Intelligent Instrumentation

Principles and Applications

Intelligent Instrumentation

Principles and Applications

Manabendra Bhuyan

CRC Press is an imprint of the
Taylor & Francis Group, an **informa** business

MATLAB® is a trademark of The MathWorks, Inc. and is used with permission. The MathWorks does not warrant the accuracy of the text or exercises in this book. This book's use or discussion of MATLAB® software or related products does not constitute endorsement or sponsorship by The MathWorks of a particular pedagogical approach or particular use of the MATLAB® software.

CRC Press
Taylor & Francis Group
6000 Broken Sound Parkway NW, Suite 300
Boca Raton, FL 33487-2742

First issued in paperback 2017

ISBN 13: 978-1-138-11435-7 (pbk)
ISBN 13: 978-1-4200-8953-0 (hbk)

Visit the Taylor & Francis Web site at
http://www.taylorandfrancis.com

and the CRC Press Web site at
http://www.crcpress.com

God has blessed me with a wonderful family and a

wonderful teacher in Instrumentation

I dedicate this book to my wife Nanti, daughter Pahi, and son Pol, and

to

Prof. Manoj Kumar Ghosh, retired professor of IIT Kharagpur, India

Contents

Preface

Classical sensors have been traditionally used in various measurement and process control applications for a variety of parameters. A signal-conditioning circuit when interfaced to a sensor enhances the performance of the sensors manyfold. Signal-conditioning operations are very common in instrumentation systems and have been used since long in the field of measurement and process control applications. With the advent of microprocessors and digital-processing technologies, such signal conditioning operations have been developed rapidly and the technologies have been found to have a good rapport with instrumentation systems. Many such sensors with microprocessor-based signal-conditioning devices have attracted a high volume of consumers.

The applications of instrumentation and process control have grown rapidly requiring medium-to-extremely complicated measurement systems. During the last decade, many new types of process parameters have evolved requiring new technologies of sensor or signal-conditioning systems. While classical sensors target general types of measurement systems, the newer sensor technologies focus on more specialized process parameter measurements. In many situations, such a sensor has not only to measure a process parameter, but has also to take additional decisions and perform many other nonconventional operations such as validation, compensation, and classification. This new category of sensors carries the tag *intelligent* and has expanded the scope of incorporating *intelligence* to instrumentation systems.

Why Do We Need a Textbook on Intelligent Instrumentation?

Incorporation of intelligence to classical sensors has been done by researchers and sensor manufacturers in various ways. Based on the design approach of such sensors with added intelligent features, several varieties of intelligent sensors are modeled, implemented, and even marketed commercially. Due to the nonavailability of specific definitions, there is no straightforward indication in such innovations about the requisite features of an intelligent instrumentation system. Very often, a sensor integrated with a digital processor in a single chip is also termed an intelligent sensor, but it is not indicative of any intuitive ability of the sensor's functionality. There are many texts on intelligent sensors and instrumentation, but none of them define the technologies and services in a categorial manner. Moreover, many of the intelligent sensors developed so far are commercially viable due to the important services they offer, while many of them are not due to inappropriate design methodology.

Texts on intelligent instrumentation and intelligent sensors abound in various research articles and manufacturers' application notes; however, general texts on their design approaches are still far less than expected. *Intelligent Instrumentation: Principles and Applications* is designed as a textbook for a first course on intelligent instrumentation.

Why do we need a classroom course on intelligent instrumentation? Over the past decade or so, many universities have included topics on intelligent instrumentation in their courses on classical instrumentation. These topics mostly cover *integrated smart sensors* and the broad topics covering the entire family of intelligent instrumentation are found missing. Conventional instrumentation has rapidly shifted to intelligent instrumentation over the last decade. Researchers are continuously trying to add intelligence to sensors using state-of-the-art methodologies, but researching for a target service is different from understanding the underlying principles and design methodologies of intelligent instrumentation.

This author has taught conventional instrumentation with varying patterns of course structures for the last 30 years. Since five years or so, I have tried to cover some topics on 'intelligent instrumentation' in the classical instrumentation course. However, the students' confusion as to which book to follow motivated me to write a textbook on intelligent instrumentation covering the design methodologies and their relevant applications.

Who Will Benefit Most from This Book?

This textbook is not self-contained and neither does it try to be not to go too much beyond its scope. It is intended as a classroom course for engineering graduates and covers the theories and applications of intelligent instrumentation or an elective course. The contents of this book can also be spread over two semesters. Apart from its usefulness in the classroom, this book will also be useful for practicing engineers and manufacturers. Besides theory on intelligent instrumentation, it includes many applications as case studies and, hence, can also be useful for researchers. The readers would also need to take a course on instrumentation as a prerequisite for this book, though Chapters 1 and 2 do cover the basics of sensors, transducers, and their performance characteristics.

How Is This Book Different from Others?

The basic feature of this book is that it explains the underlying design methodologies of intelligent instrumentation for researchers and manufacturers in a textbook-like language, translates these methodologies to numerical examples, and provides applications in case studies. There are at least 80 solved numerical examples and 14 case studies in this book. The major features of this book are as follows:

1. **Prerequisite chapters**: To understand the design methodologies of intelligent instrumentation, readers need to be familiar with the concepts of sensor devices and their performance characteristics, and signals and system dynamics. Chapters 1 through 3 cover these topics.
2. **Design emphasis**: The basic design principles of intelligent sensors are emphasized in Chapters 4 and 6 and their applications are shown using numerical examples and case studies. This approach helps the students to use the principles in real-world problems.
3. **Intelligent processing**: Intelligent sensors rely on signal processing operations such as calibration, linearization, and compensation. Chapter 5 deals exclusively with intelligent signal processing operations and provides a wide range of numerical examples.
4. **Artificial intelligence**: Artificial intelligence is one of the major components of intelligent sensors. Use of artificial neural networks (ANNs) in sensor signal processing is very useful nowadays and can solve many real-world problems. A chapter is included to explain such issues (Chapter 6).
5. **Integral use of MATLAB®**: MATLAB programs have been provided throughout the book to validate the design approaches. MATLAB can be used not only to prove the design methods, but is also an essential tool for many signal preprocessing and statistical measurements.

Organization

Chapter 1 provides a brief introduction to the basic concepts of process, process parameters, sensors and transducers, and classification of transducers, with examples ranging from radio-isotopic sensors to biosensors. The aim of this chapter is to provide a review of classical sensors and transducers.

Although a basic course on instrumentation is a prerequisite for this book, this chapter will serve as a refresher course.

Chapter 2 deals with the performance characteristics of instrumentation and measurement systems that discuss the static and dynamic characteristics. Since the intelligent processing of sensors focuses on enhancing their performance, the topics covered in this chapter will be an essential component of the book.

Intelligent signal processing deals with various types of sensor signals and the readers must therefore understand the concepts of signal representations, various transforms, and their operations in both static and dynamic conditions. Chapter 3 intends to provide such an understanding and knowledge to the readers.

Intelligent sensors developed so far by various researchers use different technologies and provide different services. The nomenclature of intelligent sensors is a complex task since, in most cases, the technologies and services are overlapping. Chapter 4 provides a unified approach to classify the intelligent sensors with their underlying design principles. It describes smart sensors, cogent sensors, soft sensors, self-validating sensors, VLSI sensors, temperature-compensating sensors, microcontrollers and ANN-based sensors, and indirect measurement sensors.

While discussing intelligent sensors in Chapter 4, the basic signal conditioning techniques were not elaborately explained. Chapter 5 addresses the issues dealing with intelligent sensor signal conditioning such as calibration, linearization, and compensation. A wide variety of calibration and linearization techniques using circuits, analog-to-digital converters (ADCs), microcontrollers, ANNs, and software are discussed in this chapter. Compensation techniques such as offset compensation, error and drift compensation, and lead wire compensation are also discussed here.

Chapter 6 deals with intelligent sensors that rely on ANN techniques for pattern classification, recognition, prognostic diagnosis, fault detection, linearization, and calibration. The chapter begins with the basic concepts of artificial intelligence and then moves on to ANN applications.

Interfacing of intelligent sensors to the processor and the users is a major issue. In order to achieve higher efficiency, uniformity, and flexibility of intelligent sensors, various interfacing protocols have been developed either in wireless platforms or on the Internet. Chapter 7 discusses a few important interfacing protocols in the wireless networking platform.

At the end of every chapter, a reference list is included to aid the reader consult the original text wherever necessary. Questions and problems for practice are also provided in a separate chapter as Chapter 8.

An Advice to Course Instructors

This book covers topics more than are required for a semester. Course instructors may organize the topics in the following manner:

Option 1 (when the students have not taken a course on instrumentation)

Semester 1: Course—Introduction to intelligent sensors (Chapters 1 and 2, Sections 4.1 through 4.7)

Semester 2: Course—Signal processing for intelligent sensors (Sections 4.8 and 4.9, Chapter 5)

Option 2 (when the students have already taken a course on instrumentation)

Semester 1: Course—Intelligent instrumentation (Chapter 4, Sections 6.1 through 6.2.2)

Semester 2: Course—Signal processing for intelligent sensors (Chapter 5, Section 6.2.3)

Chapter 7 is a supporting chapter; course instructors may therefore include one or two topics from this chapter depending on the requirement.

As a final word, the applications described in the case studies may be referred to by researchers for designing their sensors for a particular application.

Suggestions, feedback, and comments from course instructors, students, and other readers are welcome for the improvement of this book.

For MATLAB® product information, please contact

The MathWorks, Inc.
3 Apple Hill Drive
Natick, MA, 01760-2098 USA
Tel: 508-647-7000
Fax: 508-647-7001
E-mail: info@mathworks.com
Web: www.mathworks.com

Manabendra Bhuyan
Tezpur, India

Acknowledgments

I would like to thank a number of people who have made this book a better work with their support and constructive suggestions. I gratefully acknowledge the help from my colleagues, Riku Chutia and Durlov Sonowal, in solving many MATLAB® programs. I am also thankful to my research students, Madhurjya Pratim Das, Nimisha Dutta, and Awadhesh Pachauri, for their dedicated help in completing the manuscript. The help and guidance from my editors Nora Konopka, Amber Donley, Brittany Gilbert, Glenon Butler, and Arunkumar Aranganathan at CRC Press are also appreciated.

I would also like to express my gratitude to my wife Nanti, daughter Pahi, and son Pol for their patience and understanding throughout the preparation of this book.

Manabendra Bhuyan
Tezpur, India

Author

Dr. Manabendra Bhuyan is a professor of electronics and communication engineering at Tezpur (Central) University, India. He has been teaching instrumentation for more than 30 years. Currently, his teaching and research areas include instrumentation, sensor signal processing, intelligent systems, and biomedical signal processing. He has authored more than 50 research publications in various journals and conferences and holds 2 patents. He has authored a book, *Measurement and Control in Food Processing,* published by Taylor & Francis, Boca Raton, Florida, and has contributed chapters to several books. Professor Bhuyan received his BE from Dibrugarh University, India; his MTech from the Indian Institute of Technology, Kharagpur, India; and his PhD from Gauhati University, India.

1

Background of Instrumentation

1.1 Introduction

Due to necessity and curiosity, man tries to learn and understand the surroundings where he lives. The necessity comes from the urge to make man's life comfortable, whereas the curiosity leads to exploring unknown facts. In the scientific and technological world, learning and understanding of various phenomena of nature, universe, space, or man-made objects necessitates understanding the state, amount, or value of various factors that affect their phenomenon. Acquiring the knowledge of the state, amount, or value of various factors is termed as *measurement*. The factors cannot be explored fully unless the need or requirement can be quantified.

Similarly, a sense of relief cannot be obtained by a curious mind unless the facts can be explored. Exploration needs quantification of the information. However, accurate, quick, and intelligent quantification is always appreciated and, therefore, man always strives to do so.

The concept of measurement of physical factors is not new. The sundial used to measure time in Egypt dates back to the fifteenth century. In the medieval age, man learned to measure length by hand, palm, or finger. Some of the older concepts are translated into newer forms to present higher accuracy and efficiency, and many newer technologies have evolved with the advancement of science and technology with older concepts too. On the other hand, necessity and curiosity continue to flourish and have resulted in thousands of newer measurement parameters and, thereby, newer measuring instruments.

Human endeavor to quantify or measure a physical quantity has resulted in an added development to science and technology. As technology advances, measurement technology is also bound to expand. From a fire alarm to an electronic nose, from the laboratory pH meter to a counter to measure the number of sharks passing under the sea—in every sphere of our life, measurement systems are becoming increasingly common. However, measurement systems continue to develop by associating computing devices, which present them with a new feature—*intelligence*. This book will present step-by-step concepts that will help the reader to understand what those intelligent instruments are, how they work, and how such instruments can be developed for application.

1.2 Process

A process is a unit where a series of continuous or regularly occurring action takes place in a predefined or planned manner. However, we often encounter many systems where the process operation is random and cannot be modeled. Nonetheless, a system is regular or random, it experiences various forms of physical, chemical, or biological changes. The causes of such changes are the variations of some parameters that get reflected in some other parameters. These parameters are called process parameters or process variables. It is evident that the process dynamics mainly depends on the process variables. The process variables that indicate the state of the process action is called process outputs, whereas those that change the process action are called process inputs. An example of the process input and output parameters is explained with the help of a tea drier, as shown in Figure 1.1 [1].

The process of manufacturing black tea, which takes place inside the tea drier, is both a physical and a biochemical change of state of the fermented tea. The biochemical change in the tea is a complex process and difficult to model. Considering only the process of the physical changes of drying of the tea in the drier, the quality of product can be defined by the process output variables such as the moisture content and color of the black tea, whereas the moisture content of the input fermented tea, the feed rate, and the temperature of the drier can be defined as the input parameters. There are some other input variables, which are less responsible or not at all responsible for

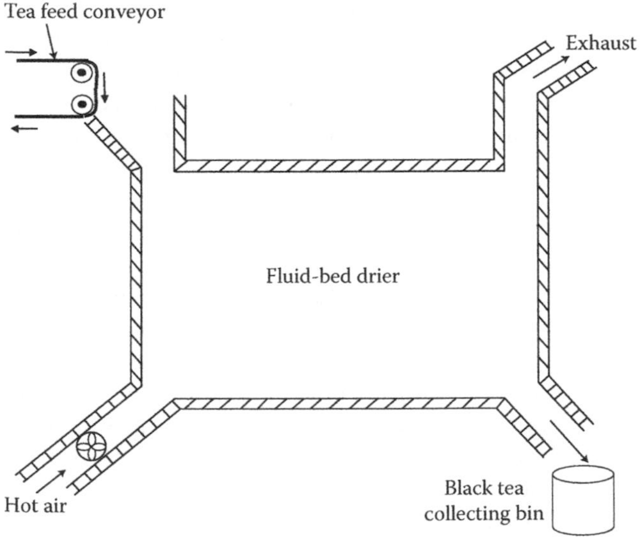

FIGURE 1.1
A tea drier.

the process action such as the density of the fermented tea. Similarly, some process output variables do not indicate any quality of the process, namely, the flow rate of the black tea in the drier. Hence, a process in a system constitutes manipulated input variables and controlled output variables. The manipulated input variables control the process dynamics, whereas controlled output variables carry the signature of the process operation.

1.3 Process Parameters

A process to be handled by measurement and instrumentation may vary widely. It may range from a simple and common type like an oven to a complex one like the fermentation of tea. Hence the parameter to be measured and controlled may also vary from the simplest like the temperature of the oven to a complex one like the flavor of tea during fermentation.

Moreover, the parameters may have to be measured under different stringent conditions—high pressure, high temperature, vibration, shocks, etc. There may be a good number of cross sub-conditions of these situations such as temperature measurement of high-pressure fluid, pressure measurement of hot gas, flavor measurement under humid condition, etc. The most important point in instrumentation under such condition is to select the right kind of sensor or to develop a special one to meet the requirement. This will need some sort of intelligence of the sensor to nullify the interfering effect inherently or by using special circuitry or computation.

A wide range of physical parameters in various systems—industrial, laboratory, biological, medical, etc.—are required to be measured. The most common physical parameters in industrial systems are time, temperature, pressure, flow, level, etc. Other less common parameters available in almost all systems are position, displacement, velocity, acceleration, weight, force, density, viscosity, etc. Some parameters are far less common and rarely need measurement or control, such as color, flavor, turbidity, sugar content, etc. The continued effort of human beings makes it possible to explore new facts and therefore newer parameters. The instrumentation will always strive to develop sophisticated techniques to address such parameters.

1.4 Classical Sensors and Transducers

Transducers and sensors are the basic devices needed to sense and convert the physical parameters to a convenient form. The convenient form of the signal is, most commonly, an electrical signal, which has many advantages compared to other forms such as mechanical, optical, fluidic, etc.

Physical parameters that need measurement in industrial, laboratory, medical, space, household, etc., are large in number. Transducers required for the most common physical parameters (almost 90%) include temperature, pressure, flow, and level.

Although it is not always specifically mentioned, a transducer may comprise two or more stages. The primary stage converts the physical parameters into other more easily measurable physical parameters, while the subsequent stages convert it to an electrical form. The first stage is called a sensor and the second stage is the transducer. A good example for a two-stage transducer is the load cell. A mechanical elastic member converts the load to a displacement or strain signal, while a resistive strain gauge converts the strain to electrical voltage on the application of a voltage. Mechanical devices like bourdon tube, bellows, diaphragm, spring, rings, levers, etc., are examples of primary sensors that convert the mechanical load or force to displacement signal.

A sensor is unique while a transducer is composite. A sensor structure gets more physically attached to the environment under operation than the transducer. Citing the same example of the load cell, the mechanical elastic member experiences the physical deformations and displacements in its molecular structure due to the application of load or force, whereas the displacement is converted to electrical signal using an electrical strain gauge. Here, the resistive gauge element does not directly react with the deformation, so it is a transducer.

1.4.1 Classification

Transducers are classified on the basis of various factors such as the type of electrical signal that a transducer develops. Hence, it can be classified as an analog or a digital transducer. Although most transducers are analog in nature, digital transducers are becoming popular due to their added features and advantages.

As explained earlier, a transducer may consist of a single unit, i.e., only the primary sensor or a combination of two stages. Thus, a transducer can be classified as a primary transducer or a secondary transducer. Transducers are designed based on different working principles; however, this author prefers to group the working principles into two classes—variable parameters type or self-modulating type, and the self-generating type. These types are interchangeably termed as active and passive transducers also. In the first type, the physical parameter causes the transducer to change an electrical parameter like resistance, capacitance, inductance, etc. These variable electrical quantities are converted to electrical voltage, current, or frequency with the help of an appropriate circuit powered by an external voltage source. Therefore, they are called as active transducers. In the other type, due to some inherent quality of the transducer, an electrical voltage is generated when the transducer interacts with the physical parameter and therefore separate power supply is not used in the transducer. This is why the transducer is termed as passive transducer. Many authors define passive and active transducers interchangeably; however, other processing circuits such as amplifiers, filters, etc., may have some active devices.

Although we commonly understand that a transducer transforms a non-electrical quantity into an electrical quantity, there is one counterpart of this, which is used for the conversion of electrical energy to mechanical or other form of energy. Taking the electrical instrumentation and measurement system as a reference, the first type of transducer is called an input transducer, while the other type of transducer is called an output transducer. The output transducers are also called *actuators.*

Most importantly, in the light of intelligent system, transducers are continuously being improved and featured with several added advantages, one of which is *intelligence.* In view of this, transducers can also be classified based on their intelligence as *dumb* transducers and *intelligent* transducers. Dumb transducers are mainly classical or conventional without any added intelligence. Intelligent transducers are further categorized into several subgroups based on their role of intelligence.

Transducers or sensors are sometimes named after the signal they handle, say, mechanical, thermal, optical, magnetic, pneumatic, radiation, biological, etc. In one sense, these sensors convert the corresponding physical signal into another form of signal; however, sometimes a transducer uses the corresponding physical energy as an intermediate signal and so such transducers are also termed by the same name. For example, an optical encoder for angular position measurement uses the optical signal, a radioactive sensor can be used to measure the level of liquid where radiation is the mediator only-calling them optical and radioactive transducer, respectively. Therefore, such nomenclature is always confusing and it is difficult to exhaustively classify the vast family of transducers. Various texts follow various methods of classifying the transducers, but this author finds the following classification as logical and optimal:

1. Self-generating type
2. Variable parameter type
3. Pulse- or frequency-generating type
4. Digital type

1.4.2 Self-Generating Transducers

Some materials have an inherent property due to which when the material is exposed to external stimulation, a voltage is developed. Transducers made of such materials are classified as self-generating transducers. Some examples of self-generating transducers are piezoelectric crystal, thermocouple, pH electrode, radioactive sensors, photocells, electrodynamic, electromagnetic, and eddy current type. Since a transducer does not have an external power supply, the voltage developed by almost all the self-generating transducers is of very low strength and cannot be directly used for displaying or actuating a control device. Therefore, voltage or current amplification is necessary before applying it to an indicator, a recorder, or a control device. Since corruption of the signal by noise is a very common problem in case of transducers generating weak signals, filtering of the signal to remove noise also becomes necessary. A self-generating transducer does not require an extra source of power supply, which is an advantage.

1.4.3 Variable Parameter Transducers

A major part of the transducers fall under this category of transducers. Unlike self-generating transducers, a variable parameter–type transducer cannot develop a voltage of its own; however, an electrical parameter of the device changes in proportion to the physical variable applied. The change in electrical parameter can be in

1. Resistance or conductance
2. Capacitance
3. Magnetic properties

1.4.3.1 Resistance or Conductance Variation

The resistance variation of a material is exploited for making these transducers. The geometrical or molecular configuration of the material is made to change causing its resistance vary, proportionately when a physical variable is applied. The variation of resistance is changed to a variation of voltage using a resistive circuit. The circuit uses a separate voltage source for the generation of the signal. Examples of resistive transducers and the physical variables that can be measured are [1]

Potentiometer—displacement, load, force, etc.

Strain gauge—strain, pressure, load, torque, etc.

Resistive thermometer—temperature, flow, etc.

Thermistor—temperature

Hygroscopic sensor—moisture content, etc.

E-nose sensor—flavor, humidity, etc.

1.4.3.2 Capacitance Variation

In capacitive transducers, the change in capacitance may be realized either by changing the dimensions of the capacitor or by changing the dielectric property of the material of the capacitor. When such variations confirm proportionality, the capacitive transducers can measure various kinds of physical parameters. The variation of the capacitance is utilized in a capacitive measuring circuit, namely, a capacitance ac bridge, an oscillator, an integrator circuit, etc. Examples of capacitive transducers are [1]

Variable area capacitor—angular displacement

Charging/discharging capacitor—rotational speed

Variable dielectric capacitor—moisture content, humidity, density, and liquid level

1.4.3.3 Magnetic Properties Variation

The working principle of a magnetic transducer relies on the fact that one or more of the following magnetic properties change in accordance with the many physical variables. These magnetic properties are self-inductance, mutual inductance, reluctance, etc. In such magnetic transducers, the inductance of a magnetic coil is allowed to change by varying either the magnetic properties of the core material or the air gap in the magnetic core. In both cases, the inductance of the transducer changes due to change in the permeability of the magnetic core.

Variable inductance transducers are mainly used for dynamic measurements of physical variables such as pressure, displacement, acceleration, force, angular position, etc. A proportional voltage signal is developed with the help of an excitation voltage applied to a circuit, which comprises the transducer. Examples of magnetic transducers are single-core reluctance pressure sensor, linear variable differential transformer, rotational variable displacement transformer, electrodynamic rotary motion transducer, synchro angle transmitter, noncontact proximity sensors, and magnetostrictive force transducer.

1.4.3.4 Pulse or Frequency-Generating Type

When self-generating types of transducers produce a train of pulses, frequency of which is proportional to the input physical variable, however, with constant amplitude, the transducer is called a pulse-generating transducer. The output pulses of such a transducer are applied to a digital counter, which determines the number of pulses during a specific time period, which is finally calibrated in terms of the input physical signal. Some examples of pulse- or frequency-generating transducers are optical disc type of rotational transducer, turbine flowmeter, radioactive flowmeter, shaft speed meter, pressure-sensing oscillator, shaft position transducer, and capacitive controlled humidity-sensing oscillator [1].

Digital transducers can generate a digital signal proportional to the physical variable, which can be conveniently interfaced to microprocessors or computers. Digital transducers have four distinct forms:

1. Direct digital encoding
2. Pulse, frequency, and time encoding
3. Analog-to-digital encoding
4. Analog-to-digital conversion

1.4.4 Radioactive Transducer

The absorption pattern or depth of penetration of radioactive rays liberated in a medium by radioisotopes such as ^{60}Co, ^{137}Cs, ^{192}Ir, etc., is utilized in radioactive transducers. The radioactive rays liberated by a radioisotope are α, β, γ, and neutron radiations. The basic characteristics of radioactive rays that are important for instrumentation are

1. Penetrating power
2. Half-life
3. Half-distance

1.4.4.1 Penetrating Power

The penetrating power of α radiation (about 11 cm in air and 0.1 mm in fabric) is the lowest of all types of radioactive radiations. Moreover, it is easily absorbed by a sheet of writing paper and an aluminum foil of 0.006 cm thickness. Penetrating power of β radiation is higher than α radiation due to its smaller mass (10 m in air and 10–12 mm in fabric) and it can be stopped by an aluminum layer of 5–6 mm and a lead sheet of 1 mm thickness. The γ radiation has similar characteristics as that of x-rays and has the highest penetrating power (several inches of lead). When a radiation travels through a medium, its radioactive strength in the medium can be expressed by

$$I = I_0 e^{-\mu_L d} \tag{1.1}$$

where

I is the strength of radiation after absorption
I_0 is the strength of radiation before absorption
μ_L is the linear absorption coefficient of the medium
d is the distance traveled in the medium

$$= \rho \mu_m$$

where

ρ is the density of the medium
μ_m is the mass absorption coefficient of the medium

The unit of radioactivity or strength of radiation is Curie (C).

The absorption coefficient varies as some of the parameters of the medium such as density, compactness, moisture content, impurities present, etc. vary. These parameters or some of their dependent variables can be measured based on this absorption principle of the radioactive rays. Two important properties of radioactive materials for instrumentation are *half-life* and *half-distance.*

1.4.4.2 Half-Life

The continuous disintegration of the radioisotope reduces its strength. The time taken by the radioisotope to fall to half of its strength is known as half-life. The half-life of a radio isotope is given by

$$t_{half} = \frac{0.693}{\lambda} \tag{1.2}$$

where λ = decay constant.

The half-life of radioactive sources varies from days to months to years. For the use of instrumentation, the half-life of a source should be high so that frequent recalibration of the detector is not required. In all radioactive sources, the date of activation and half-life is written on them by the manufacturer.

1.4.4.3 Half-Distance

Half-distance of sources depends on the source energy level as well as the material of the medium. Half-distance is the thickness of the medium, which will allow only half the value of the source' intensity entering the medium. This is important for selecting a source depending on the thickness of the medium.

A basic radioactive instrumentation scheme for detecting physical parameters requires the following components:

1. Radioactive source
2. Radioactive detector
3. Electronic processing unit
4. Indicator or recorder

Some of the radiation detectors are

1. Geiger–Muller counter
2. Scintillation counter
3. Ionization chamber
4. Proportional counter
5. Semiconductor counter

The electrical signal obtained from a radiation detector is conditioned by an electronic processing unit and the processed signal is displayed by an indicator or recorder.

1.4.5 Semiconductor Sensors

The classical transducers (macrosensors) are developed from materials such as conductors, crystals, dielectric, magnetic, and optical fibers, where the sensor principle relies on some physical or chemical properties shown by them. There is one more type of material—semiconductor—which is also used to make sensors. Semiconductor can be used in two levels—as material or as device. Whatever may be the level of use, semiconductor sensors are gaining importance for two reasons. First, they lead to microsensors, which are possible to be manufactured by micromachining in mass production with low cost. The second reason is that single-chip integration of signal processing along with the sensor is possible. The second concept leads to intelligent or smart sensor (microsensor).

1.4.5.1 Semiconductor Thermal Sensors

A good example of the semiconductor sensor is the integrated circuit (IC) thermal sensor. The basic concept of a semiconductor thermal sensor is that the forward characteristic of a *p–n* junction diode is temperature sensitive and of the order of −2mV/°C for silicon diode. A better dependency in respect of linearity is observed in base emitter voltage v_{BE} of a transistor supplied with a constant collector current I_C. The expression for v_{BE} is given by

$$v_{BE} = \frac{KT}{q}\ln\frac{I_C}{BT^3} + V_{q0} \tag{1.3}$$

where

K is Boltzmann's constant = 1.3807×10^{-23} J/K
V_{q0} is the band gap voltage = 1.12V at 300K for silicon
B is the constant that depends on geometry and doping level and is independent of temperature
T is the absolute temperature
q is the charge on an electron
I_C is the collector current

Equation 1.3 shows the dependence of temperature; however, it also depends on collector current. This is the reason why a single transistor is not attractive to be used as a temperature sensor. The alternative is a single sensor, having two identical transistors with different collector currents but

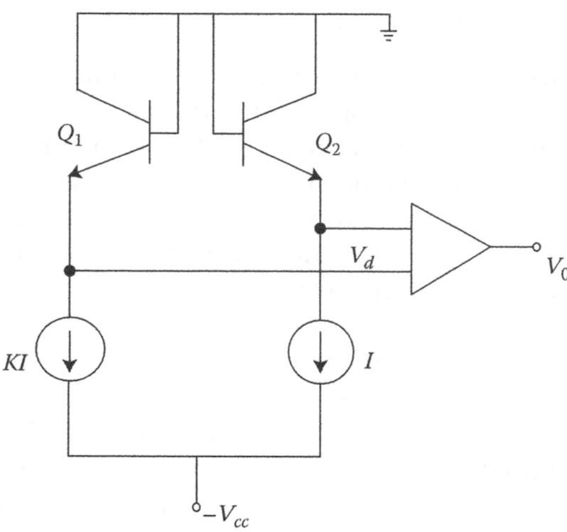

FIGURE 1.2
Dual transistor thermal sensor.

constant emitter current densities, as shown in Figure 1.2. At temperature T for both transistors, the difference voltage is given by

$$v_d = v_{BE1} - v_{BE2} = \frac{KT}{q}\ln\frac{I_{C1}}{BT^3} + V_{q0} - \frac{KT}{q}\ln\frac{I_{C2}}{BT^3} + V_{q0} \tag{1.4}$$

$$v_d = \frac{KT}{q}\ln\frac{I_{C1}}{I_{C2}} \tag{1.5}$$

If (I_{C1}/I_{C2}) is constant, v_d is directly proportional to only T.

IC thermal sensors are commercially available in a variety of forms in the range of −55°C to 150°C. These are generally available in metal (TO-52) or plastic (TO-92) packages. Analog Devices (Norwood, MA) manufactures the AD590 and AD592, which are two terminal currents proportional to absolute temperature (IPAT) sensors. The supply voltage for the AD series of thermal sensors is 4–30 V and when properly calibrated, outputs a current of 273.15 μA at 0°C and 298.15 μA at 25°C. By using an output load resistor of 1 kΩ, the sensitivity obtained is 1 mV/°C; however, an offset of 0.273 mV (at 0°C) is required to be nullified using an operational amplifier.

Similar ICs are manufactured by the National Semiconductor Corporation (Santa Clara, CA); however, these ICs are designed as voltage proportional to absolute temperature (VPTAT). LM34, LM35, LM134, LM135, LM234, LM235, and LM335 series of ICs of the National Semiconductor Corporation provides a sensitivity of 10 mV/°C over a range −55°C to 150°C. In these ICs, the offset of 2.73 mV (at 0°C) should be nullified while amplifying. These IC sensors are suitable for ambient temperature sensing and temperature compensation in other sensors.

1.4.5.2 Semiconductor Pressure Sensors

Most pressure sensors are resistive strain gauge based, where the strain gauges are bonded over a diaphragm strained due to the application of pressure. The strain gauges are connected in Wheatstone bridge configuration and the output leads of Wheatstone bridge are connected to an external circuitry. But semiconductor pressure sensors utilize strain gauge technology fabrication in miniature sizes. In a design of IC Sensors Inc., Milpitas, CA, four piezoresistive strain gauges are diffused into the surface of a single crystal diaphragm of silicon to form the Wheatstone bridge.

Motorola Semiconductors manufactures another IC, MPX series of pressure sensors where a single *p*-type diffused silicon strain resistor is deposited on an etched single crystal silicon diaphragm. Additionally, a patented X-shaped four-terminal resistor is deposited with two current taps and two voltage taps. In this configuration, when a current is passed through the current terminals and a pressure is applied at a right angle to the current direction, a transverse electric field is developed across the voltage terminals giving rise to an emf. Motorola μPX 3100 series sensors comprise of an additional signal conditioning circuit with four operational amplifiers and laser-trimmed resistors on the margin of the silicon wafer base. In a further enhanced series, μPX 2000, five laser-trimmed resistors and two thermistors are also deposited on the margin of the silicon chip that provides temperature effect.

1.4.5.3 Semiconductor Magnetic Sensors

When a magnetic field is placed perpendicular to the direction of charge carrier in a *p*-*n* junction diode, the carriers are deflected by the magnetic field due to Lorentz force. If the carriers can be deviated to a high recombination region, the I–V characteristic of a diode can be controlled by magnetic field intensity. This makes the diode a magnetodiode, providing magnetic sensitivity. Similarly, this magnetic sensitivity can be obtained from a magnetotransistor, which consists of a base, an emitter, and two collectors. A magnetic field unbalances the two collectors resulting in unbalanced collector currents. The difference between the collector currents is proportional to the applied magnetic field intensity.

However, the above devices are found commercially unsuitable due to poor repeatability, poor sensitivity, and high offset error.

1.4.5.4 Hall-Effect Sensors

Lorentz force is defined as the force on electrons when a magnetic field H is applied to a current-carrying conductor and is given by

$$F = ev \times H \tag{1.6}$$

where

e is the charge on electron
v is the electron velocity

e is given by

$$e = -q = -1.6\times10^{-19}\,\text{C}$$

The force causes some electrons to deviate from their paths and the electrons drift to one side of the conductor. This phenomenon gives rise to a noticeable increase in electronic resistance, termed as "magnetoresistive effect." In semiconductors, this magnetoresistive effect can be manifold, say, 10–10^6 on application of a magnetic field of a few tesla.

On the basis of the above effect, Edwin H. Hall demonstrated in 1879 that an electric potential (V_H) is generated in a current-carrying semiconductor, while a magnetic field was applied perpendicular to the direction of current (Figure 1.3). The Lorentz force can be shown by a vector equation as

$$F = q(V \times B) = \{q\,|v|\,|B|\sin\theta\}u \quad \text{Nw} \tag{1.7}$$

The force direction is given by right-hand screw rule. If electrons are the carriers, then

$$-qE_y = -qv_nB_z \tag{1.8}$$

$$E_y = v_nB_z \tag{1.9}$$

In *n*-type semiconductor, the average drift velocity v_n of electron is given by

$$v_n = -\frac{J_x}{qn} \tag{1.10}$$

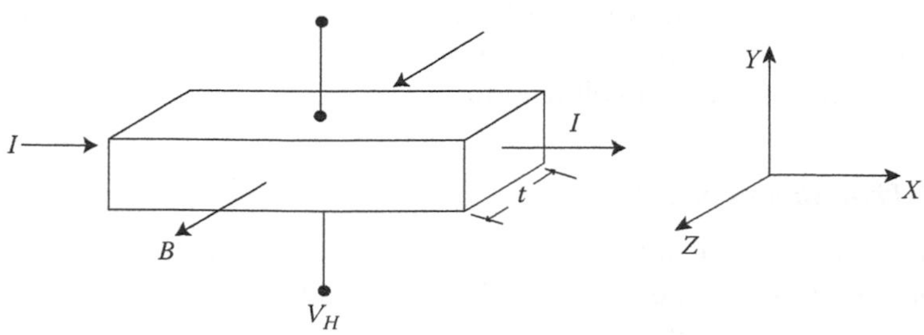

FIGURE 1.3
A Hall effect sensor.

where

J_x is the current density in x-direction

n is the electron doping density

From Equations 1.9 and 1.10 we can write

$$E_y = -\frac{J_x B_z}{qn} \tag{1.11}$$

The current density in x-direction

$$J_x = \frac{I}{t} \tag{1.12}$$

The Hall voltage developed

$$V_H = E_y = -\frac{IB_z}{(qnt)}$$

$$= R_H \frac{IB_z}{t} \tag{1.13}$$

where $R_H = -(1/qn)$ is called Hall coefficient.

Similarly, for p-type semiconductor

$$R_H = +\frac{1}{qn} \tag{1.14}$$

Hence, the Hall voltage depends on the current I, the applied magnetic field B, the thickness t of the material, and the Hall coefficient. The Hall coefficient describes the electrical properties of the material and doping density.

Other interfering factors to the Hall sensor are temperature and piezoresistivity. The temperature effect can be compensated by using a constant current source instead of a voltage source.

1.4.5.5 Photodiodes and Phototransistors

The energy of optical radiation can raise the electrodes of a p–n junction semiconductor from valence to conduction band. Thus, it can generate an electric voltage of its own, which is called a photovoltaic effect. When the radiation energy is greater than band gap, it generates an additional electron–hole pair. The accumulation of electron in n region and of holes in

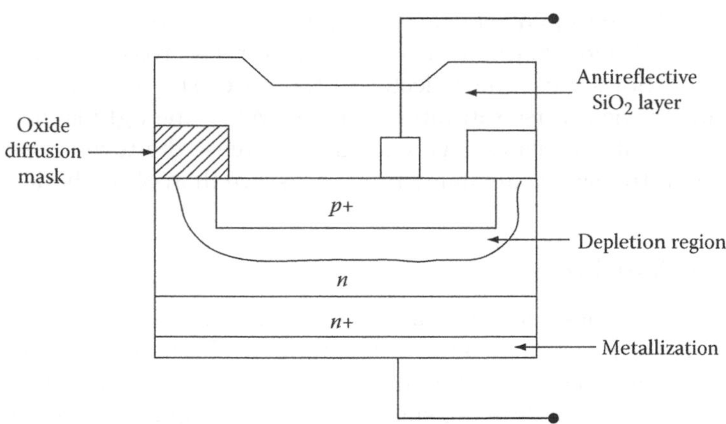

FIGURE 1.4
Layout of a photodiode.

p region causes a potential across the load connected to output leads of the *p–n* junction. Higher the radiation intensity, greater is the potential with a limitation to the band gap energy. This device is a self-generating device; however, an external reverse bias voltage is applied to increase the width of the depletion layer to provide a faster response and a current multiplication proportional to the incident radiation. Figure 1.4 shows the layout of a photodiode.

The sensitivity and spectral bandwidth of photodiode can be improved by placing an intrinsic semiconductor layer between *p* and *n* regions forming a *p–i–n* diode. The incident photons are absorbed in the intrinsic region causing a lower recombination rate.

Spectral selectivity of a photodiode can be obtained by using different materials in the absorption window for the semiconductor material. For example, silicon is transparent to radiation with a wavelength higher than 1100 nm, but a short wavelength radiation penetrates only to a low depth; therefore, the *p*-doped zone is made very thin.

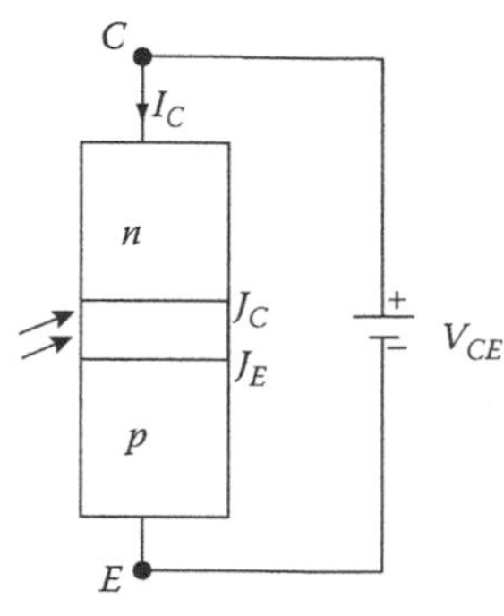

FIGURE 1.5
A phototransistor.

On the other hand, the window can be used as a filter such that, say, plastic window for $850\,\text{nm} < \lambda < 100\,\text{nm}$, and germanium for $800\,\text{nm} < \lambda < 1800\,\text{nm}$. Most commercial photodiodes are available for wavelength ranging for 0.2–2 μm.

The phototransistor (also called photodiode) is obtained by combining a photodiode and an *n–p–n* transistor. The phototransistor is usually connected in common-emitter configuration with the base open for radiation to strike it. Figure 1.5 shows a phototransistor biased in common-emitter configuration. The junction J_E is slightly forward biased; due to the open base and junction, J_C is reverse biased. Under unexcited condition, thermally generated electrons cross the base to the collector and holes cross the collector to the base. This produces collector current due to the reverse saturation current. When the light is incident on the base, minority carriers are generated to increase the reverse saturation current. Phototransistors are many times sensitive than photodiodes.

1.4.6 Array-Based Sensors

Till now, the sensors discussed above are considered to work alone or in tandem to perform a particular sensing operation. These sensors are mainly used to generate a signal function proportional to the measurand; however, they are not capable of deriving statistical metrics, signal features, and patterns as any human or animal sentience can do.

The human or animal sentience—sight, hearing, smell, touch, and taste—perform differently to a stand-alone and dumb sensor. For example, human olfactory system is an array of hundreds of olfactory receptor neurons of the nose and produces a pattern of signal that is transmitted to the brain. Each olfactory receptor neuron contains 8–20 cilia that receive the molecular odorants and convert it to electrical stimuli. The brain recognizes the pattern of the signal rather than their magnitudes. The pattern becomes a signature of the odorant because the olfactory receptors have different sensitivity to different odors. This technique of human or animal olfactory system is mimicked by the electronic nose. E-nose comprises of an array of gas sensors, each with different sensitivity to different odorant molecules. The signal generated is fed to a computer (or a microcomputer) to discriminate the class with the help of an intelligent software such as artificial neural network (ANN) or fuzzy logic.

Signal detection using an array of sensor elements is an attractive solution to overcoming sensitivity and the dimensionability limitation of a single sensor element. The three important stages of array-based signal processing and pattern recognition are

1. Preprocessing
2. Feature extraction
3. Classification and decision making

1.4.6.1 Preprocessing

Different sensors in an array respond to the physical signal in different ways, say, in case of chemoresistive and polymer sensors, there is a change in conductivity, whereas in the case of a quartz crystal microbalance (QCM) or a surface acoustic wave (SAW), sensors are frequency sensitive. Therefore, in the first type of sensors, resistive circuits are used and in the second type of sensors, oscillator circuits are used. The preprocessing unit performs smoothing, normalization, and drift correction of the data. The preprocessed data is then stored in the computer for analysis.

1.4.6.2 Feature Extraction

This stage adopts software techniques to extract some hidden information from the data. Features are some unique signature of a class of data. Supervised linear transformations such as principal component analysis (PCA) are mostly used for feature extraction. Nonlinear transforms such as Kohonen self-organizing maps are also used in many cases.

1.4.6.3 Classification and Decision Making

In supervised learning, known features are assigned to known classes and the unknown sample is placed on the class assignment. The classification stage assigns to the set of data a class label to identify the signal by comparing its features with those compiled during training. The tools available for performing classification are K-nearest neighbors (KNN), Bayesian classifiers, and ANN. Accordingly, the classifier takes a decision whether the data "falls within the class" or "does not fall," etc.

1.4.7 Biosensors

Micro/nanofabricated biochemical sensor technology has become popular recently. Such devices couple a molecular recognition process to a physiochemical microsensor that results in a sensitive signature of a molecular event [2]. With the advent of micro/nanofabricated technology, the miniaturization of the macroscale biosensors has become possible. Major applications of such devices are in low-cost medical diagnostic equipment, drug recovery, and proteomics research. The advantages of micro/nanofabricated biosensors are smaller in size, lower manufacturing cost per unit, improved sensitivity, batch manufacturing, lower energy consumption, capacity to handle smaller sample volume, scalability to large arrays, label-free detection, shorter analysis time, extended measurement bandwidth, differential measurement, and suitable construction of experimental controls.

Examples of such microsensors are ISFETS, micro/nanomechanical cantilevers, magnetic bead technologies, carbon nanotube sensors, and holographic sensors [1].

1.4.8 Actuating Devices

An actuator is a kind of inverse transducer since it converts back an electrical control signal to a mechanical form to produce a physical effect or action to actuate the final control element, such as a pneumatic or hydraulic effect such as pneumatic and hydraulic actuators. Actuators are designed and manufactured in a variety of sizes to generate force for different sizes of actions. Hence, it is important to select the right type of actuator for a specific application. The following points should be kept in mind to make the selection of the actuator simpler [1]:

1. Type of power source (electrical, pneumatic, or hydraulic)
2. Reliability required
3. Mechanical power (torque or thrust) required
4. Control functions (on/off, PID, etc.)
5. Cost

The following two types of actuators are very commonly used.

1.4.8.1 Electrical Solenoids

Solenoids can generate mechanical energy using the principle of magnetodynamics. In its basic form, a coil is excited by the electrical signal, which produces electromagnetism. The magnetic force moves a plunger where the plunger may be freestanding or spring loaded. The electrical voltage or current signal applied to the coil generates a rectilinear motion in the plunger to make a valve open or close. A solenoid is designed based on voltage or current it receives and the force required to move the assembly connected to the plunger. The duty cycle of the plunger motion (percentage of motion time to total time) is also important for considering plunger thermal constraints.

1.4.8.2 Electrical Motors

Various types of control machines like conveyors, blowers, stirrers, mixers, and pumps are run by electric motors. Sizes and types of such motors depend on the speed and torque required by the process operation. Moreover, all motors cannot be controlled with the same precision; therefore, the motor employed depends upon the precision of control. Any motor used to control some mechanical action can be said to be an actuating motor.

References

1. Bhuyan, M., *Measurement and Control in Food Processing*, CRC Press, Taylor & Francis, Boca Raton, FL, 2006.
2. Seshia, A., *Micro/Nano Fabricated Biochemical Sensors*, Cambridge University, Cambridge, U.K., www.ewh.ieee.org/tc/sensors

Further Readings

Anderson, N.A., *Instrumentation for Process Measurement and Control*, CRC Press, Boca Raton, FL, 1998.

Barney, G.C., *Intelligent Instrumentation: Microprocessor Applications in Measurement and Control*, Prentice Hall of India Pvt. Ltd., New Delhi, India, 1988.

Doeblin, E.O., *Measurement Systems, Application and Design*, McGraw Hill Publishing Co., New Delhi, India, 1986.

Johnson, C.D., *Process Control Instrumentation Technology* (6th edn.), Prentice Hall International Inc., Upper Saddle River, NJ, 2000.

Kuo, B.C., *Automatic Control Systems* (3rd edn.), Prentice Hall of India Pvt. Ltd., New Delhi, India, 1982.

Langton, M.A. and Say, M.G. (Eds.), *Electrical Engineering Reference Book* (14th edn.), Butterworth & Co., London, U.K., 1985.

Liptak, B.G. (Ed.), *Instrument Engineers Handbook: Process Measurement and Analysis*, Butterworth & Heinemann, Oxford, U.K., 1995.

Ogata, K., *Modern Control Engineering*, Prentice Hall of India Pvt. Ltd., New Delhi, India, 1982.

Patranabis, D., *Industrial Instrumentation*, Tata McGraw Hill, New Delhi, India, 1990.

Toro, D., *Electric Machines and Power Systems*, Prentice Hall of India Pvt. Ltd., New Delhi, India, 1985.

Van Valkenburg, M.E., *Linear Circuits*, Prentice Hall of India Pvt. Ltd., New Delhi, India, 1988.

2

Sensor Performance Characteristics

2.1 Introduction

The successful operation of a transducer depends on its performance characteristics. When a transducer is designed, the design parameters are chosen so as to meet the desired input–output amplitude relationship; however, similar attention should also be given to meet other factors like linearity, hysteresis, lag, temperature effect, etc. Many traditional sensors designed with high sensitivity (output/input) might show poor linearity, high hysteresis, high lag, or high temperature effect.

For example, low current rating of a resistive potentiometer for displacement can be obtained by choosing a high value of resistance of the potentiometer, but a high resistance decreases the linearity of the potentiometer. On the other hand, an increase in input voltage increases the sensitivity, which in turn increases the current drawn by the potentiometer. Therefore, a compromise has always to be made to achieve the optimum characteristics.

Researchers and transducer manufacturers always try to tune transducers to the best characteristics within the scope of practical design parameters. Improvement in the performance characteristics of instrumentation system is a matter of concern for instrumentation engineers. Hence, it is imperative to study the performance characteristics of a transducer.

An instrument should always work satisfactorily under various operating conditions. The conditions relate both to the measured and to the interfering inputs. The instrument should pick up the measurand with highest sensitivity but should respond least to the interfering inputs. For example, a strain gauge should vary fairly well with the applied strain, but the changes in the gauge resistance due to temperature change should be as low as possible. Some of the transducer characteristics relate to the measurand of interest, while some are related to the interfering inputs, or to both.

Since a measurand can be quantified in two domains, i.e., static domain and dynamic domain, the performance characteristics are also based on these two factors—static and dynamic. The characteristics associated or relevant to a signal under steady or static conditions are static characteristics. Hence, the characteristics associated with an instrument that works with a measurand, which does not vary with time, are called static characteristics. On the other hand, the factors that signify the functioning of an instrument for a measurand of time varying with nature are called dynamic characteristics. Although the characteristics are grouped under these two categories, some of the static characteristics may also be applicable in dynamic conditions. For example, *resolution* is a term categorized as a static performance, but when the system works under a dynamic condition, the term *resolution* is also equally applicable. Similarly, sensitivity of a transducer in a dynamic condition refers to the sensitivity over a range of input frequency, while static sensitivity is fixed for a particular transducer. In some cases, the dynamic sensitivity may be equal to the static sensitivity for a specific range of frequencies. There are still limitations for certain instrumentations that are designed for use in static conditions only and will not show good performance in dynamic condition. All such characteristics together are of distinct importance for intelligent signal processing systems.

2.2 Static Characteristics

As already stated, static characteristics are related to the amplitude of the response or the output of the system when the measurand or input does not vary with time. Compared to static, dynamic systems are less in number since most signals vary very slowly; hence, static characteristics alone suffice for analyzing the performance of these systems.

2.2.1 Accuracy and Precision

Accuracy can be defined as the capacity of an instrument system that gives a result that is near to the *true* or *ideal* value. The *true* or *ideal* value is the standard against which the system can be calibrated. The measured value of most systems fails to represent the true value either due to the effects inherent to the system or other interfering inputs such as temperature, humidity, vibration, etc. The accuracy of a system can be mathematically expressed as

$$A = 1 - \left| \frac{Y - X}{Y} \right| \tag{2.1}$$

where

X is the measured value

Y is the true or ideal value

Accuracy is generally expressed in percentage form as

$$\%A = A \times 100 \tag{2.2}$$

Example 2.1

In a strain gauge of resistance 120 Ω, the change in resistance of the gauge for three consecutive strain readings measured by a very accurate measuring instrument is as follows:

Strain $(\varepsilon)\times 10^{-6}$	100	150	200
Change in resistance (ΔR), Ω	0.025	0.037	0.047

The gauge factor of the strain gauge is 2.0. Determine the accuracy of the three readings.

Solution

The resistance of the gauge = R = 120 Ω
The gauge factor = λ = 2.0
From the equation of the gauge factor of a strain gauge

$$\lambda = \frac{\Delta R/R}{\varepsilon} = \frac{\Delta R}{\varepsilon R}, \text{ where } \varepsilon \text{ is applied strain}$$

$$\therefore \Delta R = \lambda \varepsilon R$$

From the gauge factor, the change in resistance ΔR for the given strain value (first reading) can be found as

$$\Delta R = 2.0 \times 100 \times 10^{-6} \times 120$$
$$= 0.024\Omega$$

Similarly, the values of ΔR for the other two readings are 0.036 and 0.048 Ω, respectively. These three values of ΔR are the true or ideal values, while the values given in the problem are the measured values. Therefore, the accuracy of the first reading is calculated as

$$A_1 = 1 - \left|\frac{0.024 - 0.025}{0.024}\right|$$
$$= 1 - \frac{1}{24} = \frac{23}{24}$$

and

$$\%A_1 = 95.83\%.$$

Similarly, the accuracies of the other two readings are

$$\%A_2 = 97.22\%$$

$$\%A_3 = 97.91\%$$

In the above example, it is observed that the accuracy gradually increases as the transducer is used at higher values of its operating range. Although a transducer shows different values of accuracies, at different operating points as in the above example, its accuracy is generally rated as a maximum likely deviation from the true value. Therefore, the strain gauge of the above example has an accuracy of 95.83%, which corresponds to a reading that shows a maximum percentage deviation from the true value.

If we want the percentage deviations of the three readings from their true values, they can be calculated as

$$\text{First reading:} \quad \frac{0.024 - 0.025}{0.024} \times 100 = -4.16\%$$

$$\text{Second reading:} \quad \frac{0.036 - 0.037}{0.036} \times 100 = -2.77\%$$

$$\text{Third reading:} \quad \frac{0.048 - 0.047}{0.048} \times 100 = +2.08\%$$

It is observed that the deviation may be either positive or negative with respect to the true value. Since the percentage deviation indicates the accuracy, it is also alternatively used to represent accuracy. Therefore, this accuracy of the strain gauge for the above example can also be ±4.16%.

Precision is a characteristic of a measuring system that indicates how closely it repeats the same values of the outputs when the same inputs are applied to the system under the same operating and environmental conditions. Although there is very less likelihood that the output response is exactly repeated, the closeness of repetition can be considered by taking a cluster of the repeating points. The degree of this precision is expressed as *the probability of a large number of readings falling within the cluster of closeness*. However, such closeness may not have closeness to the true value. Hence, an accurate system is also precise but a precise system may not be accurate.

Let us take N readings of the measurements of which the mean value is

$$\bar{X}_n = \frac{1}{N}\sum_{n=1}^{N} X_n \quad N = \text{Number of data} \tag{2.3}$$

The precision of a measurement is given by

$$P = 1 - \left| \frac{X_n - \bar{X}}{\bar{X}} \right| \tag{2.4}$$

Let us consider the same strain gauge discussed in Example 2.1 for each of the three strain values, following repeated sets of ΔR values obtained:

Strain $(\varepsilon)\times10^{-6}$	100									
Change in resistance (ΔR) Ω	0.025	0.0252	0.0251	0.0248	0.0247	0.0253	0.0250	0.0250	0.0251	0.0249

The mean of the readings is 0.02501 Ω. In the above set of readings, the precisions calculated using Equation 2.4 are 99.96%, 99.24%, 99.64%, 99.16%, 99.76%, 98.84%, 99.96%, 99.96%, 99.64%, and 99.56%.

If the accuracies for all readings are determined, their mean value is found as 95.41%. Hence, although the readings are not accurate, they are precise. Let us consider an accurate set of readings for repeated input strain of 200×10^{-6}.

Strain $(\varepsilon)\times10^{-6}$	200									
Change in resistance (ΔR) Ω	0.047	0.049	0.050	0.044	0.048	0.045	0.046	0.051	0.052	0.052

The mean of the readings is 0.048 Ω. The precisions calculated using Equation 2.4 for each of the readings are 97.91%, 97.91%, 95.83%, 91.66%, 100%, 93.75%, 95.83%, 93.75%, 91.66%, and 91.66%. It is evident that the accuracy for the set of readings is high, yet the precision is lower than the previous set of readings. Here we have examined the two sets of readings from which it is clear that the accuracy brings the readings closer to the true value, while precision makes the system capable of repeating the outputs more closely.

2.2.2 Error, Correction, and Uncertainty

The deviation of the output or response of a measurement system from the true or ideal value is called an error of the system. The error is calculated by taking the difference of the measured value and the true value. This is called absolute error. Sometimes, the error is calculated as a percentage of the full-scale range or with respect to the span of the instrument. Therefore, the error is expressed as

$$\varepsilon = X - Y \tag{2.5}$$

and

$$\%\varepsilon = \frac{X-Y}{Y_{FS}}\times100 \tag{2.6}$$

where, Y_{FS} = true or ideal full-scale value.

In common cases, the percentage error is expressed with respect to the true value as

$$\%\varepsilon = \frac{X - Y}{Y} \times 100 \tag{2.7}$$

This is called relative error.

Revisiting the strain gauge problem, the ideal and measured values of change in resistance (ΔR) for the second readings are

$$Y = 0.036\,\Omega$$

$$X = 0.037\,\Omega$$

Therefore, the absolute error is

$$\varepsilon = (0.037 - 0.036)\,\Omega$$

$$\varepsilon = 0.001\,\Omega$$

and the relative error is

$$\%\varepsilon = \frac{0.001}{0.036} \times 100$$

$$= 2.77\%$$

Now let us consider the third reading of 200 μ strain and the full-scale operating range of the strain gauge, this relative error with respect to the true maximal value is

$$\%\varepsilon = \frac{0.001}{0.048} \times 100$$

$$\%\varepsilon = 2.08\%$$

But if the relative error is calculated with respect to the true value of the span of the strain gauge, taking the first reading as the minimal value

$$\text{Span} = (0.048 - 0.024)\,\Omega$$

$$= 0.024\,\Omega$$

Hence

$$\%\varepsilon = \frac{0.001}{0.024} \times 100$$

$$\%\varepsilon = 4.16\%$$

During the calibration of an instrument, the error has to be compensated using a calibrating circuit, a microprocessor or a microcomputer used to implement in software. The *correction* is the value to be added with the measured value to get the true value. Hence, the correction can be expressed as

$$\text{Correction}\,(r) = Y - X$$
$$= -\varepsilon \tag{2.8}$$

Depending upon the polarity of deviation from the true value, the correction may be either positive or negative. For the first and second readings of the strain gauge, the correction $(r) = -0.001\ \Omega$, while for the third reading the correction $(r) = +0.001\ \Omega$.

Uncertainty is a term similar to error, which is used to express the deviation of the instrument from the true value. Uncertainty is the range of the deviation of the measured value from the true value. In a set of readings, uncertainty indicates the range of errors. For the strain gauge of the problem, the maximum error is 0.001 Ω and in all the three readings; however, the direction of deviation is positive in the first and second readings, while it is negative in the third reading. Therefore, the uncertainty is ±0.001 Ω. Hence, the uncertainty can be expressed as a range: $-r_{max}$ to $+r_{max}$ or $\pm r_{max}$

Uncertainty is also alternatively defined as a limiting error; however, it is expressed as a percentage of full-scale reading.

2.2.3 Repeatability, Reproducibility, and Hysteresis

Repeatability of an instrument signifies the degree of closeness of a set of measurements for the same input obtained by the same observer with the same method and apparatus under the same operating condition, but for a short duration of operation. Quantitatively, it is the minimum value by which the absolute value of the difference between two successive repeated measurements exceeds with a specific probability. If not specifically mentioned, the probability is considered to be 95%.

In the first set of readings for the strain gauge problem for a strain of 100×10^{-6}, the absolute value of difference between the successive measurements is calculated as 0.0002, 0.0001, 0.0003, 0.0001, 0.0006, 0.0003, 0, 0.001, and 0.0002 Ω.

Let the expected probability of repetition be 90%. Hence, eight out of nine pairs of successive readings should possess the difference of 0 Ω. The minimum value of deviation required to fulfill this target is 0.0003 Ω. If the expected probability is reduced to 30%, the repeatability is also reduced to 0.0002 Ω. A smaller quantitative value of repeatability increases the degree of repeatability.

Reproducibility is the same as repeatability, but the measurement operation is considered for a large span of time, carried out by different people at different places, and even with different instruments.

Many sensors with primary sensing devices made of elastic members show a difference between the two output readings for the same input, depending upon the direction of successive input values—either increasing or decreasing. This difference in output values is known as *hysteresis*. It is not that, hysteresis is a characteristic of mechanical or magnetic elements only, but many chemical and biochemical devices also show hysteresis. A ferromagnetic material shows hysteresis effect upon magnetization and subsequently demagnetization. Many chemical sensors upon being exposed to the chemicals get their sensitivity deformed and show a hysteresis effect.

2.2.4 Sensitivity, Offset, and Dead Band

When a measuring instrument is used to measure an unknown quantity *x*, we need to know how the instrument relates the amplitude of input *x* with the amplitude or output or response *y*. This input–output relationship is called *sensitivity*. Quantitatively, the sensitivity at any measuring point *i* is given by the slope

$$S_i = \frac{dy_i}{dx_i} \tag{2.9}$$

where x_i and y_i are the input and output at the measuring point *i*. It is desirable that a sensor has a constant sensitivity so that

$$\frac{dy_i}{dx_i} = K \quad \text{for } i = 1, 2, 3, \ldots, n \tag{2.10}$$

where *n* is the measuring point of the highest operating range.

Revisiting the same problem of strain gauge discussed earlier, where the input to the strain gauge is strain (ε) and output is the change in resistance (Δ*R*), the sensitivity of a strain gauge is determined by its gauge factor, which is given by

$$\lambda = \frac{\Delta R/R}{\varepsilon}$$

hence the true or ideal sensitivity is given by

$$\frac{\Delta R}{\varepsilon} = \lambda R = 2 \times 120 = 240\ \mu\Omega/\mu \text{ strain}$$

For the three readings of the strain gauge, the sensitivities at three different points are

$$S_1 = \frac{0.025\,\Omega}{100\,\mu} = 200\,\mu\Omega/\mu \text{ strain}$$

$$S_2 = \frac{0.037\,\Omega}{150\,\mu} = 246.66\,\mu\Omega/\mu \text{ strain}$$

$$S_3 = \frac{0.047\,\Omega}{200\,\mu} = 235\,\mu\Omega/\mu \text{ strain}$$

It is observed that although the strain gauge is expected to have a constant value of sensitivity, the three sensitivity values are different.

In many cases, the average or mean value of a set of sensitivity readings is taken as the working sensitivity. Alternatively, a best fit curve by least square method may also be plotted and the slope be taken as the sensitivity, as shown in Figure 2.1, for the three readings of the strain gauge. The slope of the best fit line is calculated graphically and the sensitivity is found as 200 Ω/strain.

In Figure 2.1, it is observed that the best fit line does not pass through the origin and the intercepts at 0.007 Ω of the ΔR axis. Hence, the equation of the line is given by

$$y = 200x + 0.007$$

This equation is of the form $y = mx + c$, where m is the slope and c is the intercept. The component $+c$ of the calibration equation of a measurement system is called the *offset*. Offset is the value that is developed in a measuring instrument even when no input is applied. This is also termed as zero error. Alternatively, the component $-c$ is a negative component produced in the system even when no input is applied. Due to this negative component, the system does not produce positive output up to a certain input level. This input level is given by

$$x = \frac{y + c}{S} \tag{2.11}$$

This quantity is called the *threshold*.

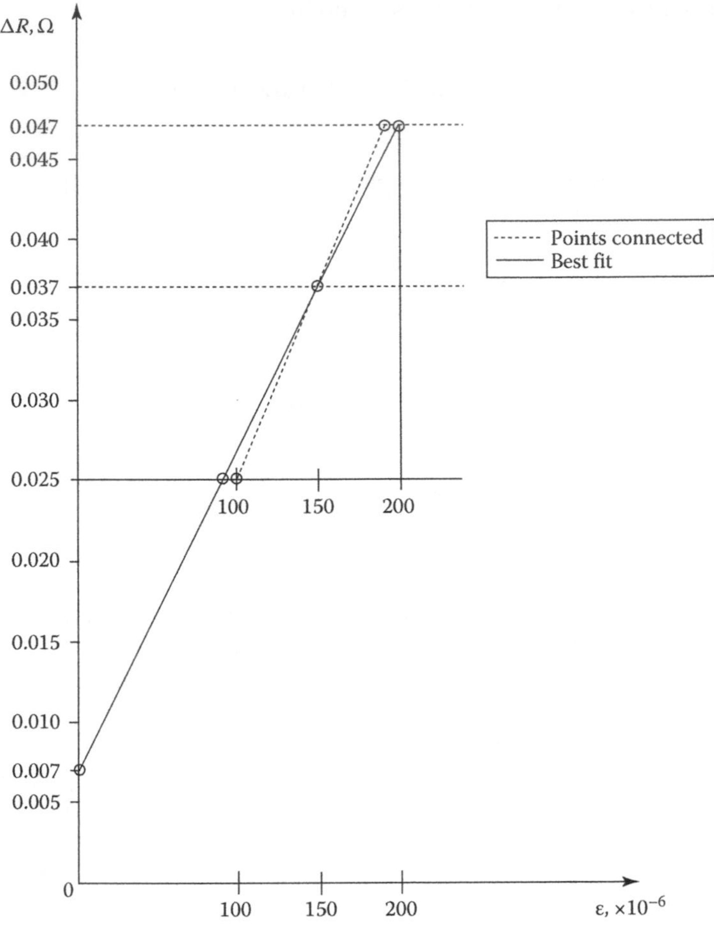

FIGURE 2.1
Sensitivity characteristics for the problem in Example 2.1.

In some instruments, the nonavailability of output right from zero level is not due to such a negative offset component, but due to some inherent reason of the system. The largest change in the input quantity to which the system does not respond is called *dead band*. The dead band is due to static friction, backlash, or hysteresis.

In the strain gauge of the problem, if the gauge operates satisfactorily within the range 100–200 μ strains, then the maximum input value of 200 μ strains is called the *range* or *full-scale* value of the gauge. The word "satisfactory" indicates that the instrument is not used beyond this range due to one or more reasons like nonlinearity, operationally not suitable, not applicable, etc.

The value of working range, i.e., 100 μ strain (200 – 100) is called the *span* of the strain gauge.

2.2.5 Resolution and Linearity

Theoretically, a measuring instrument produces the smallest output quantity on application of the smallest input; however, all smallest outputs are not practically detectable. The smallest input for which the system produces the detectable output is called its *resolution* (or discrimination). The resolution is mostly a characteristic inherent to the measuring system that depends on its geometry or structural factors.

In some measuring instruments, the input is restricted to discrete points only such as in a wire-wound potentiometer where displacement is restricted from turn to turn due to discontinuous contact of the wiper over the coil. Here, the resolution is determined by the distance between two turns. In some other instruments, although the input is linearly variable, the output is detectable only at discrete levels like a digital shaft encoder, where the detectable angle is restricted to the coded binary sequence only. Therefore, an angle between two successive binary numbers cannot be detected. Rarely, in few instruments only, both the input and the output are linearly variable to get an infinite resolution. In many measuring systems, the overall resolution is governed by the resolution of the measuring device like digital display, ADC, etc. With reference to the strain gauge discussed earlier, say, the smallest change in the resistance that can be measured by a measuring circuit is 0.001 Ω, then its corresponding strain value, i.e., 0.2×10^{-6}, is the resolution of the strain gauge. Again, let us consider that the measuring circuit used produces 1 μV for every 0.001 Ω resistance; however, the voltmeter connected to the output cannot read values smaller than 10 μV. Hence the resolution is now determined by the voltmeter and is reduced to a value of 0.01 Ω that corresponds to a strain of 2×10^{-6}.

If the output voltage of the measuring circuit is amplified using an amplifier and subsequently converted to digital output by a 12-bit ADC, the measurement system resolution will be determined by the resolution of the ADC. Assuming an amplifier amplifies the signal obtained from the strain gauge measuring circuit within the working range of 0–200 μ strain to a voltage of 0–5 V, a single bit of the ADC corresponds to $5/(2^{12})$ V, i.e., 1.22 mV that corresponds to a strain of 0.048×10^{-6}. Therefore, the resolution is increased on amplifying the signal. The resolution of many transducers are infinite; however, the indicating devices restrict the resolution of the transducer to a non-infinite value.

It was discussed in Section 2.2.1 that the measuring instruments possess some undesirable characteristics due to which the actual output deviates from the true or ideal values. The causes of deviation are various, including the inherent design characteristics and interfering inputs. Many instruments show a typical deviation from a trend of outputs even without interfering inputs making the system nonlinear. Such a characteristic of a measuring instrument is essential for calibrating the instrument by adopting various linearization techniques. In fact, when the sensitivity is constant over the operating range, the calibration characteristic is a straight line either passing through the origin or intercepting on any one of the axis. When the sensitivity changes or does not remain constant over the operating range, the instrument is said to be nonlinear. Linearity is a quantity that denotes the maximum deviation of the output from the true value as a percentage of the true value. The lesser this value is, higher is the linearity.

2.2.6 Statistical Characteristics

When an instrument is used to measure an input quantity, the output may be different from the true value due to errors. *Systematic errors* are those developed in the instruments and components of the measurement system, and are easy to be modeled, while *random errors* are produced due to the sources that cannot be specified and modeled. Due to the presence of random errors, the same instrument may show different outputs for the same inputs. The quality of such a set of measurements can be ascertained by certain statistical characteristics. Apart from the assessment of the data quality of an instrument, statistical characteristics are becoming popular for data classification using artificial neural network (ANN) techniques of sensor signal processing.

It is not that statistical analysis is useful for error analysis only. There are many systems that produce signals of random nature such as acoustics, seismology, structural vibration, etc., which need analysis by statistical methods. Statistical characteristics are used to describe all the three parameters of a signal-amplitude, frequency, and phase. The amplitude behavior is described by mean, root-mean-square, and probability density function (PDF) of the signal. These three characteristics are limited to the amplitude of the signal and are independent of frequency and phase of the signal. Similarly, the frequency behavior, which is independent of amplitude and phase of the signal, is described by the spectral density of the signal, while the phase or time behavior is described by the auto correlation function of the signal.

For a set of N readings $X_1, X_2, X_3, \ldots, X_n$ of an instrument for the same input, the mean value of the measurements is given by Equation 2.3, which is simply the arithmetic mean of the readings. Thus, the random error is eliminated to some extent by taking the mean. Higher the number of readings (N), lesser is the error. *Deviation* d_i is the departure of the ith reading from the arithmetic mean given by

$$d_i = X_i - \bar{X}_N \tag{2.12}$$

and average deviation is the mean of all the deviations obtained as

$$d_N = \frac{1}{N}\sum_{i=1}^{N} d_i \tag{2.13}$$

The measure of the extent of random error in a set of readings is stated by the root-mean-square value of the deviation given by

$$\sigma = \sqrt{\frac{1}{N}\sum_{i=1}^{N} d_i^2} \tag{2.14}$$

Mean and square values provide us the information about the central value (dc component) about which the signal fluctuates and the ac power content. These first- and second-order moments of the signal, however, do not reveal the shape of the waveform in order to know the probable time spent by each amplitude point during a finite period of time.

For a continuous time signal *x*(*t*), the mean ($\bar{x}$) and mean square ($\bar{x}^2$) values are given by

$$\bar{x} = \lim_{t\to\infty}\int_0^T x(t)\,dt \tag{2.15}$$

and

$$\bar{x}^2 = \lim_{t\to\infty}\int_0^T x^2(t) \tag{2.16}$$

A statistical way of determining this feature is to take a ratio or a proportion of the total time of stay of each amplitude with total averaging time *T*. The plot of this ratio or proportion with signal amplitude is known as the probability density function. Figure 2.2 shows the PDF of a random signal.

Thus, the probability of the signal *x*(*t*) lying between two values x_1 and x_2 is given by

$$P(x_1 < x < x_2) = \int_{x_1}^{x_2} p(x)\,dx \tag{2.17}$$

Since the total probability of the signal lying in the extended range $-\alpha$ to $+\alpha$ is unity, we can write

$$\int_{\alpha}^{+\alpha} p(x)\,dx = 1.0 \tag{2.18}$$

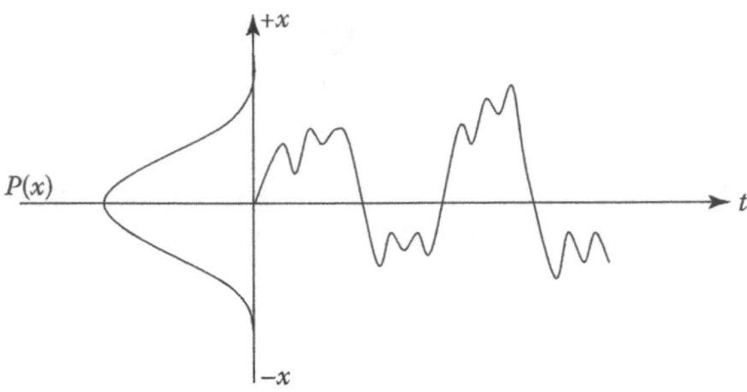

FIGURE 2.2
A Gaussian PDF function.

The expected value or the mean value of a random signal is related to PDF by

$$\bar{x} = E[x] = \int_{-\alpha}^{+\alpha} x\,p(x)\,dx \tag{2.19}$$

Similarly, the mean square value can be expressed as

$$\bar{x}^2 = E[x^2] = \int_{-\alpha}^{+\alpha} x^2 p(x)\,dx \tag{2.20}$$

and the standard deviation

$$\sigma^2 = E\left[\bar{x}^2 - x^2\right] = \int_{-\alpha}^{+\alpha} (x - \bar{x})^2 p(x)\,dx \tag{2.21}$$

The *bell-shaped* random noise shown in Figure 2.2 is known as Gaussian PDF given by

$$P_\alpha(x) = \frac{1}{\sigma\sqrt{2\pi}} e^{-(x-\bar{x})^2/2\sigma^2} \tag{2.22}$$

Most real-world system of instruments and sensors shows Gaussian noise statistics. However, strictly speaking, Gaussian PDF may be misleading unless a large number of trials or samples are used.

2.2.7 Error Modeling

A sensor or a transducer may not always produce valid measurement data. This may be due to the drift of the system parameters or interference from outside. To validate such measurements, it is necessary to analyze and model the error in the measurement. An error in general is contributed by two different components—a systematic component and a random component. The systematic component is the time-invariant offset governed by some definite rules or equations that can be modeled. Such an error may be either consistent or a linear function of the measurement. When the error is consistent or fixed in absolute value or sign, it is also called an offset. Recalling Equation 2.5 where the error is given by

$$\varepsilon = X - Y$$

with defined notations of X and Y. When the measurement includes a fixed value for all measurement points $n = 1, 2, 3, \ldots$

$$X_n = Y_n \pm C \tag{2.23}$$

Hence

$$\varepsilon = \pm C \tag{2.24}$$

Here we revisit Example 2.1 to calculate the measurement errors. Figure 2.1 shows the graphical representation of the sensitivity and the offset of the sensor. It is found that there is an almost constant offset level of 0.001 Ω with each of the three readings. Therefore, for each of the three readings, we can write

$$\begin{aligned}
\varepsilon_1 &= 0.025\,\Omega - 0.024\,\Omega \\
&= (0.024 + 0.001)\,\Omega - 0.024\,\Omega \\
&= 0.001\,\Omega \\
\varepsilon_2 &= 0.037\,\Omega - 0.036\,\Omega \\
&= (0.036 + 0.001)\,\Omega - 0.036\,\Omega \\
&= 0.001\,\Omega \\
\varepsilon_3 &= 0.048\,\Omega - 0.047\,\Omega \\
&= (0.047 + 0.001)\,\Omega - 0.047\,\Omega \\
&= 0.001\,\Omega
\end{aligned}$$

The summation inside the parenthesis indicates that the measurement reading contains a constant systematic error component. It is evident that the errors corresponding to the three measurement values are constant and equal to the offset level found in Example 2.1.

The change in resistance as calculated above corresponds to a strain of

$$\varepsilon = \frac{\Delta R}{\lambda R}$$

$$\varepsilon = \frac{0.001}{2.0 \times 120}$$

$$= 4.16 \times 10^{-6}$$

$$= 4.16\,\mu\,\text{strain}$$

There are various factors that results systematic errors in sensors. A systematic error of 4.16μ strain picked up by the strain gauge of the example discussed above may be due to the gauge being pre-strained because of the improper bonding of the gauge to the surface of the strained structure. Improper transfer of input energy from the sensor to the measurement point is also a common source of systematic error. If the sensor contact is not properly attended, sensor mis-calibration may also take place. Such situation arises in thermal sensors where the heat transfer between the medium and the sensor is a matter of concern for proper calibration. In the case of temperature measurement systems, the temperature difference between the medium and the sensor due to heat flow gap is given by

$$\Delta T = \frac{\phi d}{A\lambda} \tag{2.25}$$

where

ϕ is the heat flow (W)
d is the distance between the point of heat and sensor (m)
A is the contact surface of sensor (m^2)
λ is the thermal conductivity (W/m K)

Example 2.2

A Dallas semiconductor temperature sensor (sensor address 10F6S28E00080097) was glued to an aluminum structure of an AMS Transition Radiation Detector [1]. A calibrating platinum thermometer was glued close to the Dallas sensor. The sensors were heated in CO_2 bottle using local heaters for calibration, and measurements were recorded in a computer for calibration curve fitting by regression. Table 2.1 shows the calibration parameters for a set of Dallas sensors [1]. The calibration equation of the sensor is given by

TABLE 2.1

Calibration Parameters of Five Dallas Sensors (Statistical Errors Are Shown in Brackets)

Sensor Address	p_0 (°C)	p_1	Remarks
10F6528E00080097	−1.4(1.3)	1.05(0.07)	CO_2
10B45B8E0008007F	−2.4(2.6)	1.12(0.10)	Xe bottle
106F4A8E000800F2	−1.5(2.4)	1.05(0.10)	Mixing vessel
10A64A8E000800FB	0.7(3.6)	0.95(0.15)	
1075738E00080011	−0.4(7.1)	1.00(0.31)	Very large statistical errors

Source: Sapinski, M., *Calibration of Dallas Sensors*, INFN Sezione di Roma 1, Rome, Italy, April 2006. With permission.

$$T_{dallas} = p_1 T_{pt} + p_0$$

Find the distance between the platinum thermometer and the Dallas sensor. Take $\phi = 1$ W, $A = 10\,\text{cm}^2$, $\lambda = 10$ (W/m K).

Solution

In Equation 2.25, the systematic error ΔT for the Dallas sensor due to heat flow gap is given by

$$\frac{\phi d}{A\lambda} = \Delta T$$

The calibration table shows that the systematic error for the sensor (10F6S28E00080097) is 1.4°C. Substituting the value of ϕ, A, and λ, we get

$$1.4 = \frac{1 \times d}{10 \times 10^{-4} \times 10}$$

Hence, the distance between the platinum thermometer and the Dallas sensor is

$$d = 1.4 \times 10^{-2}\ \text{m} = 1.4\ \text{cm}$$

In many sensors, the systematic error does not remain fixed, rather it changes proportionally with some other factor of the measurement system. A good example of such a variable systematic error can be observed in resistive strain gauge. A strain gauge used for the measurement of either strain directly or other inputs such as force, load, or displacement, etc., is bonded to a mechanical structure. If necessarily the gauge is not adhesively bonded to the specimen surface, in addition to the constant systematic error due to the gauge being pre-strained, another systematic error may add to the sensor signal due to the thermal expansion of the specimen material. The thermal expansion coefficient of the gauge wire differs from that of the specimen material, hence an apparent strain is developed due to the differential thermal expansion coefficient between the two materials. If β_g and β_s be the thermal expansion coefficient of the gauge and specimen material, then the apparent strain signal developed will be

$$\varepsilon_\alpha = (\beta_g - \beta_s)\Delta T \tag{2.26}$$

Since this error in the strain measured by the measurement system is dependent on temperature variation ΔT, it can be classified as a variable systematic error.

Example 2.3

A strain gauge made of an isoelastic bonded to an aluminum surface is used for the measurement of load applied to the aluminum structure. The linear coefficient of the thermal expansion of isoelastic and aluminum at 27°C are 145×10^{-6} mm/mm °C and 23.2×10^{-6} mm/min °C, respectively. Plot the variation of apparent strain developed due to differential thermal expansion between the gauge and the specimen material over the range from 27°C to 37°C in steps of 1°C.

Solution

The apparent strain developed due to differential thermal expansion from Equation 2.26

$$\varepsilon_a = (\beta_g - \beta_s)\Delta T$$

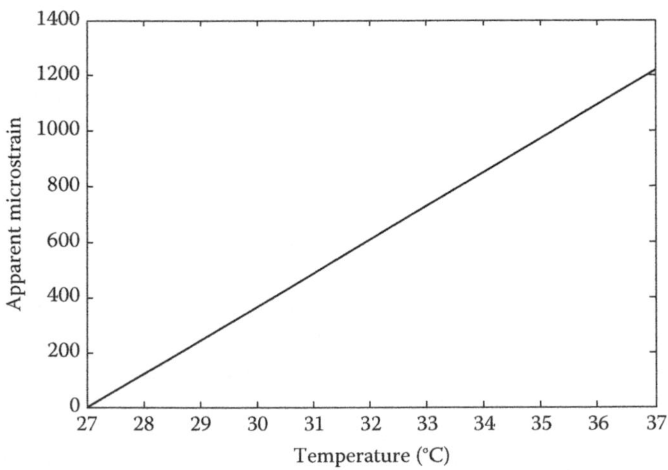

FIGURE 2.3
Apparent strain with temperature.

For the variation of temperature from 27°C to 37°C, we vary ΔT from 1°C to 10°C and substituting $\beta_g = 145 \times 10^{-6}$ and $\beta_s = 23.2 \times 10^{-6}$, we get the first value as

$$\varepsilon_{a1} = (145 - 23.2) \times 10^{-6} \times 1$$
$$= 121.8\,\mu$$

The plot of the apparent strain with temperature variation is shown in Figure 2.3.

2.3 Dynamic Characteristics

Due to the presence of energy-storing elements, such as mass, spring, and damper in mechanical and fluidic systems, thermal capacitance in thermal systems, and inductance and capacitance in electrical systems, transducers show a dynamic behavior toward the time-varying inputs than to static inputs. The behavior of such systems depends not only on their own parameters but also on the dynamic nature of the input signal.

Although a measuring instrument may have to receive input signal of varying nature, either a basic (or deterministic) or a complex (or random), it is customary to explain the dynamic characteristic of the measuring systems with respect to few common input signals, as shown in Figure 2.4.

2.3.1 Dynamic Error and Dynamic Sensitivity

When a signal is applied to a measuring system consisting of energy-storing elements, the system conserves the kinetic energy of the signal (due to its dynamic nature) for some time and releases the energy due to inherent leakage in the system. This is similar to charging a capacitor and discharging it through the leakage resistor of the capacitor. This phenomenon causes time lag between the input and output response. The behavior of the system under dynamic condition can be represented by linear

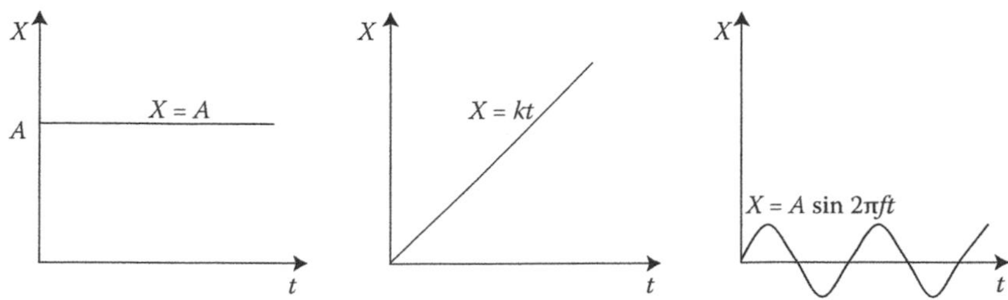

FIGURE 2.4
Common test signals.

differential equations. The dynamics of a measuring system is expressed by the order of the differential equations. A zero-order system does not have any energy-storing device such as a pure resistive network or a conductor in thermal system. First- and second-order systems are more common in measuring instruments.

When a time-varying input is applied to a zero-order system, the output tracks exactly the input, whereas in higher-order systems, the output takes some time to follow the input. Therefore, the output deviates from the input in time. The algebraic difference between the measured or indicated output and the true output at an instant of time gives the dynamic error. At that time, the dynamic error is not fixed but is a function of time. The dynamic error consists of two components—a transient component and a steady-state component. The transient component is more dominant during the initial stage of the application of input, and as soon as the output stabilizes gradually, the steady-state component becomes more dominant.

Depending upon the type of the input applied, the nature of the output for a measuring system, particularly a higher-order system, becomes different. Typical output responses of a first-order system for different types of inputs are shown in Figure 2.5.

It is clear from the figure that the error is different at different instances of time. Therefore, unlike static sensitivity, the dynamic sensitivity varies with time until the equilibrium condition is reached. The ratio between the peak values of the output signal to that of the input signal is the dynamic sensitivity.

The dynamic sensitivity of higher-order measuring systems depends on the frequency of the input signal also. For most systems, there is a range of frequency from zero within which the sensitivity is constant and beyond this range the sensitivity decreases. Figure 2.6a shows a typical amplitude-frequency response of a first-order measuring system.

Apart from such frequency-dependent amplitude error, the output undergoes a *phase error,* which is the phase difference between the output and input signal. Figure 2.6b shows the phase response of a typical measuring

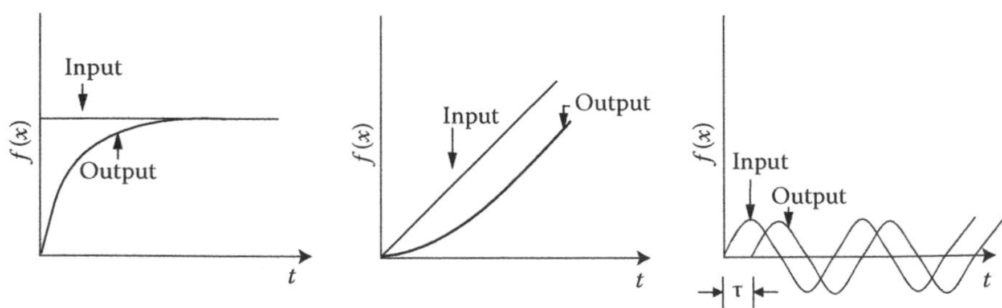

FIGURE 2.5
Output responses of a first-order system.

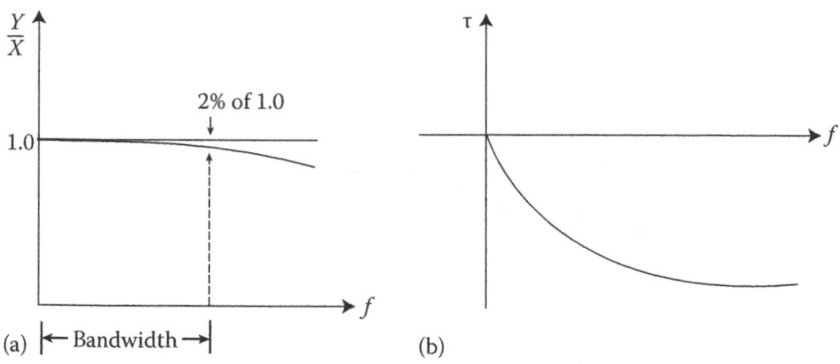

FIGURE 2.6
Frequency and phase response of a first-order system.

instrument. In Figure 2.6a, the amplitude ratio of output and input remains constant at unity (i.e., static sensitivity) for a certain frequency range, which is the *bandwidth*. In most of the measuring systems, the sharp fall of the ratio is not obtained, therefore a ±2% deviation from the static sensitivity is considered to calculate the bandwidth. Another characteristic term used for indicating the speed of response is *settling time*. Settling time is the time required for a system to reach a close range (say ±1%) of the steady-state value. The smaller the settling time, the faster is the system. The speed of response can be estimated from the knowledge of the value of the time constant, which is the time taken for the output to reach 63.2% of the steady-state value.

Example 2.4

The hot junction of a bare copper-constantan thermocouple is suddenly immersed in an oven at 200°C at constant temperature with its cold junction kept at 0°C. The static sensitivity of the thermocouple is 40 μV/°C. The response of thermocouple is shown in Figure E2.4. The step response of the thermocouple is given by $y(t) = K(1 - e^{-t/\tau})$. If the time constant of the thermocouple is 2 s, determine the dynamic error at 3 s. At what time will the dynamic error reduce to 1% of true value?

Solution
The static sensitivity of the thermocouple

$$S = 40\ \mu\text{V/°C}$$

The output at 200°C = K = 8 mV
The dynamic error is given by

$$\varepsilon(t) = K - K\left(1 - e^{-t/\tau}\right)$$

$$= Ke^{-t/\tau} \tag{E2.4.1}$$

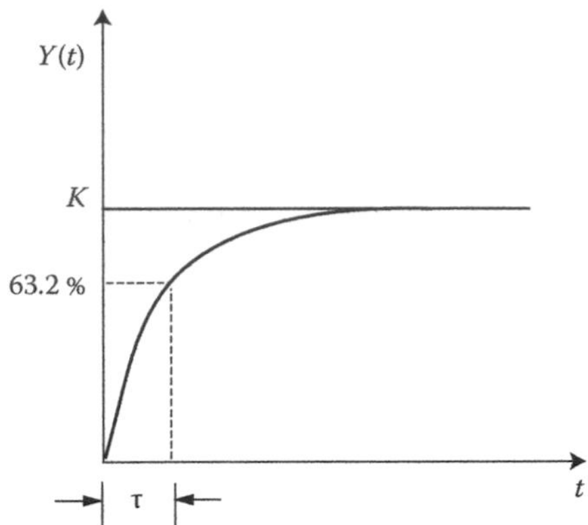

FIGURE E2.4
Step response of a thermocouple.

The dynamic error at $t = 3$ s

$$\varepsilon = 8 \times e^{-3/2}$$

$$\varepsilon = 1.78 \text{ mV}$$

Dynamic error at 1% of true value $= 8 \times 0.01 = 0.08$ mV.
From Equation 2.27

$$8 \times e^{-t/2} = 0.08$$

$$e^{-t/2} = 0.01$$

$$\therefore -\frac{t}{2} = \log_e 0.01 = -4.60$$

$$\therefore t = 9.21 \text{ s}$$

In the example of the thermocouple, the time taken by the thermocouple to reach the steady-state condition is calculated as follows:

The output equation is given by

$$y(t) = K\left(1 - e^{-t/\tau}\right)$$

The steady-state output $y(t) = K$

Taking 1% tolerance from steady-state output

$$\therefore 0.99K = K(1 - e^{-t/2})$$

$$\therefore -\frac{t}{2} = \log_e 0.01 = -4.60$$

$$\therefore t = 9.20 \text{ s}$$

∴ Settling time $(t) = 9.20$ s

Although the application of a sinusoidal temperature signal is practically not very common (it is used in heating the sensor material by using an inbuilt heater in a MOS gas sensor), for the sake of analysis, let us assume that the temperature is cycling sinusoidally with a time period of 2 s from +200°C to −200°C. We can represent the input signal with frequency

$$f = \frac{1}{2} \text{ Hz} = 0.5 \text{ Hz}$$

$$x(t) = 200 \sin 3.14t \text{ °C} \tag{2.27}$$

Now, let us assume that the thermocouple in Example 2.4 is subjected to the sinusoidal temperature input defined above and so the thermocouple output voltage will also follow the sinusoidal function with the same frequency, but the amplitude and phase will experience some dynamic error. The first-order transfer function of the thermocouple is given by

$$S(t) = \frac{Y(t)}{X(t)} = \frac{K}{1 + s\tau} \tag{2.28}$$

where

K is the static sensitivity (V/°C)
s is the Laplace operator
τ is the time constant (s)

Substituting the value of K and τ in the above equation

$$S(t) = \frac{40}{1 + 2s} = \frac{20}{s + 0.5}$$

Now taking $s = j\omega$

$$S(j\omega) = \frac{20}{j\omega + 0.5}$$

The magnitude and phase of the sensitivity of the above equation is

$$|S(j\omega)| = \frac{20}{\sqrt{(0.5)^2 + \omega^2}}$$

$$\phi = -\tan^{-1}\frac{\omega}{0.5}$$

The magnitude and phase response of the thermocouple is simulated in MATLAB® program as shown below:

```
%This program simulates the magnitude and phase response of
%the thermocouple transfer function for frequency changing
%from
% 0Hz(DC) to 100Hz.
clc;
close all;
clear all;
% Vary frequency from 0Hz to 100Hz
f=0:0.1:100;
w=2*(pi)*f;
num=20;
%Magnitude equation
den=sqrt(0.5^(2)+(w).^(2));
mag=num./den;
%Phase equation
phaser=-atan((w)./0.5);
phased=phaser*180;
figure(1);
plot(f,mag)
xlabel('Frequency, Hz');
ylabel('Magnitude');
figure(2);
plot(f,phased);
xlabel('Frequency, Hz');
ylabel('Phase, Degrees');
```

The plot of the magnitude and phase with frequencies from 0 (dc) to, say, 35 Hz is shown in Figure 2.7. For $f = 0.5$ Hz, the sensitivity is found as 12.13 μV/°C. Therefore, the output voltage can be expressed as

$$\begin{aligned} Y(t) &= Y_{max} \sin 2\pi ft \\ &= 200 \times 12.13 \sin 2\pi ft \\ &= 2.42 \sin 2\pi ft \text{ mV} \end{aligned}$$

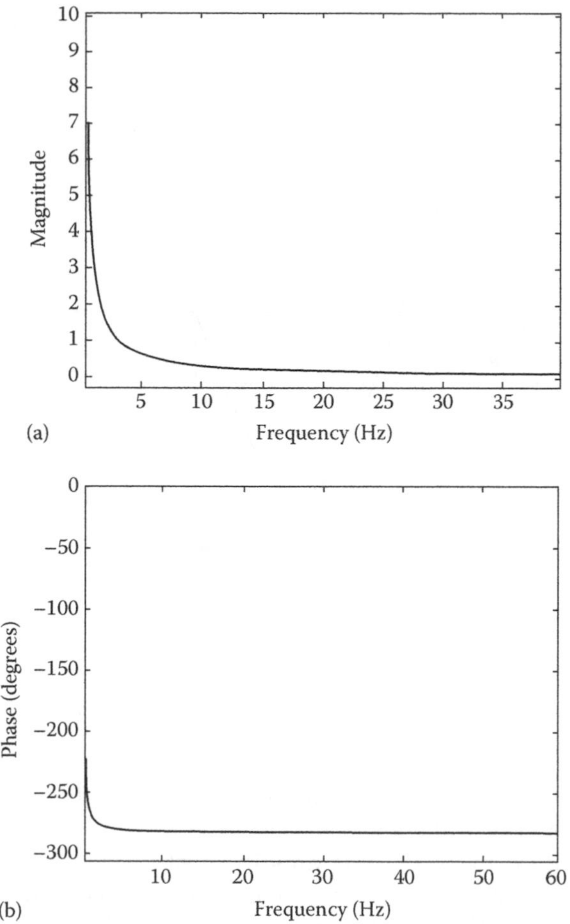

FIGURE 2.7
Frequency response of the thermocouple.

Since the frequency remains the same

$$Y(t) = 2.42 \sin 3.14t \text{ mV}$$

The plot of the input and output signals are shown in Figure 2.8.

2.4 Input–Output Impedances

Sensing a physical quantity involves some loss of energy from one stage to another. For example, in most positional sensors, some amount of input energy is lost in overcoming the frictional forces of the measuring system. Similarly,

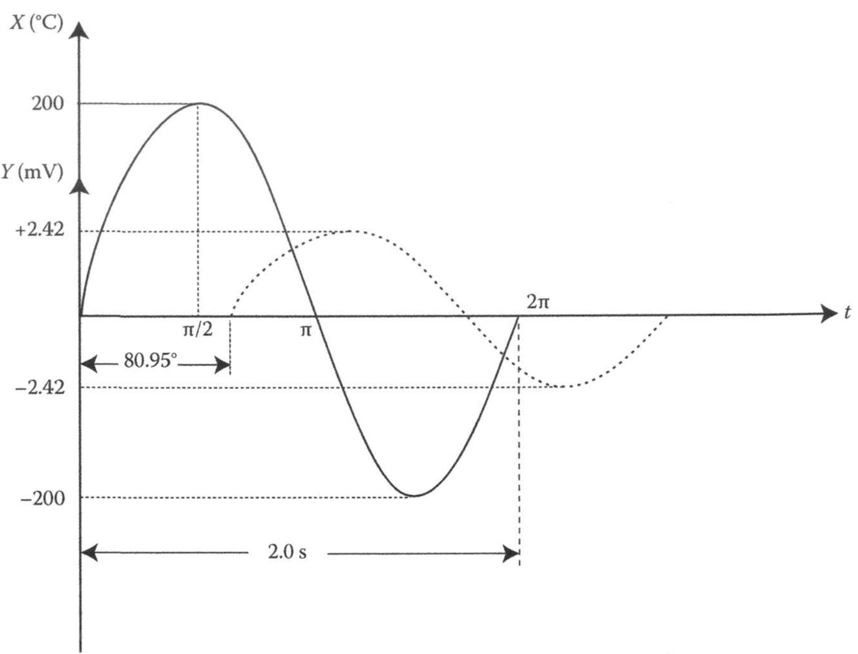

FIGURE 2.8
Input (temperature) and output (voltage) signals of the thermocouple.

in the measurement of electrical quantities like voltage by a voltmeter, some extra amount of electrical energy is consumed by the voltmeter. When such stages are depicted by block diagrams in most instrumentation texts, it is assumed that the energy transfer is maximum, i.e., least or no energy is consumed in ways other than the measurement process. This effect by which the measurement accuracy is modified by a loss of energy is called *loading effect*. However, loading effect takes place only where the quantity is measured with respect to a reference point or between two points such as pressure, voltage, temperature, etc. Such variations are called *effort variables*. On the other hand, variables that are measured at a specific reference point like electric current, heat flow, fluid flow, etc., are called *flow variables*. Contrary to the sensing of effort variables, in the measurement of flow variables, the measurement system should use maximum amount of energy from the source.

The characteristics of a sensor that restricts its input signal energy of an effort variable to be transferred to the sensor is called its input impedance (Figure 2.9). The input impedance is mathematically obtained by taking the ratio of effort variable $E(j\omega)$ and the flow variable $I(j\omega)$ in the frequency domain to give

$$Z(j\omega) = \frac{E(j\omega)}{I(j\omega)} \tag{2.29}$$

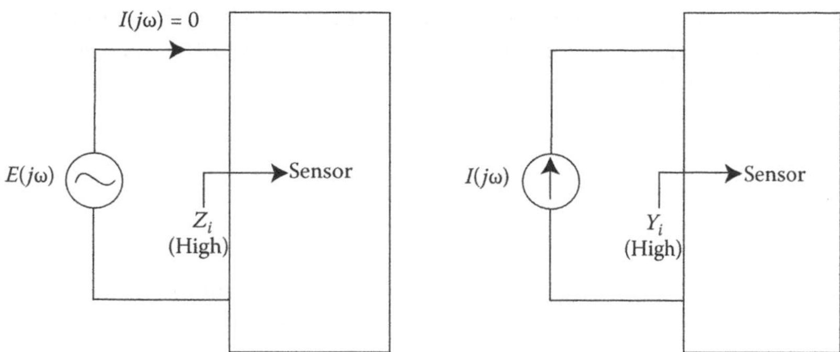

FIGURE 2.9
Equivalent impedances of a sensor (a) with high input impedance and (b) with low input impedance.

The reciprocal of impedance is admittance (Y) and is given by

$$Y_i(j\omega) = \frac{I(j\omega)}{E(j\omega)} \tag{2.30}$$

The input impedance and input admittance of a sensor measuring effort variable and flow variable, respectively, are shown in Figure 2.10.

It is desired that the sensor output is applied to an interface system for signal conditioning or directly to a meter, plotter, or other indicating or recording devices. Ideally, the interface circuit should not load the sensor. Therefore, the sensor should have low output impedance (Z_0) in case of a voltage output and low output admittance (Y_0) in case of a current output.

When the sensor produces an effort variable, i.e., a voltage signal, the voltage across the load terminal is given by

$$E_L = E\frac{Z_0}{Z_0 + Z_L} \tag{2.31}$$

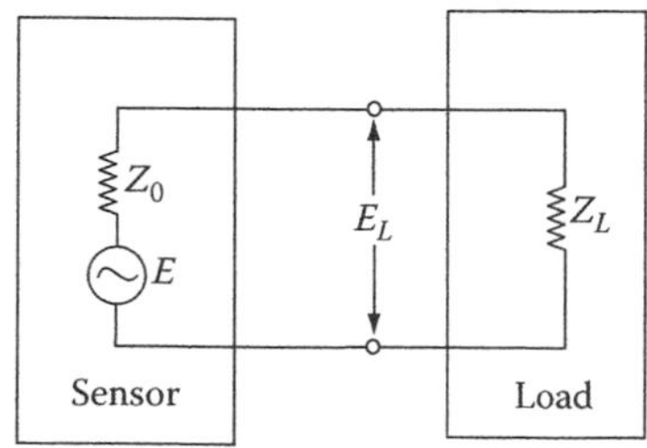

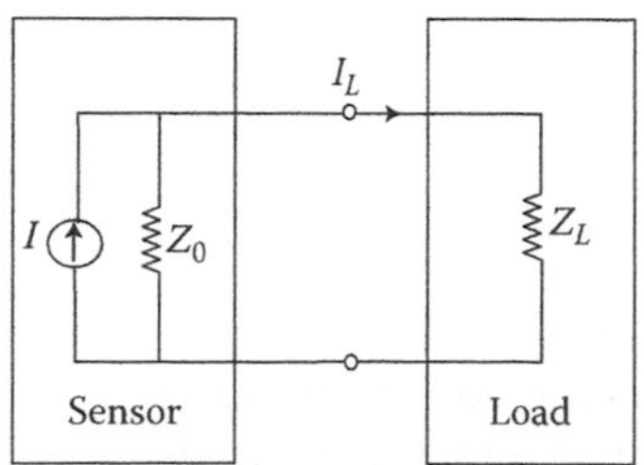

FIGURE 2.10
Impedance of voltage and current outputs (a) low sensor output and high load impedance for voltage output and (b) high sensor output and low load impedance for current output.

which becomes close to the sensor output voltage E when Z_0 is very low and Z_L is very high. Similarly, for a current output signal of the sensor, the load current I_L is given by

$$I_L = I\frac{Z_0}{Z_0 + Z_L} \tag{2.32}$$

which becomes close to the sensor output current I when Z_0 is very high and Z_L is very low. Signal source and sensor impedance of physical systems can be illustrated with the heat flow model of a temperature sensor. When a temperature sensor is installed in a medium, the conduction of heat from the medium to the sensor takes place due to the temperature difference between the medium and the sensor (Example 2.5). The higher the distance or smaller the contact surface between the measurement point and the sensor, higher is the temperature difference. In this case, temperature is the effort variable that the sensor measures and the heat flow is the flow variable involved in the process. The temperature difference for a thermal sensor is given by

$$(T - T_S) = Q\frac{d}{A\lambda} \tag{2.33}$$

where

- T is the temperature of the measurement point (K)
- T_S is the temperature of the sensor (K)
- Q is the heat flow (W)
- d is the distance between the measurement point and the sensor (mm)
- A is the surface area of the contact (mm^2)
- λ is the heat conduction coefficient (W/m K)

The above equation is analogous to that of an electrical sensor, which can be written as

$$(V - V_S) = IZ \tag{2.34}$$

$$Z = \frac{V}{I} - \frac{V_S}{I} \tag{2.35}$$

where

- V is the voltage at source
- V_S is the voltage sensed by the sensor
- I is the current
- Z is the source impedance

Comparing Equations 2.33 and 2.34, the equation of thermal impedance of the heating source can be written as

$$Z = \frac{d}{A\lambda} \tag{2.36}$$

The input impedance of the sensor is given by

$$Z_i = \frac{T}{Q} - \frac{d}{A\lambda} \tag{2.37}$$

Example 2.5

An integrated circuit (IC) temperature sensor is connected to a measurement specimen surface with a contact area of 10 cm^2 and a gap of 0.5 cm from the heating point. The heat developed is 5 W with a heat conduction coefficient of 10 W/m K. If the temperature of the specimen is 75°C, determine the thermal impedance of the specimen and input impedance of the sensor. Find the temperature at the sensor surface also.

Solution

Given:

$d = 0.5$ cm
$A = 10$ cm^2
$Q = 5$ W
$\lambda = 10$ W/m K
$T = 75$°C

Impedance of the specimen $Z = (d/A\lambda)$

$$Z = \frac{0.5 \times 10^{-2}}{10 \times 10^{-4} \times 10} = 0.5\ \text{K/W}$$

Input impedance of the sensor (from Equation 2.37) is given by

$$Z_i = \frac{T}{Q} - \frac{d}{A\lambda}$$

$$Z_i = \frac{75}{5} - 0.5 = 14.5\ \text{K/W}$$

From Equation 2.33, we can write

$$(T - T_S) = Q\frac{d}{A\lambda}$$

$$= 5 \times 0.5 = 2.5\ \text{K}$$

Temperature at the sensor surface $= T_S$

$$= 75 - 2.5$$

$$= 72.5\text{K}$$

When a sensor is interfaced to a signal processing or readout device, the connected device demands a low output impedance when the sensor produces a voltage signal. Let us consider a phototransistor which is semiconductor device that works on photo generated minority charges when radiation is applied on the base region. The phototransistor is usually connected in common-emitter (CE) configuration with the base open (Figure 2.11a). The h-parameter equivalent circuit of the CE transistor is shown in Figure 2.11b. The output impedance of the transistor is given by

$$Z_0 = \frac{1}{Y_0}$$

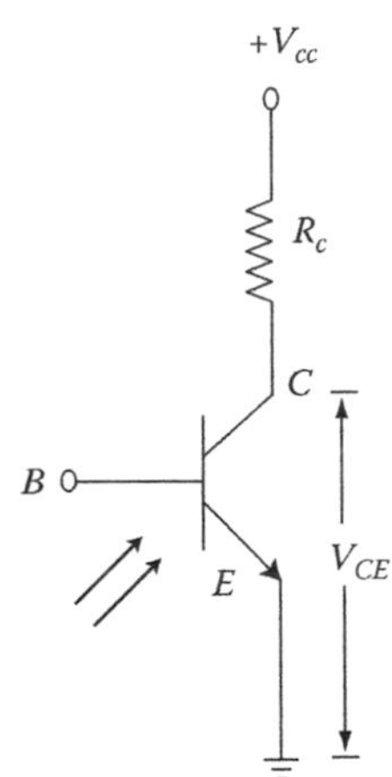

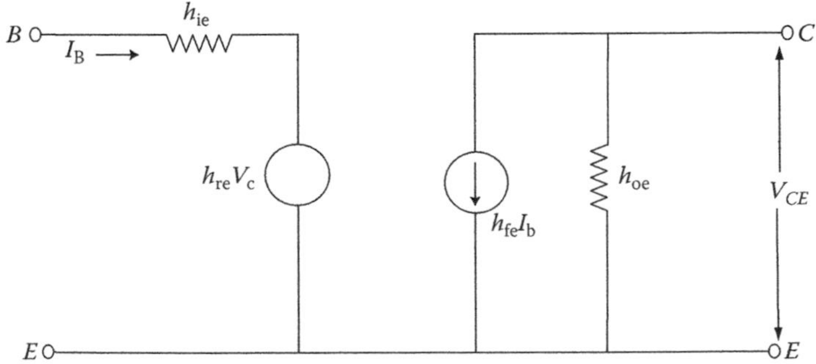

FIGURE 2.11
(a) CE configuration of phototransistor and (b) hybrid small signal model of CE transistor.

where, $Y_0 = h_{oe} - \dfrac{h_{fe}h_{re}}{h_{ie} + R_s}$

and h_{oe} = output conductance (mho)
h_{re} = reverse transverse voltage ratio (V/V)
h_{fe} = forward transverse current ratio (A/A)
h_{ie} = input resistance (ohm)

For a commercial phototransistor say MRD450 the dark currents are commonly- $I_b = 3\,\mu A$ and $I_c = 2\,mA$.

Reference

1. Sapinski, M., *Calibration of Dallas Sensors*, INFN Sezione di Roma 1, Rome, Italy, April 2006.

3

Signals and System Dynamics

3.1 Introduction

Sensors, transducers, or measuring instruments convert a physical stimulus to a convenient form of signal. Time-varying physical signals make the measuring system to behave in a dynamic mode. The behavior of the system toward time-varying signals depends on the frequency of the signal and the order of the system. Therefore, the responses of the measuring instruments can be known from the knowledge of the signal and system.

3.2 Signal Representation

A known signal can be graphically represented as a function of time, as shown in Figure 3.1.

Some signals can be represented explicitly by mathematical functions called deterministic signals. Deterministic signals may be either periodic or aperiodic. Periodic signals are those that repeat at regular intervals of time, while aperiodic signals are considered to appear only once in a broad perspective of time, as shown in Figure 3.2a and b, respectively.

Nondeterministic signals cannot be represented by a mathematical function and are called random signals. Random signals can be studied by statistical techniques such as probability or correlation theory. Figure 3.2c shows a random brain EEG signal.

3.2.1 Randomness of a Signal

Although it is not possible to say that a signal is random or periodic without observing it for a long duration of time, its behavior of excursion about a reference level with a specific interval of time may give some clue. Kendall [1] and Challis and Kitney [2] recommended a test to detect randomness based on the above criterion. Before discussing a test, let us define that a signal is said to have a turning point when the direction of excursion of the signal changes at peaks and troughs. A peak or trough is considered to have at least three consecutive points, as shown in Figure 3.3.

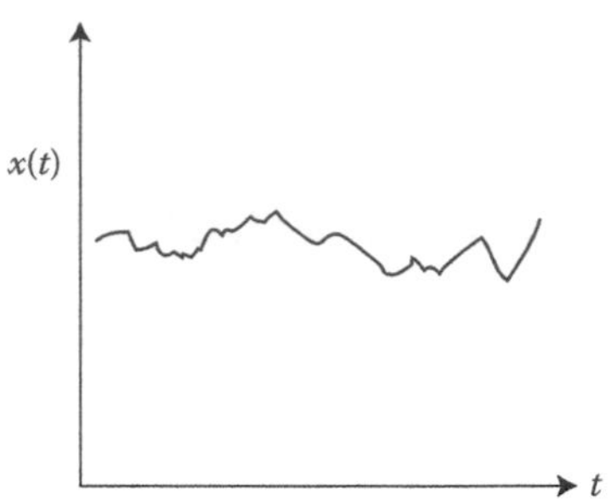

FIGURE 3.1
Representation of a signal.

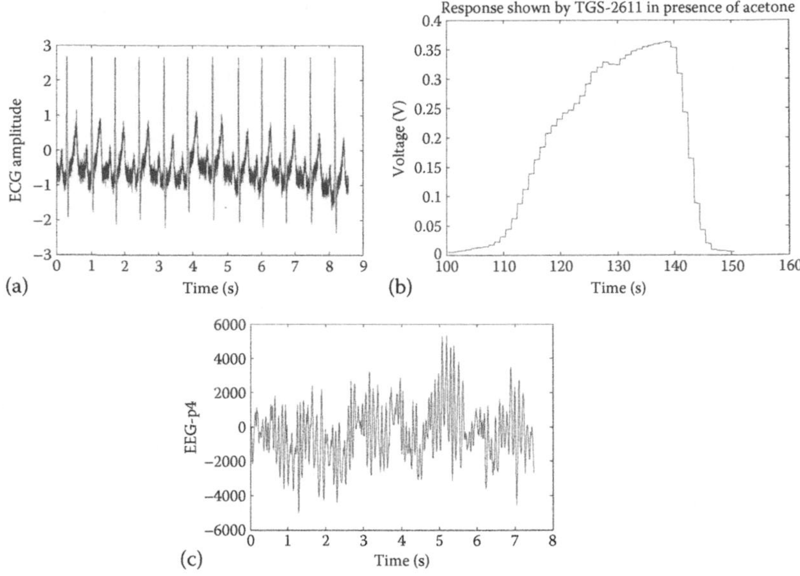

FIGURE 3.2
(a) A periodic ECG signal, (b) an aperiodic E-nose output signal, and (c) a random EEG signal.

A turning point is detected when the first-order difference of the sample of interest changes sign from the preceding sample. When a signal is sampled at N discrete levels, the signal may be called as random if the number of turning points exceeds a threshold level. Kendall and Challis and Kitney recommended that the threshold be taken as $(2/3)(N - 2)$, where N is the total number of samples [3].

3.2.2 Analog, Binary, and Ternary Signals

Signals that are continuous with time and are represented by a single value at any instant of time are called analog signals. Most physical signals are

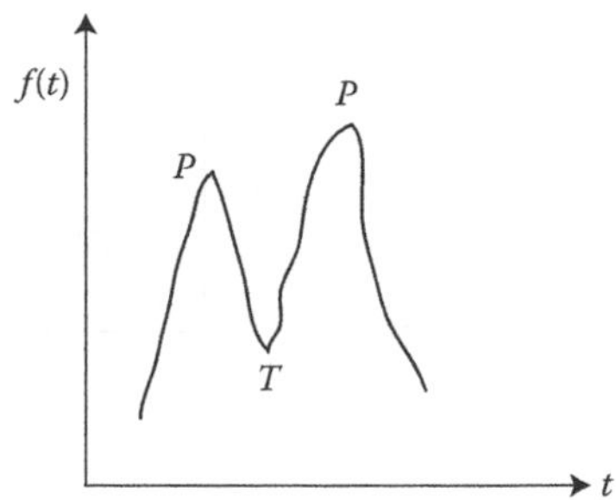

FIGURE 3.3
A signal with two peaks and one trough.

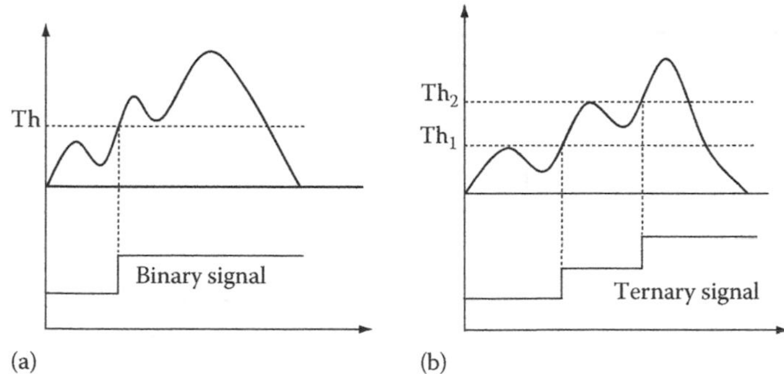

FIGURE 3.4
An analog signal conversion (a) to a binary signal and (b) to a ternary signal.

analog in nature such as the variation of temperature over a day. A binary signal assumes only two distinct values while ternary signals assume three distinct values. Signals that assume binary or ternary constant values over an instant of time are called digital signals. Unless otherwise intentionally generated by using some electrical or electronic circuit, basic measuring circuits rarely generate such signals. A digital shaft encoder develops binary signals directly, whereas an analog-to-digital converter (ADC) converts an analog signal generated by a transducer to binary. Figure 3.4 shows an analog, a binary, and a ternary form of signals.

In Figure 3.4a, the signal is divided into two levels using a threshold *Th* to make it a binary signal. In Figure 3.4b, the signal is threshold by Th_1 and Th_2 to give a ternary signal. Such type of digital signals are not used for absolute value representation of the signal, but as templates to classify or segment an analog signal.

Absolute value conversion from analog to digital is performed by digitizing the analog signals at discrete levels. Each discrete level is coded in binary format and the resulting digital number represents the analog signal. Figure 3.5 shows how an analog signal is digitized and converted to digital form.

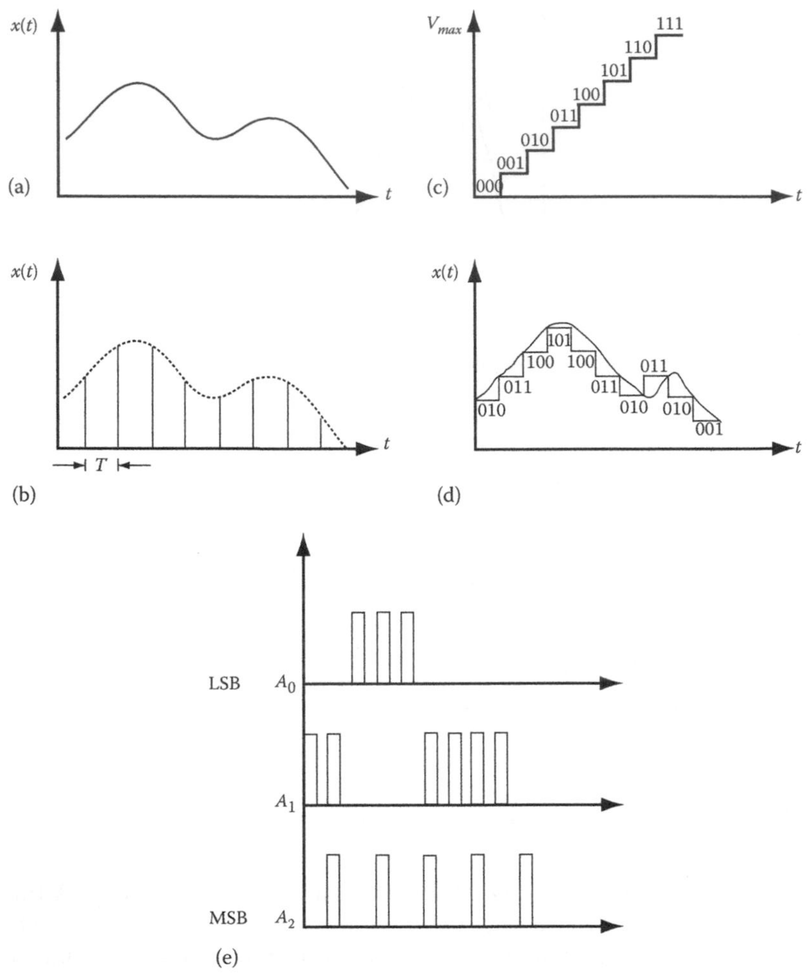

FIGURE 3.5
Discretization of analog signal: (a) analog signal, (b) discrete levels of the analog signal, (c) 8-discrete levels of a 3-bit ADC, (d) discretized signal, and (e) 3-bit digital signal.

3.3 Test Signals

The primary objective of sensors and transducers is to develop a representative signal of the input physical signals and they do so if properly designed, on application of a steady input signal. However, the physical systems of sensors and transducers experience some sort of turbulence due to the application of inputs with abrupt changes in amplitude. Therefore, measuring instruments are tested or analyzed with some typical input signals, known as the test signals. Commonly used test signals are step input, ramp input,

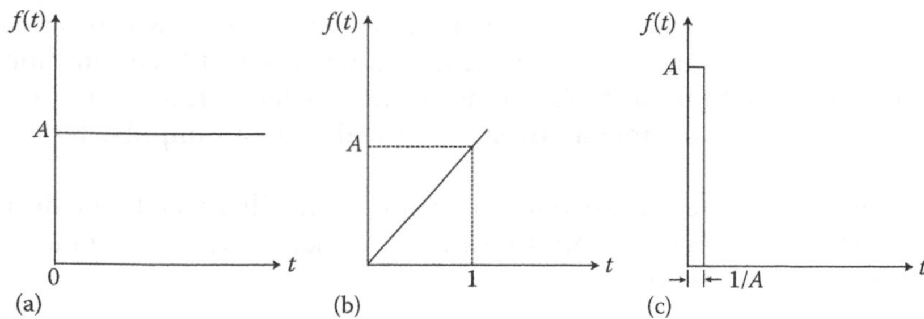

FIGURE 3.6
Test signals: (a) step, (b) ramp, and (c) impulse.

and unit impulse signals. The step input signal is shown in Figure 3.6a. The signal is represented in the time domain as

$$f(t) = A\,u(t) \tag{3.1}$$

where

$$u(t) = 1, \quad t > 0,$$
$$= 0 \quad t < 0$$

A is the amplitude of the function

Some examples of step input signals in measurement systems are placing a thermocouple to an oven temperature, which is constant at a particular level, and dipping the probe of an E-nose sensor into a chamber of volatile organic compound.

In a ramp input signal, the amplitude gradually increases at a constant rate, as shown in Figure 3.6b. The signal is represented by

$$\begin{aligned} f(t) &= At \quad t > 0, \\ &= 0 \quad\; t < 0 \end{aligned} \tag{3.2}$$

An example of a ramp input is inserting a thermocouple in an oven temperature that is gradually increased or decreased at a constant rate.

An unit impulse signal has a zero value at all times except at $t = 0$, where its magnitude is infinite. It is usually expressed by a symbol of δ function and is expressed as

$$\delta(t) = 0 \quad \text{for } t \neq 0$$

$$\int_{-\varepsilon}^{+\varepsilon} \delta(t)\,dt = 1 \quad \text{where } \varepsilon \to 0 \tag{3.3}$$

Since the impulse function has a total area of unity, and the width is infinitesimally small, it is usually assumed that it has a small time duration of $1/A$ and an amplitude of A. Applying a sudden temperature by quickly placing a thermocouple into an oven is equivalent to an impulse temperature function.

The above three special functions are mathematically related. The derivative of a step function is an impulse function, whereas derivative of a ramp function is a step function.

3.4 Fourier, Laplace, and Z-Transform

Although, we can represent a signal mathematically in the time domain, often we have to transform the signal to the frequency domain and vice versa. The reason is that many mathematical reductions are made simpler in one domain than the other. Analog signals are generally represented in Fourier and Laplace transform, while digital signals are easy to be represented in Z-transform. Laplace transform provides a systematic method for analyzing differential equations representing analog systems.

The frequency domain representation of the time domain signal dates back to the later half of the nineteenth century when Fourier suggested that any function can be resolved into an infinite series of sinusoids of appropriate amplitude and phase. Digital Fourier transformation applied to sampled signals came later with the development of digital computers only. Fourier processing is equally applicable to steady-state signals in time, space, frequency, and wave-number domains. Mathematically, Fourier transform is an integral of the product of time domain signal and a complex sinusoid of frequency, amplitude, and phase of interest. For a waveform $x(t)$ in time domain, the integration of $x(t)e^{j\omega t}\,dt$ over an infinite time results a waveform $X(\omega)$.

Therefore, any periodic function $f(t)$ in time domain can be represented by a series of trigonometric function called the Fourier series, which can be expressed as

$$f(t) = a_0 + \sum_{n=1}^{\infty} a_n \cos n\omega t + \sum_{n=1}^{\infty} b_n \sin n\omega t \tag{3.4}$$

where the coefficients

$$a_0 = \frac{1}{T}\int_{-T/2}^{+T/2} f(t)\,dt \tag{3.5}$$

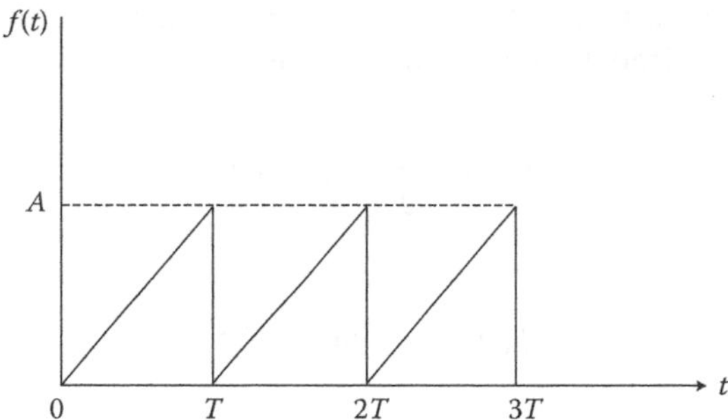

FIGURE 3.7
A periodic sawtooth waveform.

$$a_n = \frac{2}{T}\int_{-T/2}^{+T/2} f(t)\cos n\omega t\, dt \tag{3.6}$$

$$b_n = \frac{2}{T}\int_{-T/2}^{+T/2} f(t)\sin n\omega t\, dt \tag{3.7}$$

where
- T is the time period (s)
- ω is the angular velocity (rad/s)
- $n = 1, 2, 3, \ldots.$

A sawtooth wave, as shown in Figure 3.7, when transformed to a Fourier series is represented by

$$f(t) = \frac{A}{2} - \frac{A}{\pi}\sin\frac{2\pi t}{T} - \frac{A}{2\pi}\sin\frac{4\pi t}{T} \tag{3.8}$$

Fourier series representation can also be extended to an aperiodic signal considering that the signal repeats only once in an infinite time period. Before going into this discussion, let us rewrite the Fourier series in exponential function $e^{j\omega t}$ as

$$f(t) = \sum_{n=-\infty}^{\infty} d_n\, e^{jn\omega t} \tag{3.9}$$

where $j = \sqrt{-1}$ and

$$d_n = \frac{1}{T}\int_{-T/2}^{+T/2} f(t)e^{-jn\omega t}\, dt \tag{3.10}$$

Now limiting the value of T for infinity, the frequency domain representation of continuous time domain signal pair is

$$F(j\omega) = \int_{-\infty}^{+\infty} f(t)e^{-j\omega t}\,dt \tag{3.11}$$

$$f(t) = \int_{-\infty}^{+\infty} F(j\omega)e^{j\omega t}\,d\omega \tag{3.12}$$

The continuous time domain signal, when discretized into finite discrete samples, can also be represented by discrete Fourier transform (DFT). In DFT, the signal $x(t)$ is sampled into N number of samples at an interval of T_s. This permits a total time length of $T = (N-1)T_s$. In computing DFT, it is assumed that the signal $x(t)$ is periodic with a period of NT_s seconds. The DFT of N elements of the signal is given by

$$X_k = \sum_{m=0}^{n-1} x_m e^{-\left(j\frac{2\pi}{N}km\right)}, \quad \text{where } k = 0, 1, 2, 3, \ldots, (N-1) \tag{3.13}$$

The inverse DFT, i.e., IDFT is given by

$$x_k = \frac{1}{N}\sum_{m=0}^{n-1} X_m e^{j\frac{2\pi}{N}mk} \tag{3.14}$$

The IDFT is also linear and periodic in N.

Laplace transform is a powerful tool for solving or analyzing a system represented by linear differential equations. The transform uses an operator $s = \sigma + j\omega$, a complex frequency where the real part σ represents the damping of the signal. The Laplace transform pair for a function $f(t)$ is given by

$$L[f(t)] = F(s) = \int_{0}^{\infty} f(t)e^{-st}\,dt \tag{3.15}$$

$$L^{-1}F(s) = f(t) = \frac{1}{2\pi j}\int_{\sigma_1 - j\alpha}^{\sigma_1 + j\alpha} F(s)e^{st}\,ds \tag{3.16}$$

The Laplace transform of the function $f(t)$ exists only on the condition

$$\int_{0}^{\infty} |f(t)|\,e^{-\sigma t}\,dt < \infty \tag{3.17}$$

is satisfied for $\sigma > 0$.

When Fourier and Laplace transforms are related to signals in analog domain, the discrete domain counterpart is the Z-transform. The Z-transform of a continuous signal $x(t)$ is given by

$$X(z) = \sum_{n=0}^{\infty} x(nT)z^{-n} \tag{3.18}$$

where

$x(nT)$ is the sampled value of $x(t)$ at $t = nT$

T is the sampling interval in seconds

The complex variable z is defined as

$$z = e^{st} \tag{3.19}$$

Table 3.1 lists Laplace and Z-transforms for some common time functions.

TABLE 3.1

Laplace and Z-Transforms for Some Common Time Functions

Time Function ($t \geq 0$)	Laplace Transform	Z-Transform
$\delta(t)$	1	1
$\delta(t - nt)$	e^{-snt}	z^{-n}
$U(t)$	$\frac{1}{s}$	$\frac{z}{z-1}$
$t\,U(t)$	$\frac{1}{s^2}$	$\frac{T_z}{(z-1)^2}$
e^{-bt}	$\frac{1}{s+b}$	$\frac{z}{z-e^{-bt}}$
te^{-bt}	$\frac{1}{(s+b)^2}$	$\frac{T\,ze^{-bt}}{(z-e^{-bt})^2}$
$\sin \omega t$	$\frac{\omega}{s^2+\omega^2}$	$\frac{T\sin\omega T}{z^2-2z\cos(\omega T)+1}$
$\cos \omega t$	$\frac{s}{s^2+\omega^2}$	$\frac{z^2-z\cos(\omega T)}{z^2-2z\cos(\omega T)+1}$
$e^{-bt}\sin \omega t$	$\frac{\omega}{(s+b)^2+\omega^2}$	$\frac{e^{-bt}z\sin(\omega T)}{z^2-2ze^{-bt}\cos(\omega T)+e^{-2bT}}$
$e^{-bt}\cos \omega t$	$\frac{s+b}{(s+b)^2+\omega^2}$	$\frac{z^2-ze^{-bt}\cos(\omega T)}{z^2-2ze^{-bt}\cos(\omega T)+e^{-2bT}}$

Both Laplace and Z-transforms have the property of linearity and shift invariance. These properties fit well in many practical situations where the systems cannot be considered as linear and time invariant when the responses are small in amplitude and quick in time [4].

3.5 Spectral Density and Correlation Function

In Section 3.4, it was shown that any waveform whether periodic or not can be described by a series of trigonometric function indicating the frequency components present in the waveform. A diagram can be drawn to show the frequency components, consisting lines of proportionate amplitude of the components of the position of the frequencies in the horizontal axis. The frequency components are called line spectrum and the diagram is called spectral diagram. If the waveform has a frequency of f_0, the line spectrum will have components at f_0, $2f_0$, $3f_0$, … (i.e., T, $2T$, $3T$, …). Another way of describing the contribution of individual frequency components to the total signal is power spectrum. In power spectrum, the lines represent the second power of amplitude, i.e., $(\text{amplitude})^2$. If the signal is in volts, the spectrum is expressed in $(\text{volts})^2$, representing the amount of power of the signal consumed by an 1 Ω load. This also means that the total power of the signal is equal to its mean square.

When we consider Fourier transform of a random signal, we assume that the time period T of the signal approaches infinity, i.e., $1/T$ approaches zero. The power spectrum consists of infinite number of lines. Hence the line spectrum is converted to a continuous spectrum. Since the spectrum ranges to an infinite number of components, it is not possible to evaluate the total power and hence it is denoted by power per unit bandwidth having the unit of $(\text{V})^2/\text{Hz}$. This spectrum is called power density spectrum (PDS) and the value of the density is called power spectral density (PSD).

For a time domain signal $x(t)$, considering its Fourier transform over the interval $\pm T/2$ as $X(\omega)$, the spectral density is given by

$$S_X(\omega) = \lim_{T\to\infty} \frac{E\left\{|X(\omega)|^2\right\}}{T} \tag{3.20}$$

The area under the power spectrum curve gives us the total power of the signal, which is given by

$$x^{-2} = \int_{-\infty}^{+\infty} S_X(f)\,df \tag{3.21}$$

$$= 2\int_{0}^{\infty} S_X(f)df \tag{3.22}$$

PSD is an important tool for spectral analysis of random noise. A good example of its application is seismic noise analysis software developed by McNamara and Boaz using PSD and probability density function [5]. The software has been successfully implemented in seismology Data Management System, Seattle, Washington, and Advancement Seismic Data Collection Center, Golden, Colorado.

Apart from the analysis of random noise, PSD is equally useful as a feature for interpreting data and detection in many instrumentation and sensor systems. To cite an example, PSD data was used as an input feature to an ANN in an optical fiber concentration sensor system [6]. In this technique, the signal peaks sensed by an optical fiber interfaced to an optical time domain reflectometry (OTDR) were first normalized between −1 and +1, then DFT transformed using an FFT algorithm applied to the normalized data. From the resulting Fourier transform, the PSD of OTDR outputs were calculated. First six points of the PSD data were selected as the features of the ANN structure (six input nodes, three hidden nodes, and four output nodes), which were used to classify ethanol 50%, ethanol 25%, ethanol 12.5%, and distilled water. Figure 3.8 shows the PSD plots of OTDR output peaks for the four samples.

Another example of PSD analysis in sensor signal interpretation is the identification and analysis of acoustic emission signals—a technique for nondestructive testing of active defects in building structure [7]. Acoustic emission signal is strongly nonstationary, hence a time series analysis only shows how the energy of the signal is distributed during a time span. In this analysis, a test specimen of thermal load of building material was subjected to a maximum temperature of 140°C to generate an acoustic signal at this temperature.

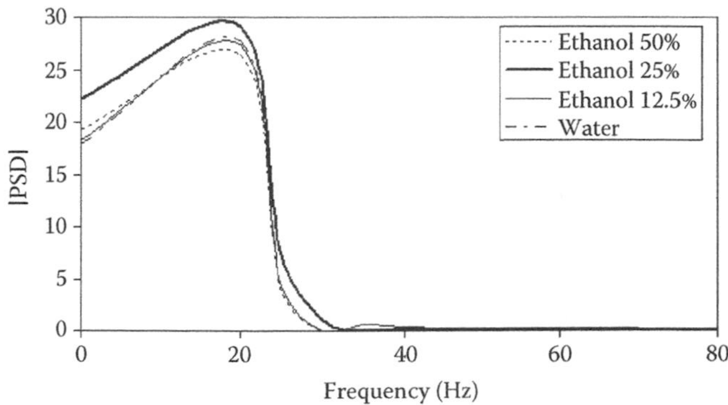

FIGURE 3.8
PSD plots of OTDR output peaks for the four samples. (Reprinted from King, D. et al., *Journal of Optics, Pure and Applied Optics*, 5, 69, 2003. Institute of Physics Publishing Ltd. With permission.)

The acoustic signal generated was sensed by an acoustic sensor (B & K 4344 accelerometer, Bruel & Kjear, Denmark) and recorded in a computer.

The analysis is not restricted only to frequency domain however; the time–frequency representation was potentially found to reveal a more realistic picture of the temporal localization of signal spectral characteristics in [8]. Margenau–Hill distribution (MHD) of time–frequency representation was applied in this technique. In this technique, time domain voltage signal of acoustic sensor, the PSD computed from the time history of the signal, and the time–frequency distribution of the signal by MHD were determined. The MHD shows the significant frequency components at 10–30, 40, and 75–95 kHz, while largest duration (0.5–1.5 ms) was associated with low-frequency components.

The evaluation of the results reveals that frequency–time distribution of PSD is a suitable tool for nondestructive studies of the effect of thermal process, which can have influence on building materials.

The *correlation* function shows statistical relationship between two signals (*cross-correlation*) or between two values of the same signals (*autocorrelation*) sampled at different instants of time. The autocorrelation function (ACF) of a signal $x(t)$ is given by

$$\Phi_{xx}(\tau) = \overline{x(t)x(t+\tau)} = \lim_{T\to\infty} \int_0^T x(t)x(t+\tau)dt \tag{3.23}$$

where

τ is the delay

$x(t+\tau)$ is the time-shifted signal

Similarly, the cross-correlation function (CCF) between two signals $x(t)$ and $y(t)$ is given by

$$\Phi_{xy}(\tau) = \overline{x(t)y(t+\tau)} = \lim_{T\to\infty} \int_0^T x(t)y(t+\tau)dt \tag{3.24}$$

In discrete domain, Equations 3.23 and 3.24 can be written as

$$\Phi_{xx}(\tau) = \lim_{N\to\infty} \frac{1}{N} \sum_{K=1}^{N} X_k(t)X_k(t+\tau) \tag{3.25}$$

$$\Phi_{xy}(\tau) = \lim_{N\to\infty} \frac{1}{N} \sum_{K=1}^{N} X_k(t)Y_k(t+\tau) \tag{3.26}$$

where

N is the number of observations

$X_k(t)$ and $Y_k(t)$ are the kth samples of the signals at time t

τ is the shifted time

The time-averaged ACF, $\Phi_{xx}(m)$, or time-averaged CCF, $\Phi_{xy}(m)$, for a delay of m samples, can be computed by taking the sum of the product $x(n)x(n+m)$ or $x(n)y(n+m)$ over the data window, respectively. Let the samples extend from $n=1$ to $n=N-1$ for N samples, then the ACF, $\Phi_{xx}(0)$, is obtained with $x(n)\cdot x(n)=x^2(n)$, but the ACF, $\Phi_{xx}(1)$, needs computation of $x(n)\cdot x(n+1)$ and there are a total of 0 to $N-2$ product terms, i.e., 1 less than required for $\Phi_{xx}(0)$. Therefore, the computation of $\Phi_{xx}(m)$ is possible with only $N-m$ data samples. Therefore, ACF is given by

$$\Phi_{xx}(m)=\frac{1}{N-m}\sum_{n=0}^{N-m-1}x(n)x(n+m) \tag{3.27}$$

In practical computations, to keep the entire ACF computations data equal, zeros are added to the signals, which fall short of m samples for the calculation of $\Phi_{xx}(m)$.

In analyzing random signals, the ACF and CCF are calculated from the expected value (i.e., sample average) of the product of two vectors. For two vectors of N data of x and y where

$$X(n)=\left[x(n)x(n-1)\quad\cdots\quad x(n-N+1)\right]^T$$

$$Y(n)=\left[y(n)y(n-1)\quad\cdots\quad y(n-N+1)\right]^T$$

Then, the CCF, $_{xy}(n)$ is

$$\Phi_{xy}(n)=E[X(n)Y^T(n)] \tag{3.28}$$

The product function is a $N\times N$ matrix. One distinct characteristic that is displayed by the ACF is peaks at intervals where the repetitive pattern of a signal is present. Similarly, the CCF shows peaks where the two signals have a similar pattern; therefore, ACF and CCF show importance to detect similarity in periodic pattern between the two signals or two segments of the same signal.

ACF and CCF have been found to be useful signal processing techniques for detecting common rhythm in biomedical signals [3], e.g., between two channels of EEG for spike and wave detection, and QRS complex detection in ECG, respectively. Figure 3.9 and 3.10 shows the p4 channel of an EEG signal and its ACF with 4.67–5.81 s segment, respectively while Figure 3.11 shows a noisy ECG signal and its CCF with the QRS template selected from the first cycle.

Another example of CCF application in sensor signal analysis is arterial pulse wave velocity determination using piezoelectric sensor [9]. In this application (Figure 3.12), the arterial pressure pulse signals $P(x_1,t)$ and $P(x_2,t)$ produced by two Polyvinylidene fluoride (PVDF) thin film piezoelectric sensors positioned at the radial artery at the wrist and over the brachial artery just below the elbow, respectively, of the subject were conditioned first and then their cross-correlation function was calculated given by the equation

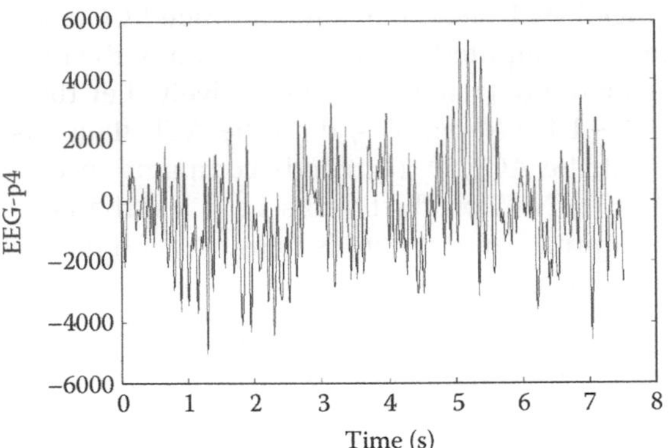

FIGURE 3.9
EEG for channel p4. (Adapted from IEEE, copyright © IEEE. With permission.)

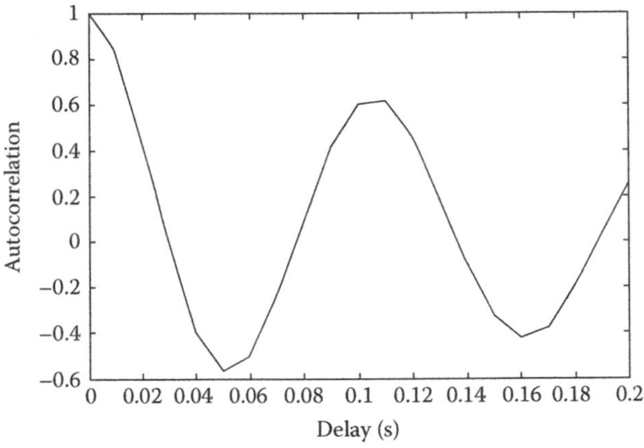

FIGURE 3.10
Autocorrelation function of 4.67–5.81 s segment with p4. (Adapted from IEEE, copyright © IEEE. With permission.)

$$\Phi_{x_1x_2}(\tau) = \frac{1}{T}\int_{-T/2}^{T/2} P(x_1,t)P(x_2,t)\,dt \tag{3.29}$$

The delay τ at which the peak CCF occurs represents the transit time Δt for the pressure wave to reach from x_1 to x_2 along the arterial segment. From the distance between the sensor positions and transit time, the velocity can be calculated as

$$v = \frac{x_2 - x_1}{\Delta t} \tag{3.30}$$

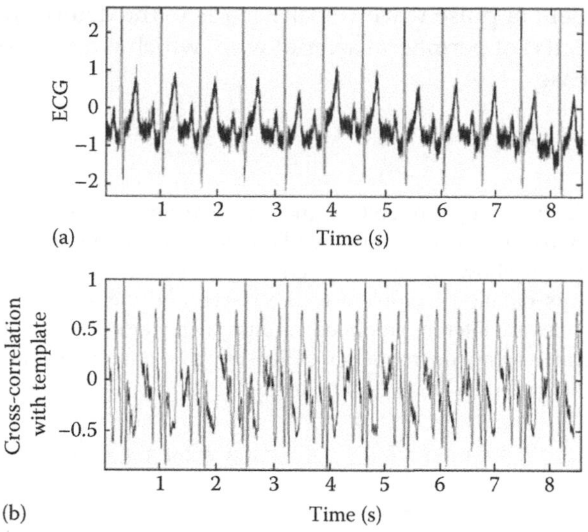

FIGURE 3.11
(a) An ECG signal with high-frequency noise and (b) CCF of QRS complex of first cycle with the ECG signal. (Adapted from IEEE, copyright © IEEE. With permission.)

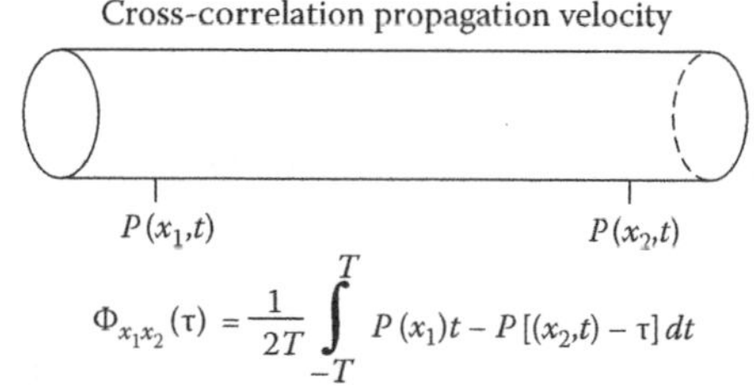

$$\Phi_{x_1x_2}(\tau) = \frac{1}{2T}\int_{-T}^{T} P(x_1)t - P[(x_2,t) - \tau]\,dt$$

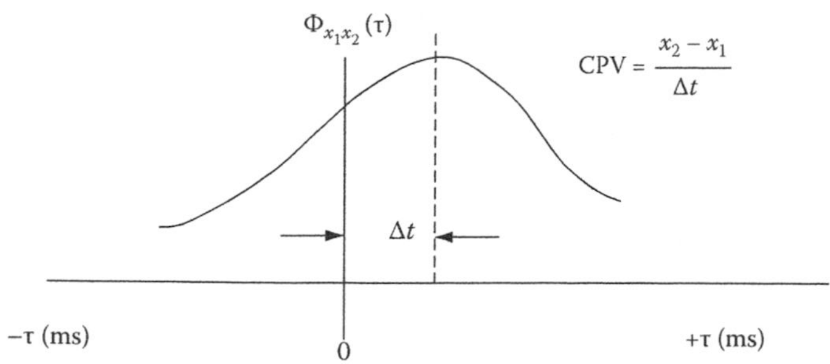

FIGURE 3.12
Arterial pulse wave velocity determination and the CCF between the two pressure pulse signals. (Reprinted from McLaughlin, et al., *Physiological Measurement*, 24, 693, 2003. Institute of Physics Publishing Ltd. With permission.)

The measurement of pulse wave velocity leads to the determination of local stiffness (elasticity) of peripheral arterial wall, which is an important feature of arterial disease.

Example 3.1

Velocity of electroneurogram (ENG) signal of *ulnar* nerve of a human subject is determined by using two concentric needle electrodes positioned over the wrist and just below the elbow (distance 62.5 cm). An electrical pulse of 100 mV amplitude and 100 μs duration is applied to the nerve and the signal sensed by the needle electrodes are filtered, amplified, and then converted to digital format by a 12-bit ADC with a sampling frequency of 500 Hz. The reference voltage is +5.0 V. Two sets of sampled data are shown below:

$$E_1\,(n) = [5.2\ 5.3\ 5.1\ 5.4\ 6.5\ 6.2\ 5.5\ 4.7\ 4.3\ 4.4\ 4.3\ 4.2\ 4.1\ 3.8\ 3.5]\ \mu\text{V}$$

$$E_2\,(n) = [3.5\ 3.8\ 4.2\ 4.9\ 5.2\ 5.5\ 5.9\ 6.1\ 6.3\ 6.4\ 5.5\ 5.2\ 5.1\ 4.8\ 4.3]\ \mu\text{V}$$

For $n = 1$–14, the peak detection and cross-correlation operation were performed on the samples data to determine the ENG velocity. Calculate

1. Cutoff frequency of LPF required to filter any high-frequency noise picked up by the electrode
2. ENG velocity by peak detection method
3. ENG velocity by CCF method
4. Amplification factor needed to amplify the signals for applying to the ADC

Solution

1. To determine the cutoff frequency of LPF, we should find the frequency components of the pulse wave by Fourier transform analysis.

   The Fourier transform of a rectangular pulse is given by

$$F(j\omega) = At_0\,\frac{\sin(\omega t_0/2)}{\omega t_0/2}\ \text{V-s}$$

   Putting $t_0 = 100 \times 10^{-6}$ s and $A = 100$ mV

$$F(j\omega) = 100 \times 100 \times 10^{-6}\,\frac{\sin\left(\omega \times 100 \times 10^{-6}/2\right)}{\omega \times 100 \times 10^{-6}/2}$$

$$F(j\omega) = 10^{-5}\,\frac{\sin(50\omega \times 10^{-6})}{50\omega \times 10^{-6}}$$

   The Fourier transform of the pulse is shown in Figure E3.1b. The cutoff frequency of the LPF is considered as the frequency at which the true component is reduced to $1/\sqrt{2}$ times.

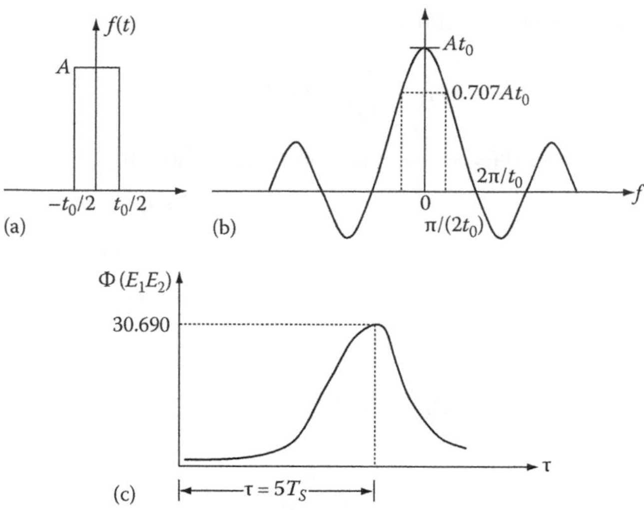

FIGURE E3.1
(a) A pulse signal, (b) Fourier transform of the pulse of Example 3.1, and (c) plot of the CCF.

The highest-frequency component at $\omega = 0$ is

$$F(0) = 10^{-5} \text{ V-s}$$

$$0.707 \times F(0) = F(0) \times \frac{\sin(50\omega \times 10^{-6})}{50\omega \times 10^{-6}}$$

By solving

$$\therefore \omega = \frac{\pi}{200} \times 10^6 \text{ rad/s}$$

$$2\pi f = \frac{\pi}{200} \times 10^6$$

$$f = \frac{1}{400} \times 10^6 \text{ Hz}$$

$$= 0.25 \times 10^4 \text{ Hz}$$

$$= 2500 \text{ Hz}$$

$$= 25 \text{ kHz}$$

2. From the sampled data

E_1 peak = 6.5 μV corresponds to $K = 5$

E_2 peak = 6.4 μV corresponds to $K = 10$

Sampling frequency = 500 sample/s

$$T_s = \frac{1}{500} = 2\times10^{-3}\ \text{s}$$

$$\begin{aligned}\text{Time difference between the peaks} &= (10-5)T_s\\ &= 5\times T_s\\ &= 5\times2\times10^{-3}\\ &= 10\times10^{-3}\\ &= 10\ \text{ms}\end{aligned}$$

$$\begin{aligned}\text{Velocity} &= \frac{\Delta x}{\Delta t}\\ &= \frac{62.5}{10\times10^{-3}}\ \text{cm/s}\\ &= \frac{0.625}{10\times10^{-3}}\ \text{m/s}\\ &= 62.5\ \text{m/s}\end{aligned}$$

3. From Equation 3.27

$$\Phi_{E_1E_2}(m) = \frac{1}{N-m}\sum_{n=0}^{N-m-1} E_1(n)E_2(n+m)$$

Here, $N = 14$, $m = 0, 1, 2, 3, \ldots, 14$ and $n = 0, 1, 2, \ldots, N-1$
Putting the values of N, m, and n

$$\Phi_{E_1E_2}(0) = \frac{1}{14}\sum_{n=0}^{13} E_1(n)E_2(n+1)$$

$$\vdots$$

$$\Phi_{E_1E_2}(14) = \frac{1}{1}\sum_{n=0}^{0} E_1(n)E_2(n+13)$$

On calculation, we get (the author used a programmable calculator to calculate the following)

$$\Phi_{E_1E_2}(0) = 23.89$$

$$\Phi_{E_1E_2}(1) = 25.864$$

$$\Phi_{E_1E_2}(2) = 27.014$$

$$\Phi_{E_1E_2}(3) = 28.017$$

$$\Phi_{E_1E_2}(4) = 30.69$$

$$\Phi_{E_1E_2}(5) = 29.283$$

$$\Phi_{E_1E_2}(6) = 29.643$$

$$\Phi_{E_1E_2}(7) = 29.935$$

$$\Phi_{E_1E_2}(8) = 29.910$$

$$\Phi_{E_1E_2}(9) = 29.058$$

$$\Phi_{E_1E_2}(10) = 27.208$$

$$\Phi_{E_1E_2}(11) = 25.442$$

$$\Phi_{E_1E_2}(12) = 24.630$$

$$\Phi_{E_1E_2}(13) = 23.875$$

$$\Phi_{E_1E_2}(14) = 22.360$$

Figure E3.1c shows the plot of the CCF. Since the CCF peak appears at $m = 4$

$$\Delta t = 5Ts$$

$$= 5 \times 2 \times 10^{-3}$$

$$= 10^{-2}$$

$$\therefore V = \frac{\Delta x}{\Delta t}$$

$$= \frac{0.625}{10^{-2}} \text{ m/s}$$

4. The reference voltage of the ADC

$$V_{ref} = +5 \text{ V}$$

The electrode potential for the peak value of ENG

$$V_{in}(\text{max}) = 6.5 \text{ µV}$$

$$\text{Amplification factor}(A) = \frac{5}{6.5 \times 10^{-6}}$$

$$= 7,69,230$$

Taking the safe value $A = 77 \times 10^4$

Ultrasonic technique is found to be an attractive approach in flow instrumentation because it is nonintrusive, does not need information on the

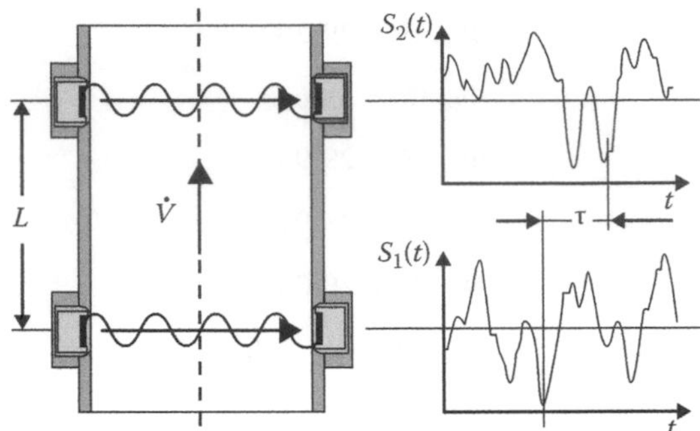

FIGURE 3.13
The ultrasonic flowmeter and the upstream and downstream ultrasonic waves. (Reprinted from Institute of Physics Publishing Ltd. With permission.)

speed of sound in the fluid, and it can be used in single and multiphase flow. Cross-correlation has been proved to be a successful tool in such ultrasonic signal analysis leading to a cross-correlation ultrasonic flowmeter [10]. In such flowmeters, two ultrasonic waves separated by an axial distance L are transmitted continuously across a pipe from wall to wall and normal to the pipe axis (Figure 3.13). The two signals received continuously in time are cross-correlated. The time delay at which the peak cross-correlation occurs is interpreted as the time required by markers in the flow to be converted from the upstream to the downstream ultrasonic wave. The distance L divided by this convection time gives the convection velocity and further calibrated in volumetric flow rate.

The ultrasonic wave was generated and sensed by two pairs of piezoelectric transducers with a resonant frequency of 220 kHz. The signals $S_1(t)$ and $S_2(t)$ were sampled at 20 kHz and then cross-correlated by the equation

$$\Phi_{12} = \frac{1}{T}\int_0^T S_1(t)S_2(t-\tau)\,dt \tag{3.31}$$

Figure 3.14a shows two signals $S_1(t)$ and $S_2(t)$ and Figure 3.14b shows the CCR, Φ_{12}. The peak of the function Φ_{12} is obtained at $\tau = 7.75$ ms. The basic principles of ultrasonic cross-correlation flowmeter are available in [11–13].

3.6 Modifying and Modulating Inputs

In ideal conditions, a sensor or a measurement system should generate an output corresponding to input physical parameters only however, no sensor or measurement system can be expected to do so; therefore, we must model

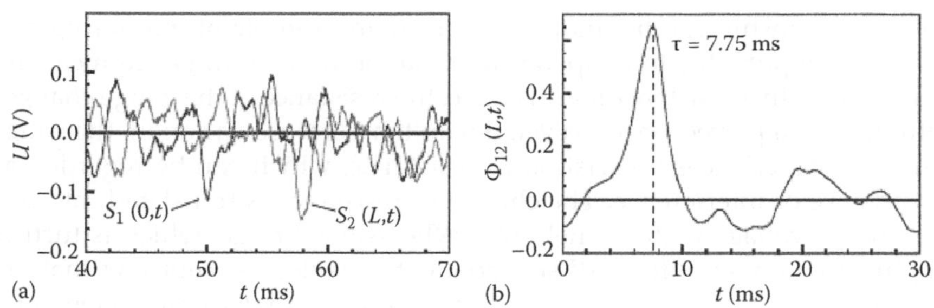

FIGURE 3.14
(a) Ultrasonic signals S_1 and S_2 and (b) CCF between S_1 and S_2. (Reprinted from Institute of Physics Publishing Ltd. With permission.)

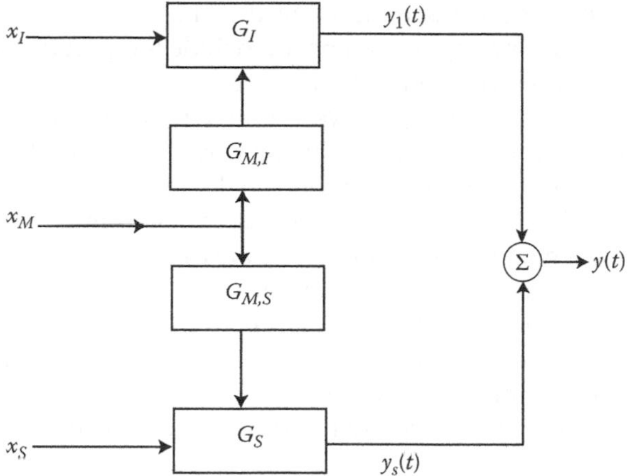

FIGURE 3.15
Effect of internal and external perturbation on measurement systems. (x_S is the signal of interest; $y(t)$ is the system output; x_I is an interference or external perturbation; x_M is a modifying input). (Redrawn from Doeblin, E.O., *Measurement System Applications*, 4th edn., McGraw-Hill, New York, 1990. With permission.)

the system in real situations with interferences. In most cases, two kinds of unwanted signals can add to the sensor output—interference or external perturbations, and a modifying input. The effect of these internal and external perturbations is pictured in a signal flow diagram [14] of Figure 3.15.

In Figure 3.15, $x_S(t)$ is the signal of interest, $y(t)$ is the system output, $x_I(t)$ is an interference or external perturbation, $x_M(t)$ is a modifying input, G_S is the actual sensor gain, and G_I, is the gain of external perturbation or interference component [14]. The sensor input x_S and interfering input x_I get transferred through the above two gains, respectively. The modifying input x_M gets transferred through $G_{M,S}$ to change G_S, while it gets transferred through $G_{M,I}$ to change G_I. The above relationships hold good for all types of gains—linear, nonlinear, varying, or random.

This relationship can be understood from an example of interfering and modifying inputs due to temperature variation in a strain gauge measurement system. In a resistance strain gauge, the resistance of the gauge changes with the strain picked up, however, due to temperature variation, the gauge resistance experiences a variation in resistance, which can be regarded as disturbance or interference. The change in resistance is translated into corresponding voltage with the help of a Wheatstone bridge, which is further amplified by an amplifier. All semiconductor devices (except few that are designed with small temperature variation) produce a drift due to temperature variation. Therefore, the variation of voltage due to the strain picked up, as well as due to the temperature variation in gauge resistance, gets amplified by a factor proportional to variation of temperature in amplifier. This second cause can be interpreted as a modifying input. This model of interfering and modifying inputs is shown in Figure 3.16.

Examples of interfering inputs in sensors include temperature and acceleration interference in pressure-sensing quartz resonators; humidity interference in resistive gas sensors and capacitive sensors; temperature interference in eddy current–based thickness gauge; linear variable differential transformer (LVDT) and semiconductor sensors; thermoelectric interference in resistive strain gauge, and electrochemical interference in electromagnetic flowmeters.

Interference external to sensor module may arise out of the measurement- or signal-conditioning circuit including the electronic devices. Such interferences include resistive interference due to differences in grounding points; capacitively coupled interference; inductive or magnetic interference; thermal interference in resistive components and semiconductor devices, etc. References [15,16] discuss many issues related to interference in measuring circuits.

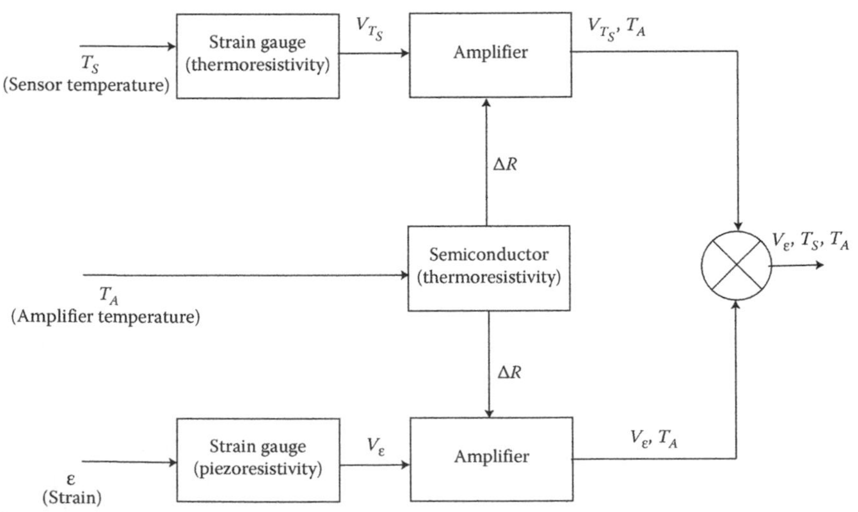

FIGURE 3.16
Model of interfering and modifying inputs of a strain gauge.

3.7 Compensation Techniques

There are several approaches by which the interference effects can be compensated, such as

1. Interference insensitive sensor design
2. Negative feedback
3. Opposing input generation

The first two approaches aim at compensating the effects at sensor level, while the last approach is basically a circuit level approach.

3.7.1 Interference Insensitive Design

It is best to eliminate the interference at the sensor level itself so that it does not propagate to other parts and gets amplified or modulates the output of interest. In the preceding example of interfering inputs of strain gauge, the temperature effect could be compensated by using a strain gauge with a low-temperature coefficient alloys such as Manganin (Cu-84%, Mn-12%, Ni-4%), temperature resistance coefficient of which is only 6×10^{-6}/K. A strain gauge intended to sense principal strain (ε_x) along the longitudinal axis of the gauge also picks up the transverse strain (ε_y) through the end loops of the gauge producing an interference input. It is difficult to compensate this in the circuit level. One approach of compensation of the transverse strain is to use resistive elements of materials insensitive to strain ($\lambda = 0$) in the end loops such as gold, silver, etc. The end loops act as a connector rather than a sensor. In another approach in foil gauge, the cross-sectional area of the end loops are made larger than that of the narrow and long longitudinal foils, thereby, reducing the transverse effect. Similarly, in magnetic sensors, thin and narrow magnetic devices are sensitive only to the magnetic field parallel to the longitudinal axis thereby cancelling all transverse magnetic fields.

Permanent magnet sensors such as accelerometers, torquers, gyroscopes, etc., when used in military and aerospace applications are greatly effected by temperature interference signals. Such a sensor made of high-energy Samarium Cobalt ($SmCo_5$ and Sm_2Co_{17}) or Neodymium–Iron–Boron (Nd Fe B) is required to operate satisfactorily under the temperature range of 10°C–50°C, however, when used in aerospace applications, are required to work at −60°C to 80°C [17]. Due to their higher negative temperature coefficient with typical values −0.04%/°C for Samarium Cobalt and −0.11%/°C for Nd Fe B, temperature interference may cause instability or even mission failure of aerospace. Therefore, conventionally, the temperature interference input is compensated by providing a thin film of thermo-magnetic alloy, usually made of Ni–Fe as a shunt over the permanent magnet. This alloy can control the leakage flux in the shunt path so that the air gap flux density remains the same in the range of operating temperature.

Interference insensitive design may include choosing a proper excitation voltage or magnetic field for the sensor. For example, in electromagnetic flowmeters, an ac exciting magnetic field eliminates the electrochemical interference produced in the electrodes and thermoelectric interference in the connecting lead wires. Another example of this approach in metal oxide gas sensors is driving the heater by a pulse of variable duty cycle rather than a fixed dc. In MGS1100 (a Motorola CO gas sensor), the manufacturer recommends to drive the heater by a 5 V pulse for 5 s first to remove the moisture from the sensor film and then 1 V for 10 s for the actual sensing period.

3.7.2 Negative Feedback

An appropriate sensor design implementing negative feedback is an attractive method to reduce the effects of modifying inputs. In a negative feedback system, the total transfer function (Figure 3.17) is given by

$$\frac{Y(s)}{X(s)} = \frac{G(s)}{1+G(s)H(s)} \tag{3.32}$$

where

- $G(s)$ is the system transfer function
- $H(s)$ is the transfer function of the negative feedback
- $X(s)$ and $Y(s)$ are the input and output of the sensor

The compensation to eliminate the effect of $x_m(s)$ is possible where $H(s)$ is insensitive to x_M. This can be achieved by designing $G(s)$ fairly larger than $H(s)$. This approximation results in changing Equation 3.32 into

$$\frac{Y(s)}{X(s)} = \frac{1}{H(s)}$$

Since $G(s)H(s) \gg 1$

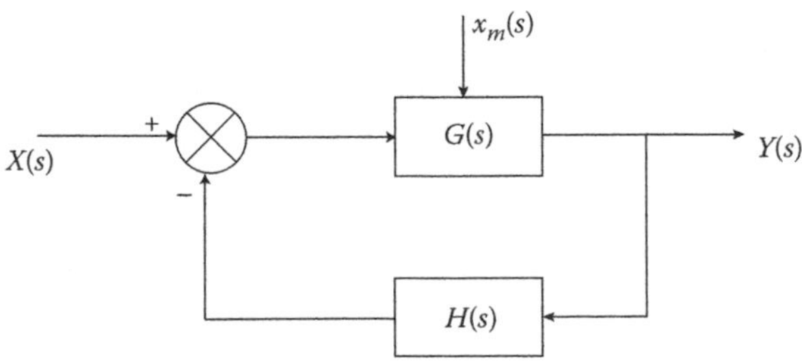

FIGURE 3.17
A negative feedback system.

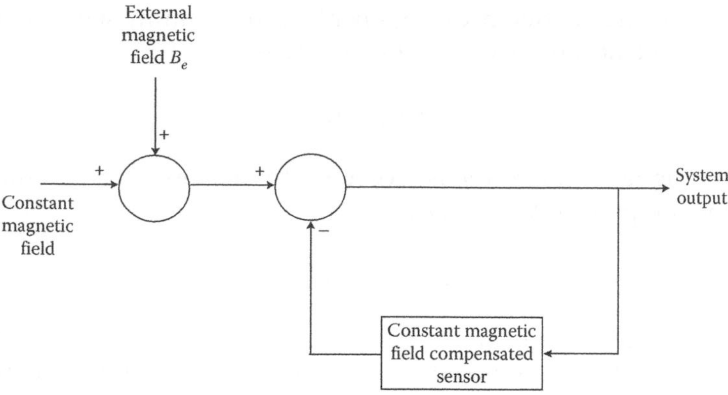

FIGURE 3.18
Principle of magnetic field generation with active compensation external magnetic field. (This compensation technique consists of a split coil, a magnetic sensor, a feedback coil and a current control circuit.)

In designing larger $G(s)$ than $H(s)$ causes lower power handling capacity, higher accuracy and sensitivity of $H(s)$, and lower energy requirement.

An example of negative feedback compensation used in measurement system is a magnetic field sensor with active compensation of active fields such as the earth's magnetic field and magnetic noise [18]. In this technique, a dual coil magnetic field generator with the function of automatically removing only the component of external magnetic fields by the generation of a constant magnetic field using negative feedback is used. The basic principle of the magnetic-field generator with negative feedback compensation of external magnetic field is shown in Figure 3.18.

This compensation technique will be discussed in Chapter 5.

3.7.3 Opposing Input Generation

When a sensor cannot inherently compensate the interfering and modifying inputs, opposing input generation approach is very common and attractive. This technique relies on developing a component equivalent to the signal developed by the interfering and modifying input by another device commonly known as *dummy* device. The signal developed by the dummy sensor may either be positive or negative in polarity with reference to the main sensor. When the signal is positive, it is subtracted from the signal of the main sensor and when it is negative, it is added. The addition or subtraction operation is performed by proper circuit arrangement with the main sensor, or separately done using an operational amplifier (OPAMP). Remembering the model of interfering and modifying inputs, let the output consist of two components—the sensor signal of interest (y_s) and the modifying or interfering input (y_{mi}) given by

$$y = y_s + y_{mi}$$

To compensate the interference component y_{mi}, the dummy sensor must produce an equivalent component y'_{mi} or $-y'_{mi}$ where

$$\pm y'_{mi} = y_{mi}$$

The sensor signal of interest y_s is recovered from y either by subtracting or adding the component $\pm y'_{mi}$ as given

$$y_s = y \mp (\pm y'_m)$$

Temperature interference in sensors is commonly compensated by such techniques, which are discussed in detail in Chapter 4. In the interference model discussed above, the sensor gain $G(s)$ is modified by the modifying input x_M through its gain $G_{M,S}$ to add a modifying component to the system output.

For a strain gauge with gauge factor λ and resistance R, which is subjected to a strain of ε, the unit change in resistance due to the applied strain is

$$\frac{\Delta R}{R} = \lambda\varepsilon$$

The gauge is made up of a material with a resistance temperature coefficient of α Ω/Ω/°C and a linear expansion coefficient of β m/m/°C. If the gauge is subjected to a temperature variation of Δt (°C), the change in resistances can be expressed as follows.

Let us define a term ε_T to denote total strain picked up by the gauge that comprises of two components ε, the applied strain and ε_t, the strain due to linear expansion for temperature. Therefore

$$\frac{\Delta R}{R} = \lambda\varepsilon_T = \lambda(\varepsilon + \varepsilon_t)$$

So

$$\Delta R = \lambda R(\varepsilon + \beta\Delta t)$$

There is another component of change in resistance ΔR_t due to temperature variation, which is given by

$$\Delta R_t = \alpha R\,\Delta t$$

Therefore, total change in resistance

$$\Delta R = \lambda R(\varepsilon + \beta\Delta t) + R\alpha\Delta t$$

$$\Delta R = R(\lambda\varepsilon) + R\Delta t(\lambda\beta + \alpha)$$

Therefore, the combined gauge factor

$$\lambda_T = \frac{\Delta R_t/R}{\varepsilon}$$
$$= \lambda + \frac{\lambda}{\varepsilon}\beta\Delta t + \frac{1}{\varepsilon}\alpha\Delta t$$
$$= \lambda + \lambda_\alpha + \lambda_\beta$$

Therefore, the combined gauge factor comprises two other components—one due to resistance temperature coefficient α and the other due to linear temperature expansion coefficient β. When a single strain gauge of above specification connected to a Wheatstone bridge configuration (with other fixed resistances R) is powered by a dc voltage V_i, the output voltages for each of the components of λ, λ_α, and λ_β can be expressed by

$$V_0 = V_i[\lambda + \lambda_\alpha + \lambda_\beta]\varepsilon$$

It is observed that the temperature interference adds components λ_α and λ_β to the gauge factor $\lambda\varepsilon$, which can be compensated by generating an equivalent component by another dummy gauge and subtracting from the sensor signal. Similarly, a negative temperature coefficient sensor such as thermistor can also be employed to compensate temperature interference. The thermistor when connected in series with the resistive sensor with positive temperature coefficient, the resulting effect can eliminate the temperature interference in spite of temperature change. Such techniques of temperature compensation in intelligent sensors have been discussed in detail in Chapter 4.

Example 3.2

A metallic strain gauge of resistance 120 Ω and gauge factor 2.0, $\alpha = 5\times10^{-3}$ Ω/Ω/°C, $\beta = 1\times10^{-3}$ m/m/°C is bonded to a beam, which is subjected to a strain of 200 μ and a temperature variation of 10°C (Figure E3.2a).

1. Determine the
   a. Gauge factor and change in resistance due to α and β.
   b. Temperature interfering output voltages for a power supply voltage of 12.0 V in the Wheatstone bridge circuit.
2. Draw the schematic circuit diagram with temperature-compensating dummy gauges.
3. Draw the schematic block diagram with temperature-compensating sensor amplifier.

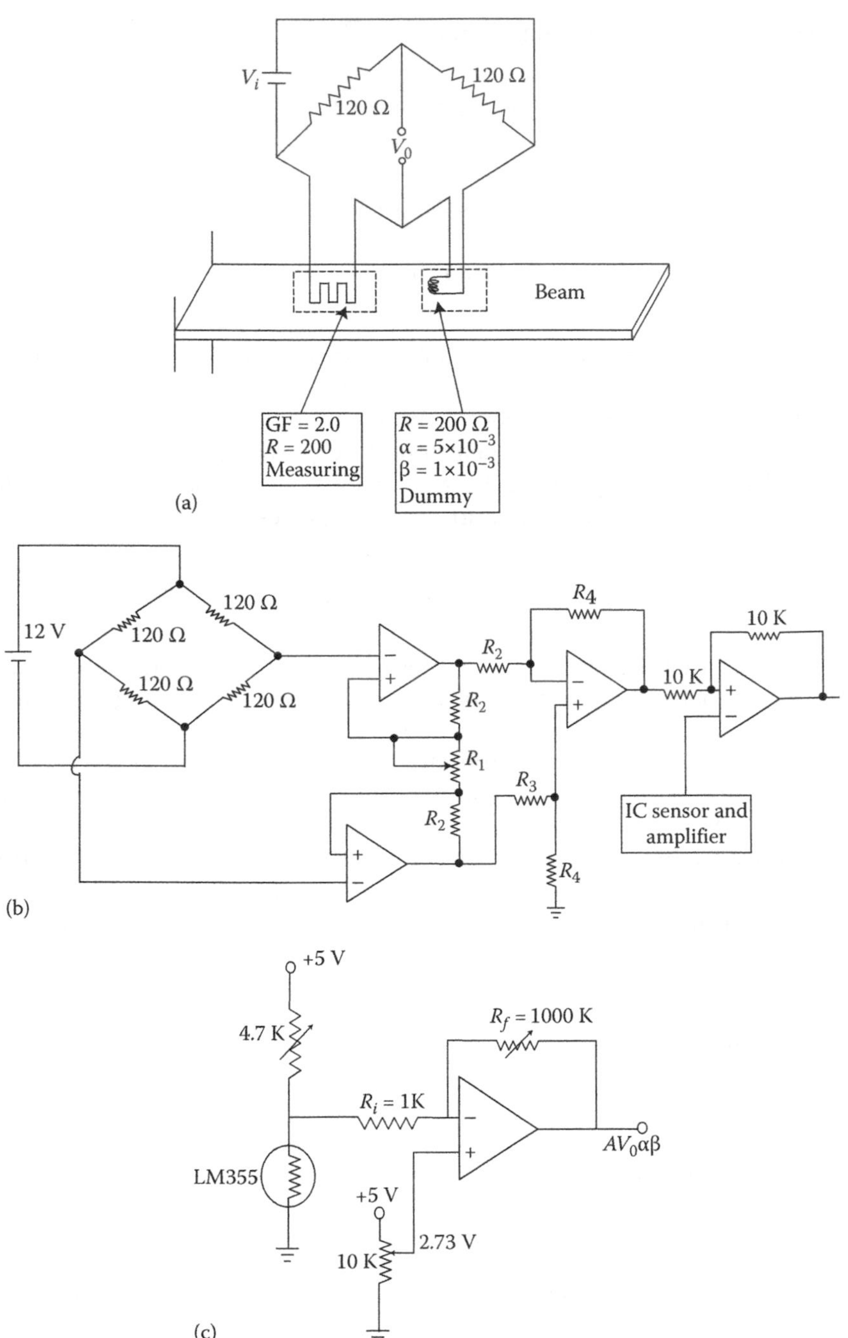

FIGURE E3.2
(a) Strain gauge with a dummy gauge, (b) temperature compensation using IC thermal sensor, and (c) IC thermal sensor circuit.

Solution
Given values

$$\varepsilon = 200\times10^{-6}$$
$$\lambda = 2.0$$
$$R = 120\ \Omega$$
$$\alpha = 5\times10^{-3}\ \Omega/\Omega/^{\circ}\mathrm{C}$$
$$\beta = 1\times10^{-3}\ \mathrm{m/m/^{\circ}C}$$
$$\Delta t = 10^{\circ}\mathrm{C}$$
$$V_i = 12.0\ \mathrm{V}$$

a. The gauge factor due to α

$$\lambda_\alpha = \frac{\alpha}{\varepsilon}\Delta t$$
$$= \frac{5\times10^{-3}\times10}{200\times10^{-6}}$$
$$= 250$$

Change in resistance

$$\Delta R_\alpha = \lambda_\alpha R\varepsilon$$
$$\Delta R_\alpha = 250\times120\times200\times10^{-6}$$
$$= 6\ \Omega$$

The gauge factor due to β

$$\lambda_\beta = \frac{\beta\lambda}{\varepsilon}\Delta t$$
$$= \frac{2.0\times10^{-3}\times10}{200\times10^{-6}}$$
$$= 100$$

Change in resistance

$$\Delta R_\beta = 100\times120\times200\times10^{-6}$$
$$\Delta R_\beta = 2.4\ \Omega$$

b. Interfering output voltages due to α

$$V_{0\alpha} = V_i\lambda_\alpha\varepsilon$$
$$= V_i\alpha\,\Delta t$$
$$= 12\times5\times10^{-3}\times10$$
$$= 600\ \mathrm{mV}$$

Interfering output voltages due to β

$$\begin{aligned} V_{0\beta} &= V_i\lambda_\beta\varepsilon \\ &= V_i\lambda\beta\,\Delta t \\ &= 12\times 2\times 10^{-3}\times 10 \\ &= 240\text{ mV} \end{aligned}$$

c. The dummy gauge insensitive to strain but having the same resistance R and coefficients α and β should be connected to the adjacent arm of the Wheatstone bridge circuit as shown in Figure E3.2b.

   The output voltages of 600 and 240 mV developed due to temperature effect can be compensated by using a temperature sensor such as semiconductor thermal sensor, as shown in Figure E3.2c.

d. The amplification factor of the instrumentation amplifier is set at 100 given by the equation

$$A_1 = \frac{R_4}{R_3}\left(1+\frac{2R_2}{R_1}\right)$$

where $R_4=10\text{ K}$, $R_3=2\text{ K}$, $R_2=100\text{ K}$, $R_1=10.52\text{ K}$

   The output voltage of the amplifier is corrupted with the two components of voltages due to temperature effect

$$\begin{aligned} A_1V_{0\alpha} &= A_1V_i\alpha\,\Delta t \\ &= 100\times 12\times 5\times 10^{-3}\,\Delta t \\ &= 6\Delta t \end{aligned}$$

$$\begin{aligned} A_1V_{0\beta} &= A_1V_i\lambda\beta\,\Delta t \\ &= 100\times 12\times 2\times 10^{-3}\,\Delta t \\ &= 2.4\,\Delta t \end{aligned}$$

and combined effect

$$\begin{aligned} A_1V_{0\alpha\beta} &= A_1(V_{0\alpha}+V_{0\beta}) \\ &= 8.4\,\Delta t \end{aligned}$$

   This effect should be subtracted from the output of the instrumentation amplifier using a differential OPAMP of unity gain. The 8.4 Δt functional voltage component is generated by an IC sensor with an amplifier, as shown in Figure E3.2c.

   Sensor output with 10 mV/°C sensitivity = $(2.73 + 10\times 10^{-3}\,\Delta t)$ V

   Offsetting 2.73 mV and amplifying by a factor A_2 the amplified output

$$= A_2[(2.73 + 10 \times 10^{-3}\Delta t) - 2.73]$$

$$= 10^{-4} A_2 \Delta t$$

By equating

$$10^{-4} A_2 \Delta t = 8.4\,\Delta t$$

$$A_2 = 840$$

$$\text{Setting } A_2 = 840 = \frac{R_f}{R_i}$$

and taking $R_t = 1$ K, $R_f = 840$ K (1000 K Pot)
R_f should be adjusted to 840 K.

3.8 System Dynamics

It has already been mentioned in Section 2.3.1 under dynamic error and dynamic sensitivity that most sensors and measurement systems comprise of energy-storing elements for which they show a different behavior to time-varying inputs than to static inputs. Such behaviors can be analyzed only when we know the input–output relationships of the system. Mathematically, the behavior of a sensor can be described by a constant coefficient linear differential equation assuming that the system is linear and time invariant. This results in an input–output transfer function of the system which can be expressed in Laplace transform. The order of the differential equation or the Laplace operator designates the order of the system as zero-order, first-order, or second-order system. Systems higher than second order are not generally used.

3.8.1 Zero-Order System

The output of a zero-order system is related to its input by a constant sensitivity K given by

$$y(t) = Kx(t)$$

The constant K is the static sensitivity that does not change with input frequency. A sensor to exhibit such a characteristic should have no energy-storing elements. Sensors comprising of simple force or load transmission systems like lever (without spring, damper, etc.) and only resistive elements will show a zero-order characteristic. However, inherent stray effects of energy-storing elements cannot be avoided practically. Example of a zero-order transducer is a potentiometer used for translational displacement measurement. The input–output relation of the potentiometer is

$$V_0 = V\frac{x_i}{x_t}$$
$$= Kx_i$$

where x_i is the applied displacement and x_t is the total displacement of the potentiometer.

3.8.2 First-Order System

The output of a first-order system is related to its input by a first-order differential equation given by

$$x(t) = a_1\frac{dy(t)}{dt} + a_0 y(t)$$

Taking Laplace transform

$$X(s) = a_1 sY(s) + a_0 Y(s)$$

Gives the transfer function

$$\frac{Y(s)}{X(s)} = \frac{K}{(1+s\tau)}$$

where
$K = (1/a_0) =$ static sensitivity
$\tau = (a_1/a_0) =$ time constant

The magnitude and phase of the transfer function in frequency domain are given by

$$|G(j\omega)| = \frac{K}{\sqrt{1+\omega^2\tau^2}}$$

and $\Phi(j\omega) = \tan^{-1}(\omega\tau)$

The response of the first-order systems to test inputs—step, ramp, and sinusoid—are shown in Figure 3.19. The dynamic error and time delay of the three test inputs can be found by the equation

$$e = y(t) - Kx(t)$$

The response equation of a first-order system to step and ramp, inputs can be found in most books of control systems.

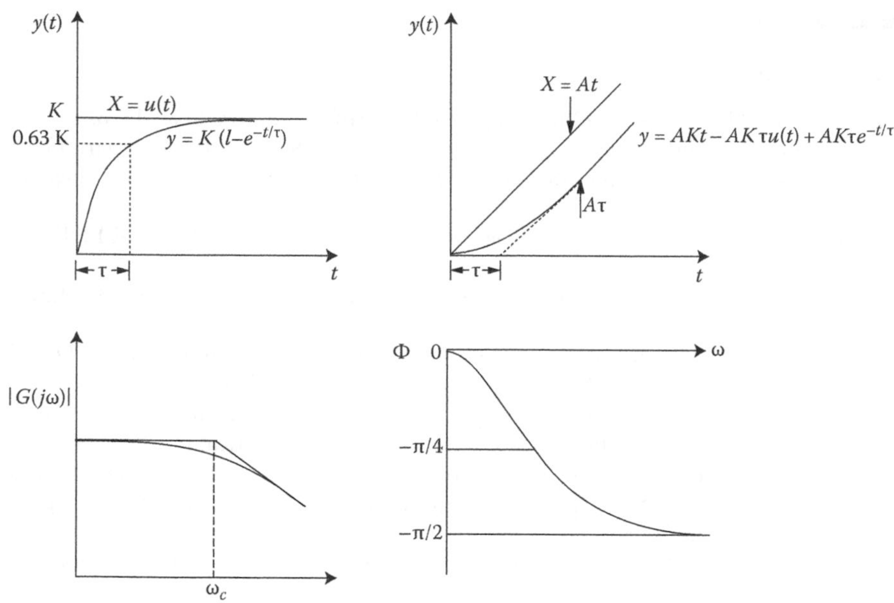

FIGURE 3.19
Response of the first-order systems to step and ramp inputs.

3.9 Electronic Noise Cancellation

Noise imposes a limit to the sensitivity of sensors and measurement systems. The complete sensor system comprises of the sensor, the signal processor, the electronic interface, and the surrounding medium where it works. Each of the components may contribute a noise component coherent with the signal of interest. Noise and interference are similar; however, interference can mostly be modeled and eliminated at the source by proper shielding and grounding. Often, the interference can be compensated using compensating techniques since the interference dynamics can be modeled. Interference is called a noise when it corrupts the signal without an exception of stopping at the source. Hence, interference is *to be compensated* while a noise is *to be filtered*. There are numerous examples of noise corrupting the signal and, therefore, an intelligent sensor can either filter out the noise or have to employ a separate technique to filter out the noise. Thermal noise in resistive sensors and resistors used in the bridge and amplifier circuit; noise due to turbulence produced in the flow system in the flow sensor; power line interference added to the ECG signal; steller noise corrupting information about outer space received by space probes on earth are a few examples.

References

1. Kendall, M., *Time Series* (2nd edn.), Charles Griffin, London, U.K., 1976.
2. Challis, R.E. and Kitney, R.I., Biomedical signal processing (in four parts) Part I. Time-domain method, *Medical and Biological Engineering and Computing*, 28, 509–524, 1990.
3. Rangayyan, R.M., *Biomedical Signal Analysis: A Case-Study Approach*, IEEE Press, 2002.
4. Swanson, D.C., *Signal Processing for Intelligent Sensor Systems*, p. 23, Marcel Dekker, 2000.
5. McNamara, D.E. and Boaz, R.I., Seismic noise analysis system using PSD and PDF function, a Stand Alone Software Package, Open File Report, 2005, 1438, USGS, 2006, U.S. Geological Survey, Reston, VA, www.pubs.usgs.gov/of/2005
6. King, D., Lyons, W.B., Flanagan, C., and Lewis, E., An optical fibre ethanol concentration sensor utilizing Fourier transform signal processing analysis and artificial neural network pattern recognition, *Journal of Optics, Pure and Applied Optics*, 5, 69–75, 2003.
7. Pazdera, L. and Smutný, J., Identification and analysis of acoustic emission signals by Cohen's class of time frequency distribution, *Proceedings of the 15th W C NDT*, Rome, Italy, 2000.
8. Smutný, J., Analysis of vibration since rail transport by using Wigner-Ville transform, *TRANSCOM 99—Third European Conference of Young Research and Science Workers in Transport and Telecommunications*, Zilina, Slovak Republic, pp. 101–104, June 1999, ISBN-80-7100-616-5.
9. McLaughlin, J., McNeill, M., Braun, B., and McCormack, P.D., Piezoelectric sensor determination of arterial pulse wave velocity, *Physiological Measurement*, 24, 693–702, 2003.
10. Schneider, F., Peters, F., and Merzkirch, W., Quantitative analysis of the cross-correlation ultrasonic flowmeter by means of system theory, *Measurement Science and Technology*, 14, 573–582, 2003.
11. Beck, M.S., *Cross-Correlation Flowmeters: Their Design and Applications*, Institute of Physics Publishing, Bristol, U.K., 1987.
12. Manook, B.A., A high resolution cross-correlator for industrial flow measurement, PhD thesis, University of Bradford, Bradford, U.K., 1981.
13. Worch, A., A clamp-on ultrasonic cross-correlation flowmeter for one phase flow, *Measurement Science and Technology*, 9, 622–630, 1998.
14. Doeblin, E.O., *Measurement Systems Applications* (4th edn.), McGraw-Hill, New York, 1990.
15. Morrison, R., *Instrumentation Fundamentals and Applications*, John Wiley & Sons, New York, 1984.
16. Ott, H.W., *Noise Reduction Techniques in Electronic Systems* (2nd edn.), John Wiley & Sons, New York, 1988.
17. Rajagopal, K.R., Singh, B., Singh, B.P., and Veda Chalaan, N., Novel methods of temperature compensations for permanent magnet sensors and actuators, *IEEE Transactions on Magnetics*, 37(4), 1995–1997, July 2001.
18. Sirai, T., A magnetic field generator with active compensation of external field, IP Publishing, *Measurement Science and Technology*, 15, 248–253, 2004.

4

Intelligent Sensors

4.1 Introduction

Classical sensors are adequately used for traditional methodological operations such as obtaining an electrical signal corresponding to a physical parameter. In comparatively complex situations, signals generated by a transducer are mathematically operated like squared or squared rooted by using a microprocessor or a microcontroller. Traditional transducers, in such situations, are termed in a way like "microcontroller based" or "microprocessor interfaced," etc. Over the years, the industrial requirement of sensors have changed tremendously in terms of cost, size, quantity, and intelligence; industrial control and instrumentation has shifted away from large distributed sensor system to integrated, small, low-cost system that could be able to perform in flexible environment conditions in real time with added signal processing capabilities. Thus, the added set of capabilities of sensor system led to a new class of sensors—intelligent sensors.

The *smartness* or *intelligence* of this new category of sensor relies on the features such as long mission duration; reliability and availability; operation in unbenign; unstructured environments real-time operation; and flexibility of use [1]. There requirements lead to components with increasingly autonomous functioning capabilities based on decentralized distribution system architecture principal. Such a system offers a wide range of quantity attribute such as ease of system integration; interoperability, scaling, portability and modularity; inherent robustness; and survivability.

Based on the method or approach of how these intelligent features are achieved in sensors, several varieties of the intelligent sensors are modeled, implemented, or even commercially marketed. However, many of them are termed interchangeably with different names by various authors or researchers. Sometimes, a sensor integrated with a simple digital processor in a single chip is also termed as "intelligent sensor," but it is not indicative of any intuitive ability of the sensor functionality. Therefore, in many texts, a reader gets mislead by the terminology used in defining various intelligent sensors.

4.1.1 Classification

Intelligent sensors may be classified via two different approaches—by the function they perform or the technology based on which they are designed. Some of the intelligent sensors function at higher hierarchical levels such as calibration, error compensation, linearization, offset elimination, self-identification, self-testing, data validation, adaptation, data recovering, signal compensation, decision making, fault detection, data classification, etc., but not limited to.

Most physical systems work with slow varying signals for which the information bandwidth requirement of the sensor is comparatively smaller than the available bandwidth of the electronics used. Therefore, the surplus bandwidth can be used for further hierarchical operation such as enhancement of reliability and reduction of power dissipation, etc. Following is a list based on the two classifications (but not exhaustive).

Based on functions

Smart sensors
Cogent sensors
Self-adaptive sensors
Self-validating sensors

Based on techniques

Soft or virtual
Mathematical
Symbolic
Computational
ANN based/fuzzy based
Intelligent integration
Very large-scale integration (VLSI)
MEMS/microsensors.

4.2 Smart Sensors

Classical sensors are said to be "dumb" in the context of intelligence or smartness needed for extended activities in flexible environmental conditions. A sensor possesses the smart sensor tag when a built-in signal processing unit is incorporated for achieving greater versatilities. In the simplest definition, smart sensors are those obtained by integrating sensors with some form of signal processing circuit.

The IEEE 1451 smart sensor interface standard defines smart sensor as an analog or digital sensor integrated or interfaced with onboard data storage/processing circuits [2]. According to prediction made by Frost and Sullivan [3], the North America smart sensor market will reach $635.2 million by 2010.

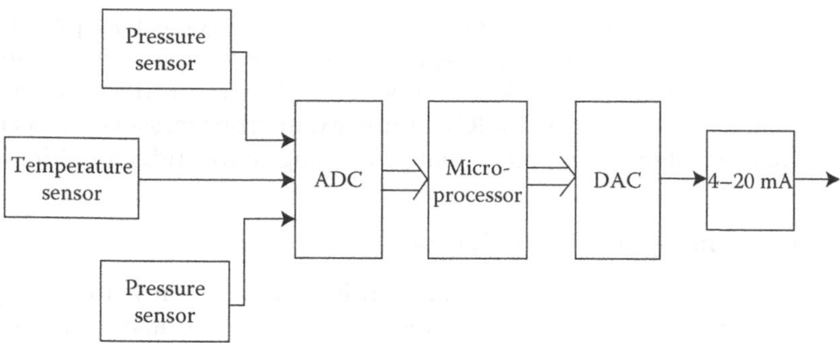

FIGURE 4.1
Block diagram of a smart pressure sensor with temperature compensation.

Remarkable growth for MEMS-based sensor, smart sensors, and sensors with bus capabilities is also expected.

Smart sensors are designed mostly to perform sensor metrological signal processing operations on board with the help of signal processing circuits, ADCs, and microprocessors. The metrological signal processing operations performed by a smart sensor are calibration, linearization, error compensation, offset elimination, A-to-D conversion, etc. A large variety of smart sensors are developed by researchers and many of them are patented and are also commercially available. While discussing intelligent sensor signal processing technique in Chapter 5, some of the smart sensors developed so far by researches will be discussed.

A smart sensor operation may be as simple as a digital communication using an on-board ADC or complex linearization and calibration algorithm implemented in a monolithic chip. The first commercial smart sensor was introduced by Honeywell in 1983. It is a smart pressure sensor consisting of two pressure sensors, generating differential and static pressure signals. The signals are multiplexed into an ADC and a microprocessor. For the compensation of error due to temperature, the temperature signal is obtained from a temperature sensor and compensation operation is performed in a microprocessor. The compensated pressure signal is converted back to analog form by a DAC and transmitted in 4–20 mA current loop. This example of smart sensor can be illustrated by a block diagram in Figure 4.1.

The various hardware techniques for the implementation of a smart sensor depends on factors such as speed needed to be maintained, possibility of integrating the sensor modules with the processing circuitry, and compatibility of the signal processing. Based on these factors, a smart sensor can be categorized as follows.

4.2.1 Monolithic Integrated Smart Sensor

In monolithic integrated smart sensors, microprocessors are integrated with signal processing circuit in a single IC package. Microsensor based on thermo-resistive, piezoresistive, magnetoresistive, piezoelectric, photoconductive, mass-sensitive, bio-sensitive, and acoustic principles are becoming popular due to small size and low cost because of mass production. Due to their similarity to semiconductor IC technology, microprocessors are becoming popular for integrating with signal processing semiconductor chips.

4.2.2 Hybrid Integrated Smart Sensors

When the sensor technology does not permit its integration with the signal processing circuit on the same chip, the sensors are mounted on the same substrate of the chip. In a variation of the hybrid integrated smart sensor, the sensor modules are completely isolated from the signal processing module. Such a hybrid and distributed smart sensor concept has been used by Aziz et al. [4] for designing an ultrasonic smart flowmeter. The block diagram of the smart flowmeter is shown in Figure 4.2.

The smart device has been designed for user-defined parameters—types of ultrasonic flowmeter, types of channel, measuring dimensions, flow range, fluid pressure, etc., and the configuration modules take care of this. The measurement modules perform correction, consider hydraulic coefficient and hydrodynamic conditions, digital filtering, etc.

Many other intelligent sensors apparently possess the characteristics of a smart sensor by the virtue of integrated sensor and mathematical computational circuit in the same chip; however, those sensors can be most suitably termed as computational VLSI sensors, which will be discussed in Section 4.7. Smart sensors basically impart smartness to the classical sensors only by performing inbuilt metrological processing circuits. Different terminologies such as cogent, computational, self-adaptive, self-validity, etc., are

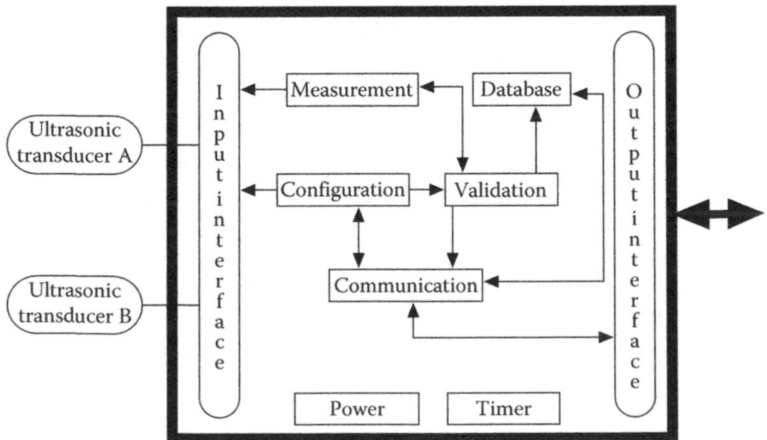

FIGURE 4.2
The minimum functionalities of the ultrasonic flowmeter. (Reprinted from IEEE, copyright © IEEE. With permission.)

used for sensors performing higher-level hierarchical operations, which will be discussed in the following sections.

4.3 Cogent Sensors

It has already been mentioned that smart sensors providing higher-level hierarchical functions are classified as per the services provided. Smart sensors provide raw data although linearized, error compensated, offset removed, calibrated, etc. When a smart sensor cannot convert data into information, a cogent sensor reduces data to decision, classification, inference, semantic transforms, etc., therefore, a cogent sensor includes an additional decision-making module in order to convert raw data into information.

Formally, a cogent sensor is a sensor that includes signal processing and a decision-making module in order to convert the raw data from the sensor hardware to the particular form required by the application. It has the capability of performing semantic transformations if necessary by removing unwanted data and produces inferred and derived information [1]. The functionalities of a cogent sensor is depicted by the block diagram shown in Figure 4.3.

It has been observed from our experience that most measurement systems under nonlinear environments are different to be modeled by linear mathematical relations and are best represented by ANN systems. Cogent sensors' capabilities like decision making, semantic transformation, and generating inference and information are based on experience and heuristic rules. Therefore, cogent sensors are commonly designed with ANN techniques employing array-based sensors fulfilling other integrated microsensor requirements. Cogent sensors can be categorized into two classes based on the dimensionality of the sensor array: monotype sensor array and multi-type sensor array [1]. Monotype sensor array exploits a large sensor array system, each component of which performs measurement of the same quantity at different localities of the field. A good example of monotype sensor is the E-nose sensor array where each sensor of the array performs gas detection. In multi-type sensor array, the sensors perform different measurement operations such as in water quality monitoring array, the sensors perform measurement of CO_2, pH, and conductivity of water.

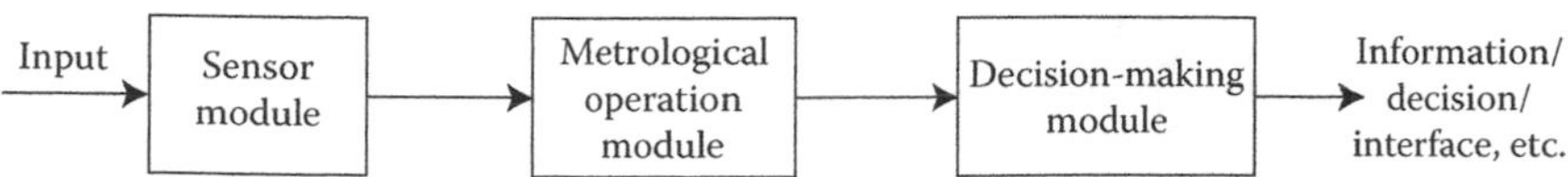

FIGURE 4.3
Functionalities of a cogent sensor.

4.4 Soft or Virtual Sensors

Sensors are key monitoring devices to keep watch on the behavior and performance of a plant. In many situations in a plant, certain measurements may not be possible or unavailable due to the following reasons:

1. Failure of the sensor
2. Removal or maintenance of the sensor
3. Insufficient sampling frequency of the sensor
4. Unavailability of the sensor or high cost
5. Unavailable due to time sharing with another measurement point

Since the measurement becomes unavailable either due to complete or temporary absence of the sensor during a certain period of time, the plant performance will be uncertain and unpredictable. A soft sensor is a solution to the above situations that can replace the unavailable physical sensor.

The principal to the soft sensor working is estimating the unavailable measurements by using a model of the plant or a part of it that correlates the measurement of interest with other variables. As the name indicates, for a soft sensor, the model is implemented in a software. Figure 4.4 shows the basic idea of how a soft sensor works.

The secondary input variables of the sensor model are the control inputs, the disturbances, and other intermediate variables of the plant. Figure 4.4 shows that, when the physical sensor is not available for measurement, the switch at the sensor output is placed in position S when the control loop uses the primary variable Y_m generated by the soft sensor. When the physical sensor is available, the switch is positioned at P and the model starts estimating its parameter for the next operation.

In an another situation, when a single physical sensor is time shared between two plants, the plant without the hard sensor is provided with soft sensor, as shown in Figure 4.5.

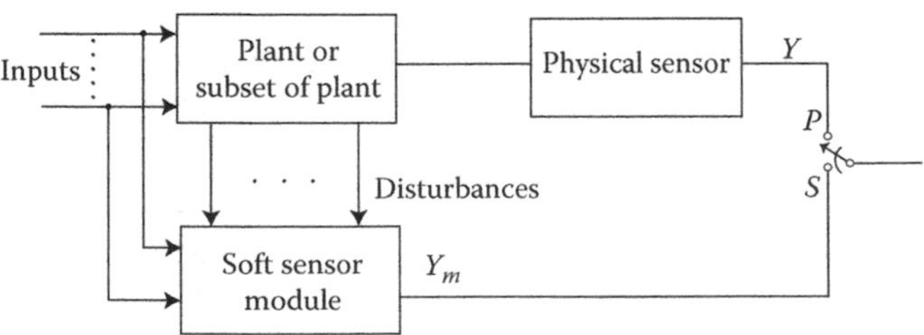

FIGURE 4.4
A soft sensor replacing a physical sensor.

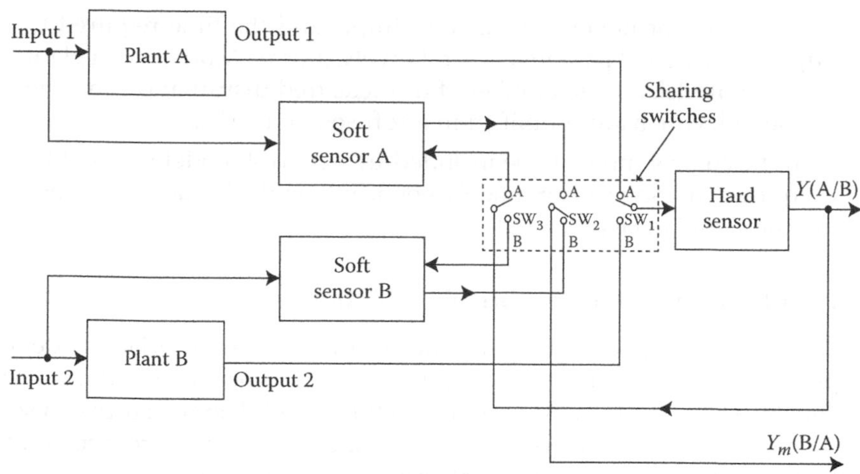

FIGURE 4.5
Time sharing of hard sensor and soft sensors.

When the hard sensor is to connect to the output of plant A, SW_1 is positioned at A and at the same time the soft sensor B is connected to the output of plant B by SW_2 giving Y_m. During this time, soft sensor A performs estimation of the plant-A model parameters using hard sensor output with the switch position SW_3 at A. Similar sharing of the hard sensor to plant-B and soft sensor to plant-A can be obtained by changing the position of SW_1, SW_2, and SW_3 to B, A, and B, respectively.

The basic requirement in designing a soft sensor is assigning a full or partial model of the plant for generating estimated measurement of a measurand of interest to replace a physical sensor. The model is determined by the method of parameter estimation. The model should be capable of generating the unavailable measurements due to the reasons discussed above. The basic principal of soft sensor model is similar to model reference adaptive control approach where the model parameters of the plant are estimated in prior to the measurement and updated continuously; however, the following are the basic differences between the model of a model-based control and a soft sensor model [5]:

1. Control model of a plant is commonly used for predicting a signal relatively for a short period, whereas in a soft sensor, the prediction of the signal should last for a long period, e.g., slow varying disturbances can be considered as constant for a control model, but it may need to be represented within a much longer duration in a soft sensor model.
2. Control model of a plant is relatively elaborated and should cover the entire plant, whereas the soft sensor model may be partial to the plant covering the part that relates to the signal of interest. Hence,

the soft sensor model is relatively simple and the time required to update the model parameters is relatively shorter. Moreover, in plant control model, a large number of unmeasured disturbances may be included, which are simplified in a soft sensor model.

3. All the control inputs to be included in a control model may not be required in a soft sensor model depending on the influence of those to the signal to be predicted.

4.4.1 Soft Sensor Secondary Variable Selection

One of the most important tasks in finding the most appropriate model of the plant for the soft sensor is to find the most significant secondary measurements, which can be used as inputs to the model. The structure of a soft sensor model is determined based on the plant behavior with respect to the primary measurement, i.e., the signal or parameter of interest.

The selection of secondary measurement variables from a large set of measurable variables, which have the best functional relationship with the primary variable, is a primary concern in soft sensor modeling. A smaller number of secondary variables are always preferable from cost and maintenance point of view. From an outward inspection, it apparently looks like all measurable variable influence the primary variable estimate; however, in many cases, all variables do not contribute to the estimate of the measurement of the signal of interest. The following sections discuss the soft sensor implementation by intelligent techniques.

4.4.1.1 Rough Set Theory

Rough set theory (RST) will be discussed here, which has become a popular tool for nonlinear system modeling. RST is a mathematical tool for relating imperfect input–output data of a system. The philosophy of the theory is based on the assumption that with every object of the universe of discourse, some information (such as data, knowledge, etc.) is associated. Objects resulting from some information are *indiscernible* (similar) based on the information available. The *indiscernibility relation* is used in RST to relate the data. Terms and mathematical relation:

Elementary set—any set of all indiscernible (similar) objects.

Crisp (precise) set—any union of some elementary sets.

Rough (imprecise) set—any elementary set that cannot be with certainty classified by employing the available knowledge.

Lower approximation set–a rough set that contains objects that *surely* belong to the set.

Upper approximation set—a rough set that contains objects that *possibly* belong to the set.

Boundary region—the difference between the lower and upper approximation.

Decision table—a table, the columns of which are labeled *attributes*, rows by *objects* of interests, and entries of table are *attribute values*. Attributes of the decision table are divided into two disjoint groups called *condition* and *decision attributes*.

Decision rule—the rule formed by each row of the decision table. If the decision rule uniquely determines the decision based on the condition, the rule is said to be *certain* otherwise it is *uncertain*.

Certainty coefficient—it expresses the condition of probability of an object to belong to the decision class determined by the decision rule.

Coverage coefficient—it expresses the condition of probability of reasons for a given decision.

An information system is a pair $S=(U,A)$, where U is the nonempty set of universe and A is the set of attributes. In information system when we distinguish the condition and decision attributes, the system can be designed by

$$S=(U,C,D) \tag{4.1}$$

where C and D are disjoint sets of *condition* and *decision* attributes, respectively. The rows of the decision table correspond to the sequence

$$C_1(x),\ldots,C_n(x);\quad d_1(x),\ldots,d_m(x),\quad \text{where } [C_1,\ldots,C_n]=C \text{ and } [d_1,\ldots,d_n]=D.$$

The sequence connectivity $[C_1(x),\ldots,C_n(x)] \rightarrow [d_1(x),\ldots,d_m(x)]$ is called a *decision rule*. Some of the coefficients that characterize the decision table are [6]

The *strength* of the decision rule is given by

$$\sigma_x(C,D)=\frac{|C(x)\cap D(x)|}{|U|} \tag{4.2}$$

A *certainty factor* of the decision table may be defined as conditional probability that belongs to $D(x)$ given y belongs to $C(x)$, and is given by

$$Cer_x(C,D)=\frac{|C(x)\cap D(x)|}{|C(x)|} \tag{4.3}$$

For a $Cer_x(C,D)=1$, then the $C_x \rightarrow D$ will be called a *certain decision rule* and for $0<Cer_x(C,D)<1$, the decision rule is referred to as *uncertain decision rule*.

The *coverage factor* of the decision rule is same as the inverse decision rule given by

$$Cov_x(C,D) = \frac{|C(x) \cap D(x)|}{|D(x)|} \tag{4.4}$$

The $Cov_x(C,D)$ factors are used for drawing inference from the information.

The use of RST for drawing inference and rules from measurable quantities of a plant control system will be discussed with the help of the following. The author found an interestingly similar case study in his research, to that of the solution shown by Zdzislaw Pawlak [6]. In his paper, Pawlak solves a problem of "client churn modeling" for mobile phone provider.

Case Study 4.1

In the tea industry, the drying of the fermented tea is carried out in drying machines. The extent of drying level of the tea is determined by the heat transfer mechanism of the dryer. From experience, it is found that the extent of drying of the tea is directly proportional to the temperature of the inlet hot air applied to the dryer and inversely proportional to the feed rate of the fermented tea. However, there are many other disturbance factors to the system and it is difficult to model the heat transfer mechanism of the dryer. From experience, the tea industry personnel maintain certain levels of parameters so that the tea is optimally dried. It is generally referred to by three terms—normally dried, overdried, and underdried. When the tea is overdried, it becomes difficult to control because the whole lot gets spoiled quickly.

RST was applied to develop rules to ascertain which variable mostly correlates the level of drying and their ranges. Table 4.1 shows conditions that were summarized after taking readings of inlet air temperature (°C), feed rate of tea (number of trays/15 min), and dryer outlet temperature (°C) at Chinnamara Tea Estate, India. These three variables are considered as

TABLE 4.1

Tea Dryer Data

Cases	Inlet Temperature Range (°C)	Feed Rate Range (Trays/15 min)	Outlet Temperature Range (°C)	Drying Status	Nos. of Reading
1	85–90	3–5	65–70	Normal	20
2	90–100	5–10	65–70	Normal	25
3	100–115	3–5	65–70	Normal	14
4	100–115	3–5	65–70	Over	12
5	90–100	3–5	60–65	Over	22
6	85–90	10–15	70–80	Under	21
7	85–90	5–10	65–70	Under	25

condition attributes. The tea drying status is described as normal, overdrying, and underdrying, which are considered as decision attributes of the system.

The temperature and feed rate readings are shown within a range of operation of the dryer. The readings were taken during a tea-manufacturing season over 112 days. Each row of the table apparently gives a decision rule; however, the certainty of the rules cannot be determined readily, e.g., Cases 3 and 4 are difficult to be decided since both have the same conditional attributes with different decision attributes. The following classification of the sets of cases can be identified:

1. Cases 1 and 2 are sets of readings where drying level is *certainly normal.*
2. Case 5 is a set of readings where drying level is *certainly overdried.*
3. Cases 6 and 7 are sets of readings where drying level is *certainly underdried.*
4. Cases 3 and 4 are sets where the drying level *cannot be decided.*

Therefore, the sets {1,2} and {5} are the lower approximation of the sets {1,2,3} and {4,5}, respectively. The sets {1,2,4} and {3,5} are the higher approximation of the sets {1,2,3} and {4,5}, respectively. The set {3,4} is the boundary region of the sets {1,2} and {5}. Table 4.1 is converted to Table 4.2 where the strength, certainty, and coverage coefficient for each case of the decision table are shown.

The decision rules can be derived from the above table by selecting a set of mutually exclusive and exhaustive decision rule that covers all facts in the information sets so that it preserves the indiscernibility relation. The ranges of inlet temperature, feed rate, and outlet temperature of the dryer are grouped as shown below:

Inlet temperature (T_i):

Low: 85°C–90°C; medium: 90°C–100°C; high: 100°C–115°C.

Feed rate (M_i):

Low: 3–5 (trays/15min); medium: 5–10 (trays/15min); high: 10–15 (trays/15min).

TABLE 4.2

Coefficient of the Decision Rules

Case	Strength	Certainty	Coverage
1	0.145	1.0	0.34
2	0.181	1.0	0.42
3	0.094	0.55	0.24
4	0.086	0.44	0.35
5	0.160	1.0	0.65
6	0.152	1.0	0.46
7	0.181	1.0	0.54

Outlet temperature (T_o):
Low: 60°C–65°C; medium: 65°C–70°C; high: 70°C–80°C.

The following rules are formulated from the seven cases of the information and their certainty values are shown below:

1. Removing M_i = low and T_o = medium from {1, 4}
   Rule 1: if T_i = low then drying = normal; Cer = 1.0
2. Removing M_i = medium and T_o = medium from {2,7}
   Rule 2: if T_i = medium then drying = normal; Cer = 1.0
3. Removing M_i = low from {3, 5}

The above rules are shown in Table 4.3.

From the decision rules, the following decision algorithms are concluded:

1. Normal drying is implied with certainty by
   Medium inlet temperature or low inlet temperature
2. Overdrying is implied with certainty by
   Medium inlet temperature and low outlet temperature
3. Underdrying is implied with certainty by
   High feed rate and high outlet temperature
4. High inlet temperature gives a certainty for overdrying with certainty of 0.44
5. High inlet temperature and medium outlet temperature gives a certainty of normal drying with certainty of 0.56

4.4.1.1.1 Rough Set Theory with Continuous Attributes Values

Luo and Shao [7] used continuous attributes values of a distillation column in refinery to identify secondary measurement variables to estimate propylene concentration. The continuous attributes values were discretized

TABLE 4.3

Decision Rules

Rule	T_i	M_i	T_o	Drying	Certainty
1	Low	—	—	Normal	1.0
2	Medium	—	—	Normal	1.0
3	High	—	Medium	Normal	0.56
4	High	—	—	Over	0.44
5	Medium	—	Low	Over	1.0
6	Low	High	High	Under	1.0
7	Low	—	—	Under	1.0

first using fuzzy discretization and equal-width-interval method. The discretized attributes were tabulated and superfluous condition attributes were removed using RST.

4.4.1.1.1.1 Fuzzy Discretization For each condition attribute x, $\left(x \in \left|x^{min}, x^{max}\right|\right)$, the range $|x^{min}, x^{max}|$ is partitioned into m parts. Each partition point of the attributes (x) is described by a Gaussian membership function of fuzzy set ($A_j(j=1,2,\ldots,m)$) such that the point belongs to the mean attribute value with membership function value of 1.

Let x_j be the value of the attribute x, then

$$\mu_{A_s}(x_j) = \underset{k\in\{1\ldots m_i\}}{MAX}\{\mu_{A_k}(x_j)\} \tag{4.5}$$

where A_k is the fuzzy subset of x corresponding to discrete value k, where A_s holds the maximal membership function value and S can be attributed as the discrete value of X_j. To illustrate this, let us take a numerical example.

Example 4.1

A set of attribute values for a variable x shown below is to be discretized using fuzzy discretization technique in four discrete levels: 1, 2, 3, and 4. Use Gaussian membership functions for the discretization of the data. $X=[10,8,9,4,4.5,3,5,0,2,7,6]$

Solution

For the given attribute values of x

$$x_{min} = 0 \quad \text{and} \quad x_{max} = 10$$

Total numbers of discrete levels $(m)=4$

Mean attribute values of the membership functions (A_1,A_2,A_3, and A_4) are

$$\begin{aligned} A_1 : \bar{x}_1 &= x_{min} + \frac{(x_{max} - x_{min})}{4} \\ &= 0 + \frac{(10-0)}{4} \\ &= 0 + 2.5 \\ &= 2.5 \end{aligned}$$

$$A_2 : \bar{x}_2 = \bar{x}_1 + 2.5 = 5.0$$

$$A_3 : \bar{x}_3 = \bar{x}_2 + 2.5 = 5 + 2.5 = 7.5$$

$$A_4 : \bar{x}_4 = \bar{x}_3 + 2.5 = 7.5 + 2.5 = 10.0$$

The Gaussian membership function for an attribute value of x is given by

$$\mu_{A_m}(x) = \frac{1}{6\sqrt{2\pi}} e^{-(x-\bar{x})^2/2\sigma^2}$$

where
  x is the attribute value of the set
  $\bar{x}$ is the mean of the partition
  σ is the standard deviation

Here we take the Gaussian distribution of σ equal to the partition value of the attributes, i.e., $\sigma = 2.5$. Figure 4.6 shows the membership functions. Now the maximum of the membership values for each of the attributes are found by calculating μ_{A_1}, μ_{A_2}, μ_{A_3}, and for each attribute values

$$\mu_{A_1}(10) = \frac{1}{\sigma\sqrt{2\pi}} e^{-(x-\bar{x}_1)^2/2\sigma^2}$$

$$= \frac{1}{2.5\sqrt{2\pi}} e^{-(10-2.5)^2/2(2.5)^2}$$

$$= 0.00176$$

$$\mu_{A_2}(10) = \frac{1}{\sigma\sqrt{2\pi}} e^{-(x-\bar{x}_2)^2/2\sigma^2}$$

$$= \frac{1}{2.5\sqrt{2\pi}} e^{-(10-5)^2/2(2.5)^2}$$

$$= 0.0215$$

$$\mu_{A_3}(10) = \frac{1}{\sigma\sqrt{2\pi}} e^{-(x-\bar{x}_3)^2/2\sigma^2}$$

$$= \frac{1}{2.5\sqrt{2\pi}} e^{-(10-7.5)^2/2(2.5)^2}$$

$$= 0.0964$$

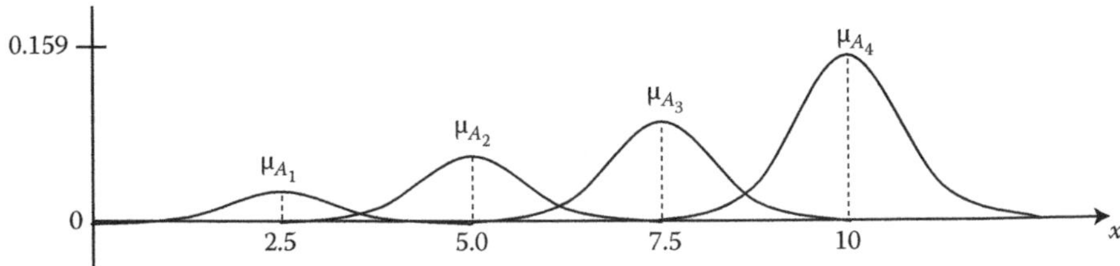

FIGURE 4.6
Gaussian membership functions for fuzzy sets.

$$\mu_{A_4}(10) = \frac{1}{\sigma\sqrt{2\pi}} e^{-(x-\bar{x}_4)^2/2\sigma^2}$$

$$= \frac{1}{2.5\sqrt{2\pi}} e^{-(10-10)^2/2(2.5)^2}$$

$$= 0.159$$

The maximum of these four membership values is μ_{A_4}, hence the discretized attribute for the attribute $x=10$ is 4.

The above calculation of μA shows that an attribute x possesses the maximum membership value μ_{A_s} for the function A_s, mean of which $(\bar{x}_s)$ is closest to the attribute. Hence, the discretized attributes(s) can be tabulated as below:

Attribute Value (x)	Closest Mean ($\bar{m}_s$)	Discretized Attribute
10	$\bar{x}_4=10$	4
8	$\bar{x}_3=7.5$	3
9	$\bar{x}_4=10$	4
4	$\bar{x}_2=10$	2
4.5	$\bar{x}_2=10$	2
3	$\bar{x}_1=10$	1
5	$\bar{x}_2=10$	2
0	$x_1=10$	1
2	$\bar{x}_1=10$	1
7	$\bar{x}_3=10$	3
6	$\bar{x}_2=10$	2

4.4.1.1.1.2 Equal-Width-Interval Discretization For each attribute x, ($x \in |x^{min}, x^{max}|$, the range is partitioned into n_i parts and each partition point holds a value x_j where $j \in (1,2,3,\ldots,n_i)$. If x is an attribute value, the following condition is obtained:

$$W_k = |x - x_k| = \min\{|x - X|\}, \quad k \in (1,2,\ldots,n_i) \tag{4.6}$$

then k is attributed as the discrete value of x.

Example 4.2

A pressure sensor developed the following set of readings that is required to be discretized. Use equal-width-interval method for discretization.

P (psi): [100.24, 94.67, 88.95, 64.75, 105.25, 102.65, 99.27, 98.53, 103.85, 101.24, 80.11, 73.65]

Solution
From the given data

Maximum pressure $= P_{max} = 103.85$ psi
Minimum pressure $= P_{min} = 64.65$ psi
Let total number of discrete levels $= n_i = 4$

The attribute points are

$$P_1 = P_{min} = 64.65\,\text{psi}$$

$$P_2 = P_{min} + \frac{(P_{max} - P_{min})}{4}$$

$$= 64.65 + \frac{(103.85 - 64.65)}{4}$$

$$= 64.65 + 9.8$$

$$= 74.45\text{psi}$$

$$P_3 = P_2 + 9.8$$

$$= 74.45 + 9.8$$

$$= 84.25\text{psi}$$

$$P_4 = P_3 + 9.8$$

$$= 84.25 + 9.8$$

$$= 94.05\text{psi}$$

The value $|x - X_k|$ for $k \in (1,2,3,4)$ are calculated for each attribute values as

For reading: 100.24 psi:

$$W_1 = |100.24 - 64.65| = 35.59$$

$$W_2 = |100.24 - 74.45| = 25.79$$

$$W_3 = |100.24 - 84.25| = 15.99$$

$$W_4 = |100.24 - 94.05| = 6.19$$

The minimum of the above four values are for $k = 4$.

For reading: 94.67 psi:

$$W_1 = |94.67 - 64.65| = 30.02$$

$$W_2 = |94.67 - 74.45| = 20.22$$

$$W_3 = |94.67 - 84.25| = 10.42$$

$$W_4 = |94.67 - 94.05| = 0.62$$

The minimum is for $k=4$

88.95 psi:

$$W_1 = |88.95 - 64.65| = 24.3$$

$$W_2 = |88.95 - 74.45| = 14.5$$

$$W_3 = |88.95 - 84.25| = 4.7$$

$$W_4 = |88.95 - 94.05| = 5.1$$

Similarly $k=3$.

Similarly, the discretized values for other reading are obtained as

64.75 psi	$k = 1$
105.25psi	$k = 4$
102.65psi	$k = 4$
99.27psi	$k = 4$
103.85psi	$k = 4$
101.24psi	$k = 4$
80.11psi	$k = 3$
73.65psi	$k = 2$

4.4.2 Model Structures

Satisfactory performance of a soft sensor depends on the selection of the best model suitable for representing the plant. The criterion for choosing the best model and its updation is normally defined in terms of the mean square error between the model output and the soft sensor output. The various basis of selecting the model are as follows [5].

4.4.2.1 ARMAX Model

Autoregressive moving average models with exogenous input (ARMAX) models are used for the estimation of the order and the model structure of the plant using all relevant information such as the measurable input/output variables, internal plant variables, measurable disturbance, and even the phenomenological information of the plant. The knowledge about which of the input/output variables are most correlated to the plant performance obtained from RST technique discussed earlier may also be incorporated. A multivariable linear, stochastic model is represented by a linear differential equation as

$$Y(k) = \sum_{i=1}^{n_a} a_i y(k-i) + \sum_{i=1}^{n_b} b_j u(k-i) + \sum_{i=1}^{n_c} c_k e(k-i) \tag{4.7}$$

where n_a, n_b, and n_c are, respectively, AR, X, and MA orders. The function a_i is the pulse response function from the past outputs to the present output $y(k)$, and b_i and c_i are the pulse response from the past and present inputs $u(k)$ and noise $e(k)$, respectively, to the present outputs. ARX is a parametric models with $c_i=0$ and with exogenous inputs (ARX) whereas ARMAX is a autoregressive moving average model with (ARMAX) and all the coefficients a_i, b_i, and c_i are estimated.

There are various estimating methods of ARMAX models such as pseudo linear regressive, correlation methods, subspace methods, etc. [8].

In multivariable regression method of estimation, the state-space representation is

$$Y(i) = B^T x(i) + e(i) \tag{4.8}$$

where

y is the predicted variable at ith instant $(n_y \times 1)$
x is the regression variable at ith instant $(n_x \times 1)$
e is the noise variable at ith instant $(n_y \times 1)$

The noise assumed to be normally distributed with zero mean with a constant covariance matrix C for N measurement of x and y the Equation 4.8 can be written as

$$Y = XB + E \tag{4.9}$$

where

Y is the predicted variable matrix $(N \times n_y)$
X is the regression variable matrix $(N \times n_y)$

$$E \in \left\{ y^T, x^T, e^T \right\}$$

The ordinary least square (OLS) estimation of B is

$$\hat{B}_{\text{OLS}} = (X^T X)^{-1}(X^T Y) \tag{4.10}$$

The OLS assumes that $C=I$, and the solution can be inaccurate if $(X^TX)^{-1}$ does not exist. This is because the OLS solution ignores that there are many interactions between predicted variables Y. To address such cases where there is high correlation on both Y and X matrices, other methods such as principle component regression (PCR), ridge regression, curds and whey, canonical correlation regression (CCR), and partial least square (PLS) are used.

Subspace identification method is a recent development in system identification. This method relies on a stochastic state-space model given by

$$X(k+1) = Ax(k) + Bu(k) + Ke(k) \quad (4.11)$$

$$Y(k) = Cx(k) + Du(k) + e(k) \quad (4.12)$$

where

x is the state variable vector $(n_x \times 1)$
u is the measured inputs $(n_x \times 1)$
y is the measured output vector $(n_y \times 1)$

In subspace method, the state vector $x(k)$ is assumed to be a linear combination of past inputs and outputs given by

$$x(k) = Jp(k) \quad (4.13)$$

where

$$p(k) = [u(k-1), \ldots, u(k-N), y(k-1), \ldots, y(k-N)]^T$$

The state vector $x(k)$ is formed by the data and cannot be specified a priori. Therefore, to estimate the state vector, J has to be determined. Once, the state vector is estimated, the state-space model parameters can be estimated via linear least square regression as shown below [8]:

$$\begin{bmatrix} \hat{A} & \hat{B} \\ \hat{C} & \hat{D} \end{bmatrix} = CoV\left(\begin{bmatrix} x(k+1) \\ y(k) \end{bmatrix}\begin{bmatrix} x(k) \\ u(k) \end{bmatrix}\right) \times CoV^{-1}\left(\begin{bmatrix} x(k) \\ u(k) \end{bmatrix}\begin{bmatrix} x(k) \\ u(k) \end{bmatrix}\right) \quad (4.14)$$

The value of J is derived from the canonical loading between the future output conditional on the future inputs and the past. In weighted singular value decomposition (SVD) strategy, J can be expressed as

$$svd(W_1 \theta W_2) = [U_1 U_2]\begin{bmatrix} \delta_1 & 0 \\ 0 & \delta_2 \end{bmatrix}\begin{bmatrix} V_1 \\ V_2 \end{bmatrix} \quad (4.15)$$

where W_1 and W_2 are two weighting matrices and are oblique projection of the future output along future inputs on the past inputs and outputs. The details about deriving J explained in [9].

4.4.2.2 Artificial Neural Network Models

Neural network systems has become increasingly popular in modeling of dynamic processes due to their extremely powerful adaptive capabilities in response to nonlinear behaviors. Neural systems can be used as a tool for mimicking cognitive tasks of complexity like human behavior. When a system is not well defined and needs to operate in a time-varying environment, neural network works well. A neutral system has the following distinct capabilities [10]:

1. Learning from the interaction with the environment without restriction to capture any kind of functional relationship between information patterns if enough training information is provided.
2. Generalized the learned information to similar situations never seen before.
3. Possessing a good degree of fault tolerance, mainly due to their intrinsic massive parallel layout.

ANN Structures

ANN structures have been devised based on a number of mathematical paradigm on interconnection and learning such as feedforward neural network, backpropagation neural network, Hopfield network, radial basis function, linear vector quantization, etc. It is found that feedforward multilayer structure is very common and popular in most applications of soft sensor. The implementation of a soft sensor by modeling the plant using a simple ANN structure will be discussed below in Case Study 4.2, which is an extension of Case Study 4.1.

Case Study 4.2

In Case Study 4.1, RST was applied to determine the correlation of the three secondary measurement variables of a tea dryer—inlet air temperature, tea federate, and exhaust air temperature, and one primary variable—tea drying status. Three decision rules were derived for the implication of normal drying, overdrying, and underdrying conditions.

Let us design an ANN model for the same problem to infer the drying status of the tea dryer using the already available input/output patterns described in Case Study 4.1. Since the drying status cannot be measured or predicted based on first principal models, the application of a neural model will be a suitable solution. We will use a feedforward backpropagation neural network in this case.

Selection of layers and nodes

Most of the classification problems can be solved by three layers—input layer, hidden layer, and output layer.

The attributes

The secondary and primary measurement attributes are (Table 4.1)

Inlet air temperature (T_i): low, medium, and high

Tea feed rate (M_i): low, medium, and high

TABLE 4.4

Discrete Codes Assigned to the Non-Numeric Attribute Values

Attributes	Non Numeric Value	Discrete Numeric Values		
		$x(0,1)$	$x(0,2)$	$x(0,3)$
T_i	Low	0	0	1
	Med	0	1	0
	High	1	0	0
M_i	Low	0	0	1
	Med	0	1	0
	High	1	0	0
T_o	Low	0	0	1
	Med	0	1	0
	High	1	0	0
S	Under	0	0	1
	Normal	0	1	0
	Over	1	0	0

Outlet exhaust temperature (T_o): low, medium, and high
Drying status (S): over, normal, and under

These attribute values are coded in Table 4.4.

Since each of the three input variables have three attribute values, the input layer needs 9 (3×3) neurons. To reduce computational complexity and time, the hidden layer is provided with two neurons, each connected to the nine inputs. Since the output corresponds to the drying status in three attributes—over, normal, and under—the output is provided with three neurons, each connected to the two hidden neurons.

It was found that training the network is difficult without biasing neurons; therefore, one biasing neuron in each layer was introduced. Hence, the total number of interconnections are $(9 \times 2) + (3 \times 2) + 2 + 3 = 29$. The network is shown in Figure 4.7.

A sigmoid function given by the equation below is chosen as the activation function of the neurons:

$$O_j = \frac{1}{1 + e^{-netj}}$$

where

$$net_j = \sum_{k=1}^{N} W_{jk} x_{jk} + b_j$$

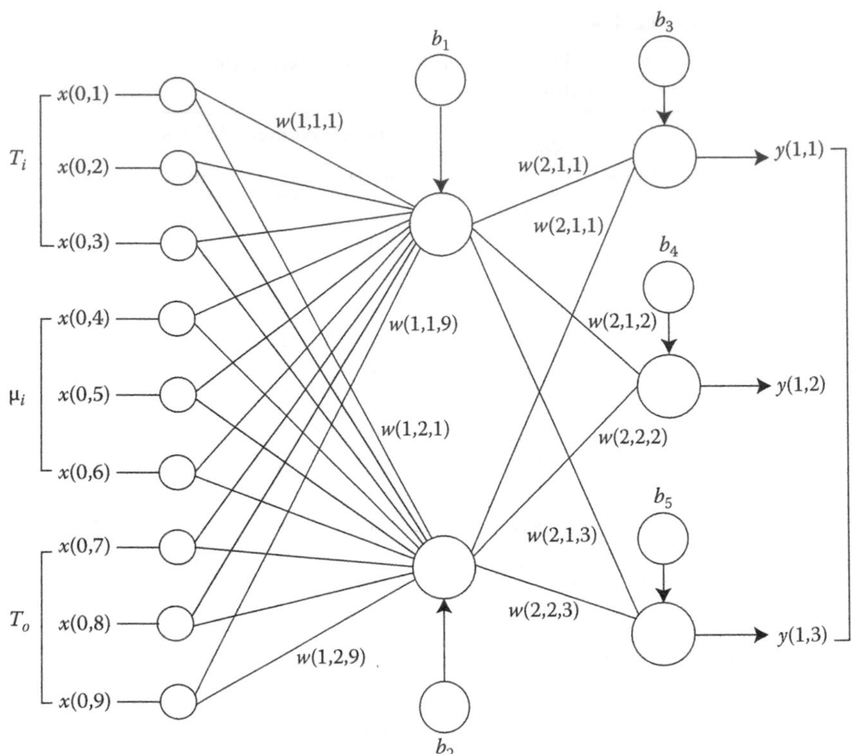

FIGURE 4.7
A multilayer feedforward neural network to determine the drying status from dryer input patterns.

In the first layer

$$net_{1,1} = \sum_{k=1}^{9} w(1,1,k)x(0,k) + b_1$$

$$net_{1,2} = \sum_{k=1}^{9} w(1,2,k)x(0,k) + b_2$$

The outputs from the two neurons

$$O_{(1,1)} = \frac{1}{1+e^{-net(1,1)}}$$

$$O_{(1,2)} = \frac{1}{1+e^{-net(1,2)}}$$

In the output layer

$$net_{2,1} = \sum_{k=1}^{2} w(2,k,1)O(1,k) + b_3$$

$$net_{2,2} = \sum_{k=1}^{2} w(2,k,2)O(1,k) + b_4$$

$$net_{2,3} = \sum_{k=1}^{2} w(2,k,3)O(1,k) + b_5$$

The outputs from the three neurons are

$$y(1,1) = \frac{1}{1+e^{-net(2,1)}}$$

$$y(1,2) = \frac{1}{1+e^{-net(2,2)}}$$

$$y(1,3) = \frac{1}{1+e^{-net(2,3)}}$$

The initial weights and the trained network weights are shown in Table 4.5.
The training data for five patterns P_1, P_2, P_5, P_7, and P_6 (Table 4.1) is shown in Table 4.6.

TABLE 4.5
ANN Weights

Weights	Initial	Trained
$w(1,2,9)$	−1.9564	−16.9922
$w(2,1,1)$	3.4435	28.4042
$w(2,1,2)$	−9.0677	−2.3560
$w(2,1,3)$	5.5480	1.6014
$w(2,2,1)$	−7.9561	36.1189
$w(2,2,2)$	−7.0232	−6.1438
$w(2,2,3)$	−6.6899	−565.989
b_1	−3.6892	−4.6031
b_2	2.7349	−0.9897
b_3	−2.0377	−12.3653
b_4	1.2041	−18.0276
b_5	2.0068	15.1686

TABLE 4.6

Training Data for Five Patterns of the Dryer

Pattern	$x(0,1)$	$x(0,2)$	$x(0,3)$	$x(0,4)$	$x(0,5)$	$x(0,6)$	$x(0,7)$	$x(0,8)$	$x(0,9)$
P_1	0	0	1	0	0	1	0	1	0
P_2	0	1	0	0	1	0	0	1	0
P_5	0	1	0	0	0	1	0	0	1
P_6	0	0	1	1	0	0	1	0	0
P_7	0	0	1	0	1	0	0	1	0
Pattern	$y(1,1)$	$y(1,2)$	$y(1,3)$						
P_1	0	1	0						
P_2	0	1	0						
P_5	1	0	0						
P_6	0	0	1						
P_7	0	0	1						

The network was trained with random initial values of the weights.

After the network is trained, it was tested with the inputs corresponding to the five patterns, which were used for training (P_1, P_2, P_5, P_6, P_7) and a few more unknown patterns. The results returned by the network are shown in Table 4.7.

It is observed that all the trained patterns P_1, P_2, P_5, P_6, and P_7 were correctly estimated by the ANN. Patterns P_3 and P_4 were not used for training, but when used for testing, ANN estimated it as normal (N), which conforms to RST decision for higher certainty (0.56).

Twenty-one test patterns (T_1–T_{21}) were applied with unknown outputs, which were estimated as shown in the table. Estimated outputs of the input patterns were compared with the RST decisions. ANN outputs of T_3 and T_{16} were approximated to U with an attribute strength of 0.28 and 0.54 returned by ANN. Out of 27 patterns, 16 fall under RST domain and for the remaining 11, RST is not able to make a decision. Out of the 16 patterns (which can be estimated by both ANN and RST), 13 decisions conform to both RST and ANN, giving a classification of 81%.

4.4.2.2.1 Dimensionality Reduction of Data

For an ANN model to be successful in soft sensor design, there must be a higher degree of correlation among the secondary measurement variables to generate the primary variable. In most cases, not all input measurement variables are highly correlated. In such cases, secondary measurement variables that explain most of the variation of the primary variable are needed to be identified. This kind of processing of the measurement data to reduce the dimension can be done by principal component analysis (PCA). In PCA, an $n \times m$ measurement matrix X is formed where each row vector $x^T(j)$ is the

TABLE 4.7

Results Obtained from the ANN on Applications of Test Patterns and Comparison with the RST Decision

Inputs $[T_i\ M_i\ T_0]$	Patterns	Decision Outputs ANN	Decision Outputs RST	Conformity ANN vs. RST Decision
001 001 001	T_1	N	—	—
001 001 010	P_1	N	N	Yes
001 001 100	T_2	N	—	—
001 010 001	T_3	U(0.28)	—	—
001 010 010	P_7	U	U	Yes
001 010 100	T_4	U	—	—
001 100 001	T_5	U	—	—
001 100 010	T_6	U	—	—
001 100 100	P_6	O	U	Yes
010 001 001	P_5	N	O/N	Yes
010 001 010	T_7	O	N	Yes
010 001 100	T_8	O	N	No
010 010 001	T_9	N	N/O	Yes
010 010 010	P_2	N	N	Yes
010 010 100	T_{10}	N	N	Yes
010 100 001	T_{11}	O	N/O	Yes
010 100 010	T_{12}	N	N	Yes
010 100 100	T_{13}	N	N	Yes
100 001 001	T_{14}	O	O(0.44)	Yes
100 001 010	P_3, P_4	N	N(0.56)	Yes
100 001 100	T_{15}	N	O(0.44)	—
100 010 001	T_{16}	U(0.54)	O(0.44)	—
100 010 010	T_{17}	U	N(0.56)	No
100 010 100	T_{18}	U	O(0.44)	—
100 100 001	T_{19}	U	O(0.44)	—
100 100 010	T_{20}	U	N(0.56)	No
100 100 100	T_{21}	U	O(0.44)	—

Note: The figure within parenthesis indicates the certainty value and else indicates a certainty value of 1.0. The output symbols: O, over dry; N, normal dry; U, under dry.

measurement data at sampling instant $j=1, 2, 3, \ldots, n$ and each column vector $x(i)$ is the measurement data for the variables $i=1, 2, 3, \ldots, m$. In PCA, the m eigenvalues λ_p and eigenvectors p^i of the symmetrical matrix X^TX are found. The subset of the largest eigenvalue explains most of the variation of the data. This subset covers the principal components space (PCS) and the remaining eigenvectors cover the residual space (RS).

Case Study 4.3

In a water treatment plant several sensors are used to measure the parameters of raw water such as color, pH, turbidity, conductivity, alkalinity, and flow. These parameters are important for chemical coagulation and pH control in the water treatment process. But due to poor environmental factors such as failure, bias, drift, spike, short circuit, open circuit, and cyclic noise, these sensors are unable to give correct measurement. A soft sensor for the water treatment process was designed by Wang [11] using ANN model so that the measurements could be validated based on knowledge-based data of several years. The sensor design is described below.

Knowledge-based data: It was observed that there is a definite trend in the change in each parameter of the water treatment plant at the same month in different years. This correlation of variation of data was the basis of the ANN. The parameters were combined in two groups—pH, conductivity, and alkalinity in one and color, turbidity, and flow in another. This grouping was made for the fact that a definite correlation was observed in the parameters of the groups separately rather than combined.

ANN model: An auto-associated neural network (AANN) model was found to be suitable for this application. The 3-10-2-10-3 structures of the AANN are shown in Figure 4.8. The network can be trained by backpropagation or any other supervised training algorithms.

4.4.2.2.2 Data Pre-processing

Instead of using the data in actual form, the data was scaled to the range [0,1]. The scaling operation was performed using the following equation:

$$\frac{1-e^{-\alpha x}}{1+e^{\alpha x}} = f(x) \tag{4.16}$$

where α is a slope of scaling factor.

Dimensionality reduction of data is also a major concern in ANN implementation, which has already been discussed earlier. PCA was applied to the data obtained from the water treatment process sensors and it was found that the data are highly correlated in two groups rather than considering them together. Therefore, six variables are reduced to three in two groups.

4.4.2.2.3 Fault Classification

Field data sets for each month from 1994 to 1998 were used to train the AANN separately, i.e., for each month, separate models were used. The sensor measurement vectors are inputs to the AANN and for each set of input there is a correct or normal output with a permitted level of error between ideal and actual. If the error crosses this limit, the system alarms a fault in the sensor.

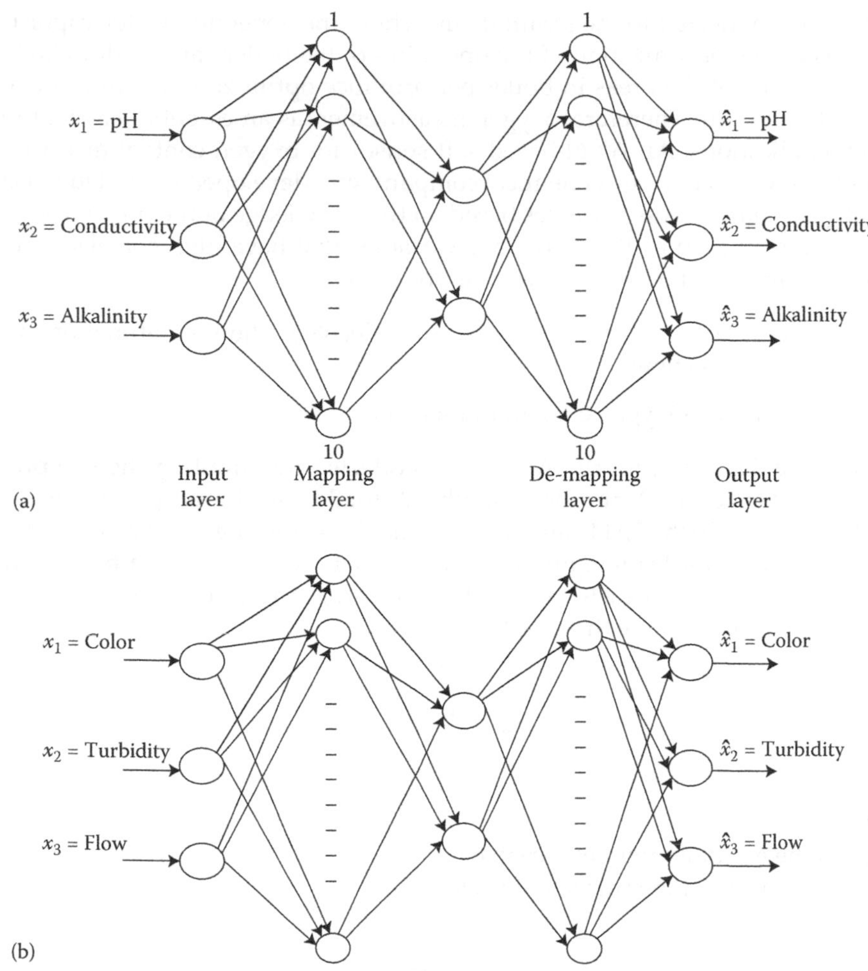

FIGURE 4.8
ANN structure for (a) pH, conductivity, alkalinity (b) color, turbidity, and flow. (Redrawn from IEEE copyright © IEEE. With permission.)

It was found in the study that in normal operation, the difference or error between the ideal and the measurement data of conductivity varies from −0.02 to +0.02, but when a sensor bias fault takes place, the error exceeds a level of −0.03.

Case Study 4.4

Boilers are found in many industries used either as power source or for processing purposes. Boilers consist of a furnace where air and fuel are combined to burn fuel to produce combination gases to a water tube system. The tubes are connected to the steam drum where the generated water vapor is removed. The optimization of the operation of the boiler can result in large savings. One of the areas in boiler performance optimization is minimization of excess air. Therefore, oxygen measurement is an essential element in such applications. An ANN-based soft sensor for oxygen content measurement of fuel gas in a petrochemical company was developed by Al-Duwaish et al. [12]. Four process variables based on the knowledge of the process engineers were chosen as the secondary variables that have high correlation to the primary variable. The process variables are

Secondary variables: Gas-burner pressure, outlet steam flow, combination airflow, and fuel gas flow.

Primary variables: Oxygen content in fuel gas.

Data—the data used for the ANN were collected by sampling the five process variables every 2 min for 2 months. A total of 18,112 samples were collected, out of which 13,144 samples were used for training the ANN and the rest were used for testing. Preprocessing of data was performed to linearly transform the data of each variable to a uniform range. The linear transformation is performed by the equation

$$\hat{x}_i^n = \frac{x_i^n - \bar{x}_i}{\sigma_i}$$

where
x_i^n is the nth data point of variable i
$\bar{x}_i$ is the mean of measurement vector

$$= \frac{1}{N}\sum_{n=1}^{N} x_i^n$$

$$\sigma_i^2 = \text{variance} = \frac{1}{N-1}\sum_{n=1}^{N}\left(x_i^n - \bar{x}_i\right)^2$$

After preprocessing, the data were used to train a multilayer feedforward ANN network with four inputs, one hidden layer, and one output layer. The number of neurons used in the hidden layer was 10 for the best estimation of the oxygen content. The network is shown in Figure 4.9.

The ANN was used to estimate the oxygen collected with test data and compared with a physical oxygen analyzer. Figure 4.10 shows the output of both the oxygen analyzer and the ANN-based soft sensor. It was observed that 92% of the errors were within the range of ±0.35.

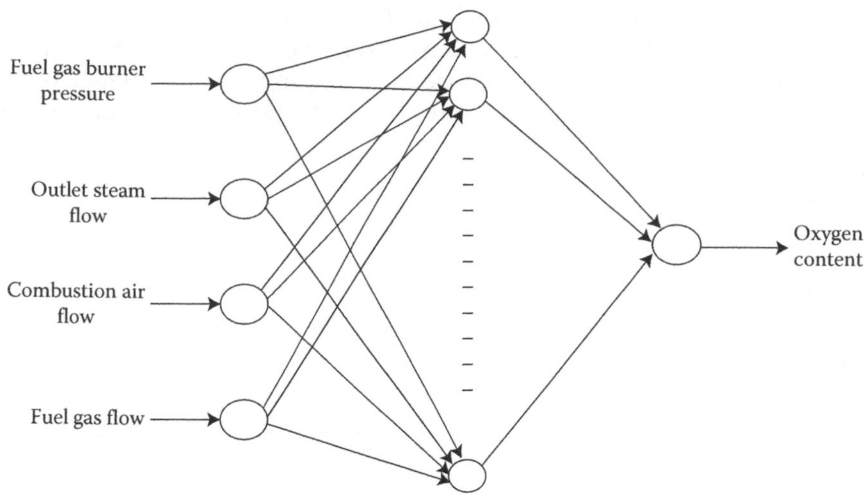

FIGURE 4.9
ANN structure for the oxygen content sensor.

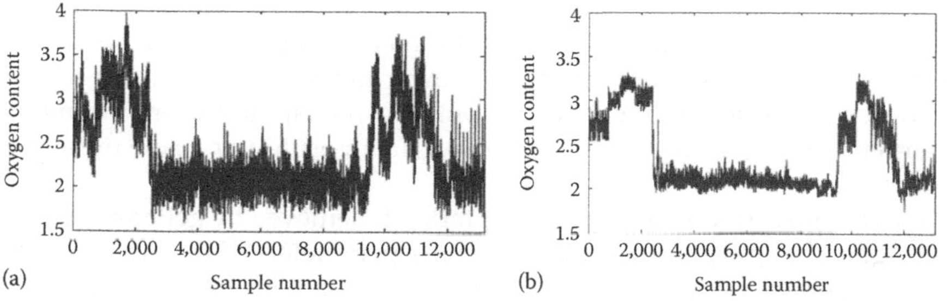

FIGURE 4.10
Measurement of O_2 content (a) by O_2 analyzer and (b) from ANN-based system. (Reprinted from Al-Duwaish, H. et al., Use of artificial neural network process analyzers: A case study, in *ESANN 2002, Proceedings of the European Symposium on ANN*, Bruges, Belgium, pp. 465–470, April 24–26, 2002. With permission.)

4.5 Self-Adaptive Sensors

Apart from performing the metrological or sensing operation, a self-adaptive intelligent sensor can perform the higher hierarchical operation of adaptation to the operational or environmental changes. A nonadaptive sensor cannot understand or judge what changes are taking place around the sensor and cannot change its parameters such that the output also changes as per requirement.

Self-adaptive sensor works on self-adaptive algorithms, which are mostly derived from digital measuring systems. Adaptive measurements are measurements where measuring procedure can be changed at the change of the properties of a signal or measuring condition [13]. The adaptation of a sensor system can be classified as

1. Adaptation to measurement accuracy
2. Adaptation to measurement time
3. Adaptation to power consumption
4. Adaptation to linearity
5. Adaptation to environmental condition

The techniques generally used for the implementation of an adaptive sensor are based on either one or combination of the following methods:

1. Algorithmic adaptation
2. Parametrical adaptation

In algorithmic adaptation, a control signal is generated in the sensor measuring system that changes or alters the measuring procedure so that the system follows a higher accuracy or a lower measurement time or a lower power consumption or linearized output or can adapt to the environmental intervention. The system basically follows an algorithm or a rule prescribed earlier.

In parametrical adaptation, the adaptation is achieved by generating a control signal that purposefully changes some parameters of the system on the basis of the current measurements. Hence, the technique basically maintains the required adaptive quality of a sensor system at change of operating factor U [14] given by

$$U \in Y_i \tag{4.17}$$

where Y_i is the number of feasible controls.

In case of parametrical adaptation

$$U = \langle P_1 P_2 \cdots P_n \rangle \tag{4.18}$$

where P_n = parameters of adaptation depending on measuring conditions resulting discrete control values D_i.

The natural trend of developing self-adaptive sensor relies on using digital or quasi-digital sensors for operational convenience. Sensor generating frequency, period, duty-cycle, or PWM outputs can conveniently be associated to the adaptive systems. This is because frequency-based digital measurement devices can easily be programmed using microprocessor or

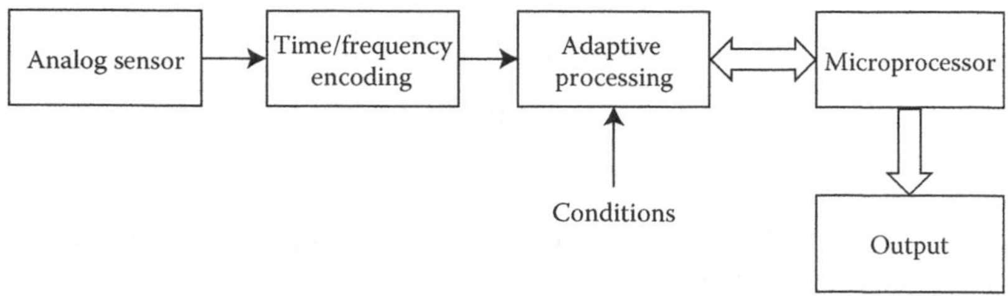

FIGURE 4.11
Block diagram of an adaptive sensor system.

microcontroller. Therefore, an analog sensor that needs adaptive features should be converted to a frequency/time generating one before the adaptation process, which is shown in Figure 4.11.

The adaptation to the need of accuracy, speed of measurement, and power consumption can be discussed in the next section.

4.5.1 Algorithmic Adaptation with Nonredundant Time of Measurement

In algorithmic adaptation, an equation of measurement explains how an operator of algorithm relates the time-dependent measurement of input action to an operation of interest for adaptation as below:

$$\lambda_j^* = \{T_s L\gamma_j(t)\} B \{\delta_s L\gamma_j(t)\} B \{P_s L\gamma_j(t)\} B \{Q_s L\gamma_j(t)\} \tag{4.19}$$

where

- T_s is the operating factor for reducing measurement time
- δ_s is the operating factor for increasing accuracy
- P_s is the operating factor for reduction power consumption
- Q_s is the operating factor for linearization output
- L is the operator of algorithmic measurement
- $\gamma_j(t)$ is the time-dependent multivariate input
- t is the moment of time for which the measurement is used such that $t \in [t_j, t_j + T]$

where

- T is the time of measurement cycle
- B is the Boolean operator

The change in the operating factor is governed by a rule based on the current result of measurement x^* given by [14]

$$\lambda_j^* = T_s L \gamma_j(t), \quad \text{if } F_x(x^*) \in I_\Gamma \tag{4.20}$$

$$= \delta_s L\gamma_j(t), \quad \text{if } F_x(x^*) \notin I_F$$
$$\text{at } I_F \in I \tag{4.21}$$

where

$F_x(x^*)$ is the characteristic of active or measuring condition that determines the change of measuring algorithm

I_F is the subset of I of possible values of $F_x(x^*)$, the set membership to which the change in operating factor is meant for

In this method, accuracy is increased at the cost of more computational time. Here, accuracy and speed of measurement are mutually exclusive of each other, i.e., when highest accuracy is needed the speed of measurement decreases and vice versa.

The steps for adaptive measurement algorithm with nonredundant time measurement according to Equations 4.20 and 4.21 are as follows [15].

Step 1: The limiting values of frequency $F_{X\,\text{limit}}$ is setup at which the algorithm change should be made.

Step 2: The limiting frequency $F_{X\,\text{limit}}$ is used for the programming for the highest accuracy of measurement.

Step 3: The frequency measurement of F_X is performed with highest accuracy.

Step 4: Measured F_X is compared with limiting F_X limit.

Step 5: If $F_X \leq F_{X\,\text{limit}}$, the control returns back to Step 2 for measurement with highest accuracy.

Step 6: The frequency measurement of F_X is performed with lower accuracy.

Step 7: Measured F_X is compared with limiting F_X limit. If $F_X \leq F_{X\,\text{limit}}$, the control returns back to Step 2 for measurement with highest accuracy.

4.5.2 Nonredundant Reference Frequency Measurement-Based Adaptation

Similar to Equations 4.20 and 4.21, an increase in accuracy and decrease in power consumption operation based on the method of nonredundant reference frequency can be obtained as [14]

$$\lambda_j^* = P_s L\gamma_j(t); \quad \text{if } F_X(x^*) \in I_f \tag{4.22}$$

$$= \delta_s L\gamma_j(t); \quad \text{if } F_X(x^*) \notin I_f \tag{4.23}$$

Here, accuracy and power consumption are mutually exclusive of each other, i.e., when accuracy of measurement increases, power consumption increases and vice versa. In this technique, change of measurement accuracy is achieved by changing the reference frequency f_0 of the frequency-generating system. When a precise measurement is to be performed within a certain boundary region of measurement frequency, the reference frequency is increased and thereby power consumption is decreased. In another words, when a measurement suffices with low accuracy, the reference can be decreased to get low power consumption.

Applications of nonredundant reference frequency–based adaptive sensors are standalone, embedded, wireless sensors, etc. The frequency-to-power consumption relation of digital circuit can be explained by the basic equation of dynamic average power of a CMOS circuit given by

$$P_a = C_{eff} V_{DD}^2 f_{CLK} \tag{4.24}$$

where

V_{DD} is the supply voltage
f_{CLK} is the clock frequency (Hz)
C_{eff} is the effective capacitance of the circuit

Since the supply voltage and capacitance are fixed for a circuit, the power consumption can be reduced by using a lowest possible clock.

Adaptive sensor systems can be designed using microcontrollers with programmable clock frequencies such as the MSP 430 from Texas Instrument [16].

In this IC there is an inbuilt system clock frequency control (SCFQCTL) register, which should be loaded with a multiplication factor K_N (which ranges from 3 to 127) to get the system clock as

$$F_{system} = K_N f_{crystal}$$

where $f_{crystal} = 32{,}768\,\text{Hz}$.

The steps for adaptive algorithm with nonredundant reference frequency measurement according to Equations 4.22 and 4.23 are same as for the nonredundant time measurement except for the following:

Step 3: The frequency measurement of F_x is performed with highest accuracy programming a highest reference frequency f_0 max.

Step 5: If $F_x \leq F_x$ limit, continue with the same reference frequency f_0 max, if $F_x > F_x$ limit, change reference frequency to f_0 min to get lowest accuracy.

4.5.3 Frequency to Digital Conversion Using Microcontroller

When a frequency-generating sensor output is to be converted to digital signal, a microcontroller can be used that encodes the signal about the measurand. The basic principles of frequency measurement are

1. Measurement of input signal pulse during a known elapsed time interval.
2. By counting clock pulse of a known reference signal during a period of the input signal.

In the first method, the clock is derived from a precision crystal clock circuit and frequency divider. A microcontroller does not use frequency divider, rather it uses two programmable counters—one to count the elapsed time and one to count the input pulses.

In the 8051 microcontroller, a 16-bit timer/counter increments a register when a machine cycle is operated. When working as a counter, the register is incremented when the input pulse changes from HIGH to LOW. Timer needs 12 machine cycles for incrementing the register, whereas the counter needs 2 machine cycles for reading the HIGH to LOW transition. Hence, the maximal count rate is 1/24 of the clock frequency. At the start of a frequency measurement, the timer is set to a prefixed time $t = T$ and the input pulses are counted by the counter till the timer value reaches $t = 0$ and then counting is stopped. The value registered by the counter gives the frequency of the input signal. If the input frequency range is $f_{min} \leq f \leq f_{max}$, an n-bit counter needs a prefixed time of

$$T = \frac{N}{f_{max} - f_{min}} \tag{4.25}$$

Example 4.3

An optical rotational sensor produces a signal of frequency 1–5 kHz. Determine the measurement time and the counter sizes to measure the frequency by a microcontroller with 12-bit resolution.

Solution

Given values:

$$f_{max} = 5\,\text{kHz}$$

$$f_{min} = 1\,\text{kHz}$$

$$n = 12$$

From Equation 4.25, the measurement time

$$T = \frac{2^{12}}{(5-1)\,\text{kHz}} = 1.024\,\text{s}$$

The numbers of counts for 1 kHz input frequency

$$N_{min} = 1 \times 10^3 \times 1.024 = 1024$$

Similarly, $N_{max} = 5 \times 10^3 \times 1.024 = 5120$.
Here, $N_{max} - N_{min} = 5120 - 1024 = 4096 = 2^{12}$

The counter size required is 13 bit.

In the second method using a reference frequency (f_c), the pulse counting of the reference frequency (clock) is made during a number of K periods of the input signal. The counted number of pulses

$$N = (f_c)KT_x \tag{4.26}$$

where T_x is the time period of the input signal.

In microcontroller implementation of this technique, the elapsed time over K input pulses is measured by a timer. The time period of one pulse cycle T_x is computed in terms of time required by the machine cycles T_{xm}. It will give

$$T_{xm} = \frac{1/f_x}{T_m} = \frac{f_c}{nf_x} \tag{4.27}$$

where n is the number of bits = 12 for 8051 microcontroller.

If the input time period ranges from T_{xm}(max) to T_{xm}(min), the elapsed time for minimum number of pulses to get a resolution of n-bit is given by

$$K \geq \frac{2^n}{T_{xm}(\text{max}) - T_{xm}(\text{min})} \tag{4.28}$$

Example 4.4

The signal frequency of the rotational sensor of Example 4.3 is measured by a microcontroller by reference frequency method using a clock of 10 MHz. Determine the minimum number of input cycle that can be used to measure the frequency with a 12-bit resolution.

Solution

Given:

$$f_x(\text{maximum}) = 5\text{ kHz}$$

$$f_x(\text{minimum}) = 1\text{ kHz}$$

$$n = 12$$

From Equation 4.27,

$$T_{xm}(\text{maximum}) = \frac{10\text{ MHz}}{12 \times 1\text{ kHz}} = 833.3 \text{ machine cycles}$$

$$T_{xm}(\text{minimum}) = \frac{10\text{ MHz}}{12 \times 5\text{ kHz}} = 166.6 \text{ machine cycles}$$

From Equation 4.28,

$$K \geq \frac{2^{12}}{833.3-166.6}$$

$$\geq 6.14$$

Hence

$$K=7$$

Therefore, the minimum number of input signal cycles is 7.

Example 4.5

In an automobile, the wheel speed is measured by a magnetic incremental positional encoder using magneto resistor sensor. The maximum wheel speed is 1600 rpm and when wheels are blocked, the speed decreases to 10 rpm. A microcontroller is used to measure the sensor frequency signal. The accuracy of the measurement at normal speed should be high, while the accuracy during wheel blocking can be low. The clock used by the microcontroller is 1 MHz.

1. Determine the measuring method that should be adopted at normal speed and blocked wheel speed. Take sensor gear teeth as 180.
2. Determine the change in measurement time in changing the method.

Solution

With time measurement method

At 10 rpm (wheel blocked), the number of pulse generated in 1 s is

$$= \frac{10 \times 180}{60} = 30$$

This gives a resolution of 1/30

$$= 3.3\%$$

At 1600 rpm (normal speed), the number of pulse generated in 1 s

$$\frac{1600 \times 180}{60} = 4800$$

That gives a resolution of 1/4800

$$= 0.020\%$$

With reference frequency method using 1 MHz clock

At 10 rpm, the pulse duration is

$$= \frac{60}{10 \times 180} = \frac{1}{30} = 0.033\ \text{s.}$$

A 1 MHz clock will have pulse in this time

$$= 0.033 \times 10^6 = 33{,}000$$

This gives a resolution of 0.0033%.

At 1600 rpm, the pulse duration is

$$\frac{60}{1600 \times 180} = \frac{1}{4800}\ \text{s}$$

Number of pulses at 1 MHz clock

$$= 10^6 \times \frac{1}{4800}$$

$$= 208$$

This gives a resolution of 0.48%.

If the microcontroller uses time-measurement method at normal speed (1600 rpm) the resolution is 0.020%, so it cannot shift to reference frequency method, which is used at blocked wheel (10 rpm) when it gets a resolution of 0.0033%.

The measurement time in reference frequency method at 1600 rpm

$$\text{Duration of a pulse} = \frac{1}{4800}\ \text{s} = 0.208\ \text{ms.}$$

The measurement time in time measurement method = 1 s.
Therefore, the change in measurement time = (1 − 0.208) s = 999.792 ms.

Case Study 4.5

In this section, two cases of self-adaptive sensors will be discussed [15], where quasi digital pressure sensor system in gas pipeline and humidity sensor is provided with the adaptivity to accuracy and speed of measurement.

Self-adaptive pressure sensor system for gas pipeline

The measurement of pressure in gas pipeline can be performed using digital pressure sensors. The sensors are basically analog sensors like capacitive or magnetic. The analog voltage of the sensor is encoded to frequency using voltage to frequency converter. Pressure drop in pipelines is a quick phenomenon that has to be detected and controlled within a short duration of time. At normal pressure, the pressure sensor output is measured by a microcontroller with highest accuracy but with lowest speed. When the pressure falls, the method used by the microcontroller is automatically switched over to one that permits high speed of measurement; however, the accuracy decreases. The parametrical adaptation allows reducing the measuring time for the microcontroller to one third.

Self-adaptive temperature and humidity sensors
Temperature sensors from Maxim [17] MAX 6576 and MAX 6577 produces period and frequency outputs, respectively, with temperature. These two sensors are designed in such a way that the accuracy can be changed 2.2 times in various ranges of its working range (−40°C to 125°C).

Humidity sensor HTF3130 from Humeril is a frequency-generating sensor, which shows variation of accuracy up to 1.7 times in the whole working range from 0% to 99%. The maximum possible full-scale relative error is 5% for the module, however, from 0% to 10% and 90% to 100% of relative humidity, the module gives an error of 3%.

4.5.4 Adaptation to Linearity

In Equation 4.19, algorithm operators for relating the time-dependent measurement of input action $\gamma_j(t)$ to various factors of adaptation were shown. In Equation 4.19, Q_s is the operating factor for the linearization of the input measurement variable. The change in the operating factor based on current measurement x^* can be expressed by

$$\lambda^* = Q_s L\gamma_j(t); \quad \text{if } F_x(x^*) \in I_f \tag{4.29}$$

$$= \delta_S L\gamma_j(t); \quad \text{if } F_x(x^*) \notin I_f \text{ at } I_f \in I \tag{4.30}$$

and

$$\lambda^* = Q_s L\gamma_j(t); \quad \text{if } F_x(x^*) \in I_f \tag{4.29}$$

$$= T_s L\gamma_j(t); \quad \text{if } F_x(x^*) \notin I_f \text{ at } I_f \in I \tag{4.31}$$

In this method of adaptation, linearity of the input measurement variable can be increased at the cost of accuracy (Equations 4.29 and 4.30) and at the cost of speed of measurement (Equations 4.29 and 4.31). Here, the characteristic of input action ($F_x(x^*)$) from a subset of I of possible values determine the change of the measuring algorithm of the variable of interest.

Linearization method for calibration of sensor output from a nonlinear sensor will be discussed in Chapter 5. Linear interpolation is the simplest technique of linearization that can be implemented by a microcontroller. Most sensors produce output with different nonlinearity values during different ranges within their operating range. One such condition can be one linear range and several nonlinear ranges, which are most common in sensors: In this condition, the sensor output follows a linear output initially up

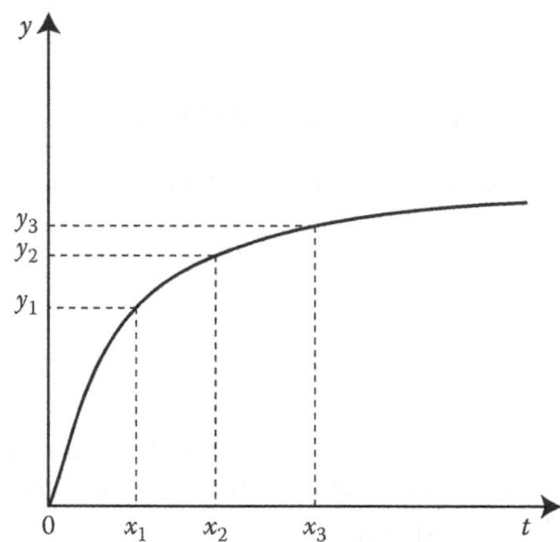

FIGURE 4.12
Input–output characteristic of a sensor with single linear range and several nonlinear ranges.

to the input measurement range $0 \leq x \leq x_1$, $x_2 \leq x \leq x_3$, etc. Figure 4.12 shows the input–output characteristic of such a sensor.

In Equations 4.28 through 4.31, the values for the operating factors of linearity and subset I_f for possible values of $F_x(x^*)$ can be written as

$$\begin{aligned}
&Q_s \in Q_{SL} && \text{for } F_x(x^*) \in I_{fL} \\
&Q_s \notin Q_{SL} && \text{for } F_x(x^*) \notin I_{fL} \\
&Q_s \in Q_{SNL1} && \text{for } F_x(x^*) \in I_{fNL1} \\
&\vdots && \\
&Q_s \in Q_{SNLm} && \text{for } F_x(x^*) \in I_{fNLm}
\end{aligned} \tag{4.32}$$

where the suffixes L and NL refer to one linear and m numbers of nonlinear ranges, respectively. The equations for accuracy (δ_S) and time (T_s) can also be expressed in a similar way

$$\begin{aligned}
&\delta_S \in \delta_{SL} && \text{for } F_x(x^*) \in I_{fL} \\
&\delta_S \notin \delta_{SL} && \text{for } F_x(x^*) \notin I_{fL} \\
&\delta_S \in \delta_{SNL1} && \text{for } F_x(x^*) \in I_{fNL1} \\
&\vdots && \\
&\delta_S \in \delta_{SNLm} && \text{for } F_x(x^*) \in I_{fNLm}
\end{aligned} \tag{4.33}$$

and

$$\begin{aligned}
&T_s \in T_{SL} && \text{for } F_x(x^*) \in I_{fL} \\
&T_s \notin T_{SL} && \text{for } F_x(x^*) \notin I_{fL} \\
&T_s \in T_{SNL1} && \text{for } F_x(x^*) \in I_{fNL1} \\
&\vdots && \\
&T_s \in T_{SNLm} && \text{for } F_x(x^*) \in I_{fNLm}
\end{aligned} \tag{4.34}$$

During the linear range of operation of the sensor, the microcontroller performs a calibration operation based on a linear calibration factor.

Giving the output by the equation

$$y(t) = kx(t) \tag{4.35}$$

Writing the Equation 4.35 in discrete form

$$y(k) = kx(k) \tag{4.36}$$

During the nonlinear region of the sensor operation, the linearization using linear interpolation gives the output by the equation

$$y(k) = y^*(k-1) + y^{*\prime}(k-1)[n(k) - n^*(k-1)] \tag{4.37}$$

where

- $y^*(k-1)$ is the sensor input value previous to the sampled instant k (stored in a lookup table (LUT) in EPROM)
- $y^{*\prime}(k-1)$ is the gradient or slop previous to the sampled instant k (stored in LUT)
- $n^*(k-1)$ is an integer calibration index factor previous to the sampled instant k corresponding to the measurement value of the sensor stored in LUT

Example 4.6

A capacitive relative humidity (RH) sensor produces the following experimental data:

V_0 (mV):	0	0.2	0.4	0.6	0.8	1.0	1.2	1.5	1.75	1.99	2.22	2.44	2.64	2.83
RH (%):	0	5	10	15	20	25	30	40	50	60	70	80	90	100

Derive a rule for adaptation to linearity and prepare a LUT for a microcontroller-based measurement. Determine the relative humidity corresponding to a sensor output voltage of 0.3 and 2.3 mV.

Solution

The slopes between each consecutive sensor sampled data are calculated and tabulated below for observation of the linearity range of the sensor. The equation used for the slope calculation is

$$\text{Slope} = \frac{RH_2 - RH_1}{V_2 - V_1}$$

For example, the slope between data nos. 2 and 3

$$\text{Slope} = \frac{10-5}{0.4-0.2} = 25$$

Data No	V_0 (mV)	RH (%)	Slope (%/RH/mV)
1	0	0	0
2	0.2	5	25
3	0.4	10	25
4	0.6	15	25
5	0.8	20	25
6	1.0	25	25
7	1.2	30	25
8	1.5	40	26.66
9	1.75	50	28.57
10	1.99	60	30.15
11	2.22	70	31.53
12	2.44	80	32.78
13	2.64	90	34.09
14	2.83	100	35.33

It is observed that for the operating range from 0% to 30% of relative humidity, the output is linear with a calibration factor of $k=25$(%/RH/mV) but, from 40% to 100%, the output is nonlinear with varying slope gradually from 26.66 to 35.33. Therefore, the following rule is formulated for adaptation:

$$y(k) = 25x(k) \quad \text{for } x(k) \in [0,1.2]$$

$$y(k) = y^*(k-1) + y^{*\prime}(k-1)[n(k) - n^*(k-1)]$$

$$\text{For } x(k) \in [1.2, 2.83]$$

The LUT for the nonlinear range is shown in the table below:

Index (n)	$y^*(k)$	$y^{*\prime}(k)$
120	30	—
150	40	0.333
175	50	0.40
199	60	0.41
222	70	0.43
244	80	0.45
264	90	0.50
283	100	0.52

For the sensor measurement range 1.2–2.83 mV, an index range 120–283 is considered.

For the sensor output voltage 0.3 mV, we use the linear equation to get

$$Y = 25 \times 0.3 = 7.5\%$$

For sensor output voltage 2.3 mV, we use the linearized equation to get –

$$n = 2.3 \times 100 = 230$$

For $n = 230$

$$x(k) = 2.3, \quad y^*(k-1) = 70, \quad y^{*\prime}(k-1) = 0.43, \quad n^*(k-1) = 222$$

Therefore

$$\begin{aligned} y(k) &= 70 + 0.43 \times [230 - 222] \\ &= 70 + 0.43 \times 8 \\ &= 73.44\% \end{aligned}$$

4.6 Self-Validating Sensors

The measurement data provided by a sensor is not always trustworthy either due to the faulty state of the sensor hardware or the inappropriateness of the estimation data. Data validation is an important task of intelligent sensor making the sensor highly sensitive to malfunctions. A data validation module included in the sensor should provide additional data indicating the qualification or confidence level of the measurement data. The self-validating function can be classified as [18]

1. Functional validation
2. Technological validation

4.6.1 Functional Validation

Functional validation aims at detecting the consistency of the measurement or estimation produced by the sensor or some resource used by the estimation algorithm. Apart from detecting the fault, the system should be able to isolate the fault, providing it a fault detection and isolation (FDI) capability. Examples of faults in sensors are a short in the thermocouple wires, removal of bonding in a bonded strain gauge, accumulation of dirt on the electrode surface of a dissolved oxygen analyzer, open heater element in a MOS gas sensor, etc.

The classical method of sensor self-validation and fault detection consists of checking the measurable variables of the plant considering certain tolerance of the nominal values under healthy conditions. Fault detection technique are basically sensor signal processing that aims at state estimation, parameter estimation, adaptive filtering, variable threshold logic, statistical decision theory, and analytical redundancy concepts. All the above approaches are basically numerical techniques that rely on mathematical modeling of the system.

4.6.1.1 Numerical Method

In order to estimate the sensor faults and its isolation, a fault model of the intelligent sensor system has to be developed.

For explaining the validation model, let us consider the system model [18] equation.

System state equation:

$$\dot{x} = A_x + Bu_r + Qd_x \tag{4.38}$$

where

x is the system state
u_r is the system control input
d_x is the deterministic disturbances affecting states

Sensor output equation can be written as

$$\text{Remote: } y_r = g_r(x, dr, \varepsilon_r) \tag{4.39}$$

$$\text{Local: } y_l = g_l(z, de, \varepsilon_l) \tag{4.40}$$

where

dr is the deterministic disturbance effecting remote measurement equation
ε is a stochastic distribution with known and normal distribution
z is an estimate of state *x*

The self-validating sensor module receives the output of the remote and local sensors y_r and y_l and the control input u_r.

4.6.1.1.1 Fault Model Based on Change in Distribution

A fault in the local sensor develops a deterministic signal δ_l giving the local sensor output

$$y_l = g_l(z, d_l, \varepsilon_l, \delta_l) \tag{4.41}$$

The fault may also reflect a change in the distribution in the measurement error ε_l from P_o^l to P_1^l when the sensor operates in normal condition, $\varepsilon_l \sim N(0, \Sigma_0)$, i.e., the measurement error is caused by a zero mean Gaussian noise and with a covariance matrix Σ_0. In faulty condition, the measurement error, $\varepsilon_l \sim N(\mu, \Sigma_1)$, i.e., the mean, is no longer zero and covariance matrix changes to Σ_1.

The local sensor output Equation 4.40 at fault condition can also be written as

$$y_l = g_l(z, de, \varepsilon_l) + \delta_l \tag{4.42}$$

Similarly, the output in the remote sensor under faulty condition can be written as

$$y_r = g_r(x, dr, \varepsilon_r, \delta_r) \tag{4.43}$$

or

$$y_r = g_r(x, dr, \varepsilon_r) + \delta_r \tag{4.44}$$

4.6.1.1.2 Fault Model Based on Estimation from Local Data

The solution of the unknown equation in the static Equation 4.40 is possible with respect to the disturbance and the process variables. Equation 4.40 can be expressed as a transformation:

$$\psi(y_l) = \begin{pmatrix} \psi_1(y_l) \\ \psi_2(y_l) \end{pmatrix} = \begin{pmatrix} \psi_1(z, \varepsilon_l) \\ \psi_2(z, \varepsilon_l, \delta_l) \end{pmatrix} \tag{4.45}$$

Any rank

$$\frac{dz_1}{dz}(z,\varepsilon_l) = \dim z \tag{4.46}$$

This leads to the estimation of *z* from

$$\psi_1(y_l) - \psi_1(z,\varepsilon_l) = 0 \tag{4.47}$$

and

$$\hat{z} = \arg\min E_o\left(\|\psi_1(y_l) - \psi_1(z,0)\|^2\right) \tag{4.48}$$

where
$\| \ \|$ is the Fuclidian norm
E_o is expectation operator under distribution P_o^l

If $\dim \phi_1(y_l) > \dim z$, there also exists a transformation

$$\phi(y_l) = \begin{pmatrix} \phi_1(y_l) \\ \phi_2(y_l) \end{pmatrix} = \begin{pmatrix} \psi_1(\varepsilon_l,\delta_l) \\ \psi_2(z,d_l,\varepsilon_l,\delta_l) \end{pmatrix} \tag{4.49}$$

where ψ is unknown input observer.

Self validation is based on fault detection using a residual under normal state

$$r_\psi = \psi_1(y_l) - \psi_1(\hat{z},0,0) \tag{4.50}$$

and

$$r_\phi = \phi_1(y_l) - \phi_1(0,0) \tag{4.51}$$

which are zero in normal deterministic operation unless otherwise the case is faulty. A self-validating system based on estimation technique will be discussed with a case study now:

Case Study 4.6

Fermentation process of tea manufacturing needs relative humidity control in a humidifier. Tea fermentation is optimally achieved with an optimal air relative humidity in the fermentation room. Humidity-controlled air is circulated and exhausted in the fermentation room by using electric fans.

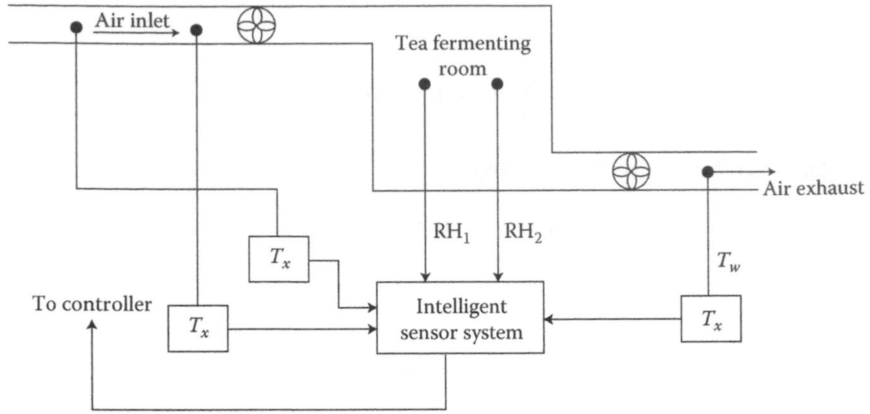

FIGURE 4.13
Humidity control of tea fermentation with self-validating sensors.

Figure 4.13 shows the schematic diagram of the humidity control system of tea fermentation process. The system is equipped with two redundant local humidity sensors RH_1 and RH_2. An estimate of the relative humidity is also done using two remote semiconductor temperature sensors to measure the dry bulb temperature T_d at the inlet and the wet bulb temperature T_w at the exhaust. Another redundant humidity sensor is installed at the outlet of the humidifier. The four linearized (piecewise) equation relating relative humidity with T_d and T_w are used to estimate the true value [19]. For example, the equation for $0°C \le T_d \le 25°C$ and $0°C \le (T_d - T_w) \le 5°C$ is given by

$$x = (0.204((T_d - T_w) + 0.072)T_d - 12.76(T_d - T_w) + 97.88 \quad (4.52)$$

In most tea industries, the dry bulb temperature T_d is maintained at 25°C, therefore the simplified equation of RH can be written as

$$x = 7.66T_w - 91.82 \quad (4.53)$$

For the self-validating sensor, the corresponding equations are

$x(t)$ is the relative humidity
$y_1(t)$ is the dry bulb temperature at inlet $= T_d$ (remote)
$y_2(t)$ is the relative humidity measured by sensor 2 (local)
$y_3(t)$ is the relative humidity measured by sensor 3 (local)
$y_4(t)$ is the wet bulb temperature at exhaust (remote)
$y_5(t)$ is the relative humidity measured at humidifier outlet by sensor 5 (remote)

The output of the remote sensors:

$$y_r = \begin{bmatrix} y_1(t) \\ y_4(t) \\ y_5(t) \end{bmatrix} = \begin{bmatrix} T_d \\ \dfrac{x}{7.66} + 11.98 \\ x + \varepsilon_3 \end{bmatrix} \tag{4.54a}$$

The output of the local sensors:

$$y_l = \begin{bmatrix} y_2(t) \\ y_3(t) \end{bmatrix} = \begin{bmatrix} x + \varepsilon_1 \\ x + \varepsilon_2 \end{bmatrix} \tag{4.54b}$$

where ε_1, ε_2, ε_3 are the measurement error.

An estimate of relative humidity can be obtained from the above equation as

$$\hat{x} = 7.66 y_4(t) - 91.76 \text{ (considering } T_d = 25°\text{C)} \tag{4.55a}$$

$$\hat{x} = 0.204 y_1^2(t) - 0.204 y_1(t) y_4(t) - 12.688 y_1(t) + 12.76 y_4(t) + 97.88 \quad \text{(considering } 0°\text{C} \le T_d \le 25°\text{C)} \tag{4.55b}$$

$$\hat{x} = y_2(t) \tag{4.55c}$$

$$\hat{x} = y_3(t) \tag{4.55d}$$

$$\hat{x} = y_5(t) \tag{4.55e}$$

A thumb rule estimate of relative humidity can be obtained with about 2% error using the equation

$$\hat{x} = 100 \left[\frac{112 - 0.1 y_1 + y_4}{112 + 0.9 y_1} \right] 8 + \varepsilon_5 \tag{4.55f}$$

where ε_5 is the estimation error.

A few structured residuals can be formulated from the above estimates as

1. $r_1(t) = y_2(t) - y_3(t)$
   if $r_1(t) \neq 0$, fault is in sensor 2 and sensor 3.
2. $r_2(t) = y_2(t) - 7.66 y_4(t) + 91.76$
   if $r_2(t) \neq 0$, fault is in sensor 2 and sensor 4, defect in exhaust fan or change in operating point $T_d = 25°$C.

3. $r_3(t) = y_3(t) - 7.66y_4(t) + 91.76$
   if $r_3(t) \neq 0$, fault is in sensor 3 and sensor 4, defect in exhaust fan or change in operating point T_d from 25°C.
4. $r_4(t) = y_2(t) - \hat{x}$ from Equation 4.55f
   if $r_4(t) \neq 0$, fault is in sensor 1 and sensor 2 or sensor 4.
5. $r_5(t) = y_3(t) - \hat{x}$ from Equation 4.55f
   if $r_5(t) \neq 0$, fault is in sensor 3 and sensor 1 and sensor 4.
6. $r_6(t) = y_2(t) - y_5(t)$
   if $r_6(t) \neq 0$, fault is in sensor 2 and sensor 5 or in inlet fan.
7. $r_7(t) = y_3(t) - y_5(t)$
   if $r_7(t) \neq 0$, fault is in sensor 3, sensor 5 or in inlet fan.

Example 4.7

During two batches of 45 min tea fermentation described in Case Study 4.6, the average readings of the sensors recorded are shown below:

Batch 1	Batch 2	Sensors
$y_1(t) = T_d = 25°C$	26°C	S1
$y_2(t) = RH_1 = 87\%$	69%	S2
$y_3(t) = RH_2 = 84.5\%$	66%	S3
$y_4(t) = T_w = 23°C$	21°C	S4
$y_5(t) = RH_3 = 85\%$	69.3%	S5

Determine the fault in the systems using the structured residue equations of Case Study 4.6. Use fault tolerance for estimate of relative humidity as 1%. Do not use the thumb rule estimate.

Solution

The residuals for batch 1 are calculated as

1. $r_1(t) = (87 - 84.5)\% = 2.5\%$
   Fault: S2 and S3
2. $r_2(t) = (87 - 7.66 \times 23 + 91.76 = 2.5\%)$
   Fault: S2, S4, Exhaust Fan (EF) change in $T_d(CT_d)$
3. $r_3(t) = (84.5 - 7.66 \times 23 + 91.76 = 0.08\%)$
   No fault: S3, S4, EF, CT_d
4. $r_4(t) =$ not used
5. $r_5(t) =$ not used
6. $r_6(t) = 87 - 85 = 2\%$
   Fault: S2, S5, inlet fan (IF)
7. $r_7(t) = 84.5 - 85 = -0.5\%$
   No fault: S3, S5, IF
   Eliminating S3, S4, IF and CT_d of (iii) from other sets

8. Fault-S2
9. Fault-S2
10. Fault-S2, S5, IF
    Eliminating S5, IF of (vii) from (viii), (ix), and (x)
11. Fault-S2
12. Fault-S2
13. Fault-S2

Therefore, there is a fault in sensor 2.

Similarly, the residuals for batch 2 are

1. $r_1(t) = 69 - 66 = 3\%$
   Fault-S2 and S3
2. $r_2(t) = 69 - 7.66 \times 21 + 91.76 = -0.1\%$
   No fault: S2, S4, EF, CT_d
3. $r_3(t) = 66 - 7.66 \times 21 + 91.76 = -3.1\%$
   Fault-S2, S4, EF, CT_d
4. $r_4(t) =$ not used
5. $r_5(t) =$ not used
6. $r_6(t) = 69 - 69.3 = -0.3\%$
   No fault-S2, S5, IF
7. $r_7(t) = 66 - 69.3 = -3.3\%$
   Fault-S3, S5 IF
   Eliminating S2, S4, EF, CT_d, S5, and IF
8. Fault: S3
   Hence, there is a fault in sensor 3.

Example 4.8

The two relative humidity sensors S2 and S3 of the system described in Case Study 4.6 produce measurement noises of zero mean Gaussian distribution with variance $\sigma_2 = 0.05\%$ and $\sigma_3 = 0.1\%$, respectively. If the sensor readings are 87% and 84.5%, respectively (batch 1 of Problem 4.7), estimate the related humidity under the following conditions: (a) no fault and (b) sensor 2 faulty.

Solution

Given:

$$y_2(t) = 87\%$$

$$y_3(t) = 84.5\%$$

$$\sigma_2 = 0.05\%$$

$$\sigma_3 = 0.1\%$$

a. Under no fault conditions

$$\hat{x} = \frac{1}{\sigma_1 + \sigma_2}(y_2(t)\sigma_3 + y_3(t)\sigma_2)$$

$$= \frac{87 \times 0.1 + 84.5 \times 0.05}{0.1 + 0.05} = 86.16\%$$

b. When sensor 2 is faulty, estimate from sensor 3

$$\hat{x} = 84.5\%$$

Case Study 4.7

Revisiting Table 4.1 of tea dryer information from which the dryer inlet temperature (T_i), dryer feed rate (M_i), exhaust temperature (T_o), and dryness of the output product values were used to model a dryer heat and mass transfer equation. The heat and mass transfer equation of a dryer is given by

$$\frac{M_a}{M_p}W_2 + \omega_1 = \frac{M_a}{M_p}W_1 + \omega_2 \tag{4.56}$$

where

(M_a/M_p) is the mass of hot air/kg of solid tea material

$W_{1,2}$ is the mass of water/kg of solid material in feeder inlet and outlet, respectively

The schematic diagram of the dryer is shown in Figure 4.14. Using the tea dryer sensor data under the three different categories of drying status—normal, overdry, and underdry—a model equation is developed based on Equation 4.56 as

$$7014\left(\frac{1}{T_o} - \frac{1}{T_i}\right) = \left(1.5 - \frac{X_0}{1 - X_0}\right)M_i \tag{4.57}$$

Giving an estimate of dryness of output tea:

$$\hat{x} = 1.5 - \frac{7014}{M_i}\left(\frac{1}{T_o} - \frac{1}{T_i}\right) \tag{4.58}$$

where

$$\hat{x} = \frac{\hat{X}_0}{1 - \hat{X}_0} \tag{4.59}$$

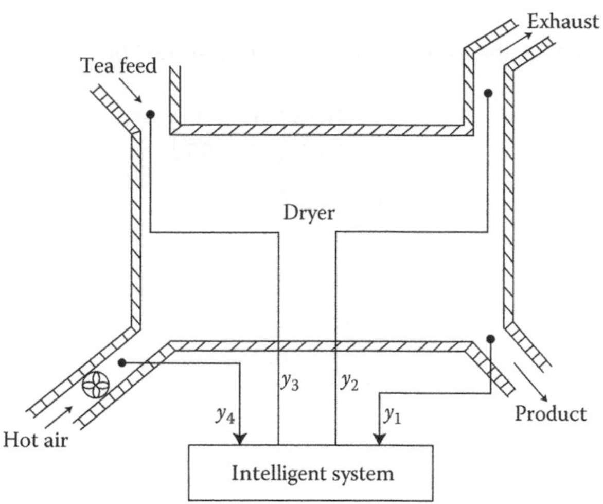

FIGURE 4.14
Schematic diagram of a tea fluid-bed dryer instrumentation.

Sensor measurement signals:

S_1 measures $y_1(t) = X_0$ (produced tea dryness fraction)
S_2 measures $y_2(t) = T_o$ (dryer exhaust temperature)
S_3 measures $y_3(t) = M_i$ (input tea feed-rate)
S_4 measures $y_4(t) = T_i$ (dryer inlet temperature)

Following residuals are formulated for sensor fault detection:

1. $$r_1(t) = y_1(t) - \left[1.5 + \frac{7014}{y_3(t)}\left(\frac{1}{y_2(t)} - \frac{1}{y_4(t)}\right)\right] \tag{4.60}$$

   If $r_1(t) \neq 0$, fault in S_1, S_2, S_3, or S_4.

2. $$r_2(t) = y_1(t) - \hat{X}_0(t) \tag{4.61}$$

   If $r_2(t) \neq 0$, fault in sensor S_1.

Example 4.9

For the Case Study 4.7, the sensor readings are given as: $T_i = 95°C$, $M_i = 7$ trays/15 min; $T_0 = 67°C$, $X_0 = 3\%$. Considering the weight of tea in a tray as 3 kg, determine the sensor linearity. Use a tolerance of 0.1% of dryness fraction.

Solution

Using Equation 4.58, the estimate of the tea dryness fraction

$$\hat{x} = 1.5 - \frac{7014}{7 \times 3}\left(\frac{1}{67} - \frac{1}{95}\right)$$
$$= 0.03 \quad \text{i.e., } 3\%$$

From Equation 4.59

$$\frac{\hat{X}_0}{1 - \hat{X}_0} = 0.03$$

Therefore

$$\hat{X}_0 = 0.029$$
$$= 2.9\%$$

$r_i(t) = 3 - 2.9 = 0.1$, which is within limit. Hence there is no fault in S_1, S_2, S_3, or S_4.

Example 4.10

In order to measure the voltage drop across a 1 kΩ resistor, three redundant meters are used, as shown in Figure E4.10. The wattmeter measures the power consumed, the voltmeter measures the voltage applied to the circuit, and the ammeter measures the current through the circuit.

1. Develop the residual equation for the meter fault detection (except short circuit or open circuit)
2. Determine the meter malfunction and error for the following sets of meter reading:
   a. 5 V, 4 mA, and 25 mW
   b. 12 V, 12 mA, and 121 mW.

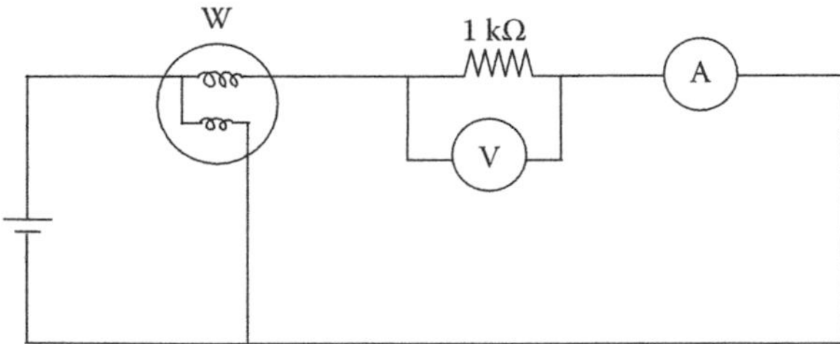

FIGURE E4.10
Circuit diagram for power measurement.

Solution
We use the following circuit equations:

$$V = IR$$

and $V = \sqrt{PR}$
Let the measurement variables be

$$y_1(t) = V + \epsilon_1$$

$$y_2(t) = I + \epsilon_2$$

$$y_3(t) = P + \epsilon_3$$

The voltage estimate equations are

$$\hat{V} = y_1(t)$$

$$\hat{V} = y_2(t)R$$

$$\hat{V} = \sqrt{y_3(t)R}$$

1. The residuals are
   a. $r_1(t) = y_1(t) - y_2(t)R$
      if $r_1(t) \neq 0$, fault in voltmeter or ammeter
   b. $r_2(t) = y_1(t) - \sqrt{y_3(t)R}$
      if $r_2(t) \neq 0$, fault in voltmeter or wattmeter
   c. $r_3(t) = y_2(t)R - \sqrt{y_3(t)R}$
      if $r_3(t) \neq 0$, fault in ammeter or wattmeter
2. For $y_1(t) = 5\,V$, $y_2(t) = 4\,mA$, and $y_3(t) = 25\,mW$

$$\begin{aligned} r_1(t) &= 5 - 4 \times 10^{-3} \times 1 \times 10^3 \\ &= 5 - 4 \\ &= 1\,V \end{aligned}$$

So, there is fault in voltmeter or ammeter

$$\begin{aligned} r_2(t) &= 5 - \sqrt{25 \times 10^{-3} \times 10^3} \\ &= 5 - \sqrt{25} \\ &= 0 \end{aligned}$$

So, there is no fault in voltmeter or wattmeter.

$$
\begin{aligned}
r_3(t) &= 4\times10^{-3}\times10^{3}\times1-\sqrt{25\times10^{-3}\times10^{3}}\\
&= 4-5\\
&= -1\,\text{V}
\end{aligned}
$$

So, there is fault in ammeter or wattmeter.

From the three logics, the fault decision is that there is a fault in ammeter only.

Similarly, for $y_1(t) = 12\,\text{V}$, $y_2(t) = 12\,\text{mA}$, and $y_3(t) = 121\,\text{mW}$

$$
\begin{aligned}
r_1(t) &= 12-12\times10^{-3}\times10^{3}\times1\\
&= 0\,\text{V, so there is no fault in voltmeter or ammeter}\\
r_2(t) &= \sqrt{12-12\times10^{-3}\times10^{3}\times1}\\
&= 12-11\\
&= 1\,\text{V, so there is fault in wattmeter or ammeter.}
\end{aligned}
$$

From the three logics, it can be ascertained that there is a fault in wattmeter only.

4.6.1.2 Artificial Intelligence for Sensor Validation

When reliable measurements are very complicated or valid mathematical models do not exist in case of complex time-varying and nonlinear plants, artificial intelligence shows importance for developing the self-validation technique in intelligent sensors. Artificial intelligence is applied broadly in two ways—expert system and artificial neural network. An expert system simulates human reasoning to solve a problem in addition to doing numerical calculation or data processing.

The reasoning is performed using heuristic knowledge rather than relationships. Expert system approaches have gradually changed from rule based using empirical reasoning to model based using functional reasoning. Model-based diagnosis uses knowledge about system structure, functions, and behavior. The advantages of model-based approaches are

1. Do not require field experience
2. Flexible to design changes
3. Less experience

However, combinations of knowledge-based techniques with numerical technique are effectively used to offer solution to many situations.

Neural network approaches also find successful application in sensor validation and fault detection due to their pattern recognition capability.

The network is trained with sensor measurement data under fault conditions and therefore can distinguish a normal and a faulty measurement. The main advantage of neural network approaches is that an approximate model can map the functional relation of the sensor. Complex rules and algorithms are avoided in neural network approach. The disadvantages of neural network are

1. Need highly accurate measurement data for online sensor validation.
2. The trained network is in the form of weight, which is difficult to comprehend.
3. Inability to explain the reasoning.

4.6.1.3 Model-Based Approach

In model-based approach, the expert knowledge is contained in a model in the expert domain. These models are used for simulation to explore hypothetical fault conditions. The RST model of tea drying status detection explained in Section 4.4.1.1 will be taken as a case study in this section to show the capability of the model to validate sensor measurements.

Case Study 4.8
The following decision rules were generated from the RST model of tea drying status monitor discussed in Case Study 4.1:

1. Normal drying is implied with certainty when there is medium inlet temperature.
2. Overdrying is implied with certainty when there is medium inlet temperature and low outlet temperature.
3. Underdrying is implied with certainty when there is low inlet temperature, high feed-rate and high outlet temperature.
4. Overdrying is implied with 0.44 certainty when there is high inlet temperature.
5. Normal drying is implied with 0.56 certainty when there is high inlet temperature.

We call the above rules as the RST model for the drying status detection. Strictly speaking, the RST model uses symbolic values such as normal, medium, and high, etc. rather than using real engineering values. Therefore, the knowledge model works on a limit theorem rather than on an equation. Now let us define the symbolic values in limit of numerical engineering units as

Inlet temperature (°C):

$$T_{i,l} \in [85, 90] \tag{4.62a}$$

$$T_{i,m} \in [90,100] \tag{4.62b}$$

$$T_{i,h} \in [100,115] \tag{4.62c}$$

Tea feed-rate (trays/15 min):

$$M_{i,l} \in [3,5] \tag{4.63a}$$

$$M_{i,m} \in [5,10] \tag{4.63b}$$

$$M_{i,h} \in [10,15] \tag{4.63c}$$

Outlet temperature (°C):

$$T_{o,l} \in [60,65] \tag{4.64a}$$

$$T_{o,m} \in [65,70] \tag{4.64b}$$

$$T_{o,h} \in [60,65] \tag{4.64c}$$

where subscripts *l*, *m*, and *h* indicate low, medium, and high, respectively.

Tea dryness fraction (%):

$$X_o \in [0,3] \tag{4.65a}$$

$$X_n \in [3,10] \tag{4.65b}$$

$$X_u \in [10,50] \tag{4.65c}$$

where subscript *o*, *n*, and *u* indicate overdry, normal dry, and underdry, respectively.

The RST rules (4.66a) through (4.66e) are written in model hypothetic equation as

$$X_n \Leftrightarrow T_{i,m} \wedge T_{i,l} \tag{4.66a}$$

$$X_o \Leftrightarrow T_{i,m} \wedge T_{o,l} \tag{4.66b}$$

$$X_u \Leftrightarrow T_{i.l} \wedge M_{i,h} \wedge T_{o,h} \tag{4.66c}$$

$$X_o \Leftrightarrow T_{i,h} \sim cer = 0.44 \tag{4.66d}$$

$$X_n \Leftrightarrow T_{i,h} \sim cer = 0.56 \tag{4.66e}$$

The estimates of the variables can be written as

Dryness fraction ($\hat{X}$):

$$X_n \Rightarrow T_{i,m} \wedge T_{i,l} \tag{4.67a}$$

$$X_n \Rightarrow T_{i,h} \sim cert = 0.56 \tag{4.67b}$$

$$X_o \Rightarrow T_{i,m} \wedge T_{o,l} \tag{4.67c}$$

$$X_o \Rightarrow T_{i,h} \sim cert = 0.44 \tag{4.67d}$$

$$X_u \Rightarrow T_{i.l} \wedge M_{i,h} \wedge T_{o,h} \tag{4.67e}$$

Inlet temperature ($\hat{T}_i$):

$$T_{i,m} \Rightarrow X_n \wedge T_{o,l} \tag{4.68a}$$

$$T_{i,m} \Rightarrow X_o \wedge T_{o,l} \tag{4.68b}$$

$$T_{i,l} \Rightarrow (X_u \wedge M_{i,h} \wedge T_{o,h}) \wedge (X_{o,n} \wedge T_{i,m}) \tag{4.68c}$$

$$T_{i,h} \Rightarrow (X_o \sim cert = 0.44) \tag{4.68d}$$

$$T_{i,h} \Rightarrow (X_n \sim cert = 0.56) \tag{4.68e}$$

Tea feed-rate ($\hat{M}_i$):

$$M_{i,h} \Rightarrow (X_u \wedge T_{i,l} \wedge T_{o,h})$$

$$M_{i,l} \Rightarrow \varnothing \tag{4.69a}$$

$$M_{i,m} \Rightarrow \varnothing$$

Outlet temperature (T_o)

$$T_{o,l} \Rightarrow X_o \wedge T_{i,m} \tag{4.70a}$$

$$T_{o,h} \Rightarrow X_u \wedge T_{i,e} \wedge M_{i,h} \tag{4.70b}$$

$$T_{o,m} \Rightarrow \varnothing$$

where $\varnothing$ indicates that the model is over-constrained for estimation of the corresponding measurement variable. One point to be remembered is that the fault detection of sensor measurement and its isolation rely on residuals that have properties that depend on the subset of faulty sensors; therefore, in knowledge-based approach also, residuals are necessary for fault isolation. The residuals on estimation of X_0, can be written as

$$r_1 = X - [(X_n \Rightarrow T_{i,m} \wedge T_{i,l}) \vee (X_n \Rightarrow T_{i,h} \sim cer = 0.56) \wedge (X_o \Rightarrow T_{i,m} \wedge T_{o,l})$$
$$\vee (X_{0,0} \Rightarrow T_{i,h} \sim cer = 0.44) \wedge (X_u \Rightarrow T_{i,l} \wedge M_{i,h} \wedge T_{0,h})]$$

If $r_1 \neq 0$, (fault is in sensor 2) or (fault is in sensor 2 ~ cer = 0.56) or (fault is in sensor 2 or sensor 4) or (fault is in sensor 2 ~ cer = 0.44) or (fault is in sensor 2 or sensor 3 or sensor 4).

The generation of residuals and validation of sensor measurements for the above example will be discussed in the following numerical problem.

Example 4.11

Four sensors are used to measure dryer inlet temperature (T_i), dryer tea feed-rate (M_i), dryer exhaust temperature (T_0), and dryness fraction of produced tea (X_0) in the tea dryer. Two sets of sensor measurement data are shown below:

Set A: $T_i = 112°C$, $M_i = 4$ trays/15 min, $T_0 = 67°C$ and $X = 5\%$

Set B: $T_i = 88°C$, $M_i = 12$ trays/15 min, $T_0 = 62°C$ and $X = 25\%$

Determine residuals and detect sensor faults for each sets of data. Use data limits to RST symbolic correspondence using Equations 4.62 through 4.65.

Solution

Data Set A:

The data values are converted to symbolic values using Equations 4.62 through 4.65 as

$$T_i = 112°C = T_{i,h}$$

$$M_i = 4 \text{ trays/15 min} = M_{i,l}$$

$$T_0 = 67°C = T_{o,l}$$

$$X = 5\% = X_n$$

The estimate of each variable from measurement data of other three variables will be performed using estimate equations:

Using Equation 4.67

$$\hat{X} = X_n \Rightarrow T_{i,h} \sim cer = 0.56 \text{ (using Equation 4.67b)}$$

Residuals

$$r_1 = X - (X_n \sim cer = 0.56)$$
$$= X_n - (X_n \sim cer = 0.56)$$
$$= 0 \sim \text{no fault in } T_i \text{ sensor with certainty } 0.56$$

Again

$$\hat{X} = X_0 \Rightarrow T_{i,h} \sim cer = 0.44 \text{ (using Equation 4.67d)}$$
$$r_2 = X - (X \sim cer = 0.44)$$
$$= X_n - (X_o \sim cer = 0.44)$$

$\neq 0 \sim$ Fault in T_i sensor with certainty 0.44. So there is a fault in T_i sensor with a certainty of 0.44.

Using Equation 4.68

$$\hat{T}_i = T_{i,h} \Rightarrow X_n \sim cer = 0.56 \text{ (using Equation 4.68e)}$$
$$r = T_i - (T_{i,h} \sim cer = 0.56)$$
$$= T_{i,h} - (T_{i,h} \sim cer = 0.56)$$
$$= 0 \sim \text{no fault in } X \text{ sensor with certainly} = 0.56$$

Using Equation 4.69

$$\hat{M}_i = \phi$$

Using Equation 4.70

$$\hat{T}_0 = \phi$$

Hence, fault decision
Fault in T_i sensor (certainty = 0.44)
No fault in X sensor
No decision for M_i and T_0 sensors

Data Set B:
Numerical data to symbolic conversion:

$$T_i = 88°\text{C} = T_{i,h}$$
$$M_i = 12 \text{ trays/15 min} = M_{i,h}$$
$$T_0 = 62°\text{C} = T_{o,l}$$
$$X = 25 = X_u$$

Using Equation 4.67

$$\hat{X} = X_u \Rightarrow T_{i,l} \wedge M_{i,h} \wedge T_{0,h(e)} \quad \text{[using Equation 4.67e]}$$
$$r = X_u - (X_u \sim T_0 \sim \in)$$

Implies that fault is in T_0 sensor.
Using Equation 4.68

$$\hat{T}_l \Rightarrow T_{i,l} \Rightarrow X_u \wedge M_{i,h} \wedge T_{0,h(l)} \quad \text{[using Equation 4.68c]}$$
$$r = T_{i,l} - (T_{i,l} \sim T_0 \sim \in)$$

Implies that fault is in T_0 sensor.
Using Equation 4.69

$$\hat{M}_1 = M_{i,h} \Rightarrow (X_u \wedge T_{i,l} \wedge T_{o,h(l)} \quad \text{[using Equation 4.69a]}$$
$$r = M_{i,h} - (M_{i,h} \sim T_0 \sim \in)$$

Implies that fault is in T_0 sensor.
Using Equation 4.70

$$\hat{T}_0 = T_{o,h} \Rightarrow (X_u \wedge T_{i,l} \wedge M_{i,h})$$
$$r = T_{o,l} - T_{o,h} \neq 0$$

So fault is in T_0 sensor only.

Example 4.12

In the tea drying status model discussed in Case Study 4.8, if the tea dryness fraction sensor (X) is 100% accurate and healthy, determine which of the sensor is faulty for the following cases:

1. $X = 15\%$; $T_i = 102°C$, $M_i = 12$ trays/15 min, and $T_0 = 74°C$
2. $X = 1\%$, $T_i = 87°C$, $M_i = 4$ trays/15 min, and $T_0 = 66°C$

Solution

Since X sensor is healthy, we use to estimate of X.

1. Numerical data to symbolic conversion (using Equations 4.62 through 4.65)

$$X = X_u$$
$$T_i = T_{i,h}$$
$$M_i = M_{i,h}$$
$$T_0 = T_{0,h}$$

From Equation 4.67e

$$X = X_u \Rightarrow T_{i,l} \wedge M_{i,h} \wedge T_{0,h} \sim \in = 0$$

From given data:

$$\hat{X} = X_u \Rightarrow T_{i,l} \wedge M_{i,h} \wedge T_{0,h} \sim \in \neq 0$$
$$r = X - \hat{X} \Rightarrow T_{i,l(h)}$$

Hence fault is in T_i sensor.

2. Numerical data to symbolic conversion (using Equations 4.62 through 4.65)

$$X = X_o$$
$$T_i = T_{i,l}$$
$$M_i = M_{i,l}$$
$$T_0 = T_{0,l}$$

From Equation 4.67c

$$X = X_o \Rightarrow T_{i,m} \vee T_{0,l}$$

From given data

$$\hat{X} = X_o \Rightarrow T_{i,l} \vee T_{0,l}$$
$$r = X - \hat{X} \Rightarrow T_{i,m(l)}$$

Hence fault is in T_i sensor.

Example 4.13

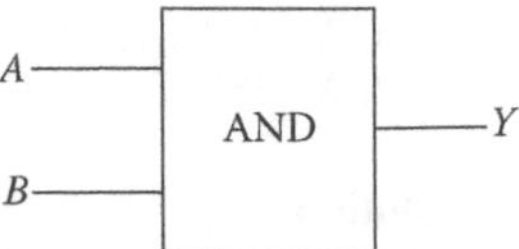

FIGURE E4.13
Input–output measurement in an AND gate.

In the digital AND system shown in Figure E4.13 three meters measures inputs A, B and output Y.

1. Develop the expert rule for estimating Y, A, and B.
2. For the following cases determine meter faults:
   a. $Y=H$, $A=H$, $B=L$, and Y meter is accurate and healthy.
   b. $Y=L$, $A=L$, $B=H$, and A meter is accurate and healthy.

Solution

1. From knowledge of digital logic system, the estimate output equations for the AND gate are

$$\hat{Y}_H \Rightarrow A_H \wedge B_H \quad [Y = AB] \tag{E4.13.1}$$

$$\hat{Y}_L \Rightarrow (A_L \vee B_L) \wedge (A_L \vee B_H) \wedge (A_H \vee B_L) \quad [Y(L) = \bar{A} + \bar{B}] \tag{E4.13.2}$$

$$\hat{A}_H \Rightarrow (B_L \wedge Y_L) \vee (B_H \wedge Y_H) \tag{E4.13.3}$$

$$\hat{A}_L \Rightarrow (B_L \wedge Y_L) \vee (B_H \wedge Y_L) \tag{E4.13.4}$$

$$\hat{B}_H \Rightarrow (A_L \wedge Y_L) \vee (A_H \wedge Y_H) \tag{E4.13.5}$$

$$\hat{B}_L \Rightarrow (A_L \wedge Y_L) \vee (A_H \wedge Y_L) \tag{E4.13.6}$$

2.
   a. Since Y meter is accurate and healthy, we estimate variable y:
   Given data:

$$Y = Y_H$$
$$A = A_H$$
$$B = B_L$$

From (E4.13.1) $\hat{Y} = Y_H \Rightarrow A_H \wedge B_H$
From given data

$$\hat{Y} = Y_H \Rightarrow A_H \wedge B_{L(H)}$$

Residual $r = Y - \hat{Y}$
$= Y_H - (Y_H \sim B \sim \in)$ implies fault in B meter.

b. Since A meter is accurate and healthy, we estimate variable A:
Given:

$$Y = Y_L$$
$$A = A_L$$
$$B = B_H$$

From (E4.13.4) $A = A_L \Rightarrow (B_L \wedge Y_L) \vee (B_H \wedge Y_L)$
From given data: $\hat{A} = A_L \Rightarrow (B_H \wedge Y_L)$
No fault in Y meter; fault in B meter cannot be ascertained.

4.6.1.4 Neural Network–Based Approach

Neural network–based soft sensor design approaches have already been discussed in Section 4.4.2.2, where basics about structure and learning methods of a network were explained; therefore, these will not be discussed here. The first important thing about a neural network–based fault detection is generating data from the sensor measurement under both healthy and faulty sensor conditions. For each set of input data there should be a definite set of output data. The input data are the input sensor measurement vectors and the output data are the output sensor estimate vectors. The second important thing is training the network. Input measurement vectors are applied to the network with specific output vectors and the weights of the network are updated during training. There are various techniques and rules adopted for training the network the descriptions of which is available in other texts. The third thing is testing the network with test measurement vectors and generating residuals. Figure 4.15 shows the block diagram of network training and residual generation.

Residuals are difference between the actual plant measurement output and neural network estimate output. Strictly speaking, the sensor blocks shown in Figure 4.15a do not represent a single sensor; rather, they represent a set of sensors. Since the plant is difficult to model due to its complexity and nonlinearity, the network is able to represent a knowledge-based

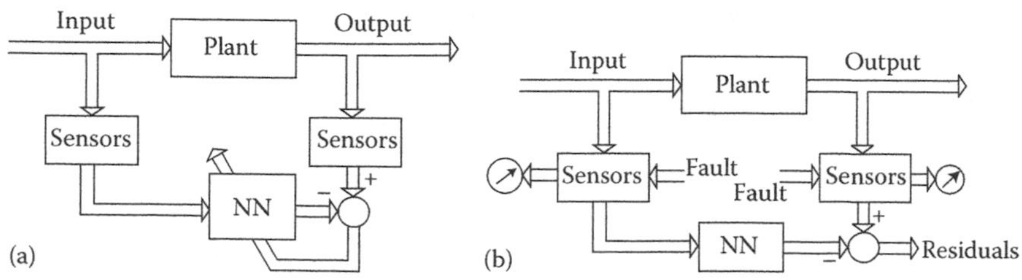

FIGURE 4.15
(a) Neural network training and (b) generating residuals.

model. An input measurement vector corresponds to a definite output measurement vector; therefore, there will be possible estimates of sensor measurements.

The next step to residual generation is the residuals evaluation for sensor validation, fault detection (Figure 4.16). Residuals are signatures of the faults; however, in most cases, a definite mathematical relation between the residuals and the faults does not exist. Therefore, a separate neural network is used to map the residuals to sensor faults. The neural network is trained with an input residual database and a target fault database.

Similar to soft sensor model using ANN discussed in Section 4.4.2.2, we will consider the same problem of tea drying status detection here for data validation and fault detection in the dryer inlet temperature (T_i), tea feed-rate (M_i), dryer exhaust temperature (T_o), and made tea dryness factor (X) sensors. Recall that, the sensor measurement variables considered in the soft sensor ANN model were discrete variables, i.e., low = 001, medium = 010, high = 100 for the input sensors; and under = 001, normal = 010, and over = 100 for the output sensors; however, in this case of sensor data validation, we will consider real numerical values. The schematic diagram and the design approach will be discussed in the following case study.

There may be various network configurations for sensor fault detection depending on the applicability. Following are some of the common configurations:

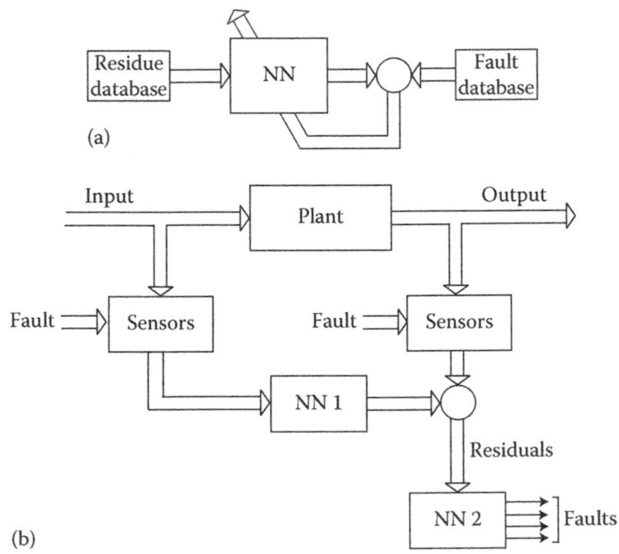

FIGURE 4.16
(a) Training the residual ANN and (b) fault detection.

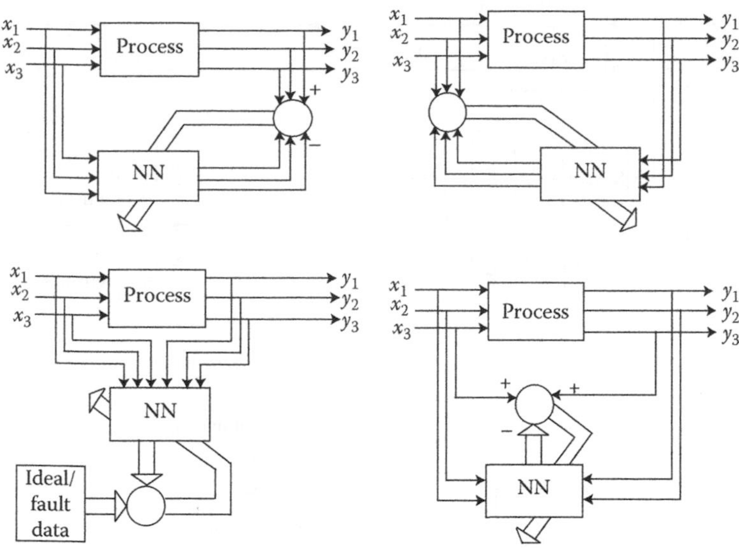

FIGURE 4.17
Various network configurations for sensor fault detection.

1. Process inputs and outputs as network inputs and outputs.
2. Process inputs and outputs, both as network input and estimated or ideal input/output values as network outputs.
3. Process inputs and outputs as network input and output, respectively.
4. A subset of process input/output as network inputs and the remaining set as network output. Example configurations are shown in Figure 4.17.

Case Study 4.9

We consider here T_i and M_i as input and T_0 and X_0 as output of the dryer, and the neural network scheme is developed as shown in Figure CS4.9. While training, the input data vectors of the ANN can be written as

$$X = \begin{bmatrix} T_i \\ M_i \end{bmatrix} = \begin{bmatrix} X_1 \\ X_2 \end{bmatrix} = \begin{bmatrix} X_{11}X_{12} & \cdots & X_{1n} \\ X_{21}X_{22} & \cdots & X_{2n} \end{bmatrix} \text{ where } n \text{ is the data points} \quad (4.70)$$

and the output data vectors of the ANN is written as

$$Y = \begin{bmatrix} T_0 \\ X_0 \end{bmatrix} = \begin{bmatrix} Y_{11}Y_{12} & \cdots & Y_{1n} \\ Y_{21}Y_{22} & \cdots & Y_{2n} \end{bmatrix} \quad (4.71)$$

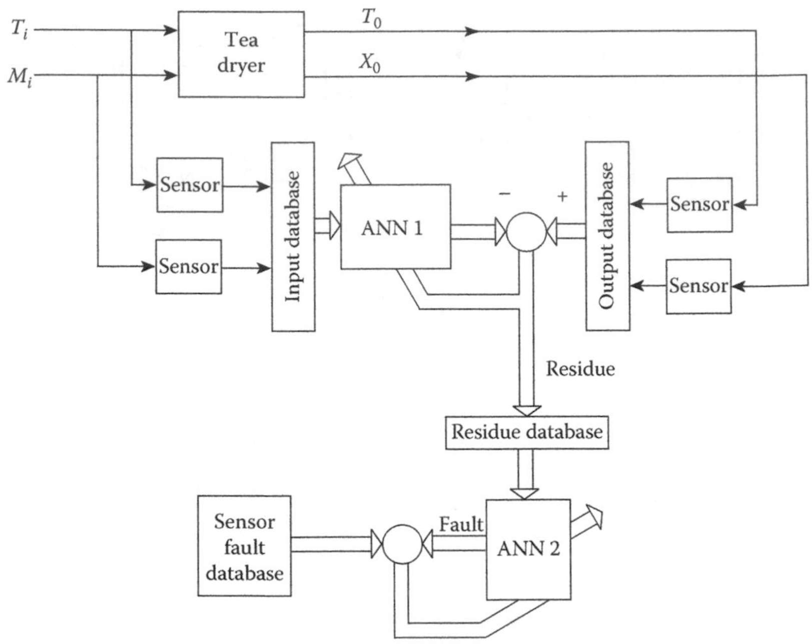

FIGURE CS4.9
Tea dryer input sensor validation by ANN training and testing.

The residue generated by the ANN is

$$R = \begin{bmatrix} T_0 - \widehat{T}_0 \\ X_0 - \widehat{X}_0 \end{bmatrix} = \begin{bmatrix} (Y_{11} - \widehat{Y}_{11}) & \cdots & (Y_{1n} - \widehat{Y}_{1n}) \\ (Y_{21} - \widehat{Y}_{21}) & \cdots & (Y_{2n} - \widehat{Y}_{2n}) \end{bmatrix} \tag{4.72}$$

where
T_0 and X_0 are actual sensor outputs
$\widehat{T}_0$ and $\widehat{X}_0$ are ANN estimated values

To train the residue ANN2, simulated fault residues are generated from ANN1. Three types of faults can be simulated:

1. Fault in T_i
2. Fault in M_i and
3. Fault in T_i and M_i both

Hence, the trained ANN1 is applied with faulty X data vectors and the faulty $\widehat{Y}$ data vectors are obtained. The faulty X and $\widehat{Y}$ vectors can be expressed as

$$X_f = \begin{bmatrix} X_{T_i(f)} & X_{M_i(f)} & X_{T_i(f),M_i(f)} \end{bmatrix} \tag{4.73}$$

$$\hat{Y}_f = \begin{bmatrix} Y_{T_i(f)} & Y_{M_i(f)} & Y_{T_i(f),M_i(f)} \end{bmatrix} \tag{4.74}$$

and the fault residue vector is

$$\begin{aligned} R_f &= Y_h - \hat{Y}_f \\ &= \begin{bmatrix} R_{T_i(f)} & R_{M_i(f)} & R_{T_i(f)M_i(f)} \end{bmatrix} \end{aligned} \tag{4.75}$$

where subscripts f and h indicate faulty and healthy, respectively.

Training ANN-1:

For training ANN-1, 67 input/output data vectors were used, which were experimentally collected from a tea dryer using healthy sensors. Example data vectors (two from each) under seven patterns are shown below:

$$X = \begin{bmatrix} T_i \\ M_i \end{bmatrix} = \begin{bmatrix} 87 & 89 & 92 & 100 & 92 & 97 & 85 & 90 & 85 & 90 \\ 3 & 5 & 7 & 9 & 4 & 5 & 12 & 9 & 7 & 8 \end{bmatrix}$$

$$Y = \begin{bmatrix} T_0 \\ X_0 \end{bmatrix} = \begin{bmatrix} 68 & 70 & 64 & 68 & 65 & 67 & 73 & 80 & 65 & 70 \\ 1 & 1 & 2 & 5 & 2 & 0 & 5 & 7 & 9 & 5 \end{bmatrix}$$

Each pair of input data of X corresponds to output data of Y.

The network structure for ANN-1 is shown in Figure CS4.10. The validity of the ANN-1 is tested by some test input and the estimated outputs are obtained. For the sample input data vectors shown above, the estimated outputs are

$$\hat{Y} = \begin{bmatrix} 67.12 & 68.21 & 68.82 & 66.77 & 66.06 & 65.57 & 71.70 & 71.16 & 71.31 & 70.67 \\ 1.50 & 2.41 & 3.05 & 1.34 & 1.01 & 0.82 & 7.19 & 6.39 & 6.62 & 5.63 \end{bmatrix}$$

Now, for generating residue for sensor fault detection, simulated fault data X were applied. The 60 data vectors were grouped as follows: 20 for T_i fault, 20 for M_i fault, and 20 for both T_i and M_i faults. For the sample data vector, the sample faulty data vector is shown below:

$$X_f = \begin{bmatrix} 60 & 65 & 82 & 100 & 92 & 97 & 60 & 70 & 60 & 75 \\ 3 & 5 & 7 & 3 & 5 & 1 & 4 & 3 & 5 & 3 \\ \multicolumn{3}{c}{\longleftarrow T_{if} \longrightarrow} & \multicolumn{4}{c}{\longleftarrow M_{if} \longrightarrow} & \multicolumn{3}{c}{\longleftarrow T_{if} \text{ and } M_{if} \longrightarrow} \end{bmatrix}$$

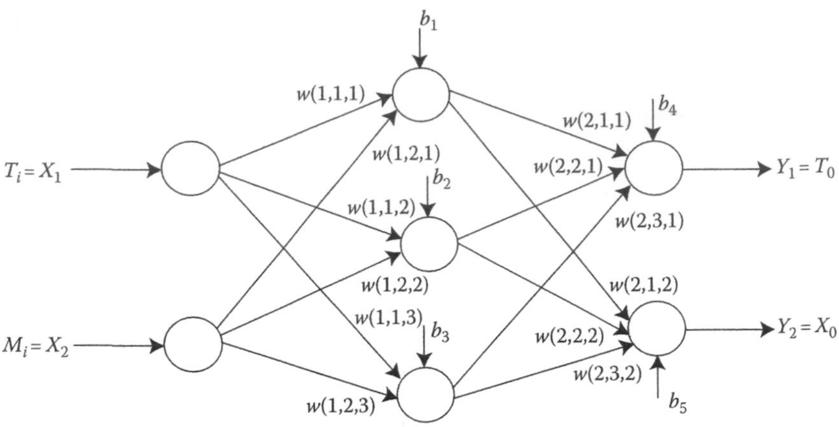

FIGURE CS4.10
ANN structure for tea dryer output estimation.

The estimated output data is obtained as

$$\hat{Y}_f = \begin{bmatrix} 71.72 & 71.72 & 71.57 & 65.30 & 66.72 & 65.29 & 71.72 & 71.68 & 71.72 & 71.5 \\ 7.2 & 7.21 & 7.00 & 0.74 & 1.31 & 0.739 & 7.21 & 7.16 & 7.21 & 6.90 \end{bmatrix}$$

The residues calculated are

$$R_f = Y - \hat{Y}_f = \begin{bmatrix} -3.72 & -1.72 & -7.57 & 2.70 & -1.72 & 1.71 & 1.28 \\ -5.20 & -2.21 & -5.00 & 0.26 & -0.31 & -0.739 & -2.21 \end{bmatrix}$$

For training residue ANN-2, 60 residue vectors were used as input to ANN-2 with target faults.

Technological validation is a partial sensor data validation approach, which is concurrent with some other hardware resources of the sensors operation such as a sensor power supply, microprocessor memory or a multiplexer, etc. Since the technological validation comes under the preview of general systems fault detection isolation, it will not be discussed in this chapter.

4.6.2 Applications of Self-Validating Sensors

A large numbers of self-validating sensor techniques find applications in sensor technology; however, a few important techniques will be discussed here.

4.6.2.1 SEVA Model

The self-validating or SEVA model developed at Engineering Science Department, Oxford University, possesses the capability of internal computing power for self-diagnostic based on the designers' expertise and of digital communication to transmit the measurement data, diagnostic, and maintenance data simultaneously [20].

The salient features of a SEVA model are

1. It exploits the manufacturer knowledge of a sensor to detect faults internally.
2. Individual impact of each fault on each measurement is possible.
3. Embedded microprocessor for self-diagnosis and correction of measurement.

The model generates the following five device-independent parameters (Figure 4.18):

1. Validated measurement value (VMV): It is the best estimate of the current measurement value. Under healthy sensor condition, it is the actual measured value; however, under faults state of the sensor, the VMV is estimated as follows:
   a. If a minor fault occurs, a correction is applied to correct the measurement. For example, say, in case of a strain gauge–based hot-gas pressure sensor, if the strain sensor is affected by temperature variation, a temperature compensation can be applied using a linear sensor resistance to temperature change coefficient. The correction term will be calculated by the microprocessor using an inbuilt semiconductor thermal sensor and the correction term is subtracted from the sensor's measurement value.
   b. If a major fault occurs—the measurement estimation is based on historical data stored in an EPROM of the microprocessor. In case of the strain gauge–based pressure sensor discussed above,

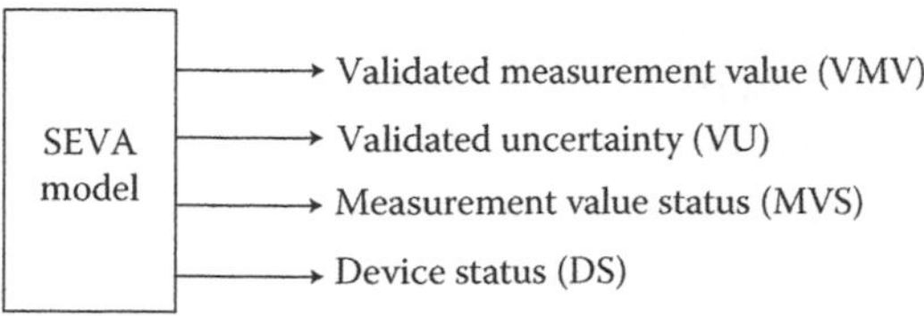

FIGURE 4.18
Parameters generated by the SEVA model. (Reprinted from Clarke, D.W. and Fraher, P.M., *Control Eng. Pract.*, 4(9), 1313, 1996, Copyright, Elsevier, 1996. With permission.)

TABLE 4.8

Measurement Status Parameters of Self-Validating Sensor

Status Parameter	Sensor Condition	Method of VMV Estimation	Correction	VU Status
CLEAR	No fault	Actual data	No	Null
BLURRED	Partial fault	Compensation technique	No	Increased
BLIND	Severe fault	Historical data		
DAZZLED	Not known temporary but severe	Historical data		
SECURE	Fault and not used	Soft a redundant sensor		
UNVALDATED	Fault and not used	Validation is not in operation		

a failure of bonding of the sensor with the test structure can be considered as a major fault. In such cases, the sensor output drastically drops and correction becomes difficult.

Historical data under various circumstances may be helpful in estimating the measurement. In this situation, the sensor works as a soft sensor with the capability of estimating based on others sensors' measurement data.

2. Validated uncertainty (VU): this is a validity index generated with each VMV. This parameter accommodates the reduced accuracy in VMV; however, it cannot indicate the cases of the reduction of accuracy. Uncertainty is defined as the deviation of a measurement from mean value of the data set expressed as percentage.

   Therefore, the microprocessor performs a computation on each measurement to calculate the uncertainty.
3. Measurement value status (MVS): this is another measurement validity index that indicates the source of the uncertainty or accuracy. The status is indicated by a discrete code. There are six MVS codes as listed in Table 4.8.
4. Device status:
   This is a discrete value summarizing the health of the sensor, which will be used mostly for maintenance purpose. For each value, any one of the following parameters is generated:
   - GOOD—there is no fault in the sensor.
   - TESTING—the intelligent system is performing fault detection during which there may be loss of measurement quantity.
   - SUSPECT—the sensor may have suffered a fault, which is yet to be diagnosed.

- IMPAIRED—there is a definite fault diagnosed, so calls a repair/maintenance of low priority.
- BAD—there is a definite fault diagnosed, so calls a repair/maintenance of high priority.
- CRITICAL—the condition of the sensor is critical and calls immediate attention.

A number of SEVA prototype devices have been developed based on commercial sensors such as dissolved oxygen [21] and mass flow [22] sensors. The first one will be discussed as a case study now.

Case Study 4.10

Dissolved oxygen sensor: Dissolved oxygen (DO_x) sensors find application in industrial and environmental monitoring. The (DO_x) sensor is basically a Clark cell; schematic diagram is shown in Figure 4.19.

When a negative potential is applied between the measurement electrode (cathode) and auxiliary electrode (anode), oxygen molecules diffuse through the membrane from the process fluid to the electrode (cathode). As a result, a steady-state current flows from the measurement electrode to the auxiliary electrode. The (DO_x) level can be expressed by the equation

$$[O_2] = G_0 G_T i_{ss} \quad (4.76)$$

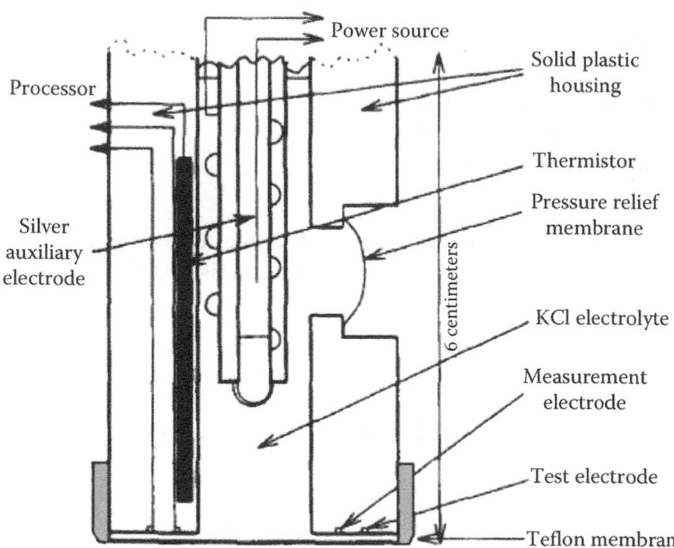

FIGURE 4.19
The Clark cell of dissolved oxygen (DO_x) sensor. (Reprinted from Clarke, D.W. and Fraher, P.M., *Control Eng. Pract.*, 4(9), 1313, 1996. Copyright, Elsevier, 1996. With permission.)

where

G_0 is the calibration constant
G_T is a temperature-dependent permeability constant
i_{ss} is the steady-state current

When the sensor membrane is fouled by deposition of dirt, etc., the flow of O_2 molecules to the electrode is retarded, the observed current i_{ss} drops down further to another equilibrium that apparently indicates a lower (DO_x) level. To detect and distinguish the act of reduced steady-state current due to membrane fouling from reduced (DO_x) level, a self-validating feature is introduced to the cell using an identical test electrode. The test electrode is not energized continuously, but it is connected to the negative potential by using a change over switch at a regular interval, say every 30 min, for a short duration, say 4 s. Since the test electrode is energized at a long interval, if the membrane is clear, sufficient (DO_x) molecule build up around the test electrode giving a pulse current. Therefore, a pre-fault indication is obtained.

Once a fault indication is obtained, it is confirmed using an oxygen transfer time τ given by

$$\tau = \frac{b^2}{D_{mo}} e^{(E_D/RT)} \tag{4.77}$$

where

b is the membrane thickness
D_{mo} is the diffusivity of membrane
E_D is the latent energy of diffusion
R is the universal gas constant
T is the temperature of fluid

At membrane foul condition, b increases, increasing τ and, thereby, a foul to clean membrane oxygen transfer time ratio τ/τ_0 is used to compare with a threshold to find a fault.

4.6.2.1.1 Compensation

When the membrane is fouled, Equation 4.76 of i_{ss} becomes

$$[O_2] = G'_D G_T i_{ss} = \frac{b'}{4FAKD'_m} G_T i_{ss} \tag{4.78}$$

where

G'_0 is the changed calibration constant
b' is the increased membrane thickness

D'_m is the fouled membrane diffusivity
F is Faraday's constant
A is the surface area of electrode
K is Henry's law constant

Since there is no any direct equation to estimate b' and D'_m, experimental data is used to estimate these two values.

In the SEVA version of the DO_x sensor, only two MVS and DS values are possible. When the membrane is clear, MVS is CLEAR and DS is GOOD. When there is a membrane foul, MVS is changed to BLURRED indicating that a compensation technique is used to estimate VMV. On the other hand, DS is changed to IMPAIRED, BAD, or CRITICAL depending upon the severity of the membrane foul. Remember that severity of membrane foul can be detected comparing τ/τ_0 with the threshold.

4.7 VLSI Sensors

With the advent of IC technology, onboard sensor data processing has been possible using VLSI circuits. Remember that smart sensor performs sensor meteorological signal processing onboard; however, VLSI circuits have the capability of handling a large number of data such as in case of an image sensor. Therefore, we will call the service provided by a VLSI sensor as data processing or data computation instead of signal processing that we generally refer to a single sensor. Depending on the type of the data processing algorithm, i.e., VLSI sensors takes the following forms –

4.7.1 Analog Numerical Computation

VLSI finds realizable applications in algorithms that are computationally complex and intensive such as those used in linear algebra-based signal processing. Matrix-based signal processing technique can also be implemented in VLSI. Singular value and eigenvalue computations are key computational Kernel of many intelligent sensors for data estimation, classification, and validation. These computations are powerful tools for many intelligent sensors; however, computational complexity is the main problem. In such cases, parallel processing is advantageous.

CORDIC (coordinate rotated digital computer)-based parallel singular value decomposition (SVD) can be implemented using a square array of processors that can handle complex value data in many signal processing applications [23]. Some basics of CORDIC algorithm and their numerical VLSI processors will be discussed here.

4.7.1.1 CORDIC Computation

It is a vector rotation–based algorithm by which all the trigonometric functions can be computed or derived. The principle of CORDIC computation is rotating the phase of a complex number by multiplying it by a succession of constant values. The multiplying factor is in powers of 2, so in binary arithmetic, it can be done by using just shift and add operation without using a real multiplier. To understand CORDIC computation, let us consider a complex number C given by

$$C = x + jy \tag{4.79}$$

which is rotated by multiplying with an another complex number given by

$$R = x_r + jy_r \tag{4.80}$$

The multiplication of C with R gives

$$C' = x' + jy' \tag{4.81}$$

where

$$x' = xx_r - yy_r \tag{4.82}$$

$$y' = yx_r + xy_r \tag{4.83}$$

to add the phase of C and R, and

$$x' = xx_r + yy_r \tag{4.84}$$

$$y' = yx_r + xy_r \tag{4.85}$$

to subtract the phase of R from the phase of C.

Example 4.14

Rotate a complex number $3 + j2$ by +90° and −90°.

Solution

To rotate by +90°, $R = x_r + jy_r = 0 + j1$

Here, $C = x + jy = 3 + j2$

$$\therefore x' = -2; \quad y' = 3 \quad \text{(using Equations 4.82 and 4.83)}$$

To rotate by −90° $R = x_r + jy_r = 0 + j1$

$$\therefore x' = -2; \quad y' = -3 \quad \text{(using Equations 4.84 and 4.85)}$$

In the above example, the complex number was rotated by full ±90°; however, for rotating by less than 90°, the rotating vector will be expressed by

$$R = 1 \pm jK \tag{4.86}$$

where K is the decreasing power of 2 giving

$$K = 2^{-L}, \quad L = 0,1,2,\ldots \tag{4.87}$$

Therefore, the rotated complex number C' can be written as

$$x' = x - 2^{-L}y \tag{4.88}$$

$$y' = y + 2^{-L}x \tag{4.89}$$

to add the phase of C and R, and

$$x' = x + 2^{-L}y \tag{4.90}$$

$$y' = y - 2^{-L}x \tag{4.91}$$

to subtract phase of R from phase of C.

Example 4.15

Rotate the complex number $5 + j10$ by an angle of ±45°.

Solution

Since $\varnothing = \pm 45°, \tan^{-1}(45)° = \pm 0.5 = \frac{y_r}{x_r}$

$$\therefore R = 1 + j0.5$$

$$\therefore K = 2^{-1}$$

The rotated complex number for +45° is

$$x' = x - 2^{-L}y = 5 - 2^{-1} \times 10$$

$$= 5 - 5 = 0$$

$$y' = 10 + 2^{-1} \times 5$$

$$= 12.5$$

$$\therefore C' = 0 + j12.5$$

TABLE 4.9
Phase and Magnitude of Vector Multiplication Process

L	K	R	Ø (degree)	$\|R\|$	CORDIC Gain
0	1.0	$1+j1.0$	45	1.41	1.41
1	0.5	$1+j0.5$	26.56	1.11	1.58
2	0.25	$1+j0.25$	14.03	1.03	1.63
3	0.125	$1+j0.125$	7.12	1.0019	1.642
4	0.0625	$1+j0.0625$	3.57	1.0004	1.645
5	0.0312	$1+j0.0312$	1.79	1.0004	1.6464
6	0.015625	$1+j0.015625$	0.89	1.0001	1.6466
7	0.007813	$1+j0.007813$	0.44	1.0000	1.6467

Similarly, for $-45°$, $x'=5+5=10$

$$y' = 10 - 2.5 = 7.5$$

$$\therefore C' = 10 + j7.5$$

Table 4.9 shows the phases and magnitude of the multiplication process for a few increasing values of L starting from 0.

CORDIC computation to determine magnitude of a complex number
The magnitude of a complex number C can be obtained from the real part after rotating it to a phase of 0°. Let the complex number

$$C = x + jy \tag{4.92}$$

be rotated to get $C'=x'+j0$.

Computationally, this can be done by rotating C successively by 90°, 45°, 22.5°, and so on till the phase becomes zero. In each rotation, we use Equations 4.88 through 4.91.

Example 4.16

Determine the magnitude of the complex number $5+j10$ by CORDIC computation.

Solution
Here, $x=5$, $y=10$, $\therefore$ $C=5+j10$

Since phase is positive, we rotate it by $-90°$
i.e., $R=0-j1$

$$\therefore x' = +10$$

$$y' = -5$$

The new number is $C_1' = 10 - j5$; $x = 10$, $y = 5$.
Now rotating by +45° (since phase is negative)

$$x' = 10 - 2^0 \times (-5) \qquad y' = -5 + 2^0 \times (10)$$
$$= 10 + 5 = 15 \qquad = 5$$

New number is $C_2' = 15 + j5$.
Now rotating by −26.56°

$$x' = 15 + 2^{-1} \times 15 = 15 + 2.5 = 17.5$$
$$y' = 5 - 2^{-1} \times 15 = 5 - 7.5 = -2.5$$

New number is $C_3' = 17.5 - j2.5$.
Now rotating by +14.03°

$$x' = 17.5 - \frac{1}{4} \times (-2.5)$$
$$= 18.125$$
$$y' = -2.5 + \frac{1}{4} \times (17.5)$$
$$= 1.875$$

New number is $C_4' = 18.125 + j1.875$.
Now rotating by −7.125°

$$x' = 18.125 + \frac{1}{8} \times (1.875)$$
$$= 18.359$$
$$y' = 1.875 - \frac{1}{8} \times (18.175)$$
$$= -0.396$$

The new number is $C_5' = 18.359 - j0.396$.
Now rotating by +3.57°

$$x' = 18.359 - \frac{1}{16} \times (-0.396) = 18.383$$
$$y' = (-0.396) + \frac{1}{16} \times (18.359)$$
$$= 0.751$$

New number is $C_6' = 18.383 + j0.751$.
Now rotating by −1.78°

$$x' = 18.383 + \frac{1}{32} \times (+0.751) = 18.40$$

$$y' = (+0.751) - \frac{1}{32} \times (18.383) = 0.176$$

The new number is $C_7' = 18.40 + j0.176$.
Now rotating by −0.89°

$$x' = 18.40 + \frac{1}{64} \times (+0.176) = 18.402$$

$$y' = (0.176) - \frac{1}{64} \times (18.40) = -0.11$$

The new number is $C_8' = 18.402 - j0.11$.
Now rotating by −0.44°

$$x' = 18.402 - \frac{1}{132} \times (+0.11) = 18.4028$$

$$y' = (-0.11) + \frac{1}{132} \times (18.402) = 0.029$$

The rotation can be further attained till y' approaches zero. Since in every step, the complex numbers are multiplied by a CORDIC gain shown in Table 4.9 the true magnitude of the complex number with zero phase can be obtained by dividing the magnitude by the CORDIC gain of 1.646.
∴ True magnitude

$$x' = \frac{18.4028}{1.646} = 11.18$$

which is same as the magnitude of the complex number given by

$$\sqrt{x^2 + y^2} = \sqrt{10^2 + 5^2}$$
$$= \sqrt{125} = 11.181$$

Since the CORDIC computation is performed in iterative method, the iterative rotation can be expressed as [24]

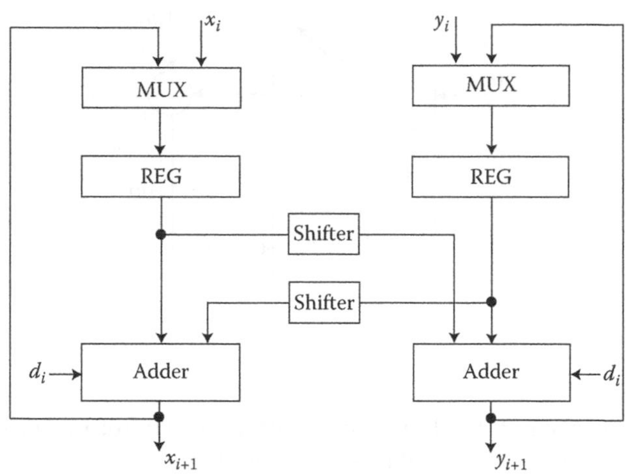

FIGURE 4.20
A 2-D CORDIC iterative processor.

$$x_{i+1} = K_i[x_i - y_i d_i 2^{-i}] \tag{4.93}$$

$$y_{i+1} = K_i[y_i + x_i d_i 2^{-i}] \tag{4.94}$$

where $K_i = 1/\sqrt{1+2^{-2i}}$ and $d_i = \pm 1$.

K_i approaches 0.6073 as the number of iteration approaches infinity. The CORDIC algorithm can be implemented in two ways. In Explicit CORDIC algorithm, the rotation angle Ø is explicitly calculated by accumulating all elementary angles $Ø_K$ to reach the rotation Ø. In Implicit CORDIC algorithm, only the control sign d_i is generated. Figure 4.20 shows a 2D CORDIC processor. Further details on CORDIC algorithm for SVD of both real and complex matrices are available in [25].

4.7.2 Adaptive Filtering

Adaptive filtering is an important area of intelligent sensor signal processing. In an adaptive filter, the coefficients are updated by an adaptive algorithm to improve or optimize the filter response. The basic structure of an adaptive filter is shown in Figure 4.21.

The input signal *x*(*n*) is filtered by the digital filter to get the output *y*(*n*), an error *e*(*n*) is generated as the difference between the filter output *y*(*n*) and the desired output *d*(*n*), which updates the filter coefficient or weight using an adaptation algorithm. Although there are various approaches to implement an adaptive filter, the least mean square (LMS) type is numerically stable,

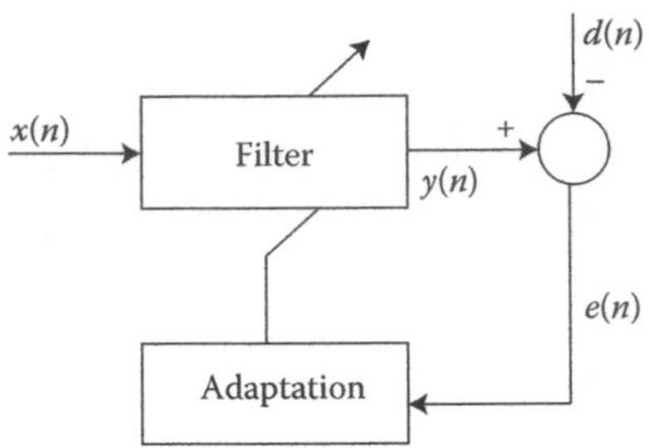

FIGURE 4.21
An adaptive filter.

fast, and convergent for VLSI implementation. Therefore, only the LMS algorithm of adaptive filter will be discussed here.

4.7.2.1 LMS Algorithm

In finite impulse response (FIR) filter operation, the filter generates the output as the weighted linear sum of the present and past K samples of the input signal given by

$$y(n) = w(n)^T x(n) = \sum_{j=0}^{N-1} (w_j^n X_{n-j}) \tag{4.95}$$

where
$x(n) = [x(n) \quad x(n-1) \quad x(n-2) \ \dots \ x(n-k+1)]^T$ is the input vector
$w(n) = [w_0(n) \quad w_1(n) \quad w_2(n) \ \dots \ w_{k-1}(n)]^T$ is the weight vector of the filter
n is the order of the filter

The adaptation algorithm uses the error signal

$$e(n) = d(n) - y(n) \tag{4.96}$$

The criteria for adaptation is the mean square error (MSE) ε given by

$$\varepsilon = E[e^2(n)] \tag{4.97}$$

where $E[\cdot]$ is the expected value.

Using Equations 4.96, 4.95, and 4.97, the MSE value can be written as

$$\varepsilon = E[d^2(n) + W^T(n)RW(n) + 2W^T(n)P] \tag{4.98}$$

where

$R = E[x(n)x(n)^T] =$ autocorrelation matrix $(N \times N)$

$P = E[d(n)x(n)] =$ cross-correlation vector $(N \times 1)$

Since the computation of R and P is difficult, a more practical and approximate solution to updating the weights by steepest descent method is used. By this method, the weight vector is increased by a value proportional to the negative gradient of MSE performance given by

$$w(n+1) = w(n) + \mu e(n)x(n) \tag{4.99}$$

where μ is the convergence factor.

The multiplication of the updated weight vector $w(n)^T$ and the sensor measurement vector $x(n)$ for the generation of the filter output $y(x)$ can be performed by a three trap bit stream multiplier (BSM) and adder, as shown in Figure 4.22 [26].

The input serial bit stream $x(n)$ and the updated serial weight stream $w(n)$ is applied to the dot product multiplier (DPM) successively through the D flip-flops (DFF) in each cycle. In each cycle, partial product terms are generated

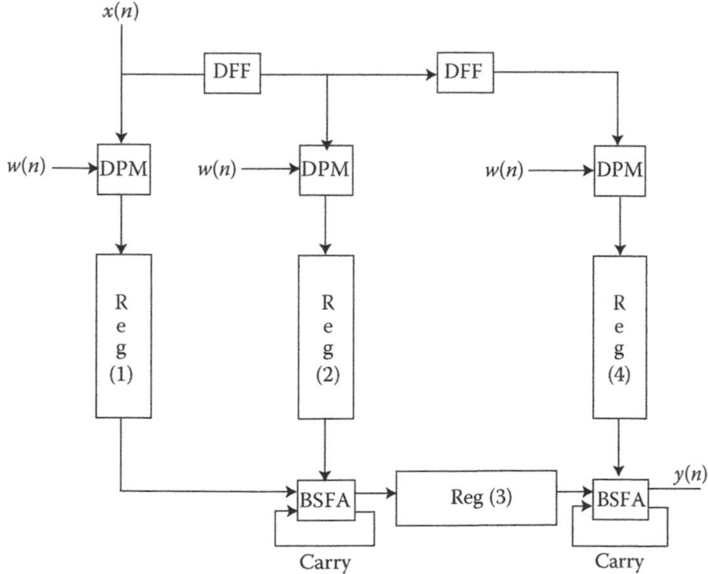

FIGURE 4.22
Three-trap bit stream multiplier. (Adapted from IEEE copyright © IEEE. With permission.)

and shifted to the bit stream full adder (BSFA) by 8-bit serial input–serial output (SISO) shift registers.

4.7.2.2 Bit Stream Multiplication

In a bit stream multiplication, partial dot products of x and w are generated using AND gates at each entry of the serial bits. The partial products are added by BSFA and successively shifted to generate the intermediate partial products. This is called triangular compression. The triangular compression for a 4-bit stream multiplication of $X = x_3x_2x_1x_0$ and $W = w_3w_2w_1w_0$ is shown below:

$$
\begin{array}{ccccccc}
 & & & \mathrm{X} = x_3 & x_2 & x_1 & x_0 \\
 & & & \mathrm{W} = w_3 & w_2 & w_1 & w_0 \\
\hline
 & & & & & & x_0w_0 \\
 & & & & & x_1w_0 & \\
 & & & & x_1w_1 & x_0w_1 & \\
\hline
 & & & & x_2w_0 & & \\
 & & & x_2w_1 & & & \\
 & & x_2w_2 & x_1w_2 & x_0w_2 & & \\
\hline
 & & & x_3w_0 & & & \\
 & & x_3w_1 & & & & \\
 & x_3w_2 & & & & & \\
x_3w_3 & x_2w_3 & x_1w_3 & x_0w_3 & & & \\
\hline
P_6 & P_5 & P_4 & P_3 & P_2 & P_1 & P_0
\end{array}
$$

The product terms P_0 through P_6 are obtained by successive addition of the dot product terms in each column. There is $(2N-1)$ product terms if N is the bit number.

Example 4.17

Show the triangular compression steps in bit stream multiplication of two 4-bit numbers 1010 and 1100.

Solution

Let $X = 1010$ and $Y = 1100$

Here $x_0 = 0$, $x_1 = 1$, $x_2 = 0$, and $x_3 = 1$

and $y_0 = 0$, $y_1 = 0$, $y_2 = 1$, and $y_3 = 1$

The bit stream dot products are

```
        X=1010
        Y=1100
      ----------
             0
            0
           00
      ----------
            0
           0
         010
         ---
         0
        0
       1
      1010
      ----------
      11110000
```

which is same as obtained from discrete multiplication as

```
        1010
        1100
      --------
        0000
       0000
      1010
     1010
    ----------
    11110000
```

Example 4.18

An 8th-order adaptive FIR VLSI filter uses a clock frequency of 5.65 kHz to filter a sensor signal, which is converted by an 8-bit ADC with a frequency of 1 MHz and a parallel to serial shift register.

1. Determine the time required to generate a filtered sample.
2. Determine the size of the buffer to be placed between the ADC and the filter.

Solution

Since the filter is 8th order

$N = 8$

Total number of product terms $= 2 \times 8 - 1 = 15$

Therefore, the filtering of a sample at an instant takes 15 clock pulses.
Given, filter clock frequency $f_c = 5.65\,\text{kHz}$

$$\text{Clock time period} = \frac{1}{5.65} \times 10^{-3}\ \text{s}$$

1. Time taken for the filtering to complete

$$= 15 \times \frac{1}{5.65} \times 10^{-3}\ \text{s} = 2.65\ \text{ms}$$

2. Frequency of ADC = 1 MHz

$$\text{ADC conversion time} = \frac{1}{100} \times 10^{-6}\ \text{s}$$
$$= 1\ \mu\text{s}$$

Number of samples converted by ADC during filtering of one sample

$$= \frac{2.65 \times 10^{-3}}{1.00 \times 10^{-6}}$$
$$= 2650$$

Therefore

$$\text{buffer size} = K$$
$$= \frac{2650}{1024}$$
$$= 2.6\ \text{kB (taking the nearest value)}$$

4.7.3 Image Processing

VLSI technology has become a major driving force for the real-time processing of image-based signals in intelligent image sensors. Development of high-performance VLSI algorithm and architecture for image compression, image enhancement, and reconstruction for real-time application find importance due to the fact that the image sensor module comprises a large number of regular structured units that need identical and concurrently operating data computations. VLSI architecture for a few image-processing applications will be discussed now.

4.7.3.1 Image Wavelets Transform

Many intelligent image sensors need to perform wavelet-based signal processing for image classification, detection, and reconstruction. Wavelet transform is a timescale representation by dilation equations as opposed to difference or differential equations. Wavelets decompose the image at one level of approximation into approximate and detailed signal at the subsequent levels, thereby resolving more information from the image. A three-level wavelet decomposition algorithm is shown in Figure 4.23. A wavelet decomposition is said to be periodic with a period of $M=2^m$ with m levels. For the wavelet shown in Figure 4.23, the period is 8. In the decomposition process, the signal is sequentially filtered with gains G and H, respectively. Moreover, each filtered output is downsampled by a factor of 2. Since after each filter output, the samples are decimated by a factor of 2, at a level-I, for a set of M samples, $M/2^i$ samples are generated at both G and H filters. For example, in the three-level wavelet structure, if there are eight input samples, v, x, z, and z' vectors are obtained as $[v(0), v(2), v(4), v(6)]$; $[w(0), w(2), w(4), w(6)]$; $[x(0), x(4)]$; $[y(0), y(4)]$; $z(0)$ and $z'(0)$, respectively. Therefore, in a level-I, total samples completed is

$$M+\frac{M}{2}+\frac{M}{4}+\cdots+2=2(M-1)$$

For simplicity, let us consider a two-tap filter transfer function as

$$G(z)=g_0+g_1z^{-1} \tag{4.100}$$

$$H(z)=h_0+h_1z^{-1} \tag{4.101}$$

In a single sample period, the wavelet decomposition algorithm performs the following computations:

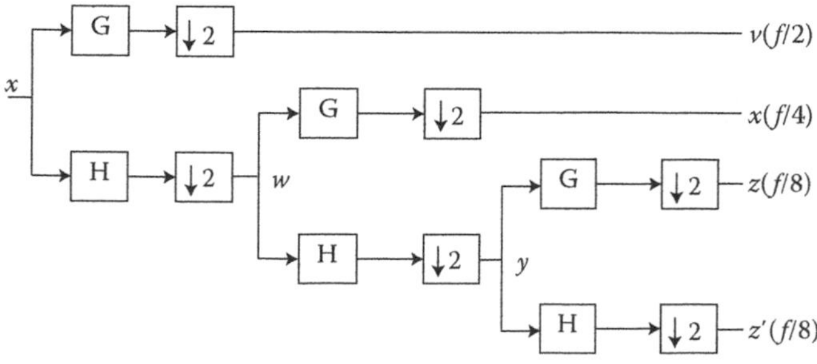

FIGURE 4.23
A three-level wavelet decomposition.

$$v(0) = g_0u(0) + g_1u(-1) \tag{4.102a}$$

$$v(2) = g_0u(2) + g_1u(1)$$

$$v(4) = g_0u(4) + g_1u(3)$$

$$v(6) = g_0u(6) + g_1u(5)$$

$$w(0) = h_0u(0) + h_1u(-1)$$

$$w(2) = h_0u(2) + h_1u(1)$$

$$w(4) = h_0u(4) + h_1u(3)$$

$$w(6) = h_0u(6) + h_1u(5)$$

$$x(0) = g_0w(0) + g_1w(-2)$$

$$x(4) = g_0w(4) + g_1w(2)$$

$$y(0) = h_0w(0) + h_1w(-2)$$

$$y(4) = h_0y(4) + h_1w(-2)$$

$$z(0) = g_0y(0) + g_1y(-4)$$

$$z'(0) = h_0y(0) + h_1y(-4) \tag{4.102n}$$

The successive development of the terms deriving a complete sequence of eight cycles is shown in Table 4.10.

The computation of the variables in (4.102) can be obtained using only one pair of G and H filters by folded architecture [27] using minimum number of registers. The minimum number of registers can be determined by analyzing the duration of live variables and forward–backward allocation of registers.

TABLE 4.10

Development Sequence of Wavelet Components

Unit Time	G-Filter	H-Filter	Available Intermediate Terms	Output Terms	Cycle
2	$v(0)$	$w(0)$	$u(-1)$, $u(0)$	$v(0)$	2
4	$v(2)$	$w(2)$	$u(1)$, $u(2)$	$v(2)$	4
5	$x(0)$	$y(0)$	$w(0)$, $w(-2)$	$x(0)$	5
6	$v(4)$	$w(4)$	$u(3)$, $u(4)$	$v(4)$	6
7	$z(0)$	$z'(0)$	$y(0)$, $y(-4)$	$z(0)$	7
8	$v(6)$	$w(6)$	$u(5)$, $u(6)$	$v(6)$	8
9	$x(4)$	$y(4)$	$w(4)$, $w(2)$	$x(4)$	9
10	$v(8)$	$w(8)$	$u(7)$, $u(8)$	$v(8)$	10
11	—	$z'(0)$	$y(0)$, $y(-4)$	$z'(0)$	11

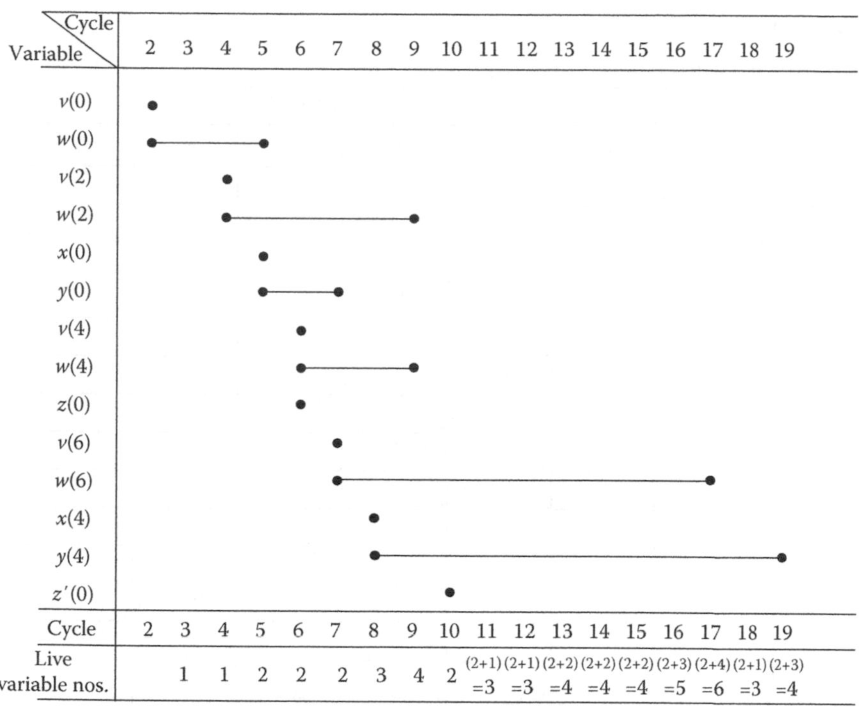

FIGURE 4.24
Duration of live variables in two-tap filter of wavelet decomposition (symbol (•) indicates generation and end of live variables).

Let us analyze the duration of live variables in the wavelet decomposition process. $v(0)$, $v(2)$, $v(4)$, and $v(6)$ are generated by the G filter and transmitted instantly; however, $w(0)$, $w(2)$, $y(0)$, $w(4)$, $w(6)$, and $y(4)$ are to be registered for a certain duration for using in other computations. For example, $w(0)$ generated in cycle 2 is used for the computation of $x(0)$ and $y(0)$ in cycle 5, so $w(0)$ should be registered as live variable till cycle 5. The live variable duration for all variables is shown in Figure 4.24.

From Figure 4.24, it is evident that at time cycle 17, a total of six variables are required for computation; therefore, the minimum number of registers needed is six. Since the generation and transmission of the variables v, x, z, and z' are performed successively, there should be an optimum allocation of registers by forward–backward method. The register allocation for the two-tap filter–based wavelet decomposition is shown in Table 4.11.

From Figure 4.24, it is observed that $w(6)$ generated in cycle 8 must be live till cycle 16; therefore, this variable is shifted forward from R_1 to R_6 and then shifted back to R_1 and then shifted forward to R_3. Similarly, $y(4)$ generated in cycle 9 is shifted forward from R_1 to R_6, then from R_1 to R_4. The architecture for the wavelet decomposition as described above is shown in Figure 4.25.

TABLE 4.11

Forward–Backward Allocation of Registers

Cycle	Generation	Registers					
		R_1	R_2	R_3	R_4	R_5	R_6
2	$W(0)$						
3		$W(0)$					
4	$W(2)$		$W(0)$				
5	$Y(0)$	$W(2)$		$W(0)$			
6	$W(4)$	$Y(0)$	$W(2)$				
7		$W(4)$	$Y(0)$	$W(2)$			
8	$W(6)$		$W(4)$		$W(2)$		
9	$Y(4)$	$W(6)$		$W(4)$		$W(2)$	
10		$Y(4)$	$W(6)$				
11			$Y(4)$	$W(6)$			
12				$Y(4)$	$W(6)$		
13					$Y(4)$	$W(6)$	
14						$Y(4)$	$W(6)$
15		$W(6)$					$Y(4)$
16		$Y(4)$	$W(6)$				
17			$Y(4)$	$W(6)$			
18				$Y(4)$			
19					$Y(4)$		

4.7.4 Neural Network

Neural network is an emerging tool for providing intelligent learning capabilities to sensors, which can respond to varying patterns. Neural network finds application mostly in speech, vision, and odor classification; however, it has expanded to many other areas also.

ANN comprises of large distributed, parallel, and simple neuron-computing elements, which can learn by a learning algorithm. Since neuron computing is nonlinear in nature, many nonlinear systems can be represented by the ANN. An ANN can be implemented in VLSI by both analog and digital approaches. Advantages of analog approaches are compactness and high speed; however, they also have the following disadvantages:

1. Susceptibility to drift and noise
2. Low precise control in synaptic weight
3. Difficult to make modifiable synapse

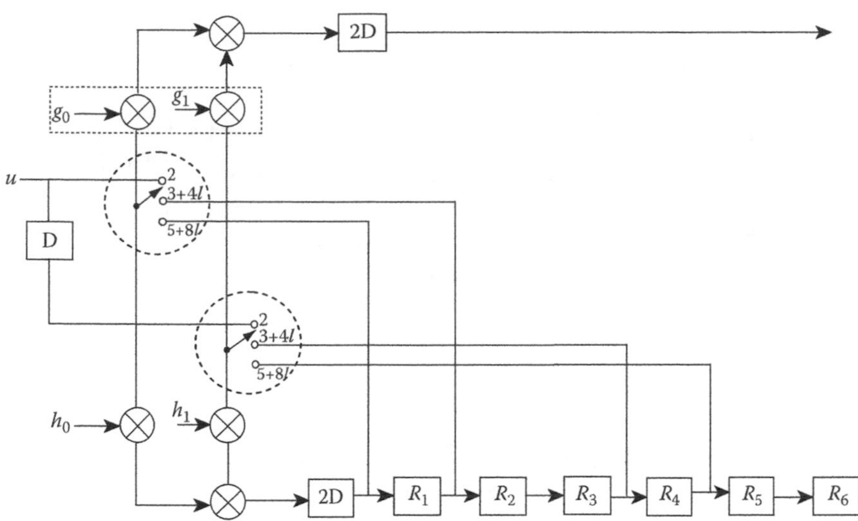

FIGURE 4.25
Architecture of wavelet decomposition.

The digital implementation of ANN possesses the following advantages:

1. Precise control of parameter and processing
2. High cascadibility
3. Relative ease of modifiable synapses
4. Increased efficiencies in terms of computational density

The disadvantages of digital approaches are large circuit size and low speed. Pioneering works for implementing ANN for specific applications both in analog and digital approaches are numerous; however, only example implemented for each of them will be discussed here.

4.7.4.1 Analog VLSI-Based ANN

Attaining maximum density in VLSI for ANN can be possible by array-based architecture. Moreover, the regularity of array-based VLSI architecture can be comfortably exploited for the ANN structure. The simple structure of neuron-to-neuron connection as shown in Figure 4.26a can be implemented by a 2D-based array computational VLSI architecture, as shown in Figure 4.26b. The 2D array grid consisting of three synapses are shared with the four inputs to generate the output y_1, y_2, and y_3. In analog computation of the synapses, each neuron computes the summation of the synaptically weighted inputs in the summer shown by the triangles. The analog circuit for a basic summer is shown in Figure 4.26c.

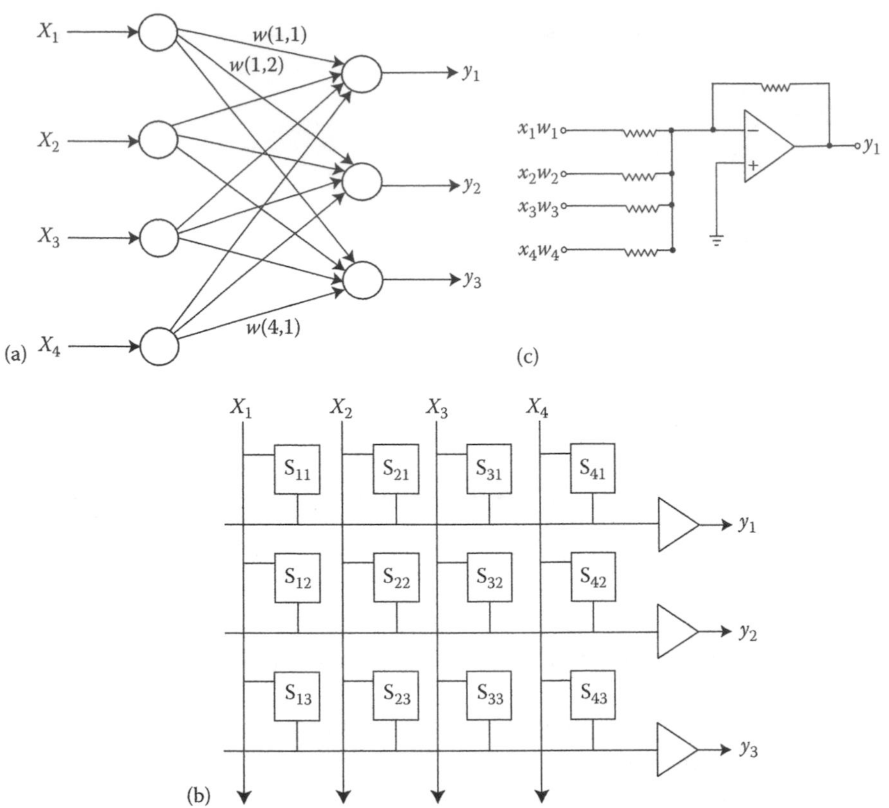

FIGURE 4.26
(a) A simple 4–3 structure ANN, (b) a basic 2D analog ANN computing array, and (c) an analog summer.

4.7.4.1.1 Array Computation

The analog computation that is required to generate the synaptically weighted inputs is performed by the blocks marked "S." Most ANNs perform dot product between an n-dimensional input vector and an $n \times m$-dimensional array of weight using multiply-accumulate operation [28]. However, in some other ANN structures like KNN, associative memory, RDF, etc., distance calculation is performed between an n-dimensional input vector and m numbers of n-dimensional reference vectors. In the case of Euclidean distance, the difference-squared-accumulate operation is performed. Figure 4.27a shows the general structure of distance computation; Figure 4.27b through d shows the dot products, Euclidean, and Manhattan distance computation structure, respectively, for a single synaptic cell. The computation variable in analog VLSI architecture may be performed by any one of the following alternative modes [28]:

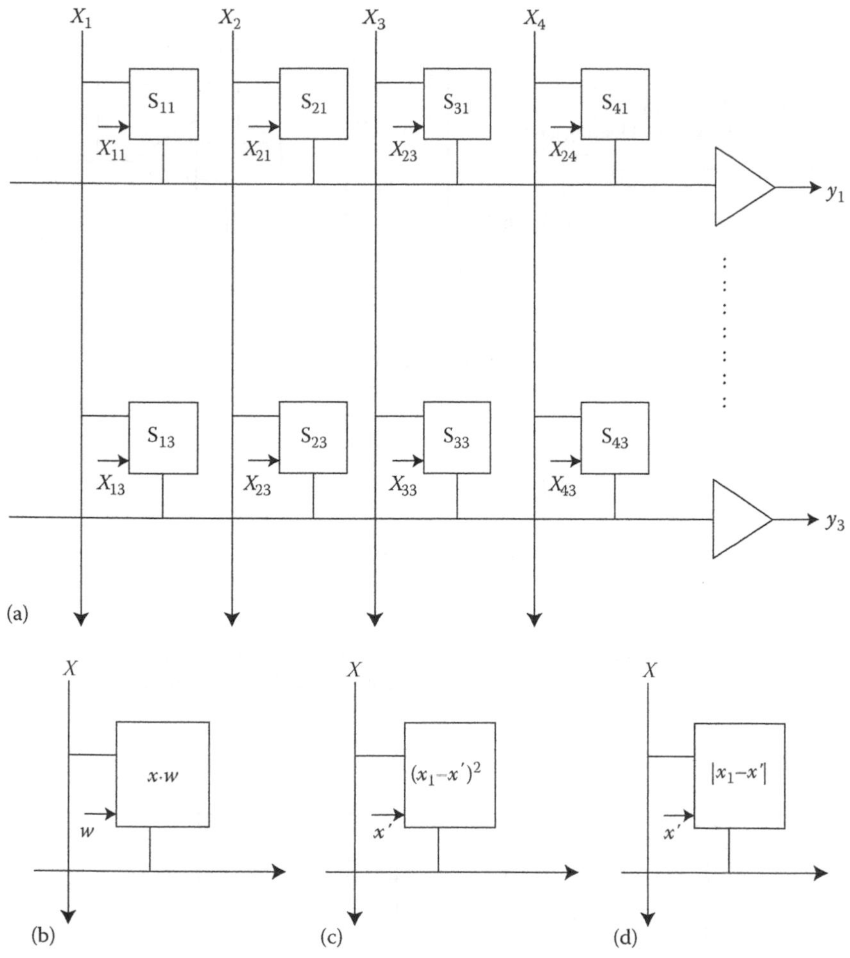

FIGURE 4.27
(a) General structure of distance calculation, (b) a dot product cell, (c) a Euclidean cell, and (d) a Manhattan cell.

1. Charge mode
2. Current mode
3. Conductance mode

4.7.4.1.1.1 Charge Mode Computation The charge mode computation is based on the use of a floating gate MOSFET, as shown in Figure 4.28a. The Q–V_g characteristic of the floating gate MOSFET reveals that, for V_g below V_T, the channel charge is zero, and when V_g exceeds V_T, the channel charge Q linearly increases with the difference between V_g and V_T. The channel charge is converted to voltage by using a charged integrator, as shown in Figure 4.28b.

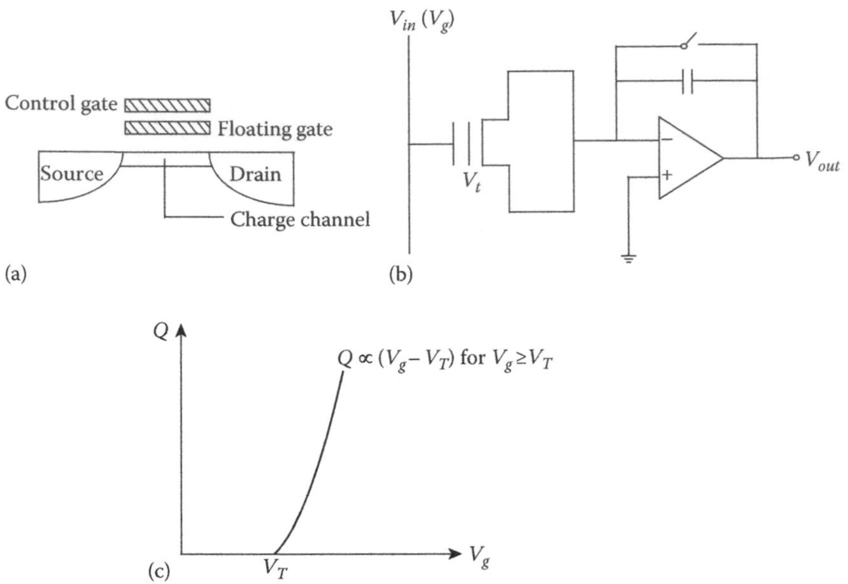

FIGURE 4.28
(a) Schematic of a floating gate MOSFET, (b) a charge integrator, and (c) channel charge vs. gate voltage characteristics. (Adapted from IEEE copyright © IEEE. With permission.)

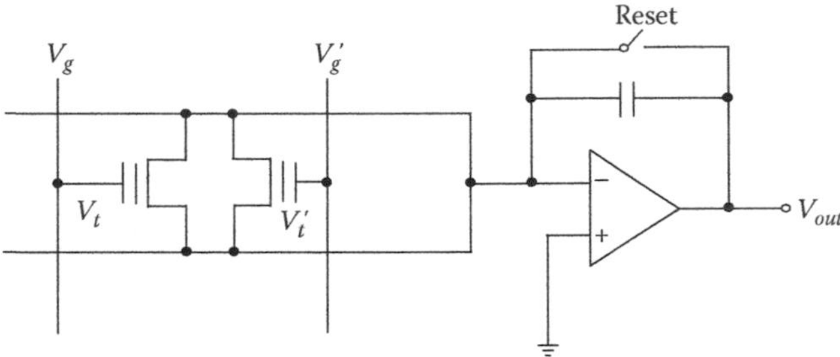

FIGURE 4.29
Difference of absolute value circuit. (Adapted from IEEE copyright © IEEE. With permission.)

For generating the difference between two analog values, V_g and V_{st}, a differential pair of floating gate MOSFET is used, as shown in Figure 4.29.

In the differential pairs (Figure 4.30), complementary inputs V_g and V_g' are applied to their gates while $+V_{st}$ are applied to their floating gates to program their threshold values. The threshold values and the gate voltages are modulated as shown in the equations

$$V_t = V_{t\,\text{min}} + V_{st}$$

$$V_{\text{out}} \propto -\left|V_{\text{in}} - V_{st}\right|$$

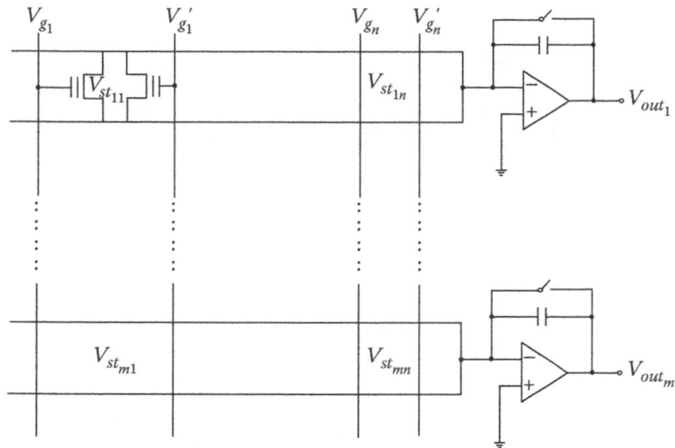

FIGURE 4.30
Distance of absolute value computing. (Adapted from IEEE copyright © IEEE. With permission.)

$$V_t' = V_{t\,\max} + V_{st}$$

$$V_g = V_{t\,\min} + V_{in}$$

$$V_g' = V_{t\,\max} - V_{in}$$

4.7.4.1.1.2 Conductance Mode In many applications of ANN, the neuron-to-neuron interconnection carries binary weighted analog signals. In such cases, a neuron computes the dot product between a digital vector of input applied to the gate and an analog vector stored in the floating gate. This dot product can be realized in conductance mode.

Conductance mode computation of dot product of an analog input with a digital bit can be performed by a floating gate MOS, as shown in Figure 4.31a. The analog signal is applied to the floating gate to program the threshold and the digital input is applied to the gate as V_g. The transistor works as a programmable switched conductance. When the gate voltage is at 0, it is below the threshold of the MOS and so the conductance is 0, while when the gate voltage is at 1, it is above the threshold and conductance is proportional to $(V_g - V_t)$.

In Figure 4.31b, a negative conductance array is connected with a common drain line, which is grounded. This facilitates to provide a differential conductance with the help of the neuron amplifier at the output. The differential conductance is further threshold to get a binary value for transmission.

4.7.4.1.1.3 Current Mode In current mode computation, only the *I–V* characteristic of a MOSFET in saturation is used for square law function implementation. The advantage of current mode computation is that it is the only method for distance squared computation in ANN. Another advantage

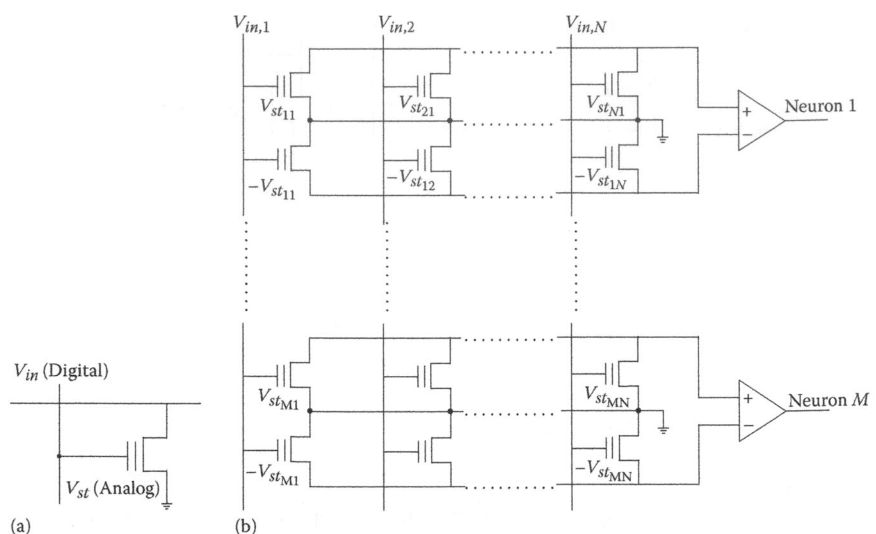

FIGURE 4.31
(a) MOS cells for conductance mode dot product computation and (b) ANN array with dot product computation. (Adapted from IEEE copyright © IEEE. With permission.)

of current mode is that, unlike charge mode or conductance mode, it uses dynamic weights instead of fixed weights.

A current mode distance squared computational cell using ten MOS transistors has been used in developing block-matching algorithm for computing relative image frame in an analog smart image sensor [28]. The input pixels are routed diagonally and the common outputs are routed horizontally.

4.7.5 Mathematical Operations

One of the important tools of smart sensors design is mathematical operation of sensor signals such as the multiplication of two signals—$v_1(t)$ and $v_2(t)$, generate terms like $v_1^2(t), v_2^2(t), v_1^3(t), v_2^3(t)$ or generate polynomials like $v_1(t)+v_1(t)v_2(t)+v_1^2(t)v_2(t)+v_1(t)v_2^2(t)$, etc. Although various multiplier or divider ICs have been developed for general purpose signal processing, smart sensor designers often need, on a case-to-case basis, mathematical circuits that fulfill the specialized needs.

A basic multiplier could be realized using programmable transconductance components [29]. The conceptual transconductance amplifier is shown in Figure 4.32a where the output current is given by

$$i_0 = G_{m1}V_1 \tag{4.103}$$

where

$$G_{m1} = G_{m1}(I_{bias1}) \tag{4.104}$$

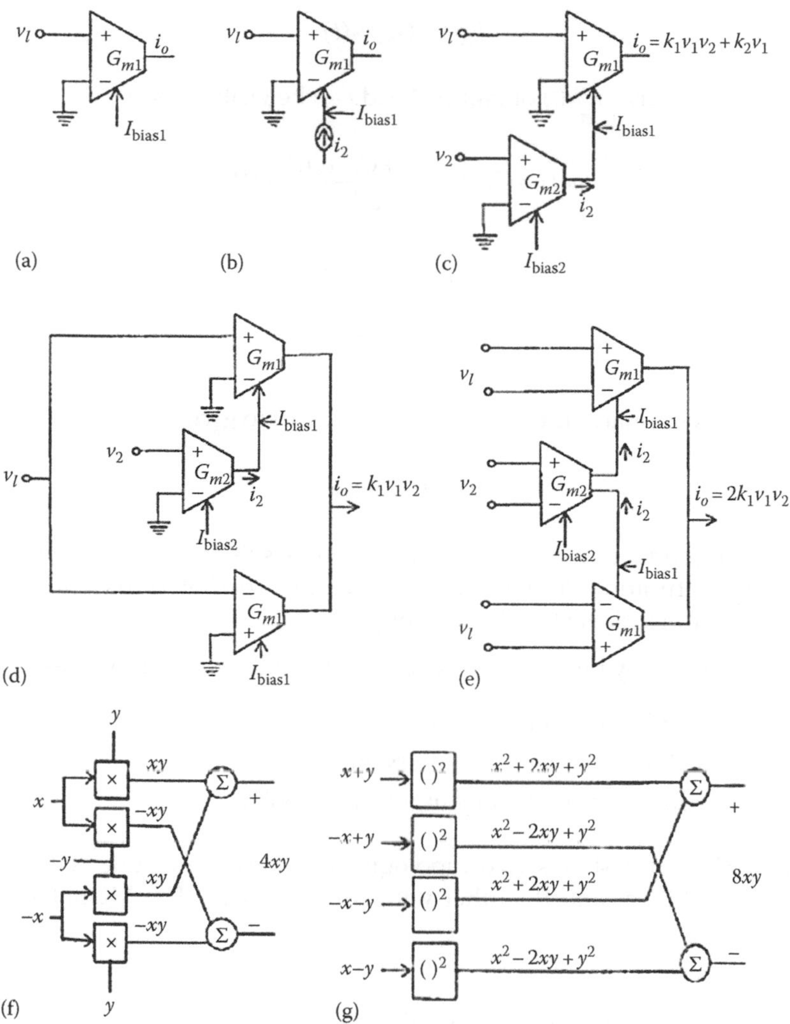

FIGURE 4.32
Transconductance multipliers. (Reprinted from IEEE, copyright © IEEE. With permission.)

In the case of a bipolar transconductor

$$G_{m1} = \frac{I_{bias1}}{2v_t} \tag{4.105}$$

where v_t is the thermal voltage (KT/q).

In Figure 4.32b, signal current i_2 is added to the bias current and, hence, the second input $v_2(t)$ can be represented by $i_2(t)$, as shown in Figure 4.32c:

$$i_2(t) = G_{m2}v_2(t)$$

The multiplier current $i_2(t)$ of Figure 4.32d can be written as

$$\begin{aligned} i_0(t) &= G_{m1}v_1 \frac{I_{bias1} + G_{m2}v_2(t)}{2V_t} v_1(t) \\ &= \frac{G_{m2}v_1(t)v_2(t)}{2V_t} + \frac{I_{bias1} + v_1(t)}{2V_t} \\ &= K_1v_1(t)v_2(t) + K_2v_1(t) \end{aligned}$$

Figure 4.32e eliminates the unwanted part $K_2v_1(t)$ to give

$$i_0(t) = K_1v_1(t)v_2(t)$$

The transconductance multipliers can be classified into two groups—linear and saturation. Table 4.12 shows eight types of transconductance multipliers under these two basic groups.

The multipliers may be classified based on the polarities of inputs as

1. Single quadrant: x and y inputs are unipolar
2. Two quadrant: x or y inputs can be bipolar
3. Four quadrant: both x and y inputs are bipolar

Figure 4.32f and g shows two topologies of multipliers—one with single quadrant multiplier and the other with square device. The MOSFET-based

TABLE 4.12

Eight Types of Transconductance Multipliers

Operating Regions	Type	Output Equation
Linear	Using V_{gs} and V_{ds}	$V_0 = -4KZ_fxy$
	Using V_{gs}^2	$I_0 = Kxy$
	Dual gate in linear region	$I_0 \approx \frac{KV_{pol}}{C}xy$
Saturation	Using V_{gs}^2 with diode connection	$I_0 = Kxy$
	Using V_{gs}^2 with gate and source injection	$I_0 = Kxy$
	Using V_{gs}^2 with substrate terminal	$I_0 \approx \frac{4K\gamma}{\sqrt{2\lvert\phi_F\rvert - Y + S}}xy$
	Using V_{gs}^2 with voltage adder	$I_0 = 4KK_0xy$
	MOS Gilbert Cell	$I_0 = 2\sqrt{2KK_3xy}$

Source: Adapted from IEEE, copyright © IEEE. With permission.

transconductance multipliers with different operating modes and types are available in [29].

The comparison of the eight categories of transconductance multipliers in respect of their linearity, voltage ranges, noise, and power consumption is available in [29]. Several multiplier architectures do not have any clear advantages over others and readers should be aware that the comparison might not hold good for all cases.

4.8 Temperature Compensating Intelligent Sensors

Temperature is a vital interfering signal in most sensors. The temperature interference gets added to the sensor signal in the following different ways:

1. From the process medium

If not thermally insulated from the process medium, the sensor will pick up the temperature signal from the process medium. In many sensor systems, this situation is unattended, as a result of which the sensor temperature deviates from the calibrated value making the measurement data erroneous. Pressure, flow, density, moisture, pH, viscosity, conductivity, flavor, and many other process sensor experiences such cross-sensitivity to temperature of the process variables. To avoid this, the sensor material should be thermally insulated from the process fluid as far as possible. Thermal insulation technique is not always feasible where a direct contact of the process medium to the sensor is essential. For example, in pH, moisture, conductivity, and flavor detection of process fluid, where the sensor material must be exposed to the process medium, the encapsulation of the sensor is not done. In some other cases like pressure, flow, density, etc., where direct contact is not mandatory, a thermal insulation becomes feasible.

2. Heating from sensor current

Resistive sensors undergo a self-heating effect due to the current flowing through the sensor. If the temperature coefficient of the sensor material is not sufficiently low, a change in resistance takes place progressively. This additional change in resistance reproduces an error in the variable being measured. The progressive increase in resistance acts as follows: due to heat developed due to i^2R loss in the sensor, the sensor resistance increases by ΔR making the sensor resistance $(R+\Delta R)$. This causes a decrease in current to $(i-\Delta i)$ giving a heat developed equal to $(i-\Delta i)^2$ $(R+\Delta R)$. If the temperature coefficient of resistance of the sensor material is not sufficiently low, $(i-\Delta i)^2$ $(R+\Delta R)$ becomes greater than i^2R. This process continues to an equilibrium condition when the heat developed in the sensor is dissipated to the surrounding. Therefore, if the sensor is not properly ventilated for the heat to be dissipated quickly, there may be a considerable change in resistance reflecting it in the variable being measured.

3. From mechanical work done

In sensors that involve a mechanical work done as a part of the sensing mechanism such as oscillation, friction, adiabatic compression or expansion, etc., heat may be developed causing sensor temperature variation. For example, in very high pressure (above 700 MPa), the measurement in fluid using resistive coil inside a kerosene-filled bellows, due to sudden pressure change, the coil experiences a transient change in resistance due to transient change in temperature caused by adiabatic expansion or compression of the fluid.

Sensors for measurement of local contact pressure between rolling elements like gear, cams, and bearing may be accomplished by bonding Manganin element of 50 μm wide and 0.08 μm thick on the contact surface. The contact pressure causes a change in resistance of the Manganin element by the principle of piezoresistivity; however, the frictional force developed between the contact materials generates a heat, which will be picked up by the sensor causing resistance variation in the sensor.

4. Secondary effect

In many cases, the variable to be measured produces a secondary effect causing temperature variation in the sensors, e.g., in light-dependent resistance (LDR), in addition to the change in LDR resistance caused by radiation energy, the heat energy of the radiation also acts simultaneously to cause increase in resistance by heating. If a semiconductor material with low-temperature coefficient is not chosen, the secondary effect of the heating may be considerable. The frictional heat developed in Manganin element of a contact pressure sensor described in (3) above can also be termed as secondary effect producing interference in the sensor.

4.8.1 Effects of Temperature on Sensors

A change in temperature in a sensor material changes its electrical characteristic in different mechanisms. In resistive sensors, the change in resistance in the sensor takes place due to the change in resistivity ρ of the material or change in the dimension, i.e., length and cross-sectional area of the conductor. Considering the equation of resistance of a resistive element

$$R = \frac{\rho L}{A} \tag{4.106}$$

where

ρ is the resistivity, Ωm
L is the length, m
A is the cross-section area, m^2

To find the change in dR in R, we differentiate Equation 4.106

$$dR = \frac{A(\rho\, dL + L\, d\rho) - \rho L\, dA}{A^2} \tag{4.107}$$

the volume of the conductor

$$V = AL$$

Therefore,

$$dV = A\, dL + L\, dA \tag{4.108}$$

and also

$$dV = L(1+\alpha\Delta T)A(1-\alpha\Delta T\upsilon)^2 - AL \tag{4.109}$$

where α is the expansion coefficient due to temperature (m/m/°C). Since α is small $(1-\alpha\Delta T)^2 \approx 1-2\upsilon\,\alpha\Delta T$.

So we can write

$$dV = AL(1+\alpha\Delta\mathrm{T})(1-2\nu\alpha T) - AL$$

$$= AL\alpha\Delta\mathrm{T}(1-2\nu)$$

From (4.108)

$$A\, dL + L\, dA = AL\alpha\Delta T(1-2\nu)$$

since $\alpha = ((dL/\mathrm{L})/\Delta T)$

$$\mathrm{L}\alpha\Delta\mathrm{T} = dL$$

Hence, $A\ dL + L\ dA = A\ dL(1-2\nu)$ and

$$L\, dA = -2\nu A\, dL$$

Substituting in (4.107) and dividing by R

$$\frac{dR}{R} = \frac{dL}{L}(1+2\nu) + \frac{d\rho}{\rho}$$

$$= \frac{dL}{L} + 2\nu\frac{dL}{L} + \frac{d\rho}{\rho} \tag{4.110}$$

From the above equation, it is evident that the change in resistance due to temperature change is due to change in length dL, change in cross-sectional area $2\upsilon dL$, and change in its resistivity $d\rho$; however, different sensor materi-also have different ratios of contribution to these three components.

Example 4.19

A given sensor material of resistance 120 Ω, length of 10 cm with a resistivity of 2×10^{-6} Ωm, resistance temperature coefficient of 0.00389(Ω/Ω)°C, and linear expansion coefficient of 0.5×10^{-8} (m/m)°C is subjected to a temperature variation of 10°C from the calibrated temperature. If the change in resistance is 12 Ω, determine the per-unit change in resistance for all three components and their percentage. Take $v=0.3$.

Solution

Given, $L=10\,\text{cm}=0.1$ m, $\rho=2\times10^{-6}$ Ωm, $\alpha_l=0.5\times10^{-8}$(m/m)°C, $\Delta R=0.1$ Ω

$$R=120\ \Omega$$

$$dL = L\alpha_l\Delta T$$

Therefore, $(dL/L)=\alpha_l\Delta T=0.5\times10^{-8}\times10=0.5\times10^{-7}$.

Putting the above value in Equation 4.110

$$\frac{120}{12}=0.5\times10^{-7}+2\times0.3\times0.5\times10^{-7}+\frac{d\rho}{\rho}$$

$$\therefore \frac{d\rho}{\rho} = \text{Per-unit change in resistance due to resistivity}$$

$$= 0.0999$$

$$\text{Percentage of change} = 0.0999\times100$$

$$= 9.99\%$$

The change due to length and cross-sectional area

$$= (100-9.99)\%$$

$$= 90.01\%$$

Although the per-unit change in the resistance due to temperature variation in resistive sensor is controlled by the three different factors mentioned above, the net effect is defined by a single coefficient called temperature coefficient of resistance, which is given by

$$\frac{dR/R}{dT}=\frac{1}{L}\frac{dL}{dT}+\frac{2\nu}{L}\frac{dL}{dT}+\frac{1}{\rho}\frac{d\rho}{dT} \tag{4.111}$$

The unit of this coefficient is Ω/Ω/°C.

For the above example, this coefficient can be found as

$$\frac{0.1/120}{10} = 8.33\times10^{-5}\Omega/\Omega/°C$$

This coefficient consists of three components given by the right-hand side of Equation 4.111. We can calculate the first two components from which the third component can be determined as

$$\frac{1}{\rho}\frac{d\rho}{dT} = \frac{dR/R}{dT} - \frac{1}{L}\frac{dL}{dT} - \frac{2\nu}{L}\frac{dL}{dT} \tag{4.112}$$

Calculating dL as

$$\begin{aligned} dL &= L\alpha_l \Delta T \\ &= 0.1\times0.5\times10^{-8}\times10 \\ &= 0.5\times10^{-8}\,\text{m} \end{aligned}$$

Using Equation 4.112

$$\begin{aligned} \frac{1}{\rho}\frac{d\rho}{dT} &= 8.33\times10^{-5} - \frac{0.5\times10^{-8}}{0.1\times10} - \frac{2\times0.3\times0.5\times10^{-8}}{0.1\times10} \\ &\cong 8.32\times10^{-5} \end{aligned}$$

Putting the values of ρ, we get

$$\begin{aligned} \frac{d\rho}{dT} &= 8.32\times10^{-5}\times2\times10^{-6} \\ &= 1.66\times10^{-10}\ \Omega\text{m}/°C \end{aligned}$$

The above value is absolute change in ρ, while the per-unit change in ρ is given by

$$\frac{d\rho/\rho}{dT} = 8.32\times10^{-5}\ \Omega\text{m}/\Omega\text{m}/°C$$

Sensor manufacturers provides a typical rating of the coefficient of temperature change in measured variable per-unit change in temperature such as Nw/°C in force sensor, microstrain/°C in a strain sensor, %RH/°C in a humidity sensor, and so on. For example, typical temperature coefficient in bulk polymer resistive humidity sensor specified in EMD-2000 and UPS-500 are −0.3%RH/°C and −0.27%/°C, respectively. Although resistance increases in most resistive sensors due to increase in temperature, the negative effect in the two humidity sensors is due to the fact that resistance decreases with humidity.

Capacitive sensors are highly stable against temperature variation compared to resistive sensors. Temperature changes the capacitance in two ways—changing the permittivity of the medium and the dimension of the capacitive element, i.e., the cross-sectional area of the charge plates and their distance. Considering a capacitance formed between two parallel plates placed in a medium with a calibrated permittivity, the capacitance is given by

$$C = \frac{\varepsilon A}{d} \tag{4.113}$$

where

ε is the permittivity of the medium
A is the cross-sectional area (m^2)
d is the distance between the plates (m)

A change in C that takes place due to change in temperature is given by

$$\begin{aligned} dC &= \frac{d(\varepsilon dA + Ad\varepsilon) - \varepsilon Add}{d^2} \\ \text{and } \frac{dC}{C} &= \frac{dA}{A} + \frac{d\varepsilon}{\varepsilon} - \frac{dd}{d} \end{aligned} \tag{4.114}$$

Equation 4.114 indicates that the per-unit change in C is contributed equally by a change in A and ε, while change in d contributes with a negative effect. In most capacitive sensors, temperature dominantly changes the permittivity only, while dimensional changes are very small.

An increase in temperature causes an increase in conductivity of the dielectric material of the capacitive sensor and the degree in variation in conductivity with temperature is a function of the ion concentration of the electrolyte. Therefore, a more conductive material experiences a smaller change in conductivity due to temperature. Due to this fact, an increase in temperature causes a relatively smaller increase in effective capacitance between a large decrease in equivalent series resistance.

4.8.2 Temperature Compensation Techniques

Although a large variety of sensors are available for temperature measurement, the following temperature sensors show greater flexibility and advantages in compensation:

1. PTC resistance temperature detectors (RTDs)
2. NTC thermistors
3. Semiconductor sensors

RTD probes with small sizes are popular for temperature compensation because of the ease of installing. The advantage of RTD due to which it is popular than other sensors are high long-term stability (0.1°C/year in industrial probes and 0.0025°C/year in laboratory probes), high sensitivity, high repeatability, high linearity, high accuracy, and fastness of response to temperature variations. Thin film platinum (Pt) RTD probes are much smaller (20–100 times) and cheaper than wire wound probes. Although their temperature range is smaller, they are popular as compensating sensor since a smaller temperature range suffices for compensation purpose and because of the advantages mentioned.

NTC thermistors are also extensively used for temperature compensation due to their advantages such as high sensitivity and high resolution. Due to the high resistivity, a smaller probe of thermistor can be used, which is advantageous for installation for compensation applications. Due to possibility of a small mass, fast response to temperature variation can be achieved. Thermistors are manufactured in various forms, shapes, and sizes such as disc, washer, and rod. Thermistor probes, with shapes and sizes which are compact and easy to be installed in the body of the sensor for which temperature compensation is necessary.

Semiconductor temperature sensors are also popular as temperature compensating sensors due to their advantages like less expensive than RTDs, more linear than thermistors, more sensitive, and less self-heating than both RTDs and thermistors. Their operating range, although limited, is sufficient for temperature compensation. One added advantage of the sensor with plastic package is that they can be installed over active elements also, which is not possible in bare RTDs and thermistors.

Although temperature compensation techniques are mostly specific to a particular sensor, the technique can be classified under three broad categories—circuit compensation, microprocessor, microcontroller-based compensation, and computer-based compensation. There is a wide variability of these techniques depending on the requirement of the application and need. Moreover, there may be a combination of two techniques, say, circuit compensation requires the help of computer algorithm, and so on.

4.8.2.1 Circuit Compensation

The advent of IC fabrication, micromachining and packaging technology has facilitated the development of fully compensated smart sensors in the form of monolithic chips combining both the sensors and the compensating circuits. In view of this, circuit compensation technique for temperature variation are more popular than microcontroller- or computer-based techniques. The basic form of circuit compensation relies on the three approaches:

1. Dummy sensor compensation
2. Half/full-bridge compensation
3. Temperature sensor–based compensation

4.8.2.1.1 *Temperature-Sensitive Dummy Sensors*

In most resistive sensors, the measuring circuit comprises typically of a Wheatstone bridge. The resistive sensor is connected to one of the branches of the bridge and fixed resistors in the others branches. The bridge is balanced by selecting proper fixed resistances so that output voltage is zero when the sensor is not excited. When the sensor is excited by a physical signal, an unbalanced output voltage is obtained, which is proportional to the physical parameters to be measured. Examples of such a configuration are applicable to potentiometric displacement sensor, strain gauge–based load, force, or pressure sensors, etc. The working principle of such sensors are discussed in Section 1.4.

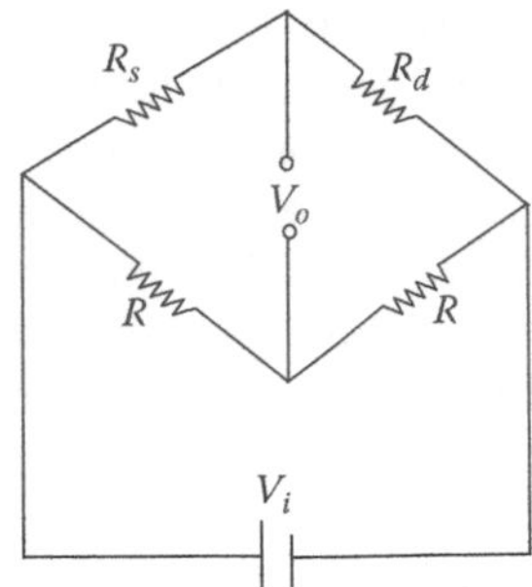

FIGURE 4.33
Temperature compensating Wheatstone bridge with dummy sensor.

In the dummy sensor approach of temperature compensation, a dummy sensor sensitive to temperature, however, that does not sense the physical variable, is connected to the adjacent arm of the sensor branches in the Wheatstone bridge. The other fixed resistances are so chosen that the bridge produces a zero voltage at unexcited condition. Such a temperature compensation circuit is shown in Figure 4.33.

Here, R_s is the sensor resistance whose value at excited condition is given by

$$R_s' = R_s(1+x) \tag{4.115}$$

where x is the input variable.

The dummy sensor is connected to the adjacent arm whose value R_d under temperature variation is given by

$$R_d' = R_d(1+t) \tag{4.116}$$

where t is the temperature variation.

In an actual system, the resistance R_s and R_d are so chosen that

$$R_s = R_d = R$$

Considering that the measuring sensor is affected by the same temperature variation as that of the dummy sensor, its resistance given by Equation 4.115 can be written as

$$R_s' = R(1+x)(1+t) \tag{4.117}$$

The output voltage V_0 is given by

$$V_0 = \frac{x}{2(2+x)} - V_i \tag{4.118}$$

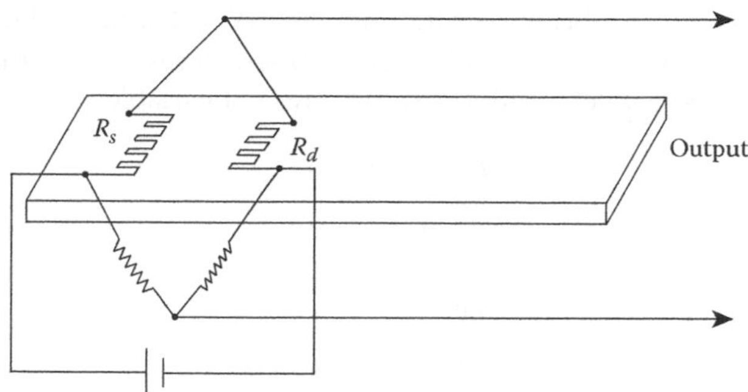

FIGURE 4.34
Dummy strain gauge installation with measuring gauge.

The output voltage shown by Equation 4.118 above is free from temperature effect that ensures temperature compensation. The dummy gauge is in fact similar to that of the measuring sensor but it is not allowed to be excited by the measuring variable. The dummy sensor is installed in such a way that the same amount of temperature variation as that of the measuring sensor takes place in it. The dummy sensor is exploited for temperature compensation by making it insensitive to the primary measurement variable using some installation methods such as by not bonding it to the strained structure in case of a strain gauge, opaquing the surface of the sensor in case of an optical sensor, blocking gas flow in case of a gas sensor, etc. Figure 4.34 illustrates a schematic diagram where a dummy gauge shown in the right-hand side is placed near to the main sensor so that the same variations take place in the dummy sensor. The dummy sensor is not bonded to the strained structure to make it ineffective to strain input.

4.8.2.1.2 Half/Full-Bridge Compensation

In a Wheatstone bridge circuit for resistive sensors, half-bridge and full-bridge configuration are often used when sensors are available with both positive and negative resistance variation for the same physical variable. Figure 4.35a and b

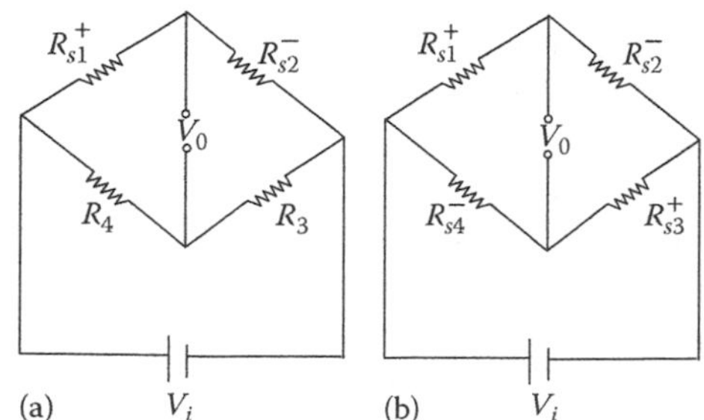

FIGURE 4.35
Wheatstone bridge circuit: (a) half bridge and (b) full bridge.

shows the half- and full-bridge configuration of Wheatstone bridge with resistive sensors. In Figure 4.35a, R_{s1}^{+} and R_{s2}^{-} are two sensor resistances that increases and decreases, respectively, with the same physical variable x, given by

$$\begin{aligned} R'_{s1} &= R_{s1}(1+x) \\ R'_{s2} &= R_{s2}(1-x) \end{aligned} \tag{4.119}$$

For simplicity, let us consider that

$$R_{s1} = R_{s2} = R_3 = R_4 = R$$

and considering the same temperature on the two sensors

$$\begin{aligned} R'_{s1} &= R(1+x)(1+t) \\ R'_{s2} &= R(1-x)(1+t) \end{aligned} \tag{4.120}$$

The output voltage is given by

$$V_0 = V_i \frac{x}{2} \tag{4.121}$$

Equation 4.121 shows that the output voltage is free from temperature effect. Similarly, for the full-bridge configuration shown in Figure 4.35b

$$\begin{aligned} R'_{s1} &= R(1+x)(1+t) \\ R'_{s2} &= R(1-x)(1+t) \\ R'_{s3} &= R(1+x)(1+t) \\ R'_{s4} &= R(1-x)(1+t) \end{aligned} \tag{4.122}$$

For which the output voltage, which is free from temperature effect is given by

$$V_0 = V_i x \tag{4.123}$$

Examples of such half-bridge and full-bridge sensor configurations are strain gauge for force and pressure measurement and potentiometer for displacement measurement. In strain gauge–based load cell, four strain gauges are used to sense tensile, and compressive strain in opposite positions of a mechanical elastic member. The strain gauges are so installed that the same temperature variation takes place in all of them.

Figure 4.36 shows a full-bridge potentiometer configuration where the rotary displacement of the wiper either increases or decreases the branch resistances configured in a Wheatstone bridge circuit. The figure shows resistance values considering a clockwise wiper rotation, i.e., $+\theta$ from the reference position

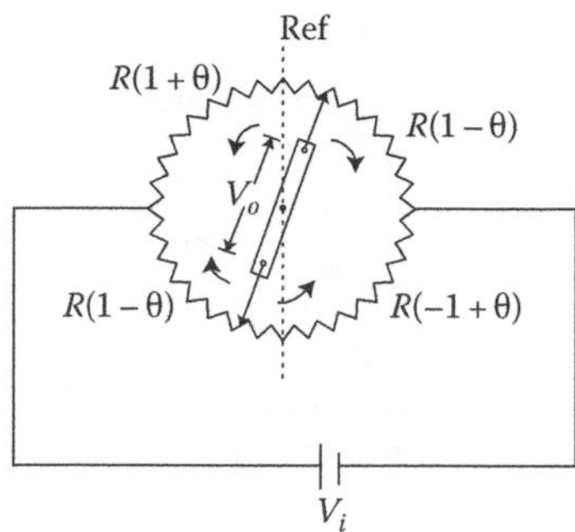

FIGURE 4.36
Rotary potentiometer in full bridge.

$$\begin{aligned}R_1 &= R(1+\theta)(1+t)\\R_2 &= R(1-\theta)(1+t)\\R_3 &= R(1+\theta)(1+t)\\R_4 &= R(1-\theta)(1+t)\end{aligned} \tag{4.124}$$

The output voltage is given by

$$V_0 = V_i\theta \tag{4.125}$$

4.8.2.1.3 Temperature Sensor–Based Compensation

In the temperature sensor–based compensation technique, a suitable temperature sensor is used to pick up the temperature variation and thereby the output signal is modified accordingly. The modification technique of the output signal may take various forms. Following are the typical methods by which the sensor output is modified to make it free from temperature effect:

1. Bridge voltage and current modification
2. Compensated amplification
3. Amplification factor modification

4.8.2.1.3.1 Bridge Voltage Modification As the temperature of a resistive sensor deviates from its calibrated reference temperature, Wheatstone bridge produces a higher voltage that includes a part generated due to temperature sensitivity of the sensor. The effect is a shift in the zero input–output voltage which is also called as offset and a change in sensitivity under input condition. Sensor manufacturers can compensate the first-order or linear effects of these changes by introducing temperature-sensitive resistance in the circuit, as shown in Figure 4.37 [30]. The bridge is excited by the voltage $+V_{ex}$ modulated by the temperature-sensitive resistor (PTC) R_{fsotc} shunted by $R_{\text{fsotc-shunt}}$.

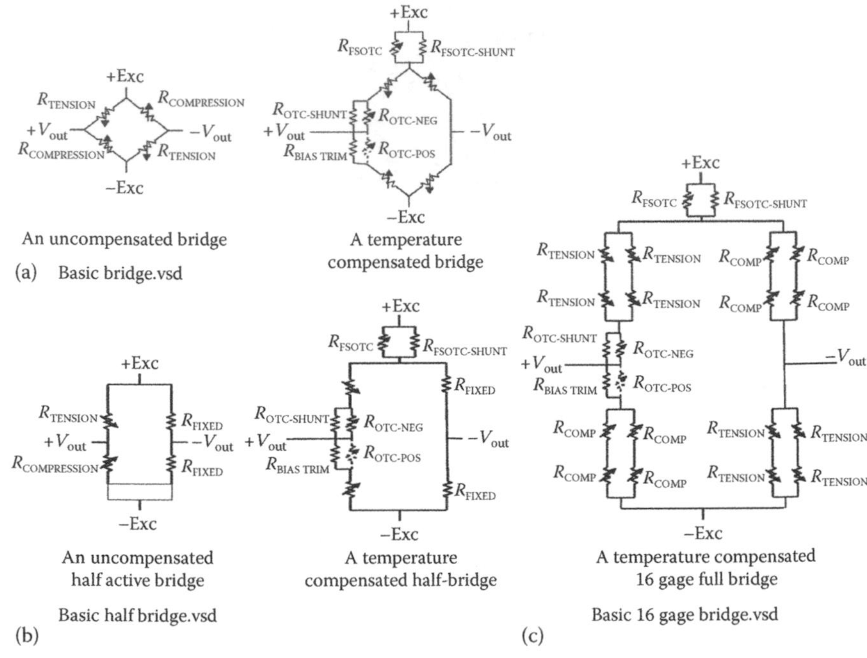

FIGURE 4.37
(a) Strain gauges wired in a Wheatstone bridge configuration, (b) strain gauges wired in a half-active Wheatstone bridge configuration, and (c) a 16-gage Wheatstone bridge configuration. (Reprinted from MAXIM, copyright © www.maxim-ic.com. With permission.)

The strain gauges—$R_{tension}$, R_{comp}, and R_{fostc}—are all subjected to the same temperature variation. As R_{FOSTC} increases, bridge excitation is proportionately reduced cancelling the inherent temperature effects by reducing the bridge output. The shunting resistor R_{SHUNT} is insensitive to strain or temperature and is used to adjust the amount of compensation required. Figure 4.38 shows a simplified circuit diagram in which the excitation voltage to the bridge is proportionately decreased as temperature effect increases the bridge output voltage. Let the bridge be excited by a voltage source V_c, then the excitation voltage to the bridge is

$$V_{ex} = \frac{V_c R}{R + R_p} \tag{4.126}$$

where

R is the bridge voltage

R_p is the parallel combination of PTC and a shunt resistance R_s so that

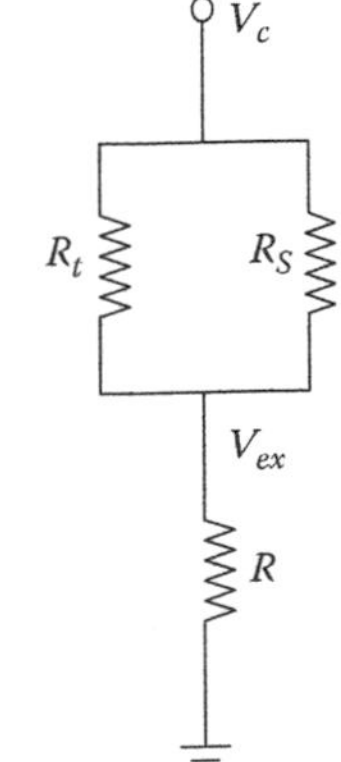

FIGURE 4.38
A simplified temperature compensation circuit.

$$R_p = \frac{R_t R_s}{R_t + R_s}$$

where R_t is the resistance of the PTC sensor. Consider that the bridge output is changed by temperature variation with a temperature coefficient of β mV/°C. The full-scale bridge output to excitation voltage ratio is denoted by a sensitivity term given by

$$S_v = \frac{V_{0(FS)}}{V_{ex}} \text{ mV/V}$$

The change in excitation voltage required to be accomplished to compensate temperature effect is

$$\Delta V_{ex} = \frac{\beta \Delta T}{S_v} \tag{4.127}$$

where ΔT is the change in temperature. This change is accomplished by using the PTC, the resistance of which increases with temperature. A constant temperature coefficient of the PTC cannot provide a linear decrease in V_{ex} with temperature; however, a variable coefficient of PTC may also be considered for a nearly linear decrease in V_{ex} as discussed in the example below.

Example 4.20

The temperature compensating bridge shown in Figure 4.38 has a temperature sensitivity of 8 mV/°C and produces an output voltage of 100 mV (full scale)/volt bridge excitation. The bridge excitation V_{ex} is 6 V when the circuit excitation is 8 V. The temperature compensation is accomplished by the PTC sensor with resistance at 25°C is 2.5 kΩ and the temperature coefficient of resistance is given by

$$\alpha_T = \frac{\alpha_{25}}{(1 + \alpha_{25}\Delta T)}$$

where

α_{25} is the 0.077 Ω/°C

ΔT is the change in temperature of the sensor (K)

Take bridge resistance as 5.6 kΩ.

Determine

1. The shunting resistance R_s
2. The required and actual excitation voltages at 27°C and 30°C

Also find the percentage error in compensation and value of compensated temperature.

Solution
The circuit is shown in Figure E4.20

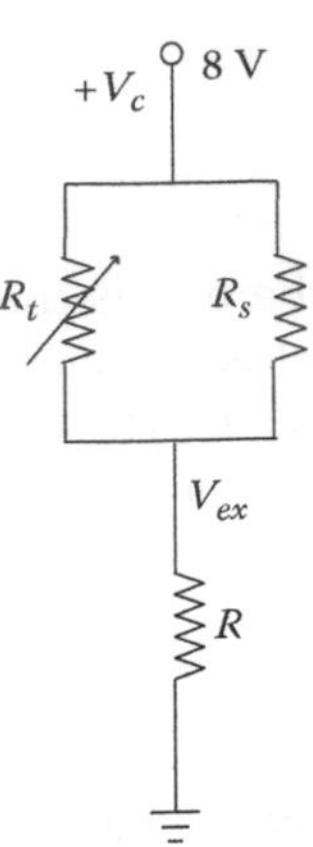

FIGURE E4.20
Temperature compensating circuit.

Given: $V_{ex} = 6\,V$, $R_t = 2.5\,k\Omega$ at 25°C
Temperature sensitivity of bridge voltage = β = 8 mV/°C
Output sensitivity = S_v = 100 mV/V (excitation)
Bridge resistance = R = 5.6 kΩ

1. The excitation voltage to bridge

$$V_{ex} = V_c \times \frac{R}{R+R_p'} \quad \text{where, } R_p = \frac{R_t R_s}{R_t + R_s}$$

$$\Rightarrow \frac{R}{R+R_p} = \frac{6}{8}$$

$$\Rightarrow \frac{5.6}{5.6+R_p} = \frac{6}{8}$$

$$\therefore R_p = 1.86\text{ k}\Omega$$

again

$$R_p = \frac{R_t R_s}{R_t + R_s}$$

$$\Rightarrow \frac{2.5 \times R_s}{2.5 + R_s} = 1.86\text{ k}\Omega$$

$$\therefore R_s = 7.28\text{ k}\Omega$$

2. For $T = 27°C$

$$\therefore \Delta T = (27 - 25)°\text{C}$$

$$= 2°\text{C}$$

Change in bridge output voltage (ΔV_0)

$$= \beta \Delta T = 8 \times 2$$

$$= 16\text{ mV}$$

Change in bridge excitation (ΔV_{ex})

$$= \frac{\Delta V_0}{S_v}$$

$$= \frac{16}{100}\text{ V}$$

New excitation voltage

$$V'_{ex} = V_{ex} - \frac{16}{100}$$

$$= 5.84 \text{ V}$$

Actual excitation voltage

$$\Delta T = (27 - 25)°\text{C}$$

$$= 2°\text{C}$$

Temperature coefficient of PTC at 27°C

$$\alpha_{27} = \frac{0.077}{(1 + 0.077 \times 2)}$$

$$= 0.066 \ \Omega/°\text{C}$$

$$R_t = 2.5 + 2.5 \times 0.066 \times 2$$

$$= 2.83 \text{ k}\Omega$$

$$R_p = \frac{2.83 \times 7.28}{2.83 + 7.28}$$

$$= 2.04 \text{ k}\Omega$$

$$\therefore V'_{ex} = V_c \frac{R}{R + R_p}$$

$$= 5.86 \text{ V.}$$

Percentage error in compensation

$$= \frac{\text{Required excitation voltage} - \text{Actual excitation voltage}}{\text{Required excitation}} \times 100\%$$

$$= \left| \frac{5.84 - 5.86}{5.84} \right| \times 100\%$$

$$= 0.34\%$$

Error in $V_{ex} = 0.02$ V.
Error in $V_0 = 0.02 \times 5$ V $= 2$ mV.

Uncompensated temperature

$$= \frac{2 \text{ mV}}{\beta}$$

$$= \frac{2 \text{ mV}}{8 \text{ mV}/°\text{C}}$$

$$= 0.25°\text{C}$$

For $T = 30°C$

$$\Delta T = (30 - 25) = 5°C$$

$$\Delta V_0 = 8 \times 5 = 40 \text{ mV}$$

$$\Delta V_{ex} = \frac{40}{100} \text{ V}$$

$$V'_{ex} = 6 - \frac{40}{100} = 5.60 \text{ V}$$

Temperature coefficient of PTC at 30°C

$$\alpha_{30} = \frac{0.077}{(1 + 0.077 \times 5)}$$

$$= 0.055 \ \Omega/°C$$

$$R_t = 2.5 + 2.5 \times 0.055 \times 5$$

$$= 3.187 \text{ k}\Omega$$

$$R_p = \frac{3.187 \times 7.28}{3.187 + 7.28}$$

$$= 2.216 \text{ k}\Omega$$

Actual excitation voltage

$$V'_{ex} = 8 \times \frac{5.6}{5.6 + 2.216}$$

$$= 5.73 \text{ V}$$

Error in $V_{ex} = |5.60 - 5.73| = 0.13$ V

Percentage error in

$$V_{ex} - = \frac{0.13}{5.6} \times 100\% = 2.35\%$$

$$\text{Error in } V_0 = 0.13 \times 100$$

$$= 13 \text{ mV}$$

Uncompensated temperature

$$= \frac{13 \text{ mV}}{8 \text{ mV/°C}} = 1.6°C$$

Frequency-based sensors
In frequency generating sensors such as piezoelectric quartz resonators, capacitive relaxation oscillator sensors, etc., the oscillation frequency can be linearly compensated for drift in frequency due to temperature. In one method [39], a thermal sensor is used to sense the temperature and produces a variables V_{cc} for a capacitive hygrometer.

Figure 4.39a shows the conceptual circuit diagram of a capacitive hygrometer, the frequency of oscillation of which is given by

$$f_0 = \frac{0.559}{RC}$$

where

$$C = C_H + C_G$$

where
C_H is the sensor capacitive
C_G is the input capacitance of the inverter gate

The sensor capacitance C_H is temperature sensitive, which causes a drift in measured relative humidity with a sensitivity of α %RH/K. The capacitance C_G depends on the supply voltage V_{cc} of the sensor. Therefore, the drift in f_0 can be compensated by varying V_{cc} accordingly. The variable V_{cc} is generated by the circuit of Figure 4.39b where the thermal sensor AD590 provides a sensitivity of 1 μA/K. The OPAMP-based current-to-voltage converter provides the adjustment of V_{cc} for the variation of temperature from the reference temperature. The temperature-compensated output frequency is divided by a factor of 100 gated by 4 MHz pulses and then counted in a counter.

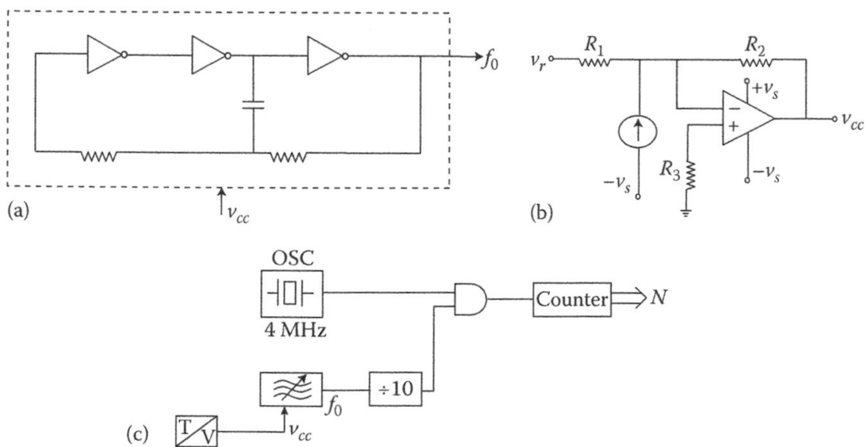

FIGURE 4.39
Capacitive hygrometer (a) conceptual circuit, (b) variable V_{cc} generation, and (c) capacitive hygrometer for Example 4.21. (Reprinted from Wiley Interscience Ltd. With permission.)

Example 4.21

The oscillation frequency of the capacitive hygrometer in Figure 4.39 is $f_0 = 0.559/RC$. The sensor capacitance C_H has a temperature drift of 0.05%RH/K. The AD590 provides a temperature current sensitivity of 1 μA/K in order to determine the dependence of C_G on V_{cc}. We keep the sensor at 20°C and 50% RH and measure the output frequency (Figure 4.39c). We obtain 8129 counts for $V_{cc} = 6\,V$, 7986 counts for $V_{cc} = 9\,V$ with a linear relationship. If the humidity sensor has capacitances of 107, 110, and 122pF at 0%, 50% and 100% RH, respectively, at 20°C, design the circuit for temperature compensation.

Solution

Given: $\alpha = 0.05\%$ RH/K

$$C_{H_0} = 107\,\text{pF}$$

$$C_{H_{50}} = 110\,\text{pF}$$

$$C_{H_{100}} = 122\,\text{pF}$$

$$N(V_{cc} = 6\text{ V}, 20°\text{C}, 50\%) = 8129$$

$$N(V_{cc} = 9\text{ V}, 20°\text{C}, 50\%) = 7986$$

From the measuring circuit

$$N = \frac{f_0 \times 0.25 \times 10^{-6}}{100}, \text{ since gated time} = \frac{1}{4}\text{ MHz} = 0.25 \times 10^{-6}$$

$$= 0.25 f_0 \times 10^{-4}$$

At 20°C and 50% RH

$$\therefore f_0(6\text{ V}) = \frac{8129}{0.25 \times 10^{-4}}\text{ Hz}$$

$$= 325.16\text{ MHz}$$

$$f_0(9\text{ V}) = \frac{7986}{0.25 \times 10^{-4}}\text{ Hz}$$

$$= 319.44\text{ MHz}$$

At 50% RH

$$C_{H_{50}} = 122\text{ pF}$$

Let at 20°C, a V_{cc} of 6 V is applied, to get output oscillation given by

$$f_0 = \frac{0.559}{RC}$$

$$\therefore \frac{0.559}{R \times 122 \times 10^{-12}}$$

$$= 325.16 \times 10^6$$

$$\Rightarrow R = \frac{0.559}{122 \times 325.16 \times 10^{-6}}$$

$$= 14.09\,\Omega$$

For a drift of 0.05% RH
Interpolating capacitance for 0.05% RH
Temperature coefficient of capacitance

$$= \frac{(122 - 107)}{100} \times 0.05$$

$$= 7.5 \times 10^{-3}\ \text{pF/K}$$

4.8.2.1.3.2 Bridge Current Modification In the case of the bridge current modification technique, the sensor Wheatstone bridge is supplied from a constant current source and an NTC temperature sensor. The resistance R_{ts} is used to stabilize the temperature sensitivity of the sensor resistance by shunting an increasing amount of excitation as temperature increases [31]. The circuit for such an application with piezoresistive pressure sensor is shown in Figure 4.40.

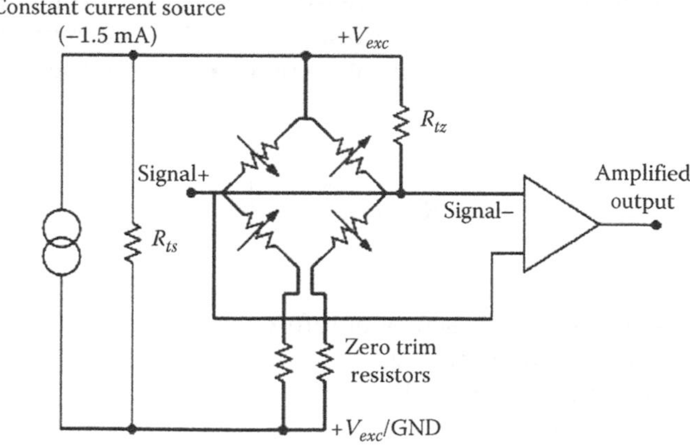

FIGURE 4.40
Resistive compensation circuit (R_{ts} for sensitivity drift, R_{tz} for offset drift, and zero-trim resistors). (Reprinted from MAXIM, copyright © www.maxim-ic.com. With permission.)

The simplified circuit diagram is shown in Figure 4.41. Let the bridge be excited by a constant current of I and the current flowing through the bridge be I_B given by

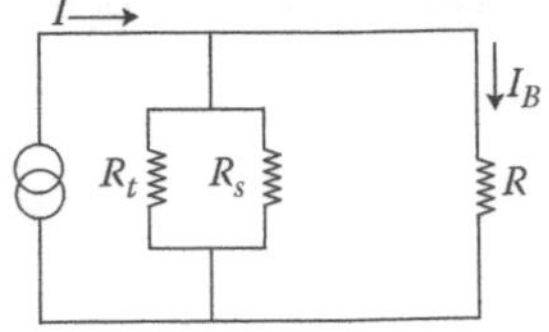

FIGURE 4.41
Constant current–based bridge current modification.

$$I_B = I\frac{R_p}{R_p + R} \quad (4.128)$$

where $R_p = R_s R_t/(R_s + R_t)$
R is the bridge resistance
R_t is the NTC temperature sensor resistance (at reference temperature)
R_s is the shunt resistance to control the current through R_t

Let the bridge output be changed due to temperature variation with a sensitivity of β mV/°C. The full-scale bridge output to bridge current sensitivity is given by

$$S_v = \frac{V_0(FS)}{I_B} \text{ mV/mA} \quad (4.129)$$

The change in bridge current required to be accomplished to compensate temperature effect is

$$\Delta I_B = \frac{\beta \Delta T}{S_v} \quad (4.130)$$

where ΔT is the temperature variation.

The NTC temperature sensor decreases as temperature increases above reference temperature, thereby the bridge excitation current gets proportionality reduced at a nearly linear rate. The resistance of the NTC sensor is given by

$$R_t = R_0 e^{-\beta\left(\frac{1}{T_0} - \frac{1}{T}\right)} \quad (4.131)$$

where
R_0 is the resistance of NTC sensor at reference temperature
T_0 is the reference temperature, K
T is the temperature of sensor, K
β is the NTC constant, K

The compensation circuit of Figure 4.40 uses another resistor R_{tz} for adjusting the offset due to temperature.

Example 4.22

The temperature compensating current mode bridge shown in Figure 4.41 has a temperature sensitivity of full-scale output voltage to bridge current coefficient of 0.0152 mV/mA/°C. The bridge excitation current at a reference temperature of 25°C is 1 mA when the constant current source is 1.5 mA. The NTC sensor has a resistance of 20 kΩ at 0°C with a constant of 4000. Take a bridge resistance value of 5.6 kΩ. Determine

1. The shunt resistance R_s.
2. The required and actual excitation current at 26°C and 30°C. Also calculate the error in compensation and value of uncompensated temperature.

Solution

The compensation circuit is shown in Figure E4.22.

Given values: $I_B = 1$ mA, $I = 1.5$ mA

$R_t = 69$ kΩ at 0°C, $\beta = 4000$ K and

$$\frac{\beta I}{V_0(FS)} = 0.0152\ (\text{mV/mA})/^\circ\text{C}$$

1. The bridge current

$$I_B = I\frac{R_p}{R_p + R}$$

where $R_p = \dfrac{R_t R_s}{R_t + R}$

At 25°C

$$R_t = R_0 e^{-\beta\left(\frac{1}{273} - \frac{1}{298}\right)}$$

$$= 69 \times e^{-4000\left(\frac{1}{273} - \frac{1}{298}\right)}$$

$$= 20.2\ \text{k}\Omega$$

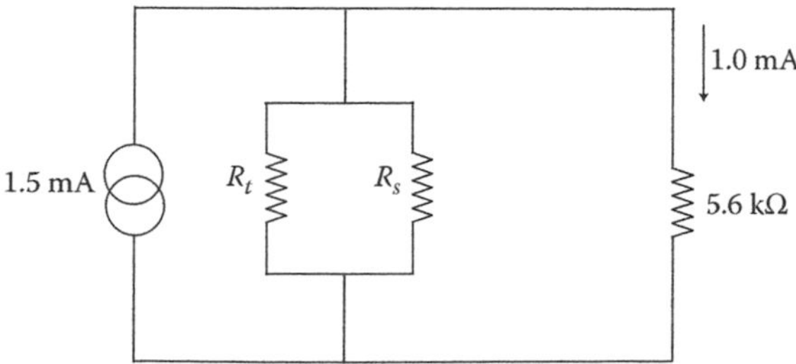

FIGURE E4.22
Constant current–based compensation circuit for Example 4.22.

$$\frac{I_B}{I} = \frac{1.0}{1.5} = \frac{R_p}{R_p + 5.6}$$

$$\Rightarrow R_p = 10.87 \text{ k}\Omega$$

Again

$$\frac{R_t R_s}{R_t + R_s} = 10.87$$

$$\Rightarrow \frac{20.2 \times R_s}{20.2 + R_s} = 10.87$$

$$\Rightarrow R_s = 23.53 \text{ k}\Omega$$

2. At 26°C

$$\therefore \Delta T = (26 - 25)°\text{C} = 1°\text{C}$$

The required change in bridge current

$$\Delta I_B = \frac{\beta I_B \Delta T}{V_0(FS)}$$

$$= \frac{\beta I_B}{V_0(FS)} \times \Delta T$$

$$= 0.0152 \times 1$$

$$= 0.0152 \text{ mA}$$

$$\therefore I'_B = (1 - 0.0152)\text{mA}$$

$$= 0.984 \text{ mA}$$

The actual bridge current

$$R_t = 69 \times e^{-4000\left(\frac{1}{273} - \frac{1}{299}\right)}$$

$$= 19.29 \text{ k}\Omega$$

$$\therefore R_p = \frac{19.29 \times 23.53}{19.29 + 23.53}$$

$$= 10.60 \text{ k}\Omega$$

$$\therefore I_B = 0.981 \text{mA}$$

Error in bridge current

$$= (0.984 - 0.981)\text{mA}$$
$$= 0.003\ \text{mA}$$

Percentage error in bridge current

$$= \frac{0.003}{0.984} \times 100$$
$$= 0.3\%$$

Value of uncompensated temperature

$$= \frac{0.003}{0.0152}$$
$$= 0.19°\text{C}$$

AT $T = 30°\text{C}$

$$T = (30 - 25) = 5°\text{C}$$
$$\therefore \Delta I_B = 0.0152 \times 5 = 0.076\ \text{mA}$$

Required I_B:

$$I'_B = (1 - 0.076)\ \text{mA}$$
$$= 0.924\ \text{mA}$$

Actual I_B:

$$R_t = 69 \times e^{-4000\left(\frac{1}{273} - \frac{1}{303}\right)}$$
$$= 16.17\ \text{k}\Omega$$

$$R_p = \frac{16.17 \times 23.53}{16.17 + 23.53}$$
$$= 9.58\ \text{k}\Omega$$

$$I'_B = 1.5 \times \frac{9.58}{9.58 + 5.6}$$
$$= 0.94\ \text{mA}.$$

$$\text{Error} = |0.924 - 0.946|\ \text{mA}$$
$$= 0.022\ \text{mA}$$

Percentage error

$$= \left(\frac{0.022}{0.924}\right) \times 100$$

$$= 2.3\%$$

Amount of uncompensated temperature

$$= \frac{0.022}{0.0152} = 1.44°\text{C}.$$

4.8.2.1.3.3 Compensated Amplification In this method, a temperature sensor senses the temperature variation in the main sensor and a proportionate amount of voltage is adjusted in the main sensor output with the help of an OPAMP. An RTD or semiconductor temperature sensor is popular for such application compared to thermocouple and thermistor. The temperature sensor circuit and the amplifier are designed so that the correct amount of voltage is developed corresponding to the temperature picked up and sensitivity of the main sensor, and the correction factor is adjusted.

The temperature sensor circuit is designed such that a voltage proportional to the temperature sensed and equal to the amount to be corrected is developed. This voltage is subtracted from the main sensor signal for temperature compensation. The designed method for temperature compensation of a strain gauge–based pressure sensor is discussed in the numerical problem below.

Example 4.23

A piezoresistive pressure sensor is composed of a bridge of one silicon strain gauge of 5000 Ω that has 1.5 μV/psi sensitivity supplied at 6 V. The other fixed resistors of the bridge are set at 5000 Ω each. The sensor has a temperature coefficient of resistance of 0.00375 (Ω/Ω)/°C. Design a temperature compensating circuit using OPAMP with the following temperature sensor: a low cost monolithic RTD sensor (AD221000 from Analog Devices) with a sensitivity of 22.5 mV/°C. The sensitivity of the transducer should be 0.3 mV/psi.

Solution

The pressure sensor amplifier and temperature compensation circuit is shown in Figure E4.23.

Pressure sensor:

Temperature coefficient of resistance $\alpha = 0.00375$ (Ω/Ω)/°C.
Original resistance $R_s = 5000\,\Omega$
Sensitivity = 1.5 μV/psi

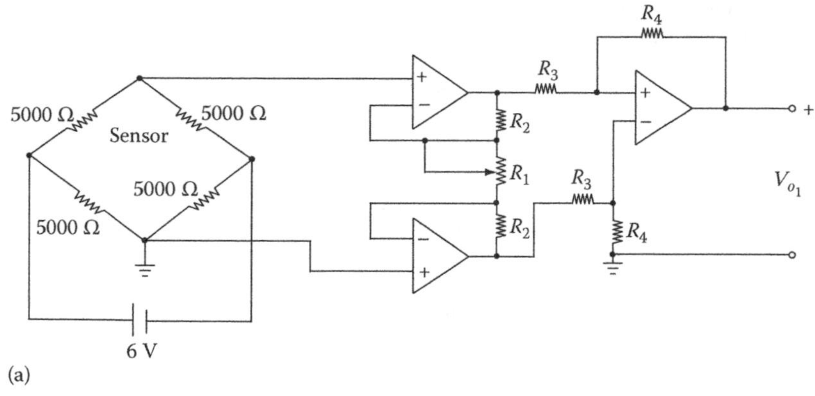

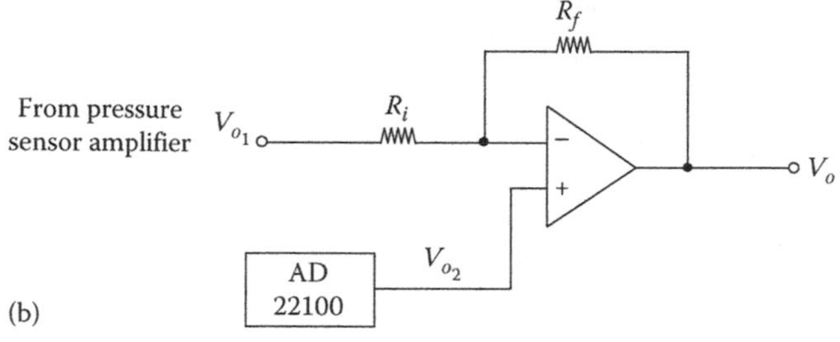

FIGURE E4.23
(a) Pressure sensor amplifier and (b) temperature compensation circuit.

Output voltage per-unit temperature

$$V_o = 6\left[\frac{5000 + 5000 \times 0.00375}{10000 + 5000 \times 0.00375} - \frac{1}{2}\right]$$
$$= 5.614 \text{ mV/°C}$$

1. For compensating with the RTD sensor

   Sensitivity = 22.5 mV/°C
   Amplification factor for the pressure sensor signal amplifier

   $$G = \frac{22.5 \text{ mV}}{5.614 \text{ µV}} = 4$$

   The gain of the instrumentation amplifier is given by

   $$G = \frac{R_4}{R_3}\left(1 + \frac{2R_2}{R_1}\right)$$

Taking $R_4 = 200\,\text{k}\Omega$
and $R_3 = 100\,\text{k}\Omega$ we get

$$\frac{R_4}{R_3} = 2$$

$$\therefore \left(1 + \frac{2R_2}{R_1}\right) = 2$$

$$\Rightarrow \frac{2R_2}{R_1} = 1$$

$$\Rightarrow \frac{R_2}{R_1} = 0.5$$

Taking $R_2 = 100\,\text{k}\Omega$

$$\therefore R_1 = 50\ \text{k}\Omega.$$

The sensitivity of the pressure sensor

$$= 1.5\ \mu\text{V/psi}$$

The overall sensitivity needed = 0.3 mV/psi

$$\text{Total amplification needed} = \frac{0.3\ \text{mV}}{1.5\ \mu\text{V}}$$

$$= 200$$

Amplification factor of the instrumentation amplifier = 4
Amplification factor of the final stage

$$= \frac{200}{4}$$

$$= 50$$

The output of the differential amplifier is

$$V_o = -\frac{R_f}{R_i}(V_{o2} - V_{o1})$$

$$= \frac{R_f}{R_i}(V_{o1} - V_{o2})$$

where $(R_f/R_i) = 50$ (as calculated earlier)
Taking $R_f = 500\,\text{k}\Omega$

$$\therefore R_i = 10\ \text{k}\Omega$$

Example 4.24

A cantilever load sensor consisting of a 120 Ω strain gauge with a gauge factor of 2.0 is installed on the surface of the cantilever to pick up the strain developed due to application of strain. The load to strain factor is 100 μ/10 kgf. The gauge is configured in a Wheatstone bridge with other fixed resistors of 120 Ω each and supplied with 5 V dc. Design a temperature compensating circuit using IC LM335 with a sensitivity of 10 mV/°C and output of 2.98 V at 25°C. Take the resistance temperature coefficient of gauge material to be 0.005 (Ω/Ω)/°C. The sensitivity of the transducer should be 0.25 V/kg.

Solution
Strain gauge:

$$R_s = 120\ \Omega,\quad \lambda = 2.0$$

$$F = \frac{100\ \mu}{10\ \text{kgf}} = \frac{10\ \mu}{\text{kgf}}$$

$$V_i = 5\ \text{V}$$

$$\alpha = 0.005\ (\Omega/\Omega)/^\circ\text{C}$$

For 1 kg input load, strain developed

$$\varepsilon = 10\ \mu$$

The bridge output for the quarter bridge

$$V_0 = \frac{\lambda}{4} \in V_i$$

$$= \frac{2.0 \times 10 \times 10^{-6} \times 5}{4}$$

$$= 25\ \mu\text{V}$$

Sensitivity required = 0.25 V/kg
and output of load cell = 25 μV/kg
Total amplification required

$$= \frac{0.25\,\text{V}}{25\,\mu\text{V}}$$

$$= 10{,}000$$

Let the load cell output be amplified by 100 times before compensation and 100 times after compensation to get

$$G = G_1 \times G_2$$
$$= 100 \times 100$$
$$= 10{,}000$$

So, we can take

$$R_2 = 100\ \text{k}\Omega$$
$$R_1 = 1\ \text{k}\Omega.$$

For 1°C temperature variation, strain gauge output voltage

$$V_{t0} = V_i \left[\frac{R_s + \Delta R_t}{2R + \Delta R_t} - \frac{1}{2} \right]$$

where $\Delta R_t = R \Delta t \alpha$

$$= 120\,\Omega \times 1 \times 0.005$$
$$= 0.6\,\Omega$$

$$\therefore V_{to} = 5 \times \left[\frac{120.6}{240.6} - \frac{1}{2} \right]$$
$$= 6.23\ \text{mV}$$

Amplified output of first stage for load cell for 1°C temperature change

$$= 6.23 \times 100$$
$$= 623\ \text{mV}$$

Now sensitivity of LM 335

$$= 10\ \text{mV/°C}$$

Removing offset of 2.73 V at 0°C for LM335 and amplifying the signal to equalize for 1°C variation in load cell:

Amplification needed for LM 335 output

$$= \frac{623\ \text{mV/°C}}{10\ \text{mV/°C}}$$
$$= 62.3$$

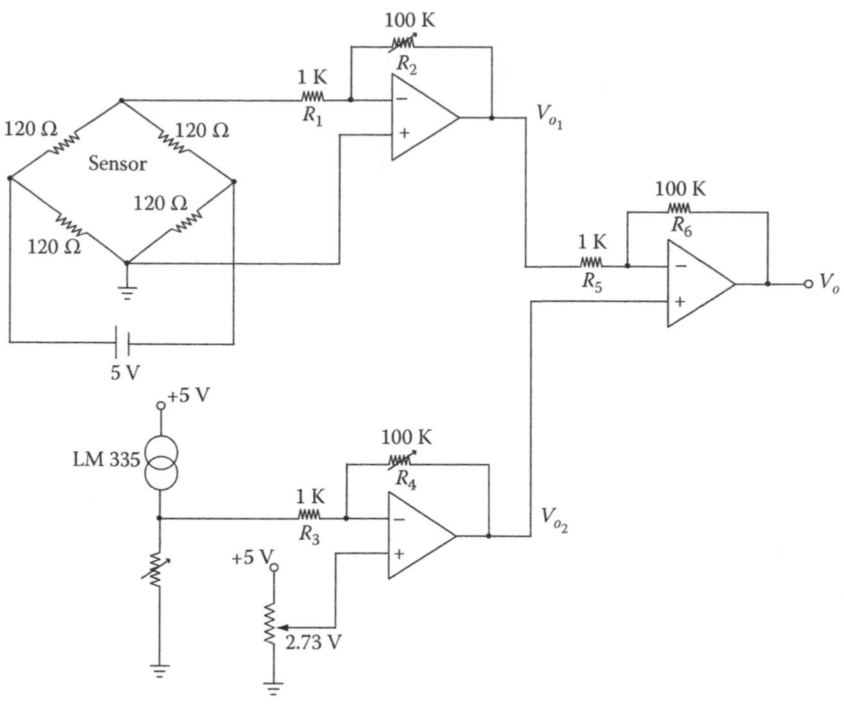

FIGURE E4.24
Temperature compensating circuit for the load cell.

For $(R_f/R_i) = 62.3$

Taking, we can choose $R_4 = 100\,\text{k}\Omega$ (potentiometer)

A final stage of differential amplifier for compensating the component proportional to temperature variation from the load cell amplifier output is shown in Figure E4.24. For an amplification factor of 100, the selected resistances are

$$R_5 = 1\,\text{k}\Omega$$

$$R_6 = 100\,\text{k}\Omega$$

4.8.2.1.3.4 Amplification Factor Modification In this technique, an OPAMP is so designed to amplify the main sensor output that the amplification factor is proportionately varied to compensate the temperature effect. This is accomplished by connecting an NTC thermistor in the input of a feedback branch of the OPAMP where an increase in temperature reduces the resistance of the feedback path changing the gain in turn. To obtain a linear change in resistance due to temperature change, a resistance is connected in parallel to the NTC thermistor.

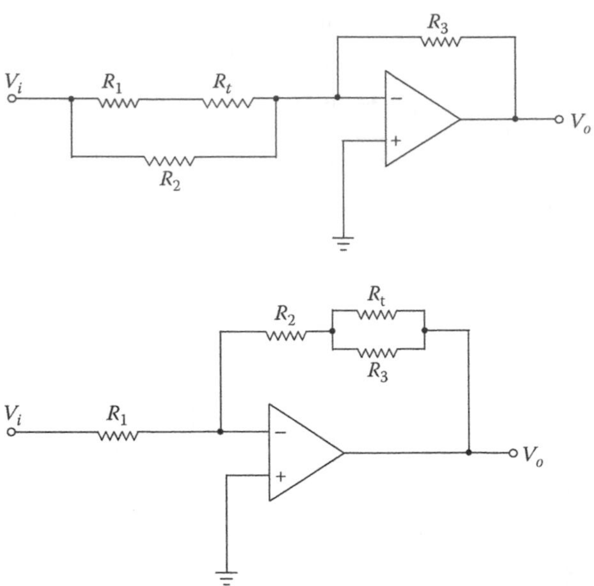

FIGURE 4.42
Amplifier gain–based compensation: (a) increasing gain and (b) decreasing gain.

In Figure 4.42a, a series-parallel combination of R_1 and R_2 with NTC thermistor R_t is connected to the inverting input of the OPAMP. The amplification factor of the amplifier is given by

$$G = -\frac{R_3}{\dfrac{R_2(R_1+R_t)}{R_1+R_2+R_t}}$$

$$= \frac{R_3}{R_2} \times \left(\frac{R_1+R_2+R_t}{R_1+R_t}\right) \tag{4.132}$$

where $R_t = R_0 e^{-B\left(\frac{1}{T_0}-\frac{1}{T}\right)}$.

In Figure 4.42b, the parallel-series combination of R_3 and R_2 with NTC thermistor R_t is connected to the feedback branch of an OPAMP whose amplification factor is given by

$$G = -\frac{R_2 + \dfrac{R_t R_3}{R_t + R_3}}{R_1}$$

$$= \frac{R_2 R_t + R_2 R_3 + R_t R_3}{R_1 R_t + R_1 R_3} \tag{4.133}$$

Example 4.25

A resistive strain gauge–based load cell consists of a single gauge of 120 Ω at 20°C with a gauge factor of 2.0 is configured in a quarter bridge supplied by 12 V. The strain gauge has a temperature coefficient of resistance of $10^{-5}(\Omega/\Omega)/°C$. The sensor is temperature compensated by reducing the gain of an OPAMP shown in Figure 4.42b. An NTC thermistor with $\beta = 4000\,K\Omega$ and $R_t = 30\,k\Omega$ at 20°C is used. Design the amplifier circuit. The readout device requires a sensitivity of 204 mV/kg load applied. Consider that a strain of 200 µ is developed per kg load applied and temperature of the sensor changes to 30°C.

Solution

$$\text{Strain gauge: } \lambda = 2.0$$

$$R_s = 120\ \Omega$$

$$\varepsilon_i = 200 \times 10^{-6}$$

$$V_i = 12\ \text{V}$$

$$\alpha = 10^{-5}\left(\Omega/\Omega\right)/°C$$

The output voltage per kg load applied at 20°C

$$V_o = \frac{\lambda \varepsilon_i V_i}{4}$$
$$= 2.0 \times 200 \times 10^{-6} \times \frac{12}{4}$$
$$= 1.2\ \text{mV}$$

Change in resistance per kg load applied

$$\Delta R = \lambda \in_i R$$
$$= 2.0 \times 200 \times 10^{-6} \times 120$$
$$= 0.048\Omega$$

Resistance of gauge ($R_s = 120.048\Omega$).

Considering 30°C temperature of the sensor

$$\Delta T = \left(30 - 20\right)$$
$$= 10°C$$

Change in resistance due to 10°C change in temperature

$$R_{st} = R_s + R_s \times \alpha \times \Delta T$$

$$= 120.048 + 120.048 \times 10^{-5} \times 10$$

$$= 120.048576\,\Omega$$

The output voltage under 1 kg load applied and 10°C temperature

$$V_o' = \left[\frac{120.048576}{120 + 120.048576} - \frac{1}{2}\right] \times 12\,\text{V}$$

$$= 1.5\,\text{mV}.$$

Sensitivity of readout device

$$= 204\,\text{mV/kg}.$$

Required gain of amplifier at 20°C

$$= \frac{204\,\text{mV}}{1.2\,\text{mV}}$$

$$= 170$$

Required gain of amplifier at 30°C

$$= \frac{204\,\text{mV}}{1.5\,\text{mV}} = 136$$

In the circuit diagram of Figure 4.42b, the thermistor (R_t) resistance at 20°C

$$= 30\,\text{k}\Omega$$

The thermistor resistance at 30°C

$$= R_t e^{-B\left(\frac{1}{T_0} - \frac{1}{T}\right)}$$

$$= 19.1\,\text{k}\Omega$$

The gain of the amplifier

$$G = \frac{R_t \parallel R_3}{R_1} + \frac{R_2}{R_1}$$

The gain at 20°C and 30°C are

$$G_{20} = \frac{30 \parallel R_3}{R_1} + \frac{R_2}{R_1}$$

$$= 170$$

$$G_{30} = \frac{19.1 \parallel R_3}{R_1} + \frac{R_2}{R_1}$$

$$= 136$$

To get a linear NTC characteristic, the parallel resistance chosen is same as R_t

$$\therefore R_3 = 30 \text{ k}\Omega$$

Hence, $$\frac{30 \parallel 30}{R_1} + \frac{R_2}{R_1} = 170 \quad \text{(E4.25.1)}$$

$$\frac{19.1 \parallel 30}{R_1} + \frac{R_2}{R_1} = 136 \quad \text{(E4.25.2)}$$

By simplifying Equations E4.25.1 and E4.25.2

$$15 + R_2 = 170\, R_1$$

$$\Rightarrow 170\, R_1 - R_2 = 15 \quad \text{(E4.25.3)}$$

$$11.67 + R_2 = 136\, R_1$$

$$\Rightarrow 136 R_1 - R_2 = 11.67 \quad \text{(E4.25.4)}$$

By solving (E4.25.3) and (E4.25.4), we get

$$R_1 = 0.0979 \text{ k}\Omega = 0.1 \text{k}\Omega$$

$$R_2 = 2 \text{ k}\Omega$$

The circuit diagram is shown in Figure E4.25

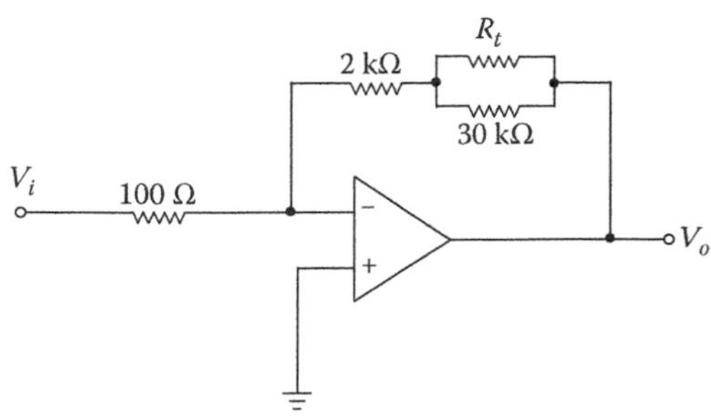

FIGURE E4.25
Compensation circuit for Example 4.25.

Example 4.26

A polymer-based resistive humidity sensor (EMD2000) configured in a resistive circuit has a sensitivity of 10 mV/%RH at 20°C and a negative temperature coefficient of 0.3%RH/°C. The sensor output is amplified by a logarithmic amplifier to linearize the signal given by

$$V_o = -\ln\left(\frac{V_i}{4.7\times10^{-4}}\right) \tag{4.134}$$

The linearized output is temperature compensated using NTC thermistor by an OPAMP of Figure 4.42a. The NTC thermistor has a resistance of 30 kΩ at 20°C and $\beta = 4000\,\text{K}$. Design the temperature compensating amplifier circuit. Consider that the readout device requires a sensitivity of 100 mV/%RH.

Solution

The circuit diagram is shown in Figure E4.26

Sensitivity of humidity sensor $S_s = \dfrac{10\text{ mV}}{\%\text{RH}}$

Temperature coefficient of the sensor $\alpha_{RH} = -0.3\%/\text{RH}/°\text{C}$

Sensitivity of the read out device $S_r = \dfrac{100\text{ mV}}{\%\text{RH}}$

Operating temperature $T_2 = 30°\text{C}$
Calibrating temperature $T_1 = 20°\text{C}$
Temperature coefficient for output voltage (α_v)

$$= \alpha_{RH} S_s$$

$$= -0.3\times10$$

$$= -3\text{ mV/°C}$$

Output voltage at 20°C/% RH

$$= 10\text{ mV}$$

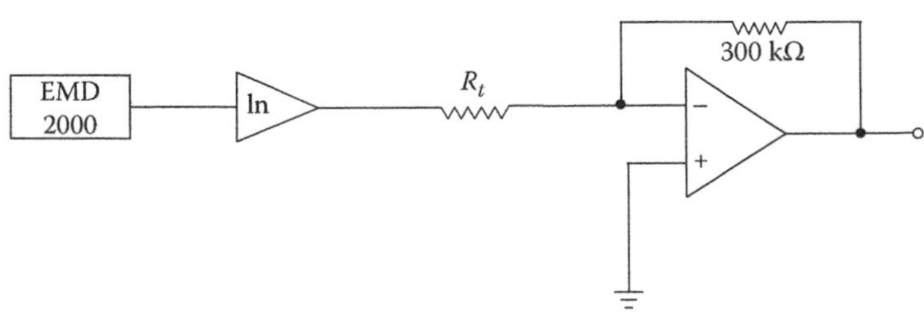

FIGURE E4.26
Compensation circuit for Example 4.26.

Output voltage at 21°C/% RH

$$= 10 - 3 \times (21 - 20)$$
$$= 10 - 3$$
$$= 7 \text{ mV}$$

Output of log amplifier at 20°C and 21°C

$$V_{20} = -\ln\left(\frac{10}{4.7 \times 10^{-4}}\right)$$
$$= -9.96 \text{ mV}$$

$$V_{21} = -\ln\left(\frac{7}{4.7 \times 10^{-4}}\right)$$
$$= -9.54 \text{ mV}$$

Required gain of compensating amplifier

$$G_{20} = \frac{S_R}{V_{20}}$$
$$= \frac{100}{-9.97}$$
$$\cong -10$$

$$G_{21} = \frac{100}{-9.54}$$
$$\cong -10.48$$

Considering only R_t in input of amplifier

$$\Rightarrow -\frac{R_3}{R_{20}} = G_{20} \quad \text{and} \quad -\frac{R_3}{R_{21}} = G_{21}$$

$$\Rightarrow \frac{R_3}{R_{30}} = 10$$

$$\Rightarrow \frac{R_3}{30e^{-4000\left(\frac{1}{293} - \frac{1}{294}\right)}}$$

$$= \frac{R_3}{28.6}$$

$$= 10.48$$

$$\Rightarrow R_3 = 299.79$$

So we choose $R_3 = 300\,\text{k}\Omega$.

4.8.3 Examples of Temperature Compensation in Sensor

The temperature compensation techniques discussed in the preceding sections have been applied in various applications with some kind of variations as well. Some of these applications will be discussed here now.

4.8.3.1 Temperature Compensation of Thermal Flow Sensors

The working principle of a thermal flow sensor is based on a heat transfer principle where a heated body is cooled by a flow system and the local rate of cooling depends on the flow velocity. Such a thermal flow sensor and its temperature compensation have been described in [32]. Here, a thermal thin flow sensor measures flow by detecting the temperature difference between a heated thin film and fluid flowing over it. Thermal sensor probes measures the rate of heat loss to the flowing fluid from a hot body—a resistive thin film held perpendicular to the fluid flow. The heat flow rate from the thin film to the fluid is proportional to the heat interchanging area A, the difference in temperature between the thin film and the fluid and the coefficient of heat transfer (h) of the thin film. The heat transfer equation at equilibrium is given by

$$I^2R = KhA(T_w - T_f) \tag{4.135}$$

where K is a constant.

The compensation of temperature effect in a thermal flow sensor was developed by Nam et al. [32]. The constant current Wheatstone bridge circuit for this technique is shown in Figure 4.43. In the circuit, R_1, R_2, and R_3 are three fixed resistors; R_h is the hot film airflow sensor, which is heated by the current flowing through it and cooled due to the difference of temperature between it and the surrounding air. This sensor is a PTC, the resistance R_h increases as temperature increases from the ambient temperature and then settles at an equilibrium state. The change in temperature of the air from the calibrated ambient value is sensed by another PTC sensor, resistance of which is R_f. The output voltage across node a and b is applied to the OPAMP,

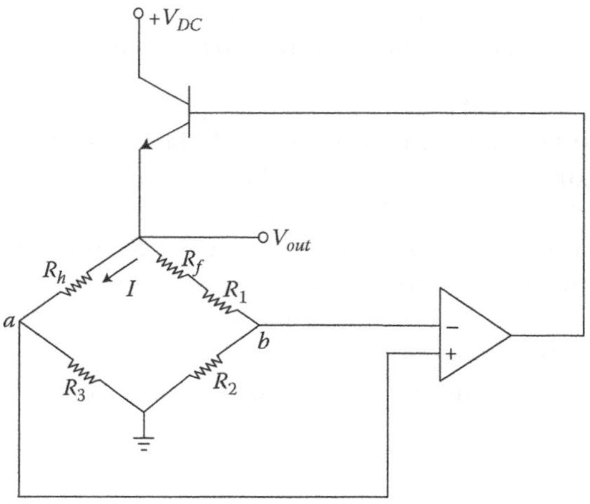

FIGURE 4.43
The temperature compensation Wheatstone bridge circuit. (Redrawn from IEEE, copyright © IEEE. With permission.)

which supplies the current I to the sensor. The current (I) produces the output voltage across R_h and R_3.

Let us consider the following definition. In the circuit, R_1, R_2, and R_3 are fixed resistors and R_{h0} is the resistance of the flow sensor at ambient condition and it is heated up to reach to a resistance value of R_h. Similarly, at normal ambient temperature, the resistance of the temperature sensor is R_{f0}, which changes to R_f after ambient temperature changes.

At normal condition the bridge is unbalanced so that

$$\frac{R_2}{R_3} \neq \frac{R_1 + R_{f0}}{R_{h0}} \tag{4.136}$$

When the sensor is heated and attains a local equilibrium, the bridge is balanced so that

$$\frac{R_2}{R_3} = \frac{R_1 + R_{f0}}{R_{h0}} \tag{4.137}$$

The node voltage at a and b become equal and the inputs of the OPAMP will attain a virtual ground condition. In this condition, the value of R_h is given by

$$R_h = (R_1 + R_f)\frac{R_3}{R_2} \tag{4.138}$$

Due to the difference between R_h and R_f, a voltage V_{out} is developed in the bridge. This voltage is proportional to the airflow rate.

Variation in air temperature

When the ambient temperature of air is fixed, the value of R_f is also constant, as shown in Figure 4.44. The difference between R_h and R_f is also constant, which is proportional to V_{out}. The characteristic for continuous change of temperature is shown in Figure 4.45. In this case, R_f continuously increases, and since the value of R_h is determined by the ratio $(R_1 + R_f)(R_3/R_2)$, the gap between R_f and R_h is not constant. This is shown by a characteristic proportional to V_{out}. To compensate the nonuniformity of the gap developed due to the varying air temperature, a compensated characteristic is shown as R_{hc}.

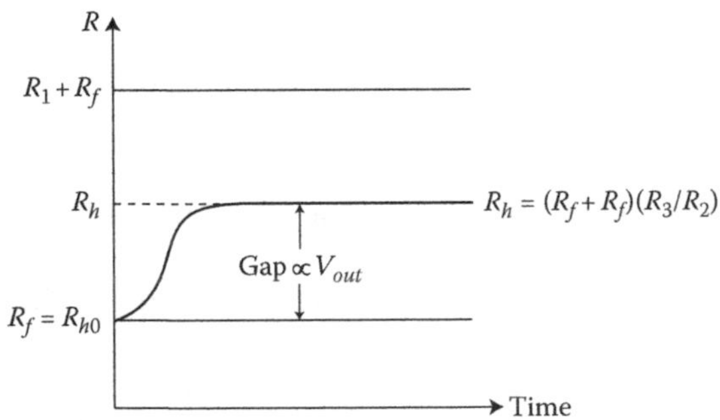

FIGURE 4.44
Characteristics when air temperature is constant. (Redrawn from IEEE, copyright © IEEE. With permission.)

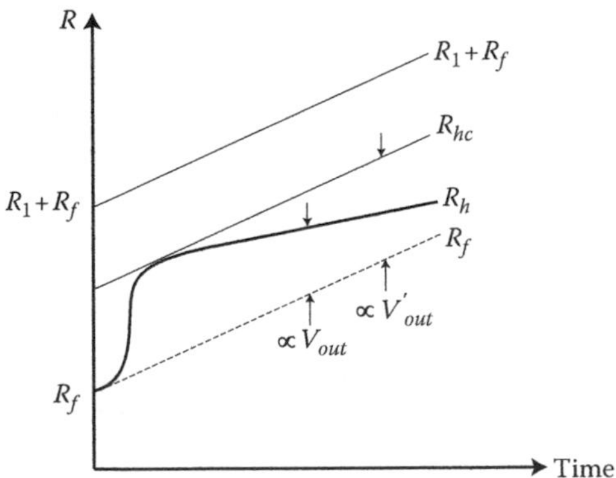

FIGURE 4.45
Characteristics when air temperature is continuously varying. (Redrawn from IEEE, copyright © IEEE. With permission.)

Compensation technique

When the value of R_{f0} changes to R_f given by equation

$$R_f + R_{f0}(1+\alpha T_f) \tag{4.139}$$

where

α is the resistance temperature coefficient
T_f is the changed temperature

Then Equation 4.138 can be written as

$$R_h = \frac{R_3}{R_2}(R_1 + R_f)$$

$$= \frac{R_3}{R_2}\{R_1 + R_{f0}(1+\alpha T_f)\}$$

$$R_h = \frac{R_3}{R_2}(R_1 + R_{f0}) + \frac{R_3}{R_2}\alpha T_f R_{f0} \tag{4.140}$$

It is seen from Equation 4.139 that the slope of R_f is $R_{f0}\alpha$, while the slope of R_h as per Equation 4.140 is $(R_3/R_2)\alpha R_{f0}$. Therefore, the slope of R_h can be made equal by multiplying by a factor (R_2/R_3) to $(R_3/R_2)\alpha R_{f0}$ as shown below

$$m_h = m_f = \frac{R_3}{R_2}\alpha R_{f0} \times \left(\frac{R_2}{R_3}\right) \tag{4.141}$$

Example 4.27

A thermal flow sensor with a resistance of 300 Ω at 20°C and a positive temperature coefficient of 0.0035 Ω/Ω/°C is connected in the Wheatstone bridge, as shown in Figure 4.43. A resistive temperature sensor with resistance 100 Ω at 20°C having a positive temperature coefficient of 0.005 Ω/Ω/°C is connected in series with R_1 in the bridge. The bridge is supplied with a constant current of 12.45 mA. A flow of 20 m/s of air with temperature of 20°C causes the sensor temperature to increase to 25°C. Take $R_3 = 200$ Ω and $R_2 = 100$ Ω so that $R_2{:}R_3 = 1{:}2$.

Determine

1. Output voltage and sensitivity
2. Output voltage if the air temperature increases to 22°C
3. Output voltages and percentage error for constant air temperature and percentage error at 22°C if $R_2 = R_3 = 100$ Ω so that $R_2{:}R_3 = 1{:}1$ and $I_T = 10$ mA

Solution
Given values:

$R_{h_a} = 300\,\Omega$; $T_a = 20°C$; $\alpha_s = 0.0035\Omega/\Omega/°C$; $R_{f_a} = 100\,\Omega$; $T_f = 20°C$;

$\alpha_f = 0.005\,\Omega/\Omega/°C$;

$I_T = 10$ mA; $\vartheta = 20$ m/s and $T_h = 25°C$.

1. The resistance of flow sensor after equilibrium condition

$$R_h = R_{h_a} + R_{h_a}\alpha_s \Delta T$$

$$= 300 + 300 \times 0.0035 \times (25 - 20)$$

$$= 305.25\,\Omega$$

Given: $R_2 = 100\,\Omega$, $R_3 = 200\,\Omega$

For balanced condition of the bridge; $(R_3/R_2) = 2$

$$R_h = \frac{R_3}{R_2}(R_1 + R_f)$$

$$R_h = 2(R_1 + 100)$$

$$\therefore R_1 = 52.626\,\Omega$$

The current through the sensor (at air temperature of 20°C)

$$I = I_T \times \frac{(R_h + R_2)}{(R_h + R_2) + (R_1 + R_{f_a} + R_3)}$$

$$= 12.45 \times 10^{-3} \times \frac{305.25 + 100}{(305.25 + 100) + (52.625 + 100 + 200)}$$

$$= 6.66 \times 10^{-3} \text{ A}$$

$$V_{out} = I \times (R_h + R_2)$$

$$= 2.7 \text{ V}$$

$$\text{Sensitivity} = \frac{2.7 \text{ V}}{20 \text{ m/s}} = 0.135 \text{ V/m/s}$$

2. When air temperature increases to 22°C $T_f = 22°C$
The temperature sensor resistance

$$R_f = R_{fa} + R_{fa}\alpha_f (T_f - T_a)$$

$$= 100 + 100 \times 0.005 \times (22 - 20)$$

$$= 101\,\Omega$$

With T_f=22°C, the sensor will heat up to

$$= 25 + 2$$

$$= 27°C$$

Hence,

$$R_h = 300 + 300 \times 0.0035 \times (27 - 20)$$

$$= 307.35\Omega$$

Sensor current

$$I = 12.45\times10^{-3} \times \frac{307.35+100}{(307.35+100)+(52.625+101+200)}$$

$$= 6.667 \text{ mA}$$

$$V_{out} = 6.667\times10^{-3}(307.35+100)$$

$$= 2.715 \text{ V}$$

$$\text{Error} = (2.715 - 2.7)$$

$$= 0.015 \text{ V}$$

$$\text{Percentage error} = \frac{0.015}{2.7}\times100$$

$$= 0.55\%$$

3. Let $R_2 = R_3 = 100\ \Omega$

$$\therefore \frac{R_2}{R_3} = 1$$

For balance of the bridge: at T_f=20°C

$$R_1 = R_h - R_{f0}$$

$$= 305.25 - 100$$

$$= 205.25\Omega$$

$$I = 0.01\times\frac{305.25+100}{(305.25+100)+(205.25+100+100)}$$

$$= 5\times10^{-3} \text{ A}$$

$$V_{out} = 5\times10^{-3}(305.25+100)$$

$$= 2.0262 \text{ V}$$

When $T_f = 22°C$, R_h heats up to $25 + 2 = 27°C$

$$R_h = 307.35\Omega \quad \text{(already calculated)}$$

$$R_f = 101\Omega \quad \text{(already calculated)}$$

Current

$$I = 0.01 \times \frac{305.35 + 100}{(307.35 + 100) + (205.25 + 101 + 100)}$$

$$= 5.006 \times 10^{-3} \text{ A}$$

$$V_{out} = 5.006 \times 10^{-3}(307.35 + 100)$$

$$= 2.039 \text{ V}$$

$$\text{Error} = (2.039 - 2.0262) \text{ V}$$

$$= 0.0128 \text{ V}$$

$$\text{Percentage error} = \frac{0.0128}{2.0262} \times 100$$

$$= 0.63\%$$

4.8.3.2 Temperature Compensation of Hot Film Crosswind Sensor

Hot film crosswind sensors are used in airflow monitoring. The basic structure of the sensor consists of a pair of PTC platinum film symmetrically deposited over a ceramic or quartz rod. The platinum films are spaced apart by a bridging material and the films are connected in series by lead wires and connected to load loop negative feedback amplifier circuit, as shown in Figure 4.46 [33]. The bridge consists of the film sensors in branch 1, an RTD

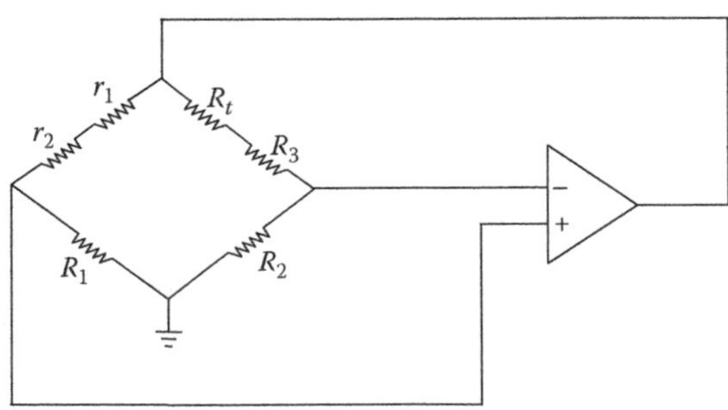

FIGURE 4.46
Constant temperature driving circuit. (Reprinted from IEEE, copyright © IEEE. With permission.)

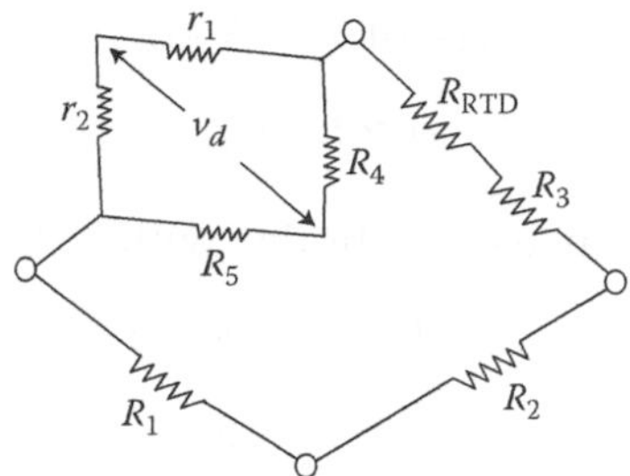

FIGURE 4.47
Wheatstone bridge with RTD sensor. (Reprinted from IEEE, copyright © IEEE. With permission.)

that measures ambient air temperature in branch 2 in series with R_3 and two fixed resistors in branches 3 and 4. When there is airflow, the temperature of the film facing the coils drops while the temperature of the film of the other side rise by the same amount. Therefore, the bridge is balanced and a voltage is developed.

A change in ambient temperature will cause the film current to change, resulting in an error in the output voltage. To prevent this, an RTD is used to sense the ambient temperature and compensate the error due to ambient temperature variation. The compensation is achieved by using the Wheatstone bridge configuration, as shown in Figure 4.47 [33]. The circuit components and sensor parameters are

r_s − Hot film sensor resistance $= r_1 + r_2$
R_1, R_2, R_3, R_4, R_5 = Fixed resistors
R_t − Resistance of RTD in series R_3
α = Temperature coefficient of Pt films
r_0 = Resistance of the Pt films at 0°C
t_1, t_2 = Temperature of the two hot film sensors, respectively (°C)
β = Temperature coefficient of RTD
R_0 = Resistance of RTD at 0°C
t_a = Ambient temperature (°C)

The resistances of the hot films after airflow are

$$r_1 = r_0(1+\alpha t_1) \tag{4.142}$$

$$r_2 = r_0(1+\alpha t_2) \tag{4.143}$$

Total resistance of the hot film branch in the bridge

$$= (r + r_2)\,||\,(R_4 + R_5) \tag{4.144}$$

$$\approx (r_1 + r_2)$$

$$= r_0(1+\alpha t_1)+r_0(1+\alpha t_2)$$

$$= 2r_0 + r_0\alpha(t_1+t_2) \quad (4.145)$$

Total resistance of the branch of the bridge with RTD is

$$R_t = R_3 + R_0(1+\beta t_a)$$

$$= R_3 + R_0 + R_0\beta t_a$$

$$= R_3 + R_0 + k_T t_a \quad (4.146)$$

The heat developed in the hot fluid due to current I dissipated by the airflow gives the equilibrium temperatures t_1 and t_2; however, the ambient air temperature t_a will interfere with the development of equilibrium temperature of the hot fluid.

Applying King's law to the hot fluids we get

$$I^2 r_1 = A + B_1\sqrt{\rho v}(t_1 - t_a) \quad (4.147)$$

$$I^2 r_2 = A + B_2\sqrt{\rho v}(t_2 - t_a) \quad (4.148)$$

where

A is the constant determined by sensor geometry
B_1, B_2 are the parameters varying with wind direction
ρ is the air density
v is the airflow rate

From Equations 4.147 and 4.148 we get

$$I^2(r_1 - r_2) = A + B_1\sqrt{\rho v}(t_1 - t_a) - \left(A + B_2\sqrt{\rho v}(t_2 - t_a)\right)$$
$$\approx (B_1 - B_2)\sqrt{\rho v}(\bar{t} - t_a) \quad (4.149)$$

where $\bar{t} = (t_1 + t_2)/2$ is the average temperature of the films.

The voltage in the inner bridge

$$V_d = \frac{I(r_1 - r_2)}{2} \quad (4.150)$$

From Equation 4.149

$$IV_d = \frac{I^2(r_1 - r_2)}{2}$$

$$= \frac{1}{2}(B_1 - B_2)\sqrt{\rho v}(\bar{t} - t_a)$$

Squaring both sides we get

$$(IV_d)^2 = \frac{1}{4}(B_1 - B_2)^2 \rho v(\bar{t} - t_a)^2 \tag{4.151}$$

Equation 4.151 can be used for measuring the flow rate v, but the equation is effected by ambient temperature t_a.

4.8.3.2.1 Temperature Compensation

From Equation 4.145, we can write

$$\frac{r_1 + r_2}{2} = r_0 + r_0\alpha\frac{t_1 + t_2}{2}$$

$$= r_0 + K_F\bar{t}$$

Therefore

$$\bar{t} = \frac{1}{K_F}\left\{\frac{r_1 + r_2}{2} - r_0\right\} \tag{4.152}$$

Now at balance, the bridge balance equation can be written as

$$(r_1 + r_2)R_2 = R_1R_t$$

Hence, $\frac{r_1 + r_2}{2} = \frac{R_1R_t}{2R_2}$

Substituting the above ratio in Equation 4.152

$$\bar{t} = \frac{1}{K_F}\left\{\frac{R_1R_t}{2R_2} - r_0\right\}$$

Substituting Equation 4.146

$$\bar{t} = \frac{1}{K_F}\left\{\frac{R_1R_3 + R_1R_0 + R_1K_Tt_a}{2R_2} - r_0\right\}$$

Hence, we get

$$\bar{t} - t_a = \frac{R_1(R_3 + R_0) - 2R_2r_0 + (R_1K_T - 2K_FR_2)t_a}{2R_2K_F} \tag{4.153}$$

To make the above equation independent of t_a we can write

$$R_1K_T - 2K_FR_2 = 0$$

which gives

$$R_2 = \frac{K_T}{2K_F}R_1 \tag{4.154}$$

where

K_F is the change in resistance of the hot film per-unit change in temperature
K_T is the change in resistance of the RTD per-unit change in ambient temperature

Equation 4.153 is a design rule that can be adopted to make the sensor self-compensating against variation in ambient temperature.

Example 4.28

A pair of hot field cross-wind sensors each of 812 Ω at 0°C resistance are used for airflow measurement. The airflow rate is 20 m/s with an ambient temperature of 25°C. The sensor temperature deviates by 5°C on airflow. Temperature compensation is performed using an RTD with resistance of 1000 Ω at 0°C in branch 2 of the Wheatstone bridge of Figure 4.46. The OPAMP supplies a constant current 10 mA to the bridge. Take K_F and K_T as 0.0312 and 4.91, respectively, and ρ as 1.2 kg/m³. Determine

1. Bridge fixed resistances with $R_1 = 3\,\Omega$
2. Output voltage V_d at 25°C and 50°C of sensor temperature
3. The corresponding airflow constant B_1 and B_2

Solution

Given values

$$r_0 = 812\,\Omega,\ R_1 = 3\,\Omega \text{ and } I = 10\,\text{mA}$$

$$t_{a_1} = 25°\text{C and } t_{a_2} = 50°\text{C}$$

$$\Delta t = 5°\text{C}$$

$$K_F = 0.0312 \text{ and } K_T = 4.91$$

$$v = 20 \text{ m/s and } \rho = 1.2 \text{ kg/m3}$$

1. From Equation 4.154

$$R_2 = \frac{K_T}{2K_F}R_1$$

$$\therefore R_2 = \frac{4.91}{2\times 0.0312}\times 3$$

$$= 236\,\Omega$$

At $t_a = 25°C$, $t_1 = 25°C$, and $t_2 = 25°C$ with no airflow

$$\therefore r_s = 2r_0 + K_F(t_1 + t_2) \quad \text{(from Equation 4.145)}$$
$$= 2 \times 812 + 0.0312 \times (25 + 25)$$
$$= 1625.56\,\Omega$$

At $t_a = 25°C$, the RTD resistance

$$R_t = R_3 + R_0 + K_T t_a \quad \text{[from Equation 4.146 with } R_3 = 0]$$
$$= R_3 + 1000 + 4.91 \times 25$$
$$R_t = (R_3 + 1122.75)\Omega$$

At balance of the bridge

$$r_s R_2 = R_t R_1$$
$$\Rightarrow 1625.56 \times 236 = R_t \times 3$$
$$\Rightarrow R_t = 12{,}7877\Omega$$
$$\therefore R_3 = 12{,}7877 - 1{,}122.75$$
$$= 126.75\,\text{k}\Omega$$

2. For ambient temperature of 25°C

$$t_1 = t_a - \Delta t$$
$$= 25 - 5 = 20°C$$

$$t_2 = t_a + \Delta t$$
$$= 25 + 5 = 30°C$$

$$r_2 = 812 + 0.0312 \times 30 \quad \text{[from Equation 4.143]}$$
$$= 812.936\Omega$$

$$r_1 = 812 + 0.0312 \times 20 \quad \text{[from Equation 4.142]}$$
$$= 812.62\Omega$$

$$V_d = \frac{I(r_1 - r_2)}{2} \quad \text{(from Equation 4.150)}$$
$$= -1.58\,\text{mV}$$

For $t_a = 50°C$ (similarly)

$$t_1 = 50 - 5 = 45°C$$

$$t_2 = 50 + 5 = 55°C$$

$$r_1 = 812 + 0.0312 \times 45$$
$$= 813.404\,\Omega$$

$$r_2 = 812 + 0.0312 \times 55$$
$$= 813.716\,\Omega$$

$$V_d = 10 \times 10^{-3} \times (813.404 - 813.716)$$
$$= -3.12\,\text{mV}$$

3. From Equation 4.149

$$I^2(r_1 - r_2) = (B_1 - B_2)\sqrt{\rho v}(\bar{t} - t_a)$$

Case I:

$$\bar{t} = \frac{t_1 + t_2}{2} = \frac{20 + 30}{2} = 25°C$$

$$\therefore t_a = 25°C$$

$$\therefore (10 \times 10^{-3})^2 \times (812.78 - 812.936) = (B_1 - B_2)\sqrt{\rho v} \times (25 - 25)$$

$$\therefore (B_1 - B_2) = \infty \text{ (Indeterministic)}$$

Case II:

$$\bar{t} = \frac{t_1 + t_2}{2} = \frac{55 + 45}{2} = 50°C$$

$$t_a = 25°C$$

$$\therefore (10 \times 10^{-3})^2 \times (813.404 - 813.716) = (B_1 - B_2)\sqrt{1.2 \times 20} \times (50 - 25)$$

$$\Rightarrow (B_1 - B_2) = 8.16 \times 10^{-7}$$

4.8.3.3 Integrated Hardware Compensation of Pressure Sensor

A half-bridge integrated compensation technique developed in [34] enables the sensor to be operated in the temperature range of −40°C to 130°C over a pressure range of 0–310 kPa. Following are the sensor parameters:

Diaphragm size: 1000 μm × 1000 μm
Diaphragm thickness: 10 μm
Piezoresistive value: 2 kΩ
Temperature coefficient of resistance: 1900 ppm/K

Substrate doping: $10^{16}/\text{cm}^3$
Sheet resistance: $200\,\Omega$/square
The piezoresistive pressure sensor errors are subdivided as

$$e_{\text{total}} = e_{\text{offset}} + e_{\text{tracking}} + e_{\text{temperature}} \tag{4.155}$$

where

e_{offset} is the zero pressure offset error
e_{tracking} is the wafer tracking error
$e_{\text{temperature}}$ is the error due to temperature sensitivity of piezoresistive sensor

The first two errors—e_{offset} and e_{tracking}—increase with the pressure value. When the piezoresistors are perfectly matched and configured in the Wheatstone bridge, the error effect should automatically cancel out due to the balanced nature of the matched resistors. However, with matched piezoresistors also the error remains.

To compensate for e_{offset} and e_{tracking}, a separate compensation bridge with a diffused resistor with positive temperature coefficient of resistance is used. The temperature signal generated is not directly used for compensation, but it is scaled depending on the pressure signal magnitude. The compensation bridge is put on the bulk part of chip and the piezoresistive sensor bridge is deposited on thin diaphragm over the chip area. The difference of the two bridge outputs is free from zero pressure offset and wafer tracking errors.

To compensate for the remaining $e_{\text{temperature}}$ error, another half-bridge temperature compensation circuit is used, which is located on the bulk part of the chip. The complete scheme of the compensation technique is shown in Figure 4.48 and the half-bridge temperature compensation circuit is shown in Figure 4.49.

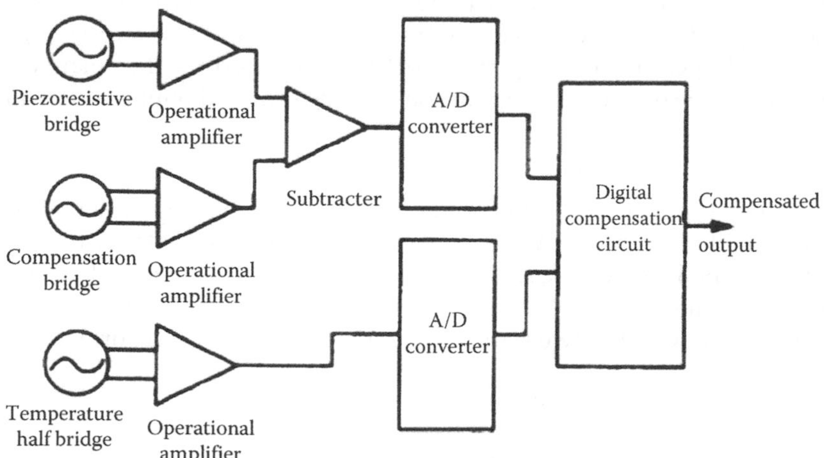

FIGURE 4.48
Block diagram of the compensation technique. (Reprinted from IEEE, copyright © IEEE. With permission.)

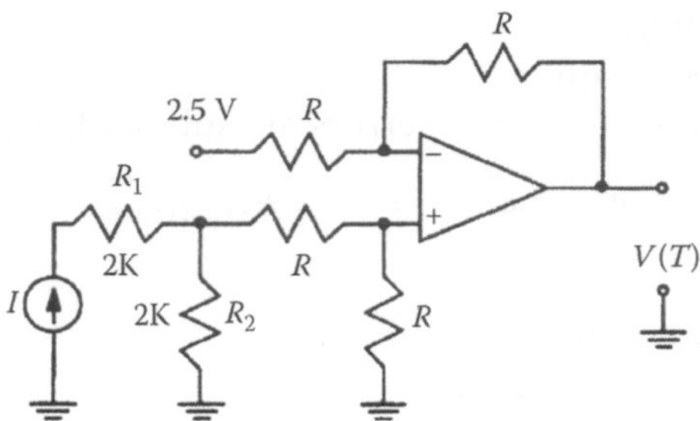

FIGURE 4.49
Half-bridge temperature compensation circuit. (Reprinted from IEEE, copyright © IEEE. With permission.)

The circuit is driven by a constant current (I) source and uses a resistance temperature detector R_2 while R_1 is used to offer a higher load resistance to the current source. The temperature detector resistance value R_2 is identical to the resistance value of the piezoresistance to ensure same wafer-to-wafer tracking error for R_2 and piezoresistance. When there is no tracking error at room temperature, the compensating resistance (R_2) generates a voltage of, say, V_{R_2} given by

$$V_{R_2} = IR_2 \tag{4.156}$$

The OPAMP generates a voltage proportional to the temperature variation beyond reference tracking error and room temperature by subtracting V_{R_2} from the fixed voltage at reference tracking error and temperature, i.e., $V_{R_{ref}}$, e.g., for a temperature sensor resistance (R_2) of $2\,\text{k}\Omega$ and constant current of 1.25 mA, the fixed voltage is 2.5 V ($2\,\text{k}\Omega \times 1.25\,\text{mA}$). The polarity of the output voltage depends on the values of the tracking errors and operating temperature range. Depending on the patterns of the tracking errors, the polarity may remain the same or it may change at some predetermined temperature value.

4.8.3.3.1 *Digital Processing of Compensation*

The analog output of the double-bridge and half-bridge compensation is converted to digital by separate ADCs. The digital signals are stored in two separate 10-bit registers. The pressure signal (P) is always positive while the half-bridge output temperature signal (T) may be either positive or negative; hence a sign bit (T_s) is added to it. The two signals are, represented, in binary forms as

$$\text{Pressure:} \quad P = P_{10}P_9P_8 \ldots P_1$$

$$\text{Temperature:} \quad T = T_sT_9T_8 \ldots T_1$$

The scales of the pressure and temperature values are shown in Tables 4.13 and 4.14, respectively.

TABLE 4.13
Digital Format of the Pressure Values

Pressure Values (Normalized)	Digital Format P_{10}	P_9	P_8	⋯	P_2	P_1	MUX Select C_1	C_0	Shifting
0	0	0	0	⋯	0	0	0	0	1
1	0	0	0	⋯	0	1	0	0	
⋮									
255	0	0	1	⋯	1	1	0	0	
512	0	1	1	⋯	1	1	1	0	2
767	1	0	1	⋯	1	1	1	0	
1023	1	0	1	⋯	1	1	0	1	3

TABLE 4.14
Digital Formats of Temperature Values

Temperature Values (Normalized)	T_s	T_{11}	T_{10}	T_9	⋯	T_3	T_2	T_1
+511	0	0	0	1	⋯	1	1	1
⋮								
0	0	0	0	0	⋯	0	0	0
⋮								
–511	1	0	0	0	⋯	0	0	0

Two control bits C_1C_0 are developed by the logic equation based on the pressure value range as shown below:

$$C_1 = \overline{P}_{10}P_9$$

$$C_0 = P_{10}$$

The temperature signal is applied to eight numbers of 3–1 MUX to shift the temperature signal based on the pressure range. This operation scales the temperature signal depending on the pressure value. For a case of maximum temperature range $\pm T_{max}$, the shifted and scaled temperature values are shown in Table 4.15.

TABLE 4.15
Scaled Temperature Values

C_1	C_0	Pressure Ranges (Normalized)	Scaled Temperature Format t_8	t_7	t_6	t_5	t_4	t_3	t_2	t_1	Temperature
0	0	0–225	T_9	T_8	T_7	T_6	T_5	T_4	T_3	T_2	$\pm T_{max}/2$
1	0	256–511	0	T_9	T_8	T_7	T_6	T_5	T_4	T_3	$\pm T_{max}/4$
0	1	512–1023	0	0	T_9	T_8	T_7	T_6	T_5	T_4	$\pm T_{max}/8$

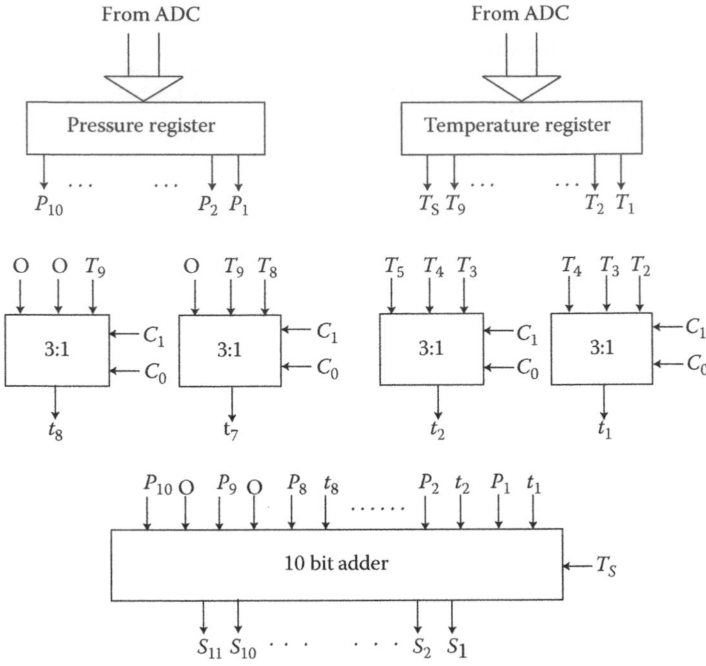

FIGURE 4.50
Signal processing logic for pressure and temperature. (Reprinted from IEEE, copyright © IEEE. With permission.)

The logic diagram of the temperature and pressure signal processing is shown in Figure 4.50. The pressure signal and the shifted temperature signal are added by a 10 bit controlled adder to generate the temperature compensated pressure signal.

Example 4.29

A piezoresistive pressure sensor, which can operate in the range of 0–310 kPa, is configured in a Wheatstone bridge circuit and the output is amplified to a range of 0–5 V. The sensor circuit shows a temperature sensitivity of –2044 mV/K. For temperature compensation of the sensor, a half-bridge circuit with a temperature sensor of resistance 2 kΩ at 20°C and temperature sensitivity of 1.22×10^{-3} Ω/Ω/K is used. The temperature signal is amplified by an OPAMP, as shown in Figure 4.49, with a gain of 10. Determine the value of constant current required for the half-bridge circuit and the fixed bias voltage.

Solution

Given values:

Pressure sensor:

$\beta = -2.44$ mV/K

Pressure range = 0–310 kPa

Full-scale output = 5 V

Temperature sensor:

$$\alpha = 1.22\times10^{-3}\ \Omega/\Omega/\text{K}$$

Resistance of temperature sensor $R_2 = 2\ \text{k}\Omega$ at 20°C
OPAMP gain = 10.

Since the pressure senor shows a temperature sensitivity of −2.44 mV/K, the half-bridge circuit output sensitivity should be 2.44 mV/K.

∴The voltage sensitivity across R_2:

$$= \frac{\text{output sensitivity}}{\text{gain}}$$

$$= \frac{2.44\times10^{-3}}{10}$$

$$= 0.244\ \text{mV/K}$$

The change in R_2 for 1°C change in temperature from reference temperature of 20°C

$$\Delta R_2 = R_2\alpha\Delta T$$

$$= 2000\times1.22\times10^{-3}\times1$$

$$= 2.44\ \Omega$$

The voltage generated by R_2 for 1°C variation

$$V_{R_2} = 2.44\times I = 0.244\ \text{mV}$$

Value of constant current

$$\therefore I = \frac{0.244\times10^{-3}}{2.44}$$

$$= 0.1\times10^{-3}$$

$$= 0.1\ \text{mA}$$

The value of fixed bias voltage in OPAMP

$$= I\times\text{Resistance at 20°C}$$

$$= 0.1\times10^{-3}\times2\times10^{3}$$

$$= 0.2\ \text{V}$$

4.8.3.4 Integrated Compensation for Pressure Sensor

In a typical application circuit developed by MAXIM, the signal conditioning IC (MAX 1450) integrates two main functional blocks—a controlled current source for driving the sensor and a programmable gain amplifier (PGA), as shown in Figures 4.51 and 4.52 [35]. In the recommended combination, a hybrid, including the sensor chip and the signal conditioning chip, results in a precise and compact measurement module. The external resistors and voltage dividers can be realized by hybrid technology and adjusted with laser trimming.

4.8.3.4.1 Sensitivity Drifts Compensation

Since the piezoresistive sensors are temperature sensitive, the bridge resistance R_B changes, which is given by the equation

$$R_B = R_{B_0}(1+\alpha_B T) \tag{4.157}$$

where

α_B is the temperature coefficient of bridge resistance (Ω/Ω/°C)
R_{B_0} is the bridge resistance at calibrated temperature
T is the temperature of the bridge under operation (°C)

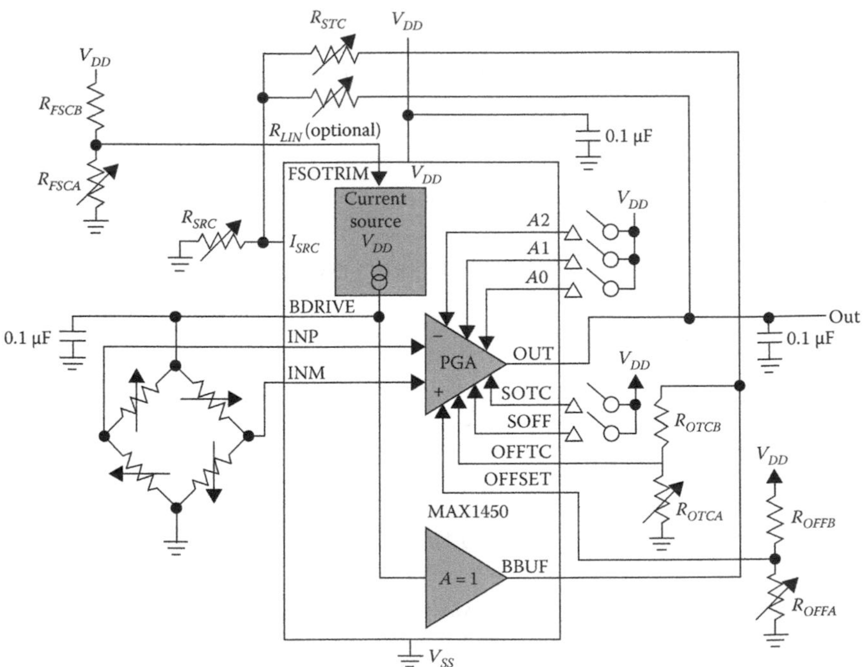

FIGURE 4.51
MAX1450 signal conditioner. (Reprinted from MAXIM, copyright © www.maxim-ic.com. With permission.)

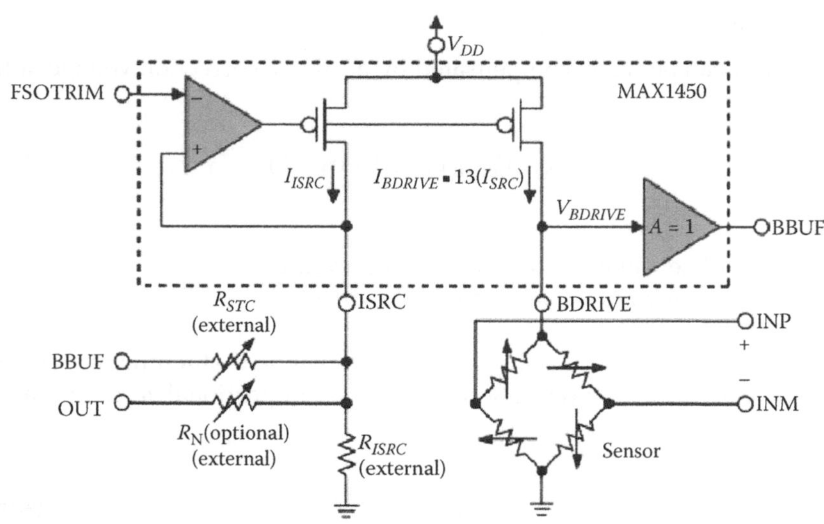

FIGURE 4.52
Controlled current source. (Reprinted from MAXIM, copyright © www.maxim-ic.com. With permission.)

The bridge output V_{DIFF} is proportional to the bridge drive voltage V_{BDRIVE} given by

$$V_{DIFF} = K_1 V_{BDRIVE}(1 + \alpha_s T) \tag{4.158}$$

where
K is a constant
α_s is the temperature coefficient of bridge sensitivity

The circuit's initial sensitivity is adjusted at the FSOTRIM pin of the IC sensor's drive voltage obtained from the V_{BDRIVE} pin to ISRC pin. Compensation of offset and offset drift is accomplished at the PGA and decoupled from the sensitivity compensation. The bridge output V_{DIFF} can also be expressed as

$$V_{DIFF} = K_2\left[1 + (\alpha_B - \alpha_S)T - (\alpha_B - \alpha_S)T^2\right] \times V_{FSOTRIM}\left(\frac{1}{R_{ISRC}} + \frac{1}{R_{STC}}\right) \times \frac{R_{B_0}}{\left[1 + K_3(1 + \alpha_B T)(R_{B_0}/R_{STC})\right]} \tag{4.159}$$

Neglecting the higher-order terms, we get

$$V_{DIFF} = \frac{K_2[1 + (\alpha_B - \alpha_s)T]}{\left[1 + K_3(1 + \alpha_B T)\left(R_{B_0}/R_{STC}\right)\right]} \tag{4.160}$$

The differential output of the sensor can be adjusted by adjusting R_{STC}.

4.8.3.4.2 *Offset Drift Compensation*

Static offset can be easily compensated by adding a correction voltage at the PGA given by

$$V_{0S} = V_{BDRIVE}V_{0S_0}(1+\alpha_{VOS}T) \tag{4.161}$$

where
V_{0S_0} is the offset voltage at reference temperature
α_{VOS} is the temperature coefficient of the offset

In sensitivity compensation, V_{BDRIVE} was independent of temperature; however, in offset drift compensation, V_{BDRIVE} is proportional to temperature given by

$$V_{BDRIVE} = \frac{K_4V_{DIFF}}{(1-\alpha_S T)} \tag{4.162}$$

Using the Taylor series expansion, we get

$$V_{BDRIVE} = K_4V_{DIFF}(1+\alpha_S T) \tag{4.163}$$

Substituting the equation for V_{BDRIVE} in Equation 4.161

$$\begin{aligned} V_{0S} &= K_4V_{DIFF}(1+\alpha_S T)V_{0S_0}(1+\alpha_{VOS}T) \\ &= S(1+\alpha_S T)V_{0S_0}(1+\alpha_{VOS}T) \\ &= SV_{0S_0}[1+T(\alpha_S+\alpha_{VOS})+\alpha_S\alpha_{VOS}T^2] \end{aligned} \tag{4.164}$$

where $S = K_4V_{DIFF}$ = Sensitivity.

Neglecting the smaller terms in Equation 4.164

$$V_{OS} = SV_{OS_0}[1+(\alpha_{VOS}+\alpha_S)T] \tag{4.165}$$

This amount of static offset can be compensated by adding (or subtracting depending on the sign of the offset temperature coefficient) a fraction of V_{BDRIVE} to PGA output (V_{out}). The IC has additional facility of linearization of the transfer curve of the pressure sensor in the PGA. This technique will be discussed in Chapter 5.

4.8.4 Microcontroller-Based Compensation

The temperature compensation technique discussed in Section 4.8.3.1 is mostly nonlinear, i.e., the signal value that can be compensated due to temperature variation is a nonlinear function of temperature. Therefore, the sensors cannot be fully compensated against temperature by circuit compensation. Microcontrollers have become increasingly popular temperature compensation of sensors for the following reasons:

1. On chip integration of CPU, memory, and sensors on the same package
2. Knowledge-based data storage in EPROM and EEPROM to adjust parameters to compensate for temperature changes
3. Can be programmed using sensor calibration data
4. Can be implemented in embedded systems requiring sensor interface having minimal memory program length, no operating systems and low software complexity
5. Flexibility in compensating as per calibrated characterization

Following are the features of a microcontroller:

1. A central processing unit (CPU) ranging from a simple 4 bit to a complex 32 or 64 bit
2. Serial input output port (UARTs)
3. Other serial communication interfaces such as IC, serial peripheral interface, and controller area network for system interconnection
4. Peripherals such as timers, event counters, PWM oscillators, etc.
5. RAM, ROM, EPROM, EEPROM, or flash memory
6. Quartz crystal oscillator, resonator, or RC circuits
7. Analog-to-digital convertor (ADC)
8. In circuit programming and debugging support
9. Programming through assembly language or high-level language

Microcontrollers are used in a feedforward mode for compensation of temperature effects of sensors as shown in the block diagram of Figure 4.53.

Voltage control
In Figure 4.53 [36], sensor voltage V_s is amplified with a high-gain amplifier A_1, with best signal-to-noise ratio and required frequency response. The voltage error compensation V_{com} is accomplished in this amplifier by the microcontroller through a DAC. In the next stage, the sensor signal is amplified by a PGA to give the necessary voltage range of the ADC. The microcontroller sends the digital value to the PGA for gain adjustment. The sensor temperature is measured by a temperature sensor producing voltage V_T, which is again amplified by an amplifier (A_2) for feeding to the ADC and then to the microcontroller. The microcontroller uses a lookup table (LUT) stored in EEPROM for error

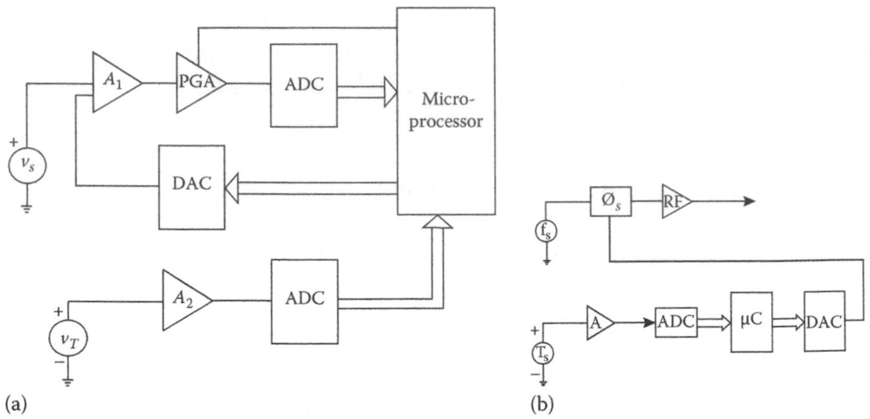

FIGURE 4.53
(a) Feedforward temperature compensation using microcontroller (voltage controlled). (b) Feedforward temperature compensation using microcontroller (frequency controlled). (Redrawn from IEEE, copyright © IEEE. With permission.)

compensation voltage V_{com}, and generates a digital value for gain adjustment of PGA corresponding to a temperature value measured as V_T.

Digital bit lengths

The voltage error of the sensor is compensated in amplifier A_1 through the K-bit DAC while the temperature drift is compensated in amplifier A_2 directly from the microcontroller. Let the sensor generate a full-scale voltage of V_{SFS} corresponding to a full-scale input of X_{FS}, hence sensitivity of the sensor is given by

$$S = \frac{V_{SFS}}{X_{FS}} \tag{4.166}$$

The sensor is sensitive to temperature and let the sensitivity be β mV/°C. The temperature drift is compensated by a signal produced by the microcontroller based on the temperature measured by a temperature sensor. The temperature sensor generates a voltage V_T with a sensitivity of α mV/°C. The temperature sensor output is amplified by amplifier A_2 and then converted to digital signal using an ADC. Let the temperature sensor be calibrated for the range $T_{ref} - T_{max}$. The output voltage range of the temperature sensor is given by

$$V_T = (V_{TFS} - V_{TR}) = \alpha(T_{max} - T_{ref}) \tag{4.167}$$

When this range is amplified by A_2, the range becomes

$$V_T' = A_2(V_{TFS} - V_{TR}) = A_2\alpha(T_{max} - T_{ref}) \tag{4.168}$$

Let the ADC with m-bit word length has a full-scale analog value of V_{FSM}, and then the word length of the m-bit number is given by

$$M = \frac{A_2(V_{TFS} - V_{TR})}{V_{FSM}} 2^m$$

$$= \frac{A_2\alpha(T_{max} - T_{ref})}{V_{FSM}} 2^m \tag{4.169}$$

The voltage error of the sensor V_e is applied to the amplifier A_1 through the K-bit DAC. In order to set the range of the DAC that outputs V_e, let us consider that

$$V_{FSK} = \text{full scale of K-bit DAC.}$$

Therefore

$$\frac{V_{FSK}}{2^K} = \text{resolution of the DAC.}$$

The voltage error V_e is amplified by A_1 and A_2 successively to get

$$V_e'' = A_1A_3V_e \tag{4.170}$$

The word length for this voltage converted by the n-bit ADC is

$$Ne = \frac{A_1A_3V_e}{V_{FSN}} 2^n \tag{4.171}$$

This voltage is fed back by the microcontroller to the DAC so that the word length of the K-bit DAC is given by

$$2^K \geq \frac{A_1A_3Ve}{V_{FSN}} 2^n \tag{4.172}$$

For setting the word length of the n-bit ADC that converts the sensor output to a digital signal, let us consider that the sensor has a temperature drift of voltage given by

$$V_d = \beta\Delta T \tag{4.173}$$

where

$\Delta T = T - T_{ref}$ and T is the temperature of the sensor (°C)
T_{ref} is the reference or calibrated temperature (°C)
β is the temperature sensitivity of the sensor (mV/°C)

The amplified drift voltage in the first stage

$$V_d' = A_1V_d$$

and after the second stage

$$V_d'' = A_1 A_3 V_d \tag{4.174}$$

The word length of the n-bit number generated from the ADC is

$$N = \frac{V_d''}{V_{FSN}} 2^n$$

$$= \frac{A_1 A_3 V_d}{V_{FSN}} 2^n \tag{4.175}$$

where V_{FSN} is the full-scale range of the n-bit ADC.

Example 4.30

The output of a strain gauge–based load sensor corrupted with a temperature drift of 0.45 mV is amplified by a first-stage amplifier with a gain of 60. The signal is then amplified by a second-stage PGA and then fed to an 8-bit ADC with a reference voltage of 5 V. The integer value equivalent of the drift voltage converted by ADC is 37. Determine the gain to be programmed for PGA.

Solution

Given

$$V_d = 0.45\,\text{mV} = 0.45 \times 10^{-3}\ \text{V}$$

$$A_1 = 60$$

$$n = 8$$

$$N = 37$$

$$V_{FSN} = 5.0\ \text{V}$$

From Equation 4.175, the word length equation

$$\frac{A_1 A_3 V_d}{V_{FSN}} 2^n = N$$

$$\Rightarrow \frac{60 A_2 \times 0.45 \times 10^{-3} \times 2^8}{5} = 37$$

$$\Rightarrow A_2 = \text{Gain of PGA} = 26.76 \approx 27$$

Example 4.31

A pressure sensor produces a voltage error of 42.5 mV, which is compensated through the amplifier and ADC of Example 4.30. The equivalent compensating voltage is generated by a DAC having a full-scale reference voltage of 5 V. Determine the integer value for the error voltage and the required word length of the DAC.

Solution
Given

$$A_1 = 60$$
$$A_2 = 26.76$$
$$V_e = 42.45\ \text{mV}$$
$$n = 8$$
$$V_{FSN} = 5\ \text{V}$$

From Equation 4.172 for the DAC word length

$$2^K \geq \frac{A_1 A_2 V_e}{V_{FSN}} 2^n$$

$$2^K \geq \frac{60 \times 26.76 \times 42.45 \times 10^{-3} \times 2^8}{5}$$

$$\geq 3489$$

$$K \geq \frac{\log 3489}{\log 2} \geq 11.76$$

$$\therefore K = 12$$

The integer value is 3489 and word length is 12.

Example 4.32

A temperature drift of 25°C–100°C in a pressure sensor is to be compensated by a microcontroller that receives the temperature signal from a PTC sensor through an 8-bit ADC with $V_{ref} = 5\,\text{V}$. The PTC sensor has a sensitivity of 0.5 mV/°C. The temperature signal is amplified by an OPAMP before feeding to the ADC. Determine the required gain of the amplifier if the integer value of the converted voltage is to be 192.

Solution
Given

$$T_{FS} = 100°\text{C}$$
$$T_{ref} = 25°\text{C}$$
$$\alpha = 0.5\ \text{mV/°C}$$
$$m = 8$$
$$M = 192$$
$$V_{FSM} = 5\ \text{V}$$

Using Equation 4.169, the integer value of the voltage

$$M = \frac{A_2 \alpha (T_{FS} - T_{ref})}{V_{FSM}} 2^m$$

$$\frac{A_2 \times 0.5 \times 10^{-3} \times (100 - 25) \times 2^8}{5} = 192$$

$$\therefore A_2 = \frac{5 \times 192}{0.5 \times 10^{-3} \times 75 \times 28}$$

$$= 100$$

Gain of the amplifier is 100.

Program routines of microcontroller

The microcontroller is programmed to carry out the calibration operation as per the following steps:

Step 1:

The main sensor is made free from input, i.e., the sensor is unstimulated and subjected to temperature variation from T_{ref} to T_{max}. The n-bit drift voltages from the ADC is continuously scanned by microcontroller. Simultaneously, the n-bit temperature signal V_T for temperature variation from $T_{ref} - T_{max}$ is also simultaneously scanned by microcontroller.

Step 2:

Using the n-bit drift voltages and m-bit temperature-dependent voltages, a LUT is prepared. For each pair of temperature output (M_T) and drift voltage (V_T), the corresponding PGA gains are estimated. The corresponding values for PGA gain control are stored in the EPROM in the form of a LUT, where the address for each EPROM entry is the corresponding temperature output (M_T). Similarly, for each value of V_e, the corresponding values to be applied to the K-bit DAC are stored in the EPROM.

4.8.4.1 Frequency Control

Sensors based on a resonant physical phenomenon generate an output frequency proportional to a measurand affecting the oscillation frequency. Such sensors are called resonant sensors such as pressure sensor quartz resonators, quartz digital thermometers, vibrating wire strain gauge, quartz microbalance, surface acoustic wave (SAW) resonators, etc. Resonant structures of single crystal silicon are particularly suitable for IC integration [37,38].

Quartz resonator sensors show good accuracy at constant temperature, but they are susceptible to drifts with temperature. Oven-controlled sensors can compensate thermal drift with a frequency stability ($\Delta f/f$) of the order of 10^{-8} in the range from 0°C to 50°C, however oven control is not possible in all cases of sensors.

Quartz digital thermometers, when precisely cut, show a very stable relationship between temperature and frequency given by

$$f = f_0[1 + a(T - T_0) + b(T - T_0)^2 + c(T - T_0)^3] \tag{4.176}$$

where

T_0 is the arbitrary reference temperature (usually 25°C)

f_0 is the frequency of oscillation at reference temperature (Hz)

a, b, and c are the constants that depend on the cutting orientation of the quartz

Such temperature sensors output frequencies ranging from about 256 kHz to 28 MHz with first-order temperature coefficient of 19×10^{-6}/°C to 90×10^{-6}/°C and temperature sensitivities up to 1000 Hz/°C.

The mass-dependent quartz microbalance sensors experience a shift in oscillation frequency f_0 by an amount

$$\Delta f \cong -f_0^2 \frac{\Delta m/A}{N\rho} \tag{4.177}$$

where

Δm is the change in mass on the crystal surface area A

N is a constant

ρ is the density of the material

Microbalance discs with a diameter of 10–15 mm and thickness of 0.1–0.2 mm provides a resonant frequency of 5–20 MHz. A vibrating wire strain gauge sensor is typically a taut string or wire of length l with a mechanical force F applied, producing a lower transverse frequency given by

$$f = \frac{1}{2l}\sqrt{\frac{F}{\mu}} \tag{4.178}$$

where μ is the longitudinal mass density (mass/length).

The oscillation frequency is measured with a variable reluctance sensor. Temperature variation causes change in wire length giving a drift in the oscillation frequency. In SAW sensors, two interleaved metal electrodes are placed on the surface of a piezoelectric material and when a voltage of frequency is applied to the electrode, the wave propagates over the surface. Another electrode pair detects the surface wave that travels at a velocity determined by the surface deformation and temperature that determines the electric property of the material.

SAW sensors are basically used in a feedback loop where the oscillation frequency is fed back to an amplifier. The total phase shift of the feedback loop is given by

$$\varnothing_T = \varnothing_0 + \varnothing_s + \varnothing_a \tag{4.179}$$

where

$\emptyset_0$ is the phase shift of the resonator

$\emptyset_s$ is the phase shift due to change in surface material properties temperature

$\emptyset_a$ is the phase shift in the amplifier input impedance

In all the sensors discussed above, temperature is a major interfering input that causes a shift in the frequency of oscillations. Temperature compensation in such sensors is achieved by the following methods.

4.8.4.1.1 *Integrating a Resonant Temperature Sensor*

Drifts in oscillation frequency due to temperature in quartz resonant sensors can be eliminated by integrating a resonant quartz temperature sensor on the same quartz surface subjected to the same temperature variation. In variable oscillation frequency generating sensors, the frequency of oscillation can be related to the variable electrical parameter (x) by

$$f = Kx \tag{4.180}$$

or

$$f = \frac{K}{x} \tag{4.181}$$

where K is a constant of proportionality. When the variable electrical parameter x changes in response to the physical measurand, the change in x can be expressed as either

$$x = x_0(1 \pm \alpha) \tag{4.182}$$

or

$$x = \frac{x_0}{(1 \pm \alpha)} \tag{4.183}$$

Substituting Equations 4.182 and 4.183 in Equations 4.180 and 4.181, we get the frequency of oscillation as either

$$f = Kx_0(1 \pm \alpha) \tag{4.184}$$

or

$$f = \frac{Kx_0}{(1 \pm \alpha)} \tag{4.185}$$

or

$$f = \frac{K}{x_0(1 \pm \alpha)} \tag{4.186}$$

or

$$f = \frac{K(1 \pm \alpha)}{x_0} \tag{4.187}$$

In Equations 4.184 and 4.187, the output frequency is directly proportional to the measurand, while Equations 4.185 and 4.186 show an inversely proportional relationship. The following cases can be used for exploiting temperature compensation:

Case 1: Considering a resonant sensor that measures a measurand variable with a sensitivity of α_s, and the sensor experiences a temperature drift with a sensitivity of α_t following the proportionality Equation 4.184, the frequency of oscillation can be written as

$$f_s = Kx_0(1+\alpha_s)(1+\alpha_t) \tag{4.188}$$

For compensation of the temperature-dependent drift in the output frequency, a resonant temperature sensor is integrated to the main sensor with identical parameters to the main sensor. The frequency output of the temperature sensor can be written as

$$f_t = Kx_0(1+\alpha_t) \tag{4.189}$$

Taking the difference of the two frequencies

$$\begin{aligned} f_s - f_t &= Kx_0(1+\alpha_t+\alpha_s+\alpha_s\alpha_t) - Kx_0(1+\alpha_t) \\ &= Kx_0(\alpha_s+\alpha_s\alpha_t) \end{aligned} \tag{4.190}$$

Neglecting the smaller term

$$f_s - f_t = Kx_0\alpha_s \tag{4.191}$$

Therefore, the difference in frequencies is free from the temperature sensitivity term. The schematic diagram of Figure 4.54 shows how the two frequencies can be subtracted to output the quantity of Equation 4.188. In this scheme, the two frequencies are counted for a duration T_N during which the difference is generated, which is given by

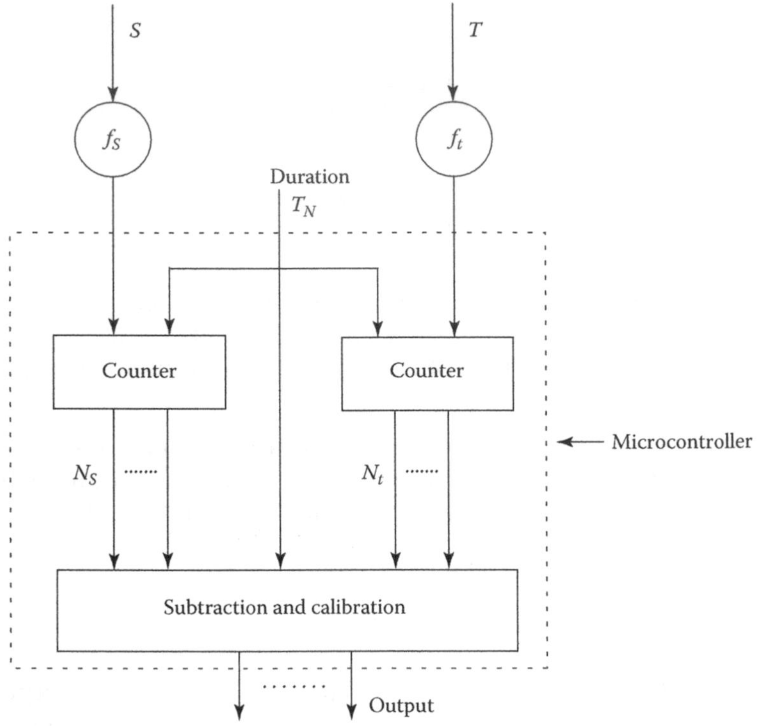

FIGURE 4.54
Microcontroller-based drift compensation for case-I.

$$\begin{aligned} N_s - N_t &= (f_s - f_t)T_N \\ &= Kx_0 T_N \alpha_s \end{aligned} \tag{4.192}$$

Alternatively, f_t can be measured first and then the counter is loaded with 2's complement of N_t and then measure f_s [39]. The microcontroller performs the subtraction of the measured counts and outputs the calibrated value.

Example 4.33

A quartz microbalance humidity sensor with a reference frequency of 5 MHz at 0% RH, which varies linearly with humidity has a sensitivity of -4.347×10^{-6}/%RH. The sensor has a temperature drift in output frequency of 50×10^{-6}/K. The temperature drift is compensated by integrating a quartz temperature sensor into the humidity sensor, having same reference frequency and temperature sensitivity.

The frequencies of both the sensors are measured by two independent counters and their difference is calculated within a duration of 500 ms. Determine the output counts for

1. A relative humidity of 70% with no temperature variation.
2. A relative humidity of 70% with a temperature variation of 100°C from calibrated temperature. Also find the error in count.
3. Error in count if the compensation scheme is not used.

Solution

Given:

$$f_{s0} = f_{t0} = 5\,\text{MHz}$$

$$\text{RH} = 70\%$$

$$\alpha_s = -4.347 \times 10^{-6}/\%\text{RH}$$

$$\alpha_t = 50 \times 10^{-6}/\text{K}$$

$$T_N = 500\text{ ms}$$

$$\Delta T = 0\text{ K (for (1)) and } 10\text{ K (for (2))}$$

1. The humidity sensor output frequency

$$f_s = f_{s0}(1 + \alpha_s \text{RH})(1 + \alpha_s \Delta T)$$

$$= 5 \times 10^6 (1 - 4.347 \times 10^{-6} \times 70)(1 + 0)$$

$$= 4{,}998{,}478\text{ Hz}$$

The temperature sensor output frequency

$$f_t = f_{t0}(1 + \alpha_t \Delta T)$$

$$= 5 \times 10^6 (1 + 50 \times 10^{-6} \times 0)$$

$$= 5{,}000{,}000\text{ Hz}$$

$$\therefore (f_t - f_s) = 1{,}522\text{ Hz}$$

Difference in count

$$= N_t - N_s = (f_t - f_s)T_N$$

$$= 1522 \times 500 \times 10^{-3}$$

$$= 761$$

2.

$$
\begin{aligned}
f_s &= f_{s0}(1+\alpha_s \text{RH})(1+\alpha_t \Delta T) \\
&= 5\times10^6(1-4.347\times10^{-6}\times70)(1+50\times10^{-6}\times10) \\
&= 5{,}000{,}977 \text{ Hz} \\
f_t &= 5\times10^{-6}(1+50\times10^{-6}\times10) \\
&= 5{,}002{,}500 \text{ Hz}
\end{aligned}
$$

$$\therefore (f_t - f_s) = 1{,}523 \text{ Hz}$$

$\therefore$ Difference in count

$$
\begin{aligned}
&= (f_t - f_s)\times500\times10^{-3} \\
&= 1523\times500\times10^{-3} \\
&= 761.38
\end{aligned}
$$

Error in count

$$
\begin{aligned}
&= 761.38 - 761.00 \\
&= 0.38
\end{aligned}
$$

3. The output frequency only due to humidity

$$
\begin{aligned}
f_{S_H} &= f_{s0}(1+\alpha_s \text{RH}) \\
&= 5\times10^6(1-4.347\times10^6\times70) \\
&= 4{,}998{,}478 \text{ Hz}
\end{aligned}
$$

The output frequency with temperature variation of 10 K

$$
\begin{aligned}
f_{S_{Ht}} &= f_{S_H}(1+\alpha_t\times\Delta t) \\
&= 4{,}998{,}478\,(1+50\times10^6\times10) \\
&= 5{,}000{,}977 \text{ Hz}
\end{aligned}
$$

The error in frequency

$$
\begin{aligned}
&= (5{,}000{,}977 - 4{,}998{,}478) \\
&= 2{,}499
\end{aligned}
$$

Error in count

$$=2499 \times T_N$$
$$= 2499 \times 500 \times 10^{-3}$$
$$= 1249.50$$
$$\cong 1,250$$

Case 2: When a sensor follows the inverse proportionality relationship between the output frequency and the electrical parameters given by Equation 4.186, the equation takes the form

$$f_s = \frac{K}{x_0(1+\alpha_s)(1+\alpha_t)} \tag{4.193}$$

Considering that the temperature drift is compensated by a resonant temperature sensor that follows the Equation 4.184, then its output oscillation is given by

$$f_t = Kx_0(1+\alpha_t) \tag{4.194}$$

Remember that the temperature sensor should be identical to the main sensor, as in Case 1. If the output oscillations are measured by two counters for a time duration of T_N, the output counts are found to be

$$N_s = f_s T_N = \frac{KT_N}{X_0(1+\alpha_S)(1+\alpha_t)} \tag{4.195}$$

and

$$N_t = f_t T_N = KX_0 T_N(1+\alpha_t) \tag{4.196}$$

The multiplication of the counter outputs gives

$$N_s N_t = \frac{(KT_N)^2}{1+\alpha_S} \tag{4.197}$$

The above expression eliminates the temperature sensitivity term providing ability to compensation of temperature drift to the sensor. The microcontroller-based measurement and compensation scheme is shown in Figure 4.55.

Example 4.34

The oscillation frequency of a vibrating wire gauge displacement sensor is given by

$$f = \frac{1}{2l}\sqrt{\frac{F}{\mu}}$$

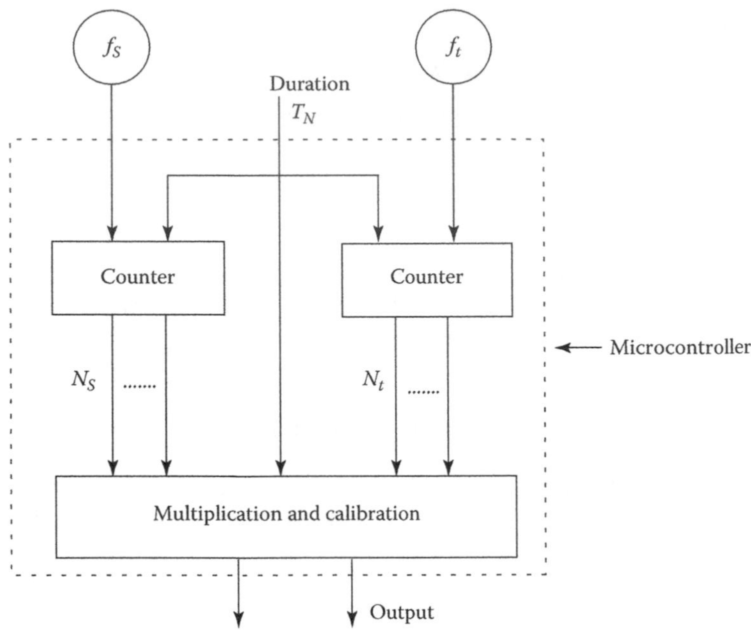

FIGURE 4.55
Microcontroller-based drift compensation for case-II.

where
l is the length of the wire, which is proportional to the displacement
F is the force applied
μ is the longitudinal mass density

The sensor outputs a reference frequency of 2 kHz, which shifts by 0.1 kHz over an operating range of 0–10 mm. The sensor wire experiences a linear thermal expansion with a frequency shift sensitivity of 3×10^{-3}/K. The temperature effect is compensated using a quartz resonant temperature sensor having the same temperature sensitivity as that of the wire gauge. The output frequencies are measured by two counters for duration of 200 ms and the count values are multiplied. Determine

1. The output counts for an input displacement of 4.5 mm at reference temperature.
2. The output counts for the same displacement of (1) and a temperature variation of 10°C.
3. The error in output counts if temperature compensation is not used.

Solution

$$\text{Given: } f_{os} = f_{ot} = 2\,\text{kHz}$$

$$x(FS) = 10\text{ mm}$$

$$\Delta F(FS) = 0.1\,\text{kHz}$$

$$\alpha_t = 3\times10^{-3}/K$$

$$x_i = 4.5\,\text{mm} \quad \text{and} \quad T = 0°\text{C for (1)}$$

$$x_i = 4.5\,\text{mm} \quad \text{and} \quad T = 10°\text{C for (2)}$$

Sensitivity of wire gauge

$$S = \frac{\Delta F(FS)}{X(FS)} = \frac{0.1\times10^3}{10} = 10\,\text{Hz/mm}$$

$$\alpha_s = \frac{s}{f_{os}} = \frac{10}{2\times10^3} = 5\times10^{-3}\ \text{mm}^{-1}$$

1.

$$f_s = \frac{f_0}{(1+\alpha_s x_i)(1+\alpha_t T)}$$

$$= \frac{2\times10^3}{(1+5\times10^{-3}\times4.5)(1+3\times10^{-3}\times0)}$$

$$= 1,955.99\ \text{Hz}$$

$$N_s = f_s T_N$$

$$= 1,955.99\times200\times10^{-3}$$

$$= 391.198$$

$$f_t = f_0(1+\alpha_t T)$$

$$= 2\times10^3(1+3\times10^{-3}\times0)$$

$$= 2,000\ \text{Hz}$$

$$N_t = 2,000\times200\times10^{-3}$$

$$= 400$$

$$\therefore N_s N_t = 391.198\times400$$

$$= 156,479$$

2.

$$f_s = \frac{2\times10^3}{(1+5\times10^{-3}\times4.5)(1+3\times10^{-3}\times0)}$$

$$= 1,899.0196\ \text{Hz}$$

$$N_s = 1899.0196 \times 200 \times 10^{-3}$$
$$= 379.8039$$
$$f_t = 2 \times 10^3 (1 + 3 \times 10^{-3} \times 10)$$
$$= 2{,}060$$
$$N_t = 2{,}060 \times 200 \times 10^{-3}$$
$$= 412$$
$$\therefore N_s N_t = 379.8039 \times 412$$
$$= 156{,}479$$

3. Ideal counts (with no temperature effect)

$$N_{s0} = 391.198$$

Nonideal counts (with temperature effect)

$$N_{st} = 379.8039$$
$$\text{Error} = 391.198 - 379.8039$$
$$= 11.39$$
$$\text{Percentage error} = \frac{11.39}{391.198} \times 100 = 2.91\%$$

4.8.4.2 Sensitivity Slope Control

Although we consider that the sensitivity of a sensor should be constant for all ambient temperature conditions, it is not true for many sensors. Such a situation arises when the electrical parameter (resistance, capacitance, or inductance) changed by the measurand variable and the temperature are the same. In such cases, the sensitivity of the sensor can be considered as a function of temperature. The sensitivity curve for the above two cases are shown in Figure 4.56.

Consider that a sensor with input measurand variable x outputs an electrical parameter y to give a sensitivity (α_s) given by

$$y = \alpha_s x \quad (4.198)$$

when the output is directly proportional to the measurand, and

$$y = \frac{\alpha_s}{x} \quad (4.199)$$

when the output is inversely proportional to the measurand.

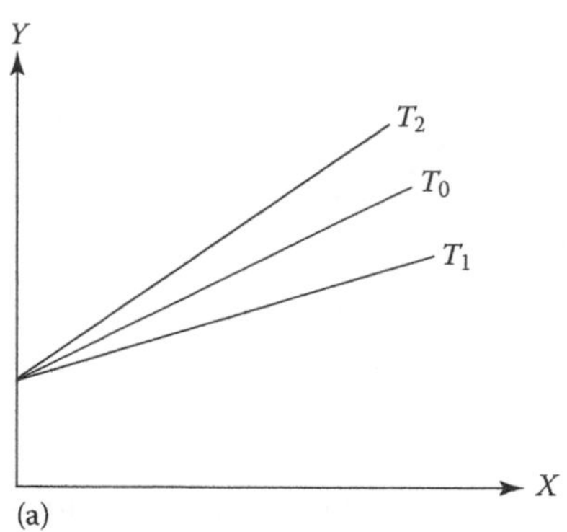

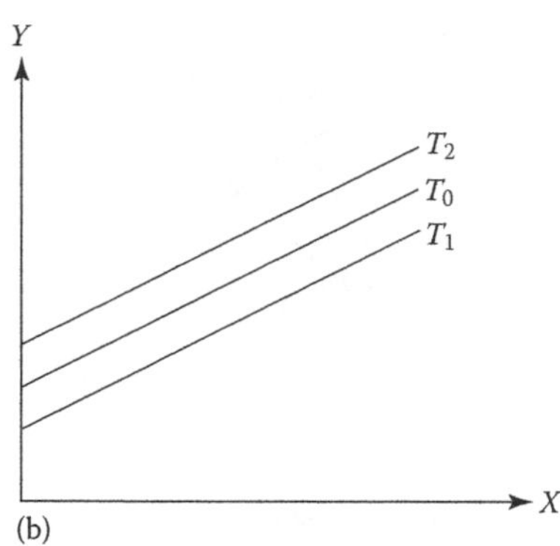

FIGURE 4.56
Sensitivity characteristics for different ambient temperatures with (a) variable slopes and constant offset and (b) constant slopes and variable offsets.

Now we will not consider α_s to be constant as we do in most sensors, instead we analyze the ways in which α_s changes with temperature. There may be two distinct ways in which α_s is varied by temperature and they are

$$\alpha_s = \alpha_{s0}(1+\alpha_t T) \tag{4.200}$$

when the sensitivity changes by a direct relationship with temperature, and

$$\alpha_s = \frac{\alpha_{s0}}{(1+\alpha_t T)} \tag{4.201}$$

when the sensitivity changes inversely with temperature.

In the above Equations 4.200 and 4.201, α_{s0} is the sensitivity of the sensor at reference temperature, α_t is the temperature coefficient of sensor sensitivity, and T is the temperature change from the reference temperature. From Equation 4.198 through 4.201, we distinctly arrive at four equations

$$y = \alpha_{s0}(1+\alpha_t T)x \tag{4.202a}$$

$$y = \frac{\alpha_{s0}x}{(1+\alpha_t T)} \tag{4.202b}$$

$$y = \frac{\alpha_{s0}(1+\alpha_t T)}{x} \tag{4.202c}$$

$$y = \frac{\alpha_{s0}}{x(1+\alpha_t T)} \tag{4.202d}$$

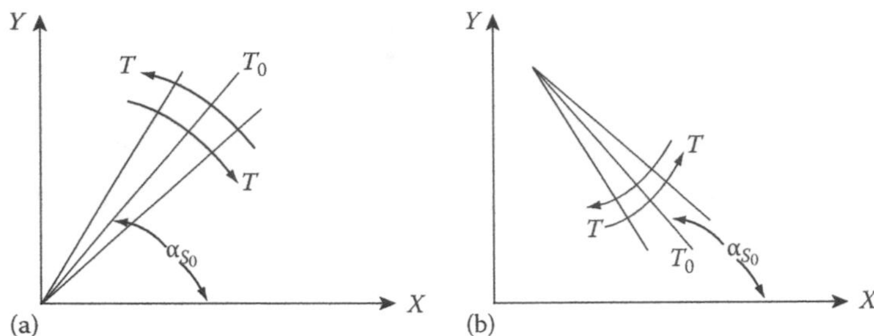

FIGURE 4.57
Sensitivity characteristics for change in temperature for (a) sensor output with direct variation with measurand and (b) sensor output with inverse variation with measurand.

The sensitivity characteristics for the four Equations 4.202a through 4.202d are shown in Figure 4.57a and b.

The above conditions are equally applicable to sensors that produce an electrical voltage, current, or pulses, the values of which are also modified by temperature. Now we will derive the temperature dependency on sensitivity in a resistive sensor for two different cases.

Case 1: The measurand (x) and temperature (T), both change the length of the resistive element.

Considering that the length of resistive sensor l is changed by the measurand x and temperature T, the increased length is given by

$$l = Kx(1+\alpha_t T) \tag{4.203}$$

where $K = l/x$ is a proportionality constant relating the length of the sensor to the measurand variable. The resistance of the sensor at stimulated condition is given by

$$R = \frac{\rho Kx(1+\alpha_t T)}{A} \tag{4.204}$$

which is similar to Equation 4.202a. Now taking the derivative of R with respect to x, the sensor sensitivity is found as

$$\alpha_s = \frac{\delta R}{\delta x} = \frac{\rho K(1+\alpha_t T)}{A} = \frac{\rho K}{A} + \frac{\rho K \alpha_t}{A} \cdot T \tag{4.205}$$

Equation 4.205 shows that the sensitivity α_s is either constant at a value $\rho K/A$ for $T=0$ or changes linearly with T with a slope $\rho K\alpha_t/A$. The sensitivity characteristics is shown in Figure 4.58a.

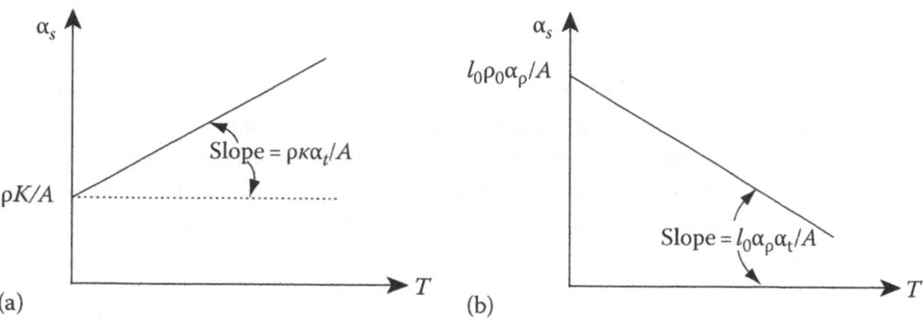

FIGURE 4.58
Sensitivity characteristics with temperature variation for change of (a) length by the measurand and (b) resistivity by the measurand.

Case 2: The measurand (x) changes the resistivity while the temperature changes the length of the resistive element.

A resistive sensor may operate on the principle of variation of resistivity (ρ) of the sensor material by the measurand variable. Now consider the change in ρ due to stimulation by x while temperature continues to act on the length as in Case 1. The equation of increased resistivity and length is given by

$$\rho = \rho_0(1+\alpha_\rho x)$$

$$l = l_0(1+\alpha_t T)$$

where
ρ_0 is the resistivity at unstimulated condition
l_0 is the length of the sensor at reference temperature
α_ρ is the sensitivity of resistivity

The resistance of the sensor is given by

$$R = \frac{\rho_0 l_0(1+\alpha_\rho x)(1+\alpha_t T)}{A} = \frac{\rho_0 l_0(1+\alpha_t T)}{A} + \frac{\rho_0\alpha_\rho l_0(1+\alpha_t T)}{A}x \tag{4.206}$$

Hence the sensitivity is

$$\alpha_s = \frac{\partial R}{\partial x} = \frac{\rho_0 l_0\alpha_\rho(1+\alpha_t T)}{A} \tag{4.207}$$

$$= \frac{l_0\rho_0\alpha_\rho}{A} + \frac{l_0\rho_0\alpha_\rho\alpha_t}{A}T \tag{4.208}$$

The above equation states that the sensitivity of the sensor is fixed at a value of $\rho_0 l_0\alpha_\rho/A$ with no temperature variation; however, it reduces with temperature with a gradient of $\rho_0 l_0\alpha_\rho\alpha_t/A$.

The sensitivity characteristics are shown in Figure 4.58b.

Example 4.35

A wire wound potentiometer with a resistance of 1260 Ω at reference temperature (20°C) is made of 100 turns of nickel-chrome wire of a total length of 1 m. The potentiometer can operate in a displacement range of 0–50 mm. For temperature compensation of the sensor, the temperature is measured and fed to a microcontroller through an ADC to adjust the slope of the potentiometer sensitivity.

Determine

1. The sensitivity of the potentiometer at temperature variation of 2°C, 5°C, and 10°C from reference temperature.
2. The resistance of the potentiometer at the above temperatures and input displacement of 20 mm.
3. The error in measurement if temperature compensation is not adopted for a displacement of 20 mm and 10°C change in temperature.

Take the resistivity and thermal expansion coefficient of the wire material as 6.8 μΩm and, 2×10^{-6}/°C respectively.

Solution

Given data:

$R = 1260\,\Omega$

$N = 100$

$l = 1$ m

$x(FS) = 50\text{ mm} = 50 \times 10^{-3}$ m

$\rho = 6.8\,\mu\Omega m$

$\alpha_t = 20 \times 10^{-6}/°C$

$T = 2°C$, 5°C, and 10°C for (1)

$X = 25\text{ mm} = 25 \times 10^{-3}$ m for (2)

1. For full stroke displacement of 50 mm, the length of sensor wire is 1 m

$$\therefore K = \frac{1\text{ m}}{50\text{ mm}} = \frac{1 \times 10^3}{50} = 20$$

We know that $R = (\rho l_0 / A)$

$$\therefore A = \frac{\rho l_0}{R} = \frac{6.8 \times 10^{-6} \times 1}{1260} = 5.39 \times 10^{-9}\text{ m}^2$$

Using Equation 4.205

$$\alpha_s = \frac{\rho K}{A} + \frac{\rho K \alpha_t}{A} T$$

Putting the values in the above equation

$$\alpha_s = \frac{6.8 \times 10^{-6} \times 20}{5.39 \times 10^{-9}} + \frac{6.8 \times 10^{-6} \times 20 \times 20 \times 10^{-6}}{5.39 \times 10^{-9}} T$$

$$= 25,232 + 0.5T\ \Omega/\text{m (displacement)}$$

From the above equation

$$\text{At } T = 0°\text{C}, \ \alpha_{s0} = 25{,}232\ \Omega/\text{m}$$

$$T = 2°\text{C}, \ \alpha_{s2} = 25{,}233\ \Omega/\text{m}$$

$$T = 10°\text{C}, \ \alpha_{s10} = 25{,}237\ \Omega/\text{m}$$

2. Using Equation 4.204

$$R = \frac{\rho K\, X}{A}(1+\alpha_t T) = \frac{\rho K\, X}{A} + \frac{\rho K\, X\alpha_t}{A}T$$

$$= \frac{6.8\times10^{-6}\times20\times20\times10^{-3}}{5.39\times10^{-9}} + \frac{6.8\times10^{-6}\times20\times20\times10^{-3}\times20\times10^{-6}}{5.39\times10^{-9}}T$$

$$= 504.63 + 0.01T$$

From the above equation

$$\text{At } T = 0°\text{C}, \quad R_0 = 504.63\,\Omega$$

$$T = 2°\text{C}, \quad R_2 = 504.65\,\Omega$$

$$T = 10°\text{C}, \quad R_{10} = 504.73\,\Omega$$

3. For $x = 20$ mm and $T = 10°$C

The actual value of resistance = 504.73 Ω
Ideal value of resistance = 504.63 Ω

$$\text{Percentage error} = \frac{504.73 - 504.63}{504.63}$$

$$\cong 0.02\%$$

Example 4.36

A semiconductor strain gauge of resistance 1200 Ω at reference temperature (20°C) and gauge factor of 200 has a temperature sensitivity to resistivity of 5 με/°C. The temperature effect is compensated by adjusting the sensitivity slope in a microcontroller based on the temperature values. Derive the expression for resistance and sensitivity of the strain gauge. Using the derived expressions, determine

1. The sensitivity when there is no temperature effect and resistance for an applied strain of 300 μ.
2. The resistance and sensitivity for an applied strain of 300 μ and temperature variation of 10°C.
3. The resistance of the gauge at a strain of 500 μ and a temperature variation of 10°C.

Solution

Given data: $R_0 = 1200\Omega$

Gauge factor $= \lambda = 200$

$$\alpha_t = 5\times10^{-6}\text{ (strain)/°C}$$

Strain $= \varepsilon = 300\,\mu$

Derivation of expressions for resistance and resistivity:

The resistance of an element is given by

$$R = \frac{\rho l}{A} \tag{E4.36.1}$$

where R and ρ are resistance and resistivity of the gauge. Since in a semiconductor gauge, the applied strain changes the resistivity, the resistivity can be expressed as

$$\rho_\epsilon = \rho_0(1+\lambda\varepsilon)$$

Since temperature also changes resistivity, the combined resistivity is given by

$$\rho_{\epsilon t} = \rho_0(1+\lambda\varepsilon)(1+\alpha_t T) \tag{E4.36.2}$$

Using Equations E4.36.1 and E4.36.2

$$R = \frac{\rho_0 l(1+\lambda\varepsilon)(1+\alpha_t T)}{A}$$

$$= R_0(1+\lambda\varepsilon)(1+\alpha_t T) \tag{E4.36.3}$$

where $\rho_0 l/A$ is the original resistance of the gauge

where ρ_0 is the original resistivity of the gauge

Expanding Equation E4.36.3

$$R = R_0(1+\alpha_t T) + R_0\lambda(1+\alpha_t T)\varepsilon \tag{E4.36.4}$$

Differentiating R w.r.t ε

$$\frac{dR}{d\varepsilon} = R_0\lambda(1+\alpha_t T)$$

$$\alpha_s = R_0\lambda + R_0\lambda\alpha_t T \tag{E4.36.5}$$

1. For $T = 0°C$ and $\varepsilon = 300\,\mu$
   Using Equation E4.36.4,

$$R = 1,200(1 + 5\times10^{-6}\times0) + 1,200\times1,200\times300\times10^{-6}(1 + 5\times10^{-6}\times0)$$

$$= 1,200 + 72$$

$$= 1,272\,\Omega$$

$$\alpha_s = 1,200\times200 + 1,200\times200\times5\times10^{-6}\times0$$

$$= 240,000\,\Omega/\text{strain}$$

$$= 0.24\,\Omega/\mu\,\text{strain}$$

2. For $T = 10°C$ and $\varepsilon = 300\,\mu$

$$R = 1,200(1 + 5\times10^{-6}\times10) + 1,200\times200\times300\times10^{-6}(1 + 5\times10^{-6}\times10)$$

$$= 1,201 + 72.0036 = 1273.0036\,\Omega$$

$$\alpha_s = 1,200\times200 + 1,200\times200\times5\times10^{-6}\times10$$

$$= 240,240\,\Omega/\text{strain}$$

$$= 0.2402\,\Omega/\mu\,\text{strain}$$

3. For $T = 10°C$ and $\varepsilon = 300\,\mu$
   Using α_s of (2)

$$\Delta R = \alpha_s\varepsilon$$

$$= 240,240\times500\times10^{-6}$$

$$= 120.12\,\Omega$$

$$\therefore R = R_0 + \Delta R$$

$$= 1,200 + 120.12$$

$$= 1,320.12\,\Omega$$

The technique of slope adjustment has been applied for temperature compensation in pH electrode systems [40], where an LM335 IC thermal sensor and PIC16F877A microcontroller is used. The thermal sensor continuously measures the temperature of the pH electrode and it is fed to the microcontroller. Based on the value of the temperature, the microcontroller calculates the sensitivity and the pH is calibrated. For example, for varying temperatures at three values, say 0°C, 25°C, and 100°C, the slope angles are shown in Figure 4.59. The slope of the pH electrode at 250°C was found as 59.16 mV/pH with 0 mV at pH of 7 (neutral). The slope varies with temperature as at 100°C, 74 mV/pH and at 0°C, 54 mV/pH.

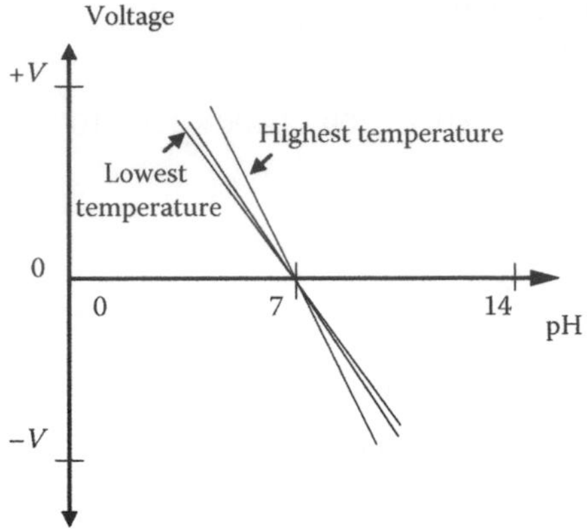

FIGURE 4.59
Slope angles of pH characteristics.

Example 4.37

A pH electrode develops the following sensitivities at two different temperatures linearly:

$$0°C : 54 mV/pH$$

$$100°C : 75 mV/pH$$

The electrode produces an output voltage of 0 mV at pH of 7 (neutral) at all temperatures.

1. Derive an equation that calculates the pH for any value of output voltage.
2. Determine the pH for an output voltage of 110 mV at 25°C.

Solution

$$T_{max} = 100°C \quad \text{and} \quad \alpha_{100} = 74 \text{ mV/pH}$$

$$T_{min} = 0°C \quad \text{and} \quad \alpha_0 = 54 \text{ mV/pH}$$

$$V_7 = 0 \text{ mV}; \quad V_0 = 110 \text{ mV} \quad \text{for (2)}$$

1. From the two values of α at two temperatures

$$\alpha = 54 + \frac{(74 - 54)}{100} \times T$$

$$= (54 + 0.2T) \text{ mV/pH} \qquad \text{(E4.37.1)}$$

At $T=0°C$, and pH$=7$ the output voltage

$$V_0 = 54\times 7\times C, \quad \text{where } C = \text{intercept at mV axis}$$

$$\therefore C_0 = -54\times 7 = -378\text{ mV}$$

At $T=100°C$, pH$=7$ the output voltage

$$V_0 = 74\times 7 + c = 0$$

$$\therefore C_{100} = -74\times 7 = -518\text{ mV}$$

∴ From the two values of C at two temperatures

$$C = -378 - \frac{(518-378)}{100}T$$

$$= -378 - 1.4T$$

$$= -(378+1.4T)\text{mV} \qquad \text{(E4.37.2)}$$

The equation of V_0 is

$$V_0 = \alpha\times \text{pH} + C$$

Putting the values of α and C from Equations E4.37.1 and E4.37.2

$$V_0 = (54+0.2T)\text{pH} - (378+1.4T)$$

$$\therefore \text{pH} = \frac{[V_0 + (378+1.4T)]}{(54+0.2T)} \qquad \text{(E4.37.3)}$$

2. Using Equation E4.37.3

$$\text{pH} = \frac{110+(378+1.4\times 25)}{(54+0.2\times 25)}$$

$$= 8.86$$

4.9 Indirect Sensing

In measurement systems, a situation often arises when some important variables are difficult to be measured due to unavailability of an appropriate sensor or high noise level. Assistance in solving the problems created by unavailability of sensor for certain process variables are important issues in intelligent sensor technology. The soft sensors discussed in Section 4.4 are one solution to this problem. A soft sensor uses a model of the process or plant and the model is simulated in the software to generate the measurand. With the help of secondary measurements (the inputs and the outputs), the model estimates the process parameters when the hard sensor is available.

Indirect sensing is another solution to the problem of unavailability of a direct sensor. A soft sensor estimates the model parameters using the secondary measurements and determines the unknown primary measurand, while indirect sensing estimates the variable through other measurable variables using one of the following techniques:

1. Least square parameter estimation
2. Spectral analysis–based parameter estimation
3. Fuzzy logic based
4. ANN based

Strictly speaking, a soft sensor estimates the model parameters using the hard sensor measurements when it is available while an indirect sensor estimates the model parameters using some other indirect measurements. Therefore, a soft sensor is a temporary solution when the hard sensor is temporarily unavailable while an indirect sensor is a permanent solution for a measurand when the hard sensor is not at all available. This section discusses the first three approaches and the ANN based approach has already been discussed in Section 4.4.

4.9.1 Least Square Parameter Estimation

Consider a system where the measurand variable y_x is related to a local set of parameter vector α_x and the corresponding indirect measurement vector $\varnothing_x$ by the vector relational equation

$$y_x = \varnothing_x^T \alpha_x \tag{4.209}$$

Now we define the system measurement variable y related to the system parameter vector α and the corresponding indirect measurement vector $\varnothing$ given by

$$y = \varnothing^T \alpha \tag{4.210}$$

where

$$\varnothing^T = [\varnothing_1 \quad \varnothing_2 \quad \varnothing_3 \quad \cdots \quad \varnothing_i]$$

$$\alpha^T = [\alpha_1 \quad \alpha_2 \quad \alpha_3 \quad \cdots \quad \alpha_i]$$

$$\text{so that } \varnothing_x \in \varnothing$$
$$\text{or} \in f(\varnothing)$$

and $\alpha_x \in \alpha$
or $\alpha_x \in f(\alpha)$

The system measurement vector Y and the indirect measurement matrix $\varnothing$ are constructed using N measurements as

$$Y = \begin{bmatrix} y(1) \\ y(2) \\ \vdots \\ y(N) \end{bmatrix} \tag{4.211}$$

$$\varnothing = \begin{bmatrix} \varnothing_1(1) & \varnothing_2(1) & \cdots & \varnothing_i(1) \\ \varnothing_1(2) & \varnothing_2(2) & \cdots & \varnothing_i(2) \\ \varnothing_1(N) & \varnothing_2(N) & \cdots & \varnothing_i(N) \end{bmatrix} \tag{4.212}$$

The least square estimate of the system parameter vector $\hat{\alpha}$ is given by

$$\hat{\alpha} = \{\varnothing^T \varnothing\}^{-1} \varnothing^T Y \tag{4.213}$$

where $\hat{\alpha} = \begin{bmatrix} \hat{\alpha}_1 & \hat{\alpha}_2 & \cdots & \hat{\alpha}_i \end{bmatrix}$
The local parameter vector α_x is constructed from $\hat{\alpha}$ given by

$$\alpha_x^T = \begin{bmatrix} \alpha_{x1} & \alpha_{x2} & \cdots & \alpha_{xj} \end{bmatrix} \tag{4.214}$$

and the local indirect measurement vector is formed

$$\varnothing_x^T = [\varnothing_{x1} \quad \varnothing_{x2} \quad \cdots \quad \varnothing_{xj}] \tag{4.215}$$

so that $j \leq i$

Now the measurand y_x is estimated from

$$y_x = \varnothing_x^T \alpha_x \tag{4.216}$$

The procedure of the least square estimation for indirect sensing is illustrated by an example discussed below.

Example 4.38

In the circuit shown in Figure 4.61, the measured variables are the supply voltage V_i, the circuit current I, and the charge accumulated on the positive plate of the capacitor q.

1. Propose a least square estimation technique to estimate the values of R and C.
2. Estimate the frequency of the input voltage from the measured data and the parameters.
3. Write a MATLAB® program for (1) and (2).

Solution

1. The circuit KVL equation can be written as

$$V_i = IR + \frac{1}{C}\int I\,dt$$

Since $\int I dt = q$

$$V_i = IR + \frac{1}{C}q$$

Writing the above equation in vector form

$$V_i = \begin{bmatrix} R & \frac{1}{C} \end{bmatrix} \begin{bmatrix} I \\ q \end{bmatrix}$$

which is of the form

$$y = \varnothing^T \alpha, \text{ where } y = V_i, \varnothing^T = [I \quad q] \text{ and}$$

$$\alpha^T = \begin{bmatrix} R & \frac{1}{C} \end{bmatrix}$$

Taking N measurements

$$\varnothing^T = \begin{bmatrix} I(1) & q(1) \\ \vdots & \vdots \\ I(N) & q(N) \end{bmatrix}$$

and

$$Y = \begin{bmatrix} V_i(1) \\ V_i(2) \\ \vdots \\ V_i(N) \end{bmatrix}$$

Using LS estimation technique, the estimated parameter

$$\hat{\alpha} = [\varnothing^T \varnothing]^{-1} \varnothing^T Y$$

2. From the circuit equation

$$V_i = IR + \frac{i}{\omega C}$$

$$\Rightarrow \omega = \frac{1/C}{(V_i - IR)}$$

Here, the measured variables are V_i and I.

3. The MATLAB program for the estimation of circuit parameters $\alpha^T = \begin{bmatrix} R & \frac{1}{C} \end{bmatrix}$

```
% This programme simulates parameter estimation of an RC
%circuit. The
% current and the charge generated due to the applied voltages
%are
% simulated and the data is used for estimation of the
%capacitance and
% resistance.
clc;
close all;
clear all;
format long;
%The values of resistance(R) in ohm, capacitance(C) in
%microfarad
% and frequency(f)in Hz
% is used in this example to calculate the impedance %%%
R=500;
C=0.000001;
f=1000;
Xc=1./(2*pi*f*C);
Z=sqrt(R.^2+Xc.^2); %Z= impedance
%The applied voltage (Vi) is varied from 0.4V to 10V in steps
of 0.01V
% Then the circuit current and the charge is calculated %%%%
Vi=[0.4:0.01:10]
i=Vi./Z;          % i=current
Vc=Vi.*(Xc./Z); % Vc=voltage across the capacitor
q=C.*Vc; % q=charge across the capacitor
%Parameter estimation
phiT=[i;q];       % Formation of measurement matrix and its
%transpose
```

```
Y=Vi';              % Formation of the input variable vector
phi=phiT';
alpha=((inv(phiT*phi))*phiT*Y) % least square parameter
%estimation
```

A practical example of least square estimation method of indirect sensing is discussed in Case Study 4.11.

Case Study 4.11

An intelligent indirect dynamic torque sensor for permanent magnet brushless DC drives [41].

Direct torque sensing by displacement, strain sensors or torsion bar (angular twist) are disadvantageous due to bulky attachment and cost of sensor and signal processing. Indirect torque sensor of dc drives can be performed with the help of motor parameter estimation and some measurable quantities. Motor parameter estimation can be performed either offline or online. In the developed method, the torque equation is proposed as

$$T_e = \frac{3}{2}\frac{P}{2}\left[\left\{i_d\frac{d}{d\theta}l_{dd} + i_q\frac{d}{d\theta}l_{dq} + i_f\frac{d}{d\theta}l_{df} + \lambda_q\right\}i_d + \left\{i_q\frac{d}{d\theta}l_{qq} + i_d\frac{d}{d\theta}l_{qd} + i_f\frac{d}{d\theta}l_{qf} + \lambda_d\right\}i_q\right] \tag{4.217}$$

where

l = inductances

θ = rotor angles

In this application, the direct-axis current i_d was separately controlled to be at zero at all times giving Equation 4.217 as

$$T_e = \frac{3}{2}\frac{P}{2}\left[i_q\frac{d}{d\theta}l_{qq} + i_f\frac{d}{d\theta}l_{qf}\right]i_q \tag{4.218}$$

It was further assumed that the significant harmonic components of inductances vary with rotor angle θ. Further higher harmonic components were ignored since their contributions to inductance variations are small. Therefore, mutual inductances are expressed as

$$l_{qf} = L_{qf6}\sin 6\theta + L_{qf12}\sin 12\theta + \cdots \tag{4.219a}$$

$$l_{df} = L_{df0} + L_{df6}\cos 6\theta + L_{df12}\cos 12\theta + \cdots \tag{4.219b}$$

Moreover, the contribution of stator winding self and mutual inductances to torque are smaller such that

$$l_{qq} = l_{qd} = 0 \tag{4.219c}$$

Therefore, the approximation of torque equation is given by

$$T_e = \frac{3}{2}\frac{P}{2}[L_{df0}i_f + \{6L_{qf6} + L_{df6}\}i_f\cos\theta + \{12L_{qf12} + L_{df12}\}i_f\cos 12\theta]i_q \quad (4.220)$$

$$= \frac{3}{2}\frac{P}{2}[K_0 + K_6\cos 6\theta + K_{12}\cos 12\theta]i_q \quad (4.221)$$

where K_i is the ith torque harmonic coefficient, which is multiple of 6.

Now for relating the torque equation with measurement variable, the quadrature axis voltage equation is given as

$$\nu_q = Riq + l_{qq}\frac{d}{dt}i_q + l_{qd}\frac{d}{dt}i_d + \omega\left[iq\frac{d}{d\theta}l_{qq} + i_d\frac{d}{d\theta}l_{qd} + i_f\frac{d}{d\theta}l_{qf} + \lambda_d\right] \quad (4.222)$$

Under the same assumptions of Equations 4.218 through 4.219, the above equation can be written as

$$\upsilon_q = Riq + l_{qq}\frac{d}{dt}i_q + l_{qd}\frac{d}{dt}i_d + \omega[K_0 + K_6\cos 6\theta + K_{12}\cos 12\theta] \quad (4.223)$$

Writing Equation 4.223 in vector form

$$\upsilon_q = \left[i_q \quad \frac{diq}{dt} \quad \omega \quad \omega\cos 6\theta \quad \omega\cos 12\theta\right]\begin{bmatrix} R \\ l_{qq} \\ K_0 \\ K_6 \\ K_{12} \end{bmatrix} \quad (4.224)$$

which is in the form

$$y = \varnothing^T\alpha$$

where

$$\varnothing^T = \left[i_q \quad \frac{diq}{dt} \quad \omega \quad \omega\cos 6\theta \quad \omega\cos 12\theta\right] \quad (4.225)$$

$$\alpha^T = \left[R \quad l_{qq} \quad K_0 \quad K_6 \quad K_{12}\right] \quad (4.226)$$

It is possible to estimate the motor parameters α from the measurement vectors $\varnothing^T$ and Y by the least square estimation:

$$\hat{\alpha} = [\varnothing^T \varnothing]^{-1} \varnothing^T Y \tag{4.227}$$

where

$$Y^T = \begin{bmatrix} v_q(1) & v_q(2) & \ldots & v_q(N) \end{bmatrix} \tag{4.228}$$

Once the parameter α is estimated, Equation 4.221 can be used to measure the torque T_e. The following steps should be adopted in the indirect sensing of torque:

1. Perform dynamic measurements of quadrature axis current i_q, its rate of change (diq/dt), angular frequency ω, rotor angle θ, and quadrature axis v_q.
2. Construct measurement matrix $\varnothing^T$ using Equation 4.225 and Y^T using Equation 4.228.
3. Estimate parameter vector $\hat{\alpha}$ using Equation 4.227.
4. Estimate torque T_e for each measurement value of i_q using Equation 4.221.

4.9.2 Spectral Analysis–Based Parameter Estimation

Consider a dynamic system with input $x(t)$ and output $y(t)$ defined by the transfer function $G(j\omega)$ as shown in Figure 4.60 given by [42]. Spectral density functions $S_{xx}(j\omega)$ and $S_{xy}(j\omega)$ can be obtained from N pairs of input–output data and the transfer function estimation can be performed by

$$\hat{G}(j\omega_k) = \frac{S_{xy}(j\omega_k)}{S_{xx}(j\omega_k)}, \quad k = 1, 2, \ldots, N \tag{4.229}$$

The system transfer function can be denoted by

$$G(S) = \frac{b_0 + b_1 s + \cdots + b_m s^m}{1 + a_1 s + \cdots + a_n s^n} = \frac{B(s)}{A(s)} \tag{4.230}$$

where the parameter vector is

$$\theta = \begin{bmatrix} b_0 & b_1 & \cdots & b_m & a_1 & \cdots & a_n \end{bmatrix}^T \tag{4.230}$$

For approximating the estimated transfer function $\hat{G}(j\omega)$ to the actual transfer function $G(j\omega)$, a loss function is defined as

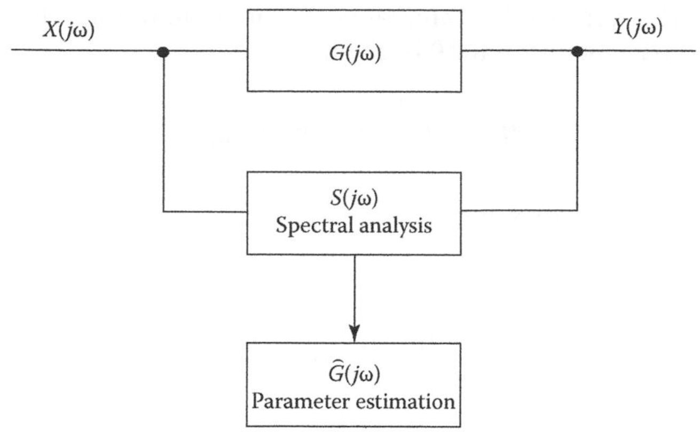

FIGURE 4.60
Spectral analysis–based parameter estimation. (Redrawn from IEEE, copyright © IEEE. With permission.)

$$J = \sum_{K=1}^{N} \left| q(K)\left[G(j\omega_K) - \hat{G}(j\omega_K)\right]\right|^2 \tag{4.231}$$

Using nonlinear optimization method, the weighting multinomial $q(k)$ is considered as

$$q(k) = 1$$

The loss function takes the form

$$J = [G(\theta) - \hat{G}(\theta)]^T [G(\theta) - \hat{G}(\theta)] \tag{4.232}$$

where

$$G(\theta) = \left[G(j\omega_1,\theta) \quad G(j\omega_1,\theta) \quad \cdots \quad G(j\omega_N,\theta)\right]^T \tag{4.233}$$

$$\hat{G}(\theta) = \left[\hat{G}(j\omega_1,\theta) \quad \hat{G}(j\omega_2,\theta) \quad \cdots \quad \hat{G}(j\omega_N,\theta)\right]^T \tag{4.234}$$

The iterative approximation of $G(\theta)$ is

$$G(\theta) \cong \hat{G}(\hat{\theta}_L) + \frac{\partial \hat{G}(\hat{\theta}_L)}{\partial \theta}(\theta - \hat{\theta}_L) \tag{4.235}$$

where $\widehat{\theta}_L$ is the estimated parameter vector at Lth iteration. The new value of estimated parameter vector θ is given by

$$\widehat{\theta}_{L+1} = \widehat{\theta}_L - [H_{\widehat{\theta}_L} + \mu_L I]^{-1} g_{\widehat{\theta}_L} \quad (4.236)$$

where I is an unity matrix and

$$H_{\widehat{\theta}L} = R_e \left\{ \left[\frac{\partial G(\widehat{\theta}_L)}{\partial \theta^T} \right]^* \left[\frac{\partial G(\widehat{\theta}_L)}{\partial \theta^T} \right] \right\} \quad (4.237)$$

$$g_{\widehat{\theta}L} = R_e \left\{ \left[\frac{\partial G(\widehat{\theta}_L)}{\partial \theta^T} \right]^* \left[G(\widehat{\theta}_L) - \widehat{G}(\theta) \right] \right\} \quad (4.238)$$

where μ is an adjusting factor.

Consider the circuit shown in Figure 4.61 that results a transfer function given by

$$G(j\omega) = \frac{K}{1 + j\omega\tau} \quad (4.239)$$

where $\tau = RC$.

Writing the transfer function in the form of Equation 4.230

$$G(S) = \frac{K}{1 + s\tau} = \frac{A(S)}{B(S)} \quad (4.240)$$

where $b_0 = 1$, $a_1 = \tau$ to give

$$\theta = \begin{bmatrix} b_0 & a_1 \end{bmatrix}^T = \begin{bmatrix} K & \tau \end{bmatrix}^T \quad (4.241)$$

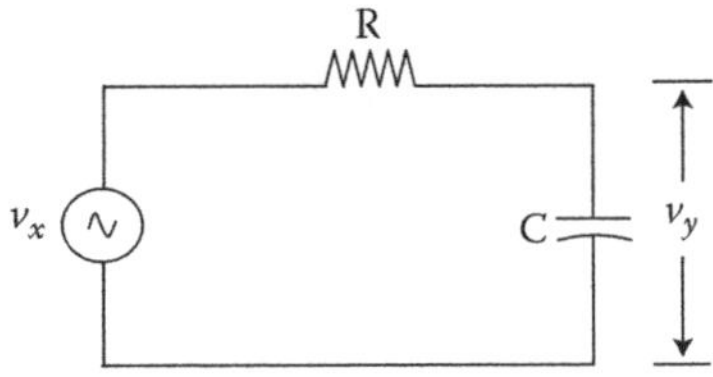

FIGURE 4.61
An RC circuit.

For N sets of υ_x and υ_y measurement data for angular frequency ω_K

$$\upsilon_x(\omega_K) = \begin{bmatrix} \upsilon_x(\omega_1) & \upsilon_x(\omega_2) & \cdots & \upsilon_x(\omega_N) \end{bmatrix} \tag{4.242}$$

$$\upsilon_y(\omega_K) = \begin{bmatrix} \upsilon_y(\omega_1) & \upsilon_y(\omega_2) & \cdots & \upsilon_y(\omega_N) \end{bmatrix} \tag{4.243}$$

From the above measurement vector, the spectral density functions are formed as

$$S_{xx}(\omega_K) = \begin{bmatrix} S_{xx}(\omega_1) & S_{xx}(\omega_2) & \cdots & S_{xx}(\omega_N) \end{bmatrix} \tag{4.244}$$

$$S_{xy}(\omega_K) = \begin{bmatrix} S_{xy}(\omega_1) & S_{xy}(\omega_2) & \cdots & S_{xy}(\omega_N) \end{bmatrix} \tag{4.245}$$

The estimated transfer function is determined from Equation 4.229 to get

$$\hat{G}(j\omega) = \begin{bmatrix} \hat{G}(j\omega_1) & \hat{G}(j\omega_{12}) & \cdots & \hat{G}(j\omega_N) \end{bmatrix} \tag{4.246}$$

For matching the spectral analysis to the transfer function, equating

$$\begin{aligned} \hat{G}(j\omega_1) &= \frac{K}{1 + A_1 a_1} \\ \hat{G}(j\omega_2) &= \frac{K}{1 + A_2 a_1} \\ &\vdots \\ \hat{G}(j\omega_N) &= \frac{K}{1 + A_N a_1} \end{aligned} \tag{4.247}$$

To form the matrix for estimation of θ, we form the equations

$$\begin{aligned} \hat{G}(j\omega_1) + \hat{G}(j\omega_1) A_1 a_1 &= K \\ \hat{G}(j\omega_2) + \hat{G}(j\omega_2) A_2 a_1 &= K \\ &\vdots \\ \hat{G}(j\omega_N) + \hat{G}(j\omega_N) A_N a_1 &= K \end{aligned} \tag{4.248}$$

By manipulation

$$\begin{aligned} K-\hat{G}(j\omega_1)A_1a_1 &= \hat{G}(j\omega_1) \\ K-\hat{G}(j\omega_2)A_2a_1 &= \hat{G}(j\omega_2) \\ &\vdots \\ K-\hat{G}(j\omega_N)A_Na_1 &= \hat{G}(j\omega_N) \end{aligned} \tag{4.249}$$

Writing in matrix form

$$\gamma\hat{\theta} = \hat{G} \tag{4.250}$$

where

$$\gamma^T = \begin{bmatrix} 1 & 1 & \cdots & 1 \\ -G(j\omega_1)A_1 & -G(j\omega_2)A_2 & \cdots & -G(j\omega_N)A_N \end{bmatrix} \tag{4.251}$$

$$\hat{G}^T = \begin{bmatrix} \hat{G}(j\omega_1) & \hat{G}(j\omega_2) & \cdots & \hat{G}(j\omega_N) \end{bmatrix} \tag{4.252}$$

$$\hat{\theta}^T = \begin{bmatrix} K & a_1 \end{bmatrix} = \begin{bmatrix} K & \tau \end{bmatrix}$$

Updating of the estimated parameters:

For updating the estimated parameter using Equation 4.236, we write the recursive equation

$$\begin{bmatrix} K_{L+1} & a_{1,L+1} \end{bmatrix} = \begin{bmatrix} K_L & a_{1,L} \end{bmatrix} - \begin{bmatrix} H_{\hat{\theta}_L} + \mu_L I \end{bmatrix}^{-1} g_{\hat{\theta}_L} \tag{4.253}$$

where $H_{\hat{\theta}_L}$ and $g_{\hat{\theta}_L}$ can be obtained from

$$H_{\hat{\theta}_L} = R_e \left\{ \left[\frac{G(\hat{\theta}_L) - G(\hat{\theta}_{L-1})}{[\hat{\theta}_L - \hat{\theta}_{L-1}]^T} \right]^* \left[\frac{G(\hat{\theta}_L) - G(\hat{\theta}_{L-1})}{\left[\hat{\theta}_L - \hat{\theta}_{L-1}\right]^T} \right] \right\}$$

$$g_{\hat{\theta}_L} = R_e \left\{ \left[\frac{G(\hat{\theta}_L) - G(\hat{\theta}_{L-1})}{\left[\hat{\theta}_L - \hat{\theta}_{L-1}\right]^T} \right]^* [G(\hat{\theta}_L) - \hat{G}] \right\}$$

4.9.3 Fuzzy Logic Based

Least square– or spectral analysis–based parameter estimation techniques involve complex mathematical computations that need fast real-time processing of the measurement data. Fuzzy logic approach is comparatively computationally simpler since it is based on rules. Nonlinear continuous functions are easy to be implemented in fuzzy models. Moreover, the memory requirement is also lower than the look-up table estimation or ANN. Fuzzy logic–based system modeling for indirect sensing can be realized by the following steps [43]:

1. Determination of the system model:

   The fuzzy model for the system can be developed from two main sources:

   a. Static input–output characteristics of the system
   b. Dynamic real-time operating effects and interferences

   Once the above two sources are available, the training input–output data can be used to model the system in a rule-based manner. Since the indirect sensing aims at the measurement of a single output variable, the fuzzy model will be considered as a single output system while input may be multidimensional. However, for reducing modeling complexity, the dimension of the input variable should be as low as possible.

   Now we define n-input-one-output pairs of measurements, where for each pair there are n input data and one output data. Each point of measurement data presented to the training systems is given by

$$\left[x_1(k), x_1(k) \quad \cdots \quad x_n(k); y(k)\right]$$

   where
   n is the number of inputs
   k is the point of data pair

2. Mapping input–output to fuzzy regions:

   It is customary in fuzzy logic to divide the input–output space into several fuzzy regions, say $N_{x1}, N_{x2}, \ldots, N_{x_n}$ and N_y, which are called fuzzy sets. Remember that more number of regions would provide greater accuracy; however, memory requirement is also to be considered. The fuzzy sets for each input–output variable are assigned membership functions $\mu_{x1}, \mu_{x2}, \ldots, \mu_{x_n}$ and μ_y. Various types of membership functions can be used in different applications and, for

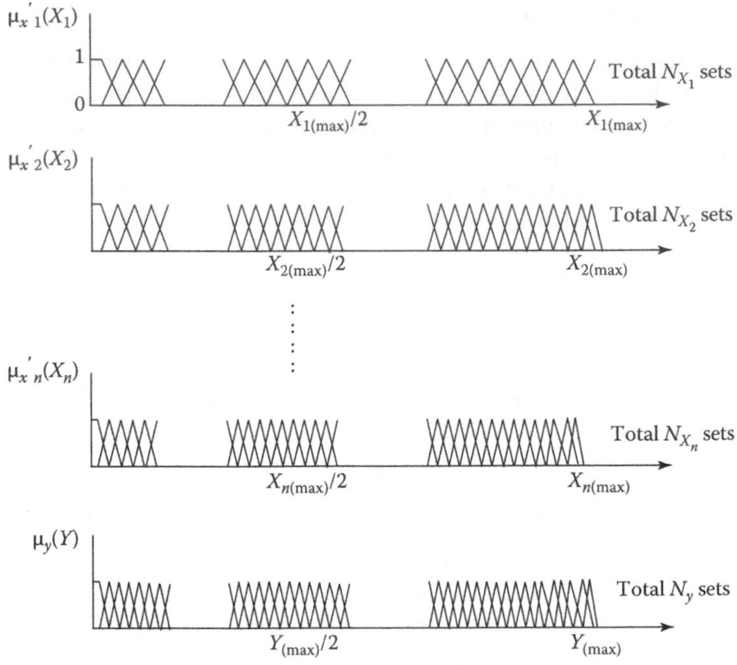

FIGURE 4.62
Fuzzy domain regions for input–output variables.

simplicity, let us consider a triangular membership function. Figure 4.62 shows the triangular membership function assigned to the input–output variables.

3. Assigning fuzzy linguistic attributes:
   For each membership region, a linguistic attribute is assigned. This assignment is based on the type of application. For example, values of variables are grouped in P groups, say
   $P = 3$, therefore

   1: "LOW" for $0 \leq x(y) < (\text{Max}/3)$
   2: "MEDIUM" for $(\text{Max}/3) \leq x(y) < (2\text{Max}/3)$
   3: "HIGH" for $(2\text{Max}/3) \leq x(y) < \text{Max}$

   Therefore, each membership region will have a unique linguistic attribute, say LOW1, MEDIUM10, HIGH12, etc.

4. Fuzzy rule:
   For each pair of input–output data set, a rule is assigned and each rule has a degree of confidence.

   For example

   Rule:

IF x_1 is SMALL10 AND x_2 is HIGH5…AND x_n is MEDIUM7, THEN y is LOW12.

$$\text{Degree: } \mu_{SMALL10}(x_1)\cdot\mu_{HIGH5}(x_2)\ldots\mu_{MEDIUM7}(x_n)\cdot\mu_{LOW12}(y)$$

where

$\mu_{SMALL10}(x_1)$: membership function of x_1 in region SMALLl10

$\mu_{HIGH5}(x_2)$: membership function of x_2 in region HIGH5

⋮

$\mu_{MEDIUM7}(x_n)$: membership function of x_n in region MEDIUM7

$\mu_{LOW12}(y)$: membership function of y in region LOW12

When the same antecedent produces different consequences in rules, the rule with the highest degree value is considered.

5. Fuzzy rule–based LUT:

   From the training data, the corresponding rules are generated and stored in the 2D table, as shown in Table 4.17. The 2D table can accommodate two antecedent linguistic variables and the entries are the linguistic consequent variables. It is not that all linguistic variables have consequents and when there is no consequent for a rule, entries are marked "×."

For example, when x_1 is low i and x_2 is medium m, there is no consequent, so entry is marked "×," and tick-marked entries indicate some linguistic variable output y.

Let us revisit Case Study 4.1 where the drying status of a tea dryer was modeled using RST. Here, we will develop a fuzzy model of the dryer for

	Input Variable x_1							
Input Variable x_2	**Low i**	**Low $(I-1)$**	**….**	**Medium j**	**Medium $(j-1)$**	**….**	**High $(k-1)$**	**High k**
Low l	✓	✓		✓	✓		✓	✓
Low $(l-1)$	✓	✓		✓	✓		✓	✓
⋮								
Medium (m)	×	✓		✓	✓		✓	✓
Medium $(m-1)$	✓	✓		✓	✓		✓	✓
⋮								
High $(n-1)$	✓	✓		✓	✓		✓	✓
High n	✓	✓		✓	✓		✓	✓

FIGURE 4.63
Example look up table for fuzzy rule base.

measuring the drying status. From Table 4.1, the patterns 1, 2, 5, 6, and 7 are chosen while patterns 3 and 4 will not be used since they were found to have certainty factors of 0.56 and 0.44, respectively. Table 4.1 shows the data corresponding to the five selected patterns.

Now we assign triangular membership functions to the input–output variables. Dryer inlet temperature (T_i) is divided into two groups—LOW and HIGH; tea feed rate (M_i) into three groups—LOW, MEDIUM, and HIGH; drier outlet temperature (T_o) into three groups—LOW, MEDIUM and HIGH and drying status into three groups—UNDER, NORMAL, and OVER. The ranges of the variables and the number of fuzzy membership regions are

$$T_i : 85°\text{C} - 100°\text{C}; 16 \text{ regions}$$

$$M_i : 3 - 15 \text{ trays}/15 \text{ min}; 12 \text{ regions}$$

$$T_0 : 60°\text{C} - 80°\text{C}; 21 \text{ regions}$$

D (dryness): 0%–10%; 21 regions

The triangular membership functions are shown in Figure 4.64.

The various linguistic attributes used for the variables are shown in Table 4.16.

For each of the input–output data, the fuzzy rules are formed and stored in the LUT, as shown in the Table 4.17.

In this table, unused boxes are marked ×, which indicates that there is no rule for that antecedent. On the other hand, in some entries, there are more than one consequent for the same antecedent. For example data Nos. 3 and 4 of pattern 5 form the following two rules, respectively:

Rule 1: IF T_i is *H*9, M_i is *L*3 and T_0 is *L*1 THEN D is 01
Rule 2: IF T_i is *H*9, M_i is *L*3 and T_0 is *L*1 THEN D is 02

In such cases, the rule with the highest membership function will be used. For example:

When T_i: 92°C, M_i: 3 trays/15 min and T_0: 65°C, then the degree of the membership functions for the two rules are

Rule 1:
$\mu_{H9}(T_i) = 1.0$, $\mu_{L3}(\mu_i) = 1.0$, $\mu_{L1}(T_0) = 1.0$, and $\mu_{01}(D) = 0.8$

Rule 2:
$\mu_{H9}(T_i) = 1.0$, $\mu_{L3}(\mu_i) = 1.0$, $\mu_{L1}(T_0) = 1.0$, and $\mu_{01}(D) = 1.0$
The degree of the two rules are

Rule 1: $1.0 \times 1.0 \times 1.0 \times 0.8 = 0.8$
Rule 2: $1.0 \times 1.0 \times 1.0 \times 1.0 = 1.0$

Therefore, rule 2, with stronger degree, is used as rule base and stored in the LUT in column *H*9 and row *L*1 as 02.

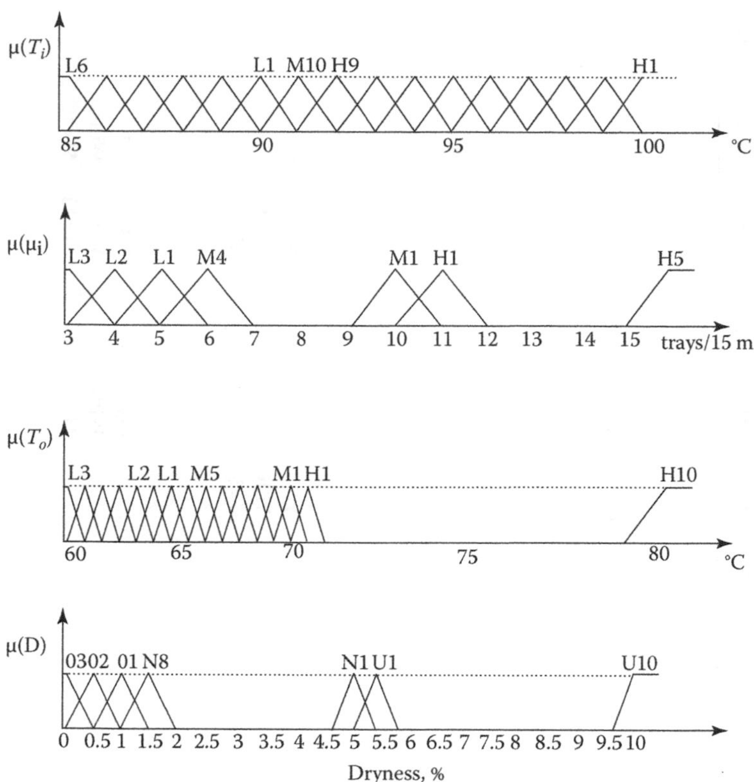

FIGURE 4.64
Regions of membership functions of the input–output variables.

TABLE 4.16
Linguistic Attributes of the Tea Dryer Parameters

Variable	Variable Range	Linguistic Region	Total Regions
T_i	85°C–90°C	L6–L1	16
	90°C–100°C	H10–H1	
M_i	3–5 trays/15min	L3–L1	17
	6–10 trays/15min	M4–M1	
	11–15 trays/15min	H1–H10	
T_0	60°C–65°C	L6–L1	21
	66°C–70°C	M5–M1	
	71°C–80°C	H1–H10	
D	0%–1%	O3–O1	21
	1.5%–5%	N8–N1	
	5.5%–10%	U1–U10	

TABLE 4.17

Lookup Table of Fuzzy Rule Distribution for Tea Dryer Status Detection

T_o \ T_i	L6	L5	L4	L3	L2	L1	H10	H9	H8	H7	H6	H5	H4	H3	H2	H1
H3	NI With M_i=13	x	x	x	x	x	x	x	x	x	x	x	x	x	x	x
L5	x	x	x	x	x	x	x	x	x	x	x	x	x	x	x	x
L4	x	x	x	x	x	x	x	x	x	x	x	x	x	x	x	x
H2	NI With M_i=13	x	x	x	x	x	x	x	x	x	x	x	x	x	x	x
L2	x	x	x	x	x	x	x	O1 With M_i=5 & N7 With M_i=7	x	x	x	x	x	x	x	x
L1	U8 With M_i=7	N7 With M_i=5 & U8 With M_i=7	x	x	x	x	x	O2 With M_i=4 & N7 With M_i=7	x	x	O1 With M_i=5	x	N3 With M_i=10	N5 With M_i=10	x	x
M5	O1 With M_i=4 & N3 With M_i=5	x	N7 With M_i=5 & U6 With M_i=7	x	x	x	x	x	x	x	x	O3 With M_i=5	x	N5 With M_i=10	N7 With M_i=10	x

M4	**U6 With** M_i**=6**	x	**N5 With** M_i**=4 & U4 With** M_i**=7**	x	x	x	x	x	x	**N3 With** M_i**=7**	**O2 With** M_i**=4 & N5 With** M_i**=7**	x	**O3 With** M_i**=5**	**O1 With** M_i**=8**	**O1 With** M_i**=9**	**O1 With** M_i**=10**
M3	**N7 With** M_i**=3 & N3 With** M_i**=5**	x	**U2 With** M_i**=6**	x	x	x	x	x	x	x	**O3 With** M_i**=3 N7 With** M_i**=6**	**N7 With** M_i**=7**	x	**O1 With** M_i**=8**	x	**O1 With** M_i**=9**
M2	x	x	**N7 With** M_i**=4 & N1 With** M_i**=5**	**N5 With** M_i**=5 & N1 With** M_i**=6**	**N1 With** M_i**=7**	x	x	x	x	x	x	**O3 With** M_i**=3 & O1 With** M_i**=6 & O1 With** M_i**=7**	x	x	x	x
M1	x	x	x	**N7 With** M_i**=4 & N3 With** M_i**=5**	**N1 With** M_i**=5 & N1 With** M_i**=7**	**N1 With** M_i**=8**	x	x	x	x	x	**O1 With** M_i**=5**	x	x	x	x
H9	x	x	x	x	**U6 With** M_i**=9**	x	x	x	x	x	x	x	x	x	x	x

(continued)

TABLE 4.17 (continued)

Lookup Table of Fuzzy Rule Distribution for Tea Dryer Status Detection

T_o \ T_i	L6	L5	L4	L3	L2	L1	H10	H9	H8	H7	H6	H5	H4	H3	H2	H1
H2	×	×	×	×	N5 With M_i=4	×	×	×	×	×	×	×	×	×	×	×
H10	×	×	×	×	×	U4 With M_i=9	×	×	×	×	×	×	×	×	×	×
H6	×	×	U10 With M_i=11	×	×	×	×	×	×	×	×	×	×	×	×	×
H5	N5 With M_i=10	×	U4 With M_i=13	U10 With M_i=13	U8 With M_i=13	×	×	×	×	×	×	×	×	×	×	×
H7	×	×	U8 With M_i=10	U6 With M_i=10	×	×	×	×	×	×	×	×	×	×	×	×
H8	U4 With M_i=9	×	×	×	U4 With M_i=10	×	×	×	×	×	×	×	×	×	×	×

References

1. Gaura, E. and Newman, R.M., Smart, intelligent and cogent MEMS based sensors, *Proceedings of the IEEE 2004, International Symposium on Intelligent Control*, Taipei, Taiwan, September 2–4, 2004.
2. Ranky, P.G., Smart sensors, *Sensor Review*, 22(4), 301–311, 2002.
3. Frost & Sullivan, North America smart sensors market, 2002.
4. Aziz, F., Kanev, Z., Barboucha, M. Maimouni, R., and Staroswiecki, M., An ultrasonic flowmeter designed according to smart sensor concept, *Electrotechnical Conference, 1996. MELECON'96, 8th Mediterranean Volume: 3*, 1996, pp. 1371–1374.
5. Gonzalez, G.D., Redard, J.P., and Barrera, R., Issues in soft sensor applications in industrial plants, *Symposium Proceedings, ISIE'94, 1994 IEEE International Symposium on Industrial Electronics*, 1994, pp. 380–385.
6. Pawlak, Z., Rough set theory and its applications, *Journal of Telecommunications and Information Technology*, 3, 7–10, 2002.
7. Luo, J.X. and Shao, H.H., Selecting secondary measurement for soft sensor modeling using rough set theory, *Proceedings of the IEEE, Fourth World Congress on Intelligent Control and Automation*, Shanghai, China, pp. 415–418, June 10–14, 2002.
8. Juricek, B.C., Seborg, D.E., and Larimore, W.E., Process control applications of subspace and regression-based identification and monitoring methods, *IEEE American Control Conference*, Portland, OR, June 8–10, 2005.
9. Van Overschee, P. and De Moor, B., *Subspace Identification for Linear Systems*, Kluwer Academic Publisher, Boston, MA, 1996.
10. Zhang, C.H., Liu, X., Shi, J., and Zhu, J.H., Neural soft sensor of product quality prediction, *Proceedings of the Sixth World Congress on Intelligent Control and Automation, IEEE*, Dalian, China, June 21–23, 2006, pp. 4881–4885.
11. Wang, P., Design and applications of soft sensor object in process control, *Proceedings of the Sixth International Conference on Intelligent System Design and Application*, 2006. ISDA'06. Sixth International Conference on 2006, pp. 107–111.
12. Al-Duwaish, H., Ghouti, H., Halawani, L. et al., Use of artificial neural network process analyzers: A case study, *Proceedings of the European Symposium on ANN (ESANN 2002)*, Bruges, Belgium, pp. 465–470, April 24–26, 2002.
13. Korobeynikov, S.A., Determination of characteristics of decision rule at adaptive measurement, Investigated in Russia, 2004, pp. 1851–1855 (http://zhurnal.gpi.ru/).
14. Pystynskiy, I.N., Titov, V.C., and Shyarabakina, T.A., *Adaptive Photoelectric Converters with Microprocessor*, Energoatomizdat, Moscow, Russia, 1990.
15. Yurish, S.Y., Self adaptive smart sensors and sensor systems, *Sensors and Transducers Journal*, 94(7), 1–4, July 2008.
16. MSP430-Ultra-Low Power Microcontroller, Texas Instruments, USA2H, 2008. http://www.ti.com
17. MAXIM, SOT Temperature Sensors with Period/Frequency Output, 2008.
18. Staroswiecki, M., Intelligent sensors: A functional view, *IEEE Transactions on Industrial Informatics*, 4, 238–249, November, 2005.
19. Bhuyan, M., An integrated PC based tea processing monitoring and control system, PhD Thesis, Gauhati University, India, 1997.

20. Henry, M.P. and Clarke, D.W., The self-validating sensor: Rationale, definitions and examples, *Control Engineering Practice*, 1(4), 585–610, 1993.
21. Clarke, D.W. and Fraher, P.M., Model based validation of a DOx sensor, *Control Engineering Practice*, 4(9), 1313–1320, 1996.
22. Henry, M.P., A SEVA sensor: The Coriolis mass flow meter, *IFAC/IMACS Symposium, Safe Process*, Espoo, Finland, 1994.
23. Ray Liu, K.J. and Yao, K., *High Performance VLSI Signal Processing Innovative Architecture and Application* (Ed.), IEEE Press, Lixouri, Greece, 2009.
24. Andraka, R., *A Survey of CORDIC Algorithm for FPGA Based Systems*, Copyright AMC Inc., New York, 1998.
25. Hsiao, S.F., Parallel singular value decomposition of complex matrices using multidimensional CORDIC algorithm, *IEEE Transactions, Signal Processing*, 44(3), 685–697, March 1996.
26. Perwaiz, A. and Khan, S.A., LMS bit stream adaptive filter design, *10th International Conference on Computer Modeling and Simulation, IEEE*, Cambridge, U.K., pp. 507–512, 2008.
27. Knowles, G., VLSI architecture for the discrete wavelet transform, *Electronics Letter*, 26(15), 1184–1185, 1990.
28. Kramer, A.H., Array based analog computation: Principles, advantages and limitations, *Proceedings of the IEEE, Micro-Neuro*, Washington, DC, pp. 68–79, 2006.
29. Han, G. and Sanchez Sinencio, E., CMOS transconductance multipliers: A tutorial, *IEEE Transactions on Circuits and Systems—II: Analog and Digital Signal Processing*, 45(12), 1550–1558.
30. MAXIM, Driving strain gauge bridge sensors with signal conditioning ICs, 2009. www.maxim-ic.com
31. MAXIM, Demystfying piezoresistive pressure sensors, 2009. www.maxim-ic.com
32. Nam, T., Kim, S., Kim, S., and Park, S., The temperature compensation of a thermal flow sensor with a mathematical method, *12th International Conference in Solid State Sensors, Actuator and Micro Systems*, Boston, MA, June 8–12, 2003.
33. Chen, A., Zu, J., and Li, B., Temperature compensation for hot film crosswind sensors, *IEEE International Conference, Instrumentation and Measurement*, Ottawa, Canada, May 19–21, 1997.
34. Akbar, M. and Shanblatt, M.A., A fully integrated temperature compensation technique for piezoresistive pressure sensor, *IEEE Transactions Instrumentation and Measurement*, 42(43), 771–775, June 1993.
35. MAXIM, A different approach to compensation, 2009. www.maxim-ic.com
36. Kolen, P.T., Self calibration/compensation technique for microcontroller based sensor arrays, *IEEE Transactions, Instrumentation and Measurement*, 43(4), 620–623, August 1994.
37. Busser, R.A., Resonant sensors, Chapter 7, in Bau, H.H., de Rooij, N.F., and Kloeck, B. (Eds.), *Mechanical Sensors, Vol 17 of Sensors, A Comprehensive Survey*, Gopel, W., Hesse, J., and Zemel, J.N. (Eds.), VCH (John Wiley & Sons), New York, 1994.
38. Stemme, G., Resonant silicon sensors, *Journal of Micromechanical and Microengineering*, 1, 113–115, 1991.
39. Pallas-Areny, R. and Webster, J.G., *Sensors and Signal Conditioning* (2nd edn.), Wiley Interscience, New York, 2001.

40. Haleem, M.A., Haque, M.Z.U., Haloi, M., Muqueet, M.A., and Shaikh, M.F., Digital pH meter with temperature compensation, *International Conference on ICEE 2008*, Budapest, Hungary, July 2008.
41. Lee, T.H. and Low, T.S., An intelligent indirect dynamic torque sensor for permanent magnet brushless DC drives, *IEEE Transactions, Industrial Electronics*, 41(2), 191–200, April 1994.
42. Xiong, S.-S. and Zhou, Z.Y., Dynamic parameter estimation of velocity sensors using an indirect measurement approach, *IEEE Transactions on Instrumentation and Measurement Technology Conference*, Budapest, Hungary, 2001.
43. Ertugral, N. and Cheok, A.D., Indirect angle estimation in switched reluctance motor drives using fuzzy-logic based model, *IEEE Transactions, Power Electronics*, 15(6), 1029–1044, December 2000.

5

Linearization, Calibration, and Compensation

5.1 Introduction

It was discussed in Sections 2.2.2 and 2.2.5 that sensors possess undesirable characteristics due to which the outputs deviate from the ideal or true values. This characteristic of a sensor is known as nonlinearity. Linearity describes how closely the sensor output relates to a specified suitable straight line, which is considered as the true or ideal input–output characteristics. It has been observed that most sensors are typically either nonlinear or linear over a limited range of interest only. Even sensors that produce approximately linear outputs may cause problems when precise measurements of the signal are required. It is unrealistic to feed nonlinear sensor signals to linear devices like meters, plotters, actuators, etc., without linearization. Strictly speaking, here a sensor does not mean the sensing device alone, but includes the complete signal conditioning system comprising the amplifier, filter, analog-to-digital converter (ADC), etc. Therefore, the nonlinearity of a sensor may also be resulted from the nonlinearity of the amplifier, filter, ADC, etc. The nonlinearity of a sensor circuit is resulted from nonlinearity of devices like metal-oxide semiconductor (MOS)-channel resistors, gate capacitor, etc. Operational amplifiers also show limited gain and pass band due to which measurement systems suffers from the problem of nonlinearity.

Therefore, a faithful representation of a physical variable is possible only when the signal is linearized before applying to the output devices. The process of linearization involves the task of mapping the sensor measurement data to a straight line or linear characteristic already determined as per experimental data. The techniques of mapping may be different for different situations. Broadly, the techniques can be classified into two categories—circuit-level linearization and software-level linearization; however, some of the specific methods may be applicable to both. The method of linearization under these two categories may be classified as

Hardware linearization:

1. Analog processing
2. Digital processing
3. Nonlinear ADC

Nonlinear analog linearization is associated with circuit complexity and dependent on environmental conditions such as temperature. On the other hand, linearization based on interpolation using digital signal processing requires complex mathematical operations, switching, and large lookup table (LUT). Nonlinear ADC is comparatively more flexible to perform linearization of low-cost sensors.

Computer-based linearization techniques are

1. Interpolation
2. Piece-wise linearization
3. LUT
4. Artificial neural network

5.2 Analog Linearization of Positive Coefficient Resistive Sensors

Most resistive sensors such as resistance temperature detector (RTD), thermistor, resistive hygrometer, etc., produce nonlinear change in resistance with the measurable value. In such sensors, the outputs can be linearized with the help of a resistive circuit connected to the sensor. One possible arrangement for resistive senor linearization can be using a parallel combination of the nonlinear sensor with fixed resistor, but such circuits can only be used at the cost of sacrificing some amount of sensitivity. Moreover, when the sensitivity of the nonlinear sensor increases with higher values of the input variable, a simple parallel combination does not work for linearization.

The following equation typically approximates the nonlinear change in resistance of common resistive temperature sensors:

$$R_t = R_0 e^{-\alpha x} \tag{5.1}$$

or

$$R_t = R_0(1+\alpha_1 x^2 + \alpha_2 x^2) \tag{5.2}$$

where the sensor resistance is denoted by R_x, R_0 is the resistance at a reference or calibrated input x_0, and α is the resistance coefficient of the sensor.

Equation 5.1 is similar to the equation of a thermistor, while Equation 5.2 is that of an RTD. We differentiate Equations 5.1 and 5.2 to determine the variation of sensitivity of the sensor, and we get

$$\frac{dR_x}{dx} = -R_0 \alpha e^{-\alpha x} \tag{5.3}$$

and

$$\frac{dR_x}{dx} = R_0(\alpha_1 + 2\alpha_2 x) \tag{5.4}$$

Equations 5.3 and 5.4 reveal that the sensitivity of a negative resistance coefficient sensor decreases exponentially, while for positive coefficient sensors it increases linearly from a constant value $R_0\alpha_1$. if we assume that Equation 5.2 does not contain the terms $\alpha_2 x^2$, the (dR_x/dx) term will have only $R_0\alpha_1$ in Equation 5.4. It means that the sensitivity remains constant at $R_0\alpha_1$ with the variation of input x.

5.2.1 Linearization by Shunt Resistance

Let a measurable variable x have a nonlinear effect on the resistance R_x of the resistive sensor given by the equation

$$R_x = R_0(1 + A_x + Bx^2) \tag{5.5}$$

The sensor can be linearized by connecting a resistance R parallel to R_x to give a total linear resistance of

$$R_{Tx} = \frac{RR_x}{R + R_x} = \frac{R_x}{1 + (R_x/R)} \tag{5.6}$$

In Equation 5.6, the variable R_x in the denominator maintains proportionality in R_{Tx} as x increases. The sensitivity of the sensor as well as the total resistance can be obtained from the following:

For the nonlinear sensor

$$\frac{dR_x}{dx} = R_0 A + 2R_0 B_x = R_0(A + 2B_x) \tag{5.7}$$

and for the linearized sensor

$$\frac{dR_{Tx}}{dx} = \frac{R^2}{(R_x + R)^2}\frac{dR_x}{dx} = \frac{AR_0 + BR_0 x}{\left[1 + (R_0/R) + (Ax + Bx^2)(R_0/R)\right]} \tag{5.8}$$

The sensitivity per unit the sensor resistance is given by

For nonlinear sensor

$$\frac{(dR_x/dx)}{R_x} = \frac{A + 2B_x}{1 + A_x + Bx^2} \tag{5.9}$$

and for the linearized sensor neglecting smaller terms

$$\frac{(dR_x/dx)}{R_{Tx}} = \frac{AR + 2BRx}{(R + R_0) + (R + 2R_0)Ax + (R + 2R_0)Bx^2} \tag{5.10}$$

Equations 5.9 and 5.10 reveal that we gained in linearity of the sensor at the cost of decreased sensitivity. Now we will verify Equations 5.5 through 5.10 for a resistive temperature sensor that produces output similar to Equation 5.5. An RTD made of platinum 385 follows a temperature resistance characteristic equation

$$R_t = R_0[1 + \alpha_1 T + \alpha_2 T^2 + \alpha_3 (T - 100)^3] \tag{5.11}$$

where
$\alpha_1 = 3.9083 \times 10^{-3}/°C$
$\alpha_2 = -5.775 \times 10^{-7}/°C^2$
$\alpha_3 = -4.183 \times 10^{-12}/°C^3$ for below 0°C and $\alpha_3 = 0$ above 0°C
R_0 = resistance at 0°C = 100 Ω
T = temperature, °C

The RTD has an operating range of −200°C to 850°C. The temperature resistance characteristics for the RTD for a temperature range of 0°C–850°C is shown in Figure 5.1a

The sensitivity of the RTD is given by (for temperature above 0°C)

$$\frac{dR_t}{dT} = R_0\alpha_1 + 2R_0\alpha_2 T \tag{5.12}$$

and sensitivity per-unit resistance of the sensor is

$$\frac{(dR_x/dT)}{R_t} = \frac{R_0(\alpha_1 + 2\alpha_2 T)}{R_0(1 + \alpha_1 T + \alpha_2 T^2)} = \frac{\alpha_1 + 2\alpha_2 T}{1 + \alpha_1 T + \alpha_2 T^2} \tag{5.13}$$

The sensitivity and per-unit sensitivity characteristic are shown in Figure 5.1b and c, respectively. The sensitivity of a resistive sensor shunted by a fixed resistance given by Equations 5.8 and 5.10 decreases as the variable x increases. Therefore, to get an effective linearization, Equation 5.6 indicates that the shunting resistor R should have a negative resistance [1].

5.2.1.1 Positive Feedback OPAMP Circuit (Current Source)

The negative resistance equivalence of the shunting resistor in Equation 5.6 can be effectively realized by an operational amplifier (OPAMP) measuring circuit shown in Figure 5.2. Consider a measurable range of the RTD where the resistance changes from an initial value R_i at temperature T_i to a final

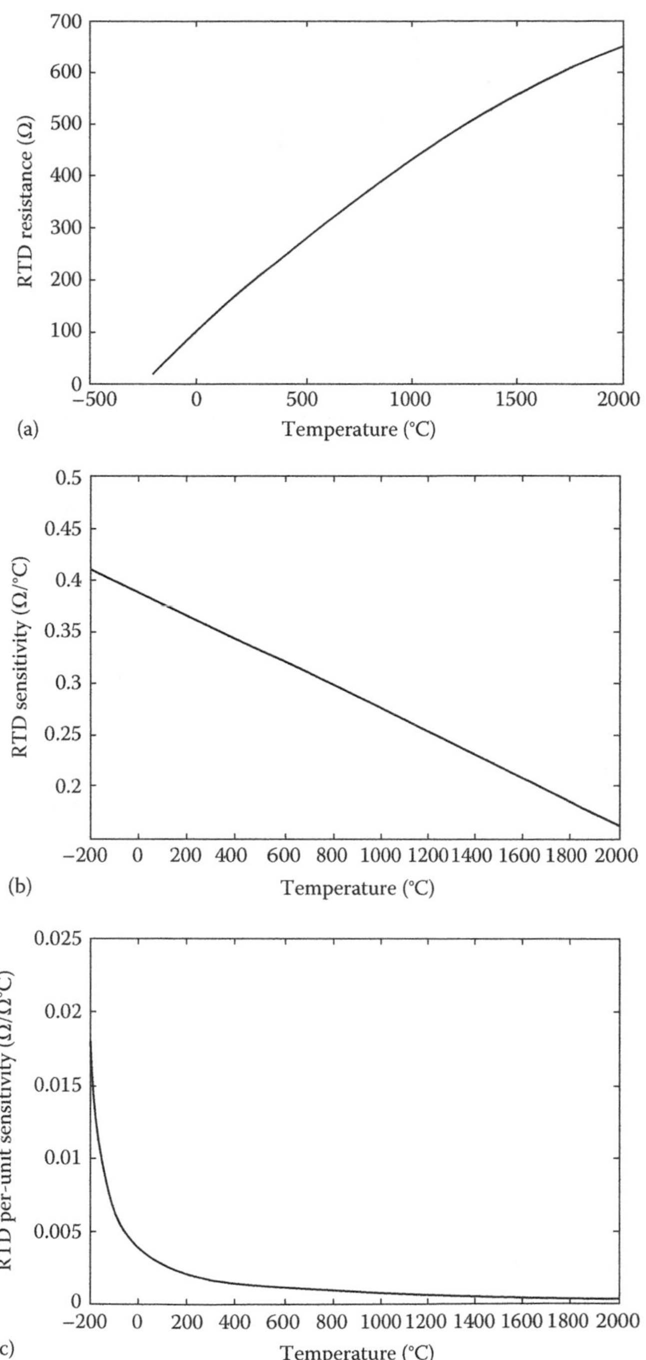

FIGURE 5.1
RTD characteristics: (a) temperature resistance, (b) sensitivity, and (c) per-unit sensitivity.

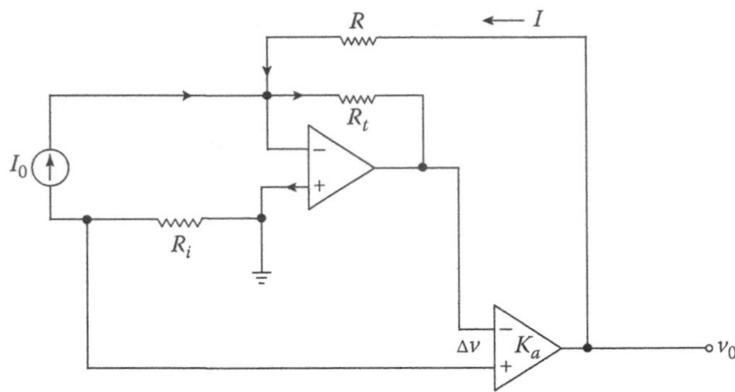

FIGURE 5.2
Positive feedback OPAMP linearization circuit (current source). (Redrawn from IEEE, copyright © IEEE. With permission.)

value R_f at temperature T_f. During this range, the resistance R_x can be specified as

$$R_t = R_i + \Delta R_t \tag{5.14}$$

The resistor with value R_i is connected to the current source I_0 and the RTD is connected in the feedback path of the OPAMP and, as a result, the output produces the differential voltage Δv. The differential voltage is further amplified by another OPAMP with a gain K_a introducing a positive feedback of current I to the current source I_0. This gives the output of the first stage:

$$\Delta v = (I_0 + I)R_t - I_0 R_i \tag{5.15}$$

and the positive feedback current by equation

$$I = \frac{v_0}{R} \tag{5.16}$$

where

v_0 is the linearized output voltage
R is the positive feedback resistor

From Equations 5.15 and 5.16, we get

$$\Delta v = I_0(R_t - R_i) + v_0 \frac{R_t}{R} \tag{5.17}$$

The output voltage from the second stage and using Equation 5.15

$$v_0 = k_a \Delta v$$

$$\Rightarrow k_a \left[I_0(R_t - R_i) + v_0 \frac{R_t}{R} \right]$$

$$\Rightarrow \frac{k_a I_0(R_t - R_i)}{1 - (R_t k_a / R)}$$

$$v_0 = \frac{k_a I_0(R_t - R_i)}{1 - (R_t / R')} \tag{5.18}$$

where $R' = (R/K_a)$.

Equation 5.18 is similar to Equation 5.6 except that the denominator has an equivalent negative resistor R' shunted to R_t.

To determine the resistance R, an optimal linearization condition is formulated between two extreme temperatures T_i and T_f with a mid value of T_m. Let the corresponding RTD resistances be R_i, R_f, and R_m, respectively. The differences of the output voltages in the two halves of the temperature ranges are

$$v_{0f} - v_{0m} = v_{0m} - v_{0i} \tag{5.19}$$

Substituting Equation 5.18 in Equation 5.19 for the corresponding output voltages we get

$$2(R_m - R_i)\left(1 - \frac{k_a R_f}{R}\right) = (R_f - R_i)\left(1 - \frac{k_a R_m}{R}\right)$$

$$\Rightarrow \frac{1}{R}[2(R_m - R_i)k_a R_f - (R_f - R_i)k_a R_m] = 2(R_m - R_i) - (R_f - R_i)$$

$$\Rightarrow R = k_a \frac{2(R_m - R_i)R_f - (R_f - R_i)R_m}{2R_m - R_i - R_f}$$

$$\therefore R = k_a \frac{R_f(R_m - R_i) - R_i(R_f - R_m)}{2R_m - R_i - R_f} \tag{5.20}$$

Example 5.1

In the RTD linearization circuit shown in Figure 5.2, the connected RTD has the following parameters:

$$R_0 = 100\,\Omega \text{ at } 0°\text{C}$$

$$\alpha_1 = 3.9083 \times 10^{-3}/°\text{C}$$

$$\alpha_2 = -5.775 \times 10^{-7}/(°\text{C})^2$$

The RTD is to be calibrated by linearization in the temperature range 0°C–100°C for generating output of 0–10 V. Consider a gain of 200 in the OPAMP.

1. Determine the sensitivity of the RTD at 10°C, 50°C, and 100°C in (Ω/°C) before linearization.
2. The optimal value of R and current I_0.
3. The output voltages at initial, mid, and end value of temperatures.

Solution

Given values:

$$R_0 = 100\,\Omega,\ \alpha_1 = 3.9083\times10^{-3}/°C,\ \alpha_2 = -5.775\times10^{-7}/(°C)^2, \text{ and } k_a = 200$$

1. The RTD resistance from Equation 5.11 is

$$R_t = R_0(1+\alpha_1 T^2 + \alpha_2 T^2)$$

Putting the values of R_0, α_1, α_2, and T, we get

$$R_{t,10} = 100[1+3.9083\times10^{-3}\times10-5.775\times10^{-7}\times10^2]$$
$$= 103.9\ \Omega$$

$$R_{t,50} = 100[1+3.9083\times10^{-3}\times50-5.775\times10^{-7}\times50^2]$$
$$= 119.39\ \Omega$$

$$R_{t,100} = 100[1+3.9083\times10^{-3}\times100-5.775\times10^{-7}\times100^2]$$
$$= 138.50\ \Omega$$

Sensitivity at a particular temperature is

$$S_t = \frac{R_t - R_0}{T - T_0}$$

$$S_{10} = \frac{103.9-100}{10-0} = \frac{3.9}{10} = 0.39\ \Omega/°C$$

$$S_{50} = \frac{119.39-100}{50-0} = \frac{19.39}{50} = 0.387\ \Omega/°C$$

$$S_{100} = \frac{138.50-100}{100-0} = \frac{38.5}{100} = 0.385\ \Omega/°C$$

2. Since the calibration is to be done in the temperature range 0°C–100°C

$$T_i = 0°C, \quad T_m = 50°C, \quad \text{and} \quad T_f = 100°C$$

The corresponding RTD resistances

$$R_i = 100\ \Omega,\quad R_m = 119.39\ \Omega,\quad \text{and}\quad R_f = 138.50\ \Omega$$

Using Equation 5.20

$$R = 200\times\frac{138.50(119.39-100)-100(138.5-119.39)}{2\times 119.39-100-138.5}$$

$$= 200\times\frac{2685.51-1911}{0.28}$$

$$= 0.553\ \text{M}\Omega$$

From Equation 5.18

$$I_0 = \frac{v_{0f}(1-K_a(R_f/R))}{K_a(R_f-R_i)}$$

Since the output voltage at end value of temperature T_f is 10 V, i.e., v_{0f}= 10 V

$$I_0 = \frac{10(1-200\times(138.5/0.553\times 10^6))}{200(138.5-100)}$$

$$= 1.23\ \text{mA}$$

3. The output voltage equation

$$v_0 = \frac{K_a I_0(R_t-R_i)}{(I-(R_t K_a/R))}$$

The output voltage at 0°C, 50°C, and 100°C are

$$v_{0,0} = \frac{200\times 1.23\times 10^{-3}(100-100)}{(1-((100\times 200)/0.553\times 10^6))}$$

$$= 0\ \text{V}$$

$$v_{0,50} = \frac{200\times 1.23\times 10^{-3}(119.39-100)}{(1-((119.39\times 200)/0.553\times 10^6)}$$

$$= 4.98\ \text{V}$$

$$v_{0,100} = \frac{200\times 1.23\times 10^{-3}(138.5-100)}{(1-((138.5\times 200)/0.553\times 10^6)}$$

$$= 9.98\ \text{V}$$

5.2.1.2 Positive Feedback OPAMP Circuit (Voltage Source)

The positive feedback OPAMP circuit using voltage source [2] is shown in Figure 5.3, which is commercially available in the integrated circuit (IC) form (MAX 4236-37A). The OPAMP circuit provides a positive feedback through R_2. The voltage applied to the non-inverting terminal of the OPAMP can be obtained by applying superposition theorem. Consider the voltage source V_s and output V_0 shorted to ground

$$V_+^s = V_s \frac{R_{2t}}{R_{2t} + R_5}$$

$$\text{where } R_{2t} = \frac{R_2 R_t}{R_2 + R_t} \tag{5.21}$$

and R_t is the RTD resistance.

Again considering the positive feedback from V_0 and V_s shorted to ground

$$V_+^0 = V_0 \frac{R_{5t}}{R_{5t} + R_2}$$

$$\text{where } R_{5t} = \frac{R_5 R_t}{R_5 + R_t} \tag{5.22}$$

Therefore, the total voltage at non-inverting terminal

$$V_+^s + V_+^0 = V_s \frac{R_{2t}}{R_{2t} + R_5} + V_0 \frac{R_{5t}}{R_{5t} + R_2} \tag{5.23}$$

The voltage at the inverting terminal of the OPAMP

$$V_- = V_0 \frac{R_4}{R_4 + R_3} \tag{5.24}$$

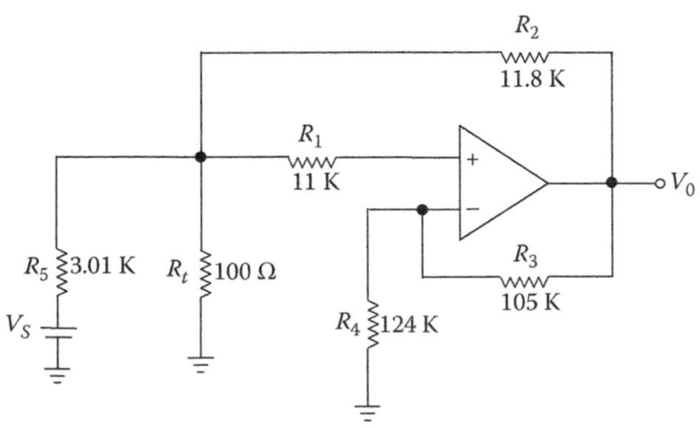

FIGURE 5.3
Positive feedback OPAMP linearization circuit (voltage source). (Reprinted from MAXIM, copyright © www.maxim-ic.com. With permission.)

For calibrating the RTD, the differential voltage can be equated to zero, such that

$$V_+^s + V_+^0 = V_- \tag{5.25}$$

Using Equations 5.21 through 5.25, we get

$$V_0 = \frac{V_s \times (R_{2t}/(R_{2t}+R_5))}{(R_4/(R_4+R_3)) - (R_{5t}/(R_{5t}+R_2))} \tag{5.26}$$

In the above transfer function, the denominator decreases due to the negative term, which is proportional to temperature and, thereby, the proportionality of output voltage with temperature is maintained. Note that Equation 5.26 resembles Equation 5.6 but with a negative term in the denominator.

Now we will calculate the output voltage of the linearization circuit with the circuit parameters shown over the temperature range of –100°C to +200°C. Let us consider the temperature points –100°C, –50°C, 0°C, 50°C, 100°C, 150°C, and 200°C.

The RTD resistance at these temperature readings can be calculated using the equation

$$R_t = R_0[1 + \alpha_1 T + \alpha_2 T^2 + \alpha_3 (T-100)^3]$$

The values of R_0, α_1, and α_2 are shown in Equation 5.11. The calculated resistances are

$$R_{-100} = 60.26\,\Omega; \quad R_{-50} = 80.31\,\Omega; \quad R_0 = 100\,\Omega; \quad R_{50} = 119.4\,\Omega; \quad R_{100} = 138.5\,\Omega;$$

$$R_{150} = 157.32\,\Omega; \quad R_{200} = 175.84\,\Omega$$

At –100°C

$$R_{2t} = \frac{11.8 \times 0.06026}{11.8 + 0.06026} = 0.0599\,\text{k}\Omega \quad \text{(using Equation 5.21)}$$

$$R_{5t} = \frac{3.01 \times 0.06026}{3.01 + 0.06026} = 0.0590\,\text{k}\Omega \quad \text{(using Equation 5.22)}$$

Again, using Equation 5.26, we get

$$V_{0,-100} = \frac{5 \times \left(\dfrac{0.0599}{0.0599 + 3.01}\right)}{\left(\dfrac{12.4}{12.4 + 105}\right) - \left(\dfrac{0.0590}{0.059 + 11.8}\right)}$$

$$= 0.969 \text{ V}$$

At −50°C

$$R_{2t} = \frac{11.8 \times 0.08031}{11.8 + 0.08031} = 0.0797\ \text{k}\Omega$$

$$R_{5t} = \frac{3.01 \times 0.08031}{3.01 + 0.08031} = 0.0782\ \text{k}\Omega$$

$$V_{0,-50} = \frac{5 \times \left(\dfrac{0.0797}{0.0797 + 3.01}\right)}{0.10562 - \left(\dfrac{0.0782}{0.0782 + 11.8}\right)}$$

$$= 1.301\ \text{V}$$

At 0°C

$$R_{2t} = \frac{11.8 \times 0.1}{11.8 + 0.1} = 0.0991\ \text{k}\Omega$$

$$R_{5t} = \frac{3.01 \times 0.1}{3.01 + 0.1} = 0.096\ \text{k}\Omega$$

$$V_{0,0} = \frac{5 \times \left(\dfrac{0.0991}{0.0991 + 3.01}\right)}{0.10562 - \left(\dfrac{0.096}{0.096 + 11.8}\right)}$$

$$= 1.633\ \text{V}$$

At 50°C

$$R_{2t} = \frac{11.8 \times 0.1194}{11.8 + 0.1194} = 0.1182\ \text{k}\Omega$$

$$R_{5t} = \frac{3.01 \times 0.1194}{3.01 + 0.1194} = 0.1148\ \text{k}\Omega$$

$$V_{0,50} = \frac{5 \times \left(\dfrac{0.1182}{0.1182 + 3.01}\right)}{0.10562 - \left(\dfrac{0.1148}{0.1148 + 11.8}\right)}$$

$$= 1.974\ \text{V}$$

At 100°C

$$R_{2t} = \frac{11.8 \times 0.1385}{11.8 + 0.1385} = 0.13682\ \text{k}\Omega$$

$$R_{5t} = \frac{3.01 \times 0.1385}{3.01 + 0.1385} = 0.1324\ \text{k}\Omega$$

$$V_{0,100} = \frac{5 \times \left(\dfrac{0.1368}{0.1368 + 3.01}\right)}{0.10562 - \left(\dfrac{0.1324}{0.1324 + 11.8}\right)}$$

$$= 2.298\ \text{V}$$

At 150°C

$$R_{2t} = \frac{11.8 \times 0.15732}{11.8 + 0.15732} = 0.1552\ \text{k}\Omega$$

$$R_{5t} = \frac{3.01 \times 0.15732}{3.01 + 0.15732} = 0.1495\ \text{k}\Omega$$

$$V_{0,150} = \frac{5 \times \left(\dfrac{0.1552}{0.1552 + 3.01}\right)}{0.10562 - \left(\dfrac{0.1495}{0.1495 + 11.8}\right)}$$

$$= 2.632\ \text{V}$$

At 200°C

$$R_{2t} = \frac{11.8 \times 0.17584}{11.8 + 0.17584} = 0.1732\ \text{k}\Omega$$

$$R_{5t} = \frac{3.01 \times 0.17584}{3.01 + 0.17584} = 0.1661\ \text{k}\Omega$$

$$V_{0,200} = \frac{5 \times \left(\dfrac{0.1732}{0.1722 + 3.01}\right)}{0.10562 - \left(\dfrac{0.1662}{0.1661 + 11.8}\right)}$$

$$= 2.965\ \text{V}$$

To analyze the linearity achieved from the circuit, the sensitivities at different temperatures must be calculated. Let us do that here:

Sensitivity between −100°C and −50°C:

$$= \frac{1.301 - 0.969}{-50 - (-100)} = 6.64 \times 10^{-3}\ \text{V/°C}$$

Sensitivity between −50°C and 0°C:

$$= \frac{1.633 - 1.301}{0 - (-50)} = 6.64 \times 10^{-3}\ \text{V/°C}$$

Sensitivity between −0°C and 50°C:

$$= \frac{1.974 - 1.633}{50 - 0} = 6.82 \times 10^{-3}\ \text{V/°C}$$

Sensitivity between 50°C and 100°C:

$$= \frac{2.298 - 1.974}{100 - 50} = 6.48 \times 10^{-3}\ \text{V/°C}$$

Sensitivity between 100°C and 150°C:

$$= \frac{2.632 - 2.298}{150 - 100} = 6.68 \times 10^{-3}\ \text{V/°C}$$

Sensitivity between 150°C and 200°C:

$$= \frac{2.965 - 2.632}{200 - 1500} = 6.66 \times 10^{-3}\ \text{V/°C}$$

The average sensitivity of the RTD linearization circuit is 6.65 mV/°C. The highest nonlinearity observed in the range 0°C–150°C with a deviation in sensitivity of 0.17 mV/°C which is 2.55% of the average sensitivity.

Example 5.2

The RTD of Example 5.1 is linearized using the circuit of Figure 5.3.

1. What should be the value of V_s to achieve a uniform sensitivity of 13.3 mV/°C?
2. Design an OPAMP to calibrate the output of the linearization circuit in the temperature range of −100°C to +200°C with corresponding output voltage range of −2.5 to 5 V. Take data of (1) above.

Solution

1. To achieve a uniform sensitivity in the operating range of −100°C to +200°C, let us consider the two extreme temperature points.
   At −100°C:
   Using Equation 5.11

$$R_{-100} = 100[1+3.908\times10^{-3}\times(-100)-5.775\times10^{-7}\times(-100)^2$$

$$= 60.26\,\Omega$$

$$R_{2t} = 0.0599\text{ k}\Omega \quad \text{(using Equation 5.21)}$$

$$R_{5t} = 0.0590\text{ k}\Omega \quad \text{(using Equation 5.22)}$$

$$V_{0,-100} = 0.1937V_s$$

Similarly at 200°C

$$R_{200} = 100[1+3.908\times10^{-3}\times(200)-5.775\times10^{-7}\times(200)^2$$

$$= 175.84\,\Omega$$

$$R_{2t} = 0.1732\text{ k}\Omega$$

$$R_{5t} = 0.1661\text{ k}\Omega$$

$$V_{0,200} = 0.5929V_s$$

The uniform sensitivity over the temperature range −100°C to +200°C is given by

$$S = \frac{V_{0,200}-V_{0,100}}{200-100}$$

$$= \frac{0.5929-0.1937V_s}{100}\text{ mV/°C}$$

$$= 1.33V_s\text{ mV/°C}$$

The required sensitivity is 13.3 mV/°C

$$\therefore 1.33V_s = 13.3$$

$$\Rightarrow V_s = \frac{13.3}{1.33} = 10\text{ V}$$

2. Design of calibrating OPAMP:
   Input to the OPAMP at the two extreme points

$$V_{0,-100} = 0.1937\times10 = 1.93\text{ V}$$

$$V_{0,200} = 0.5929\times10 = 5.92\text{ V}$$

Let us calculate the voltage at 0°C:

$R_{2t} = 0.0991 \text{k}\Omega$ (using Equation 5.21)
$R_{5t} = 0.096 \text{k}\Omega$ (using Equation 5.22)
$V_{0,0} = 3.26 \text{ V}$ (using Equation 5.26)

The temperature linearized output and the calibrated output voltage are tabulated below:

Temperature (°C)	Linearized Output (V)	Calibrated Output (V)
−100	1.93	−2.5
0	3.26	0
200	5.92	5.0

Range of linearized output = 5.92 − 1.93 = 3.99 V
Range of calibrated output = 5.00 − (−2.5) = 7.50 V
Amplification factor required = 7.5/3.99 = 1.87

The amplifier circuit is shown in Figure E5.2. The amplifier is used in the inverting mode; the non-inverting input is applied with a fixed voltage of 3.26 V from a 5 V power supply through a pot. The linearized output v_0 is applied to the inverting terminal. A gain of 18.7 is fixed by the pot (20 kΩ) connected to the feedback path.

At −100°C, using equation $v_0 = -(R_t/R_i)(v_t - v_-)$

$$v_{out} = -18.7(3.26 - 1.93)$$
$$= -2.5 \text{ V}$$

At 0°C

$$v_{out} = -18.7(3.26 - 3.26)$$
$$= 0 \text{ V}$$

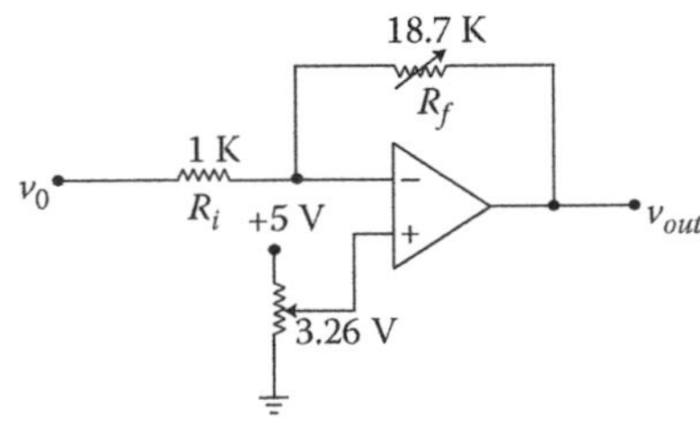

FIGURE E5.2
OPAMP.

At 200°C

$$v_{out} = -18.7(3.26 - 5.92)$$
$$= 5\ \text{V}$$

5.2.1.3 Linearization Using Feedback Amplifier

The output voltage obtained from the bridge circuit consisting of the nonlinear resistive sensor in one of the branches is also nonlinear. In such circuits, linearization is typically performed digitally using a microcontroller.

Nobbs [3] suggested a simple analog method for the linearization of RTD, where a current is passed through the RTD in such a way that it depends on its resistance, thereby getting a linear output voltage.

It was mentioned earlier in this section that linearization of a nonlinear resistive sensor demands Equation 5.6 to have a negative part in the denominator, which is of the form

$$R_{Tx} = \frac{R_x}{1-(R_x/R)} \tag{5.27}$$

An equation similar to Equation 5.27 can be obtained for the sensor [3], which is given by

$$A_t = \frac{R_t}{1-\beta R_t} \tag{5.28}$$

where
R_t is the RTD resistor
β is a constant

Equation 5.28 can be realized using an OPAMP, as shown in Figure 5.4 [3]. In the circuit, V_s is a reference voltage and V_t is the linearized output voltage. Considering the reference voltage V_s being amplified, we get

$$V_{0s} = -\frac{R_2}{R_3}V_s \tag{5.29}$$

and, the amplified output V_{ot}

$$V_{ot} = \left(1+\frac{R_2}{R_3}\right)V_t \tag{5.30}$$

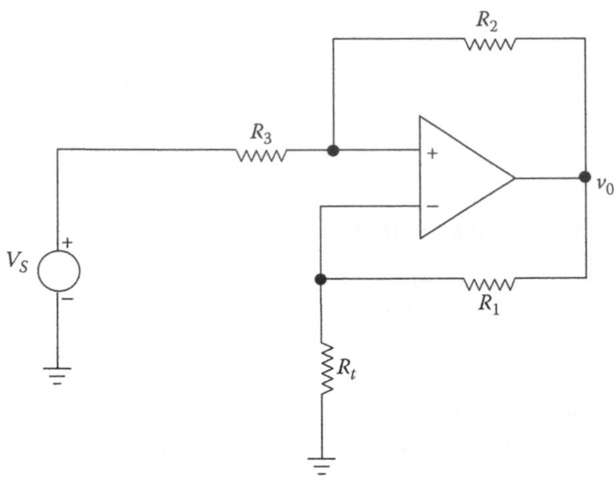

FIGURE 5.4
OPAMP linearization circuit for RTD. (Reprinted from Institute of Physics Publishing Ltd. With permission.)

and $V_0 = V_{os} + V_{ot}$

$$\therefore V_0 = \left(1 + \frac{R_2}{R_3}\right)V_t - \frac{R_2}{R_3}V_s \tag{5.31}$$

again

$$V_t = V_0 \frac{R_t}{R_1 + R_t}$$

$$= \left(1 + \frac{R_2}{R_3}\right)\frac{R_t}{R_1 + R_t}V_t - \frac{R_2}{R_3}\frac{R_t}{(R_1 + R_t)}V_s$$

$$\therefore V_t = -\frac{V_s(R_2/R_3)(R_t/(R_1 + R_t))}{1 - (1 + (R_2/R_3))(R_t/(R_1 + R_t))}$$

$$V_t = -\frac{V_s K_x}{1 - (K_x/K)} \tag{5.32}$$

where

$K_x = (R_2/R_3)(R_t/(R_1 + R_t))$

$K = R_2/(R_2 + R_3)$

With some approximation, Equation 5.32 can be written as

$$V_t = -\frac{V_s B R_t}{(1 - B R_t)} \tag{5.33}$$

where $B = R_2/(R_1 R_3)$.

Writing Equation 5.33 using Equation 5.28

$$V_t = -A_t V_s \tag{5.34}$$

To find out the optimal value of B in Equation 5.33, the value of A_t is calculated for different values of B and temperature. Let us consider the same RTD defined in Equation 5.11 and calculate V_t for various values of B and temperature at −100°C to 200°C. We consider here a reference voltage $V_s = 5$ V. The value of B was changed from 1×10^{-4} to 6×10^{-4} in steps of 1×10^{-4}. It was observed that the variation of sensitivity ($\Delta V_t/\Delta T$) is least for $B = 2 \times 10^{-4}$. Hence the V_t for $B = 2 \times 10^{-4}$ was found as follows:

At −100°C, $V_t = -0.0610$ V

At −50°C, $V_t = -0.0816$ V

At 0°C, $V_t = -0.1020$ V

At 50°C, $V_t = -0.1223$ V

At 100°C, $V_t = -0.1424$ V

At 150°C, $V_t = -0.1624$ V

At 200°C, $V_t = -0.1822$ V

From Equation 5.33

$$B = \frac{R_2}{R_1 R_3} = 2 \times 10^{-4}$$

Taking $R_1 = 10$ kΩ and $R_3 = 5$ kΩ, we get $R_2 = 10$ kΩ.

Note that Equation 5.32 was approximated to Equation 5.33 to justify that the equation resembles Equation 5.18. Now we will directly apply Equation 5.32 to calculate V_t at different temperature. Writing Equation 5.32

$$V_t = -\frac{V_s \dfrac{R_2}{R_3} \dfrac{R_t}{(R_1 + R_t)}}{1 - \left(1 + \dfrac{R_2}{R_3}\right) \dfrac{R_t}{R_1 + R_t}}$$

Putting values of $V_s = 5\,\text{V}$ and $(R_2/R_3) = 2$

$$V_t = \frac{10R_t}{10 - 2R_t}$$

At −100°C, $V_t = -0.0609\,\text{V}$
At −50°C, $V_t = -0.0816\,\text{V}$
At 0°C, $V_t = -0.1020\,\text{V}$
At 50°C, $V_t = -0.1223\,\text{V}$
At 100°C, $V_t = -0.1424\,\text{V}$
At 150°C, $V_t = -0.1624\,\text{V}$
At 200°C, $V_t = -0.1822\,\text{V}$

Hence, the voltages (V_t) calculated from the direct Equation 5.32 and the approximated Equation 5.33 are equal.

Example 5.3

An RTD with $R_0 = 200\,\Omega$ and $\alpha_1 = 3.9083 \times 10^{-3}$, $\alpha_2 = -5.775 \times 10^{-7}$, and $\alpha_3 = -4.183 \times 10^{-12}$ (for a temperature below 0°C) is linearized by the OPAMP circuit of Figure 5.4. The OPAMP uses a reference voltage of 5 V. Determine the resistance values when the output voltages V_t at different temperatures are as shown below:

T (°C):	−100	0	100
V_t (V):	−0.080	−0.121	−0.163

Solution
Given RTD data: $R_0 = 200\,\Omega$, $\alpha_1 = 3.9083 \times 10^{-3}$, $\alpha_2 = -5.775 \times 10^{-7}$, $V_s = 5\,\text{V}$.
The RTD resistances at the specified temperature are
At −100°C,

$$R_{-100} = 200[1 + 3.9003 \times 10^{-3} \times (-100) - 5.775 \times 10^{-7} \times (-100)^{-2} - 4.183 \times 10^{-12} \times (-100)^{-3}]$$
$$= 120.67\,\Omega$$

At 0°C,

$$R_0 = 200\,\Omega$$

At 100°C,

$$R_{100} = 200(1+3.9083\times10^{-3}\times100-5.775\times10^{-7}\times(100)^2$$
$$= 277\,\Omega$$

Using Equation 5.33

At −100°C,

$$-0.080 = \frac{-5\times B\times 0.120}{(1-B\times 0.120)}; \quad \text{by solving we get, } B = 0.1149$$

At 0°C,

$$-0.121 = \frac{-5\times B\times 0.20}{(1-B\times 0.20)}; \quad \text{by solving } B = 0.1008$$

At 100°C,

$$-0.163 = \frac{-5\times B\times 0.277}{(1-B\times 0.277)}; \quad \text{by solving } B = 0.1139$$

Taking the average of the three values, $B=0.1098$.
From Equation 5.33

$$B = (R_2/R_1R_3)$$

Taking $R_1 = 10\,\text{k}\Omega$, $R_3 = 5\,\text{k}\Omega$

$$\therefore R_2 = 5.5\,\text{k}\Omega$$

The amplifier circuit is shown in Figure E5.3.

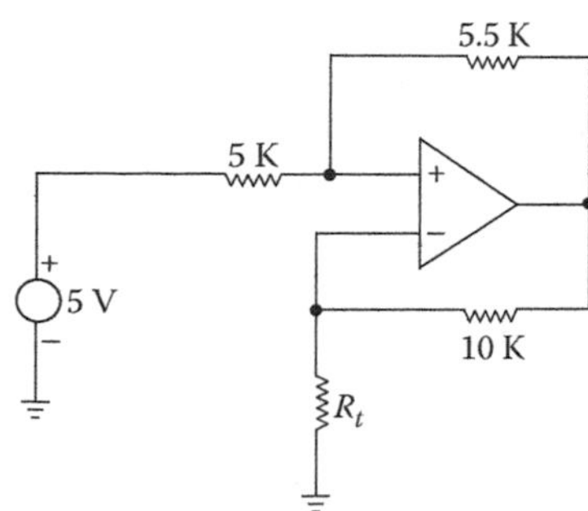

FIGURE E5.3
Amplifier circuit of Example 5.3.

5.3 Linearization of Negative Coefficient Resistive Sensors

The general equation of a negative coefficient resistive sensor and its sensitivity with input is shown in Equations 5.1 and 5.3. A typical example of negative coefficient resistive sensor is a thermistor. The change in resistance with temperature for a thermistor is given by

$$R_t = R_0 e^{\beta\left(\frac{1}{T} - \frac{1}{T_0}\right)} \tag{5.35}$$

where

- T is the thermistor temperature in K
- T_0 is the reference temperature (say 298 K)
- R_0 is the resistance of the thermistor at temperature T_0
- β is the thermistor constant or characteristic temperature in K

The usual value of β ranges from 2000 to 4000 K.

Nonlinear resistive sensor can be linearized with certain limitation by shunting it to a fixed resistor R. The parallel combination of the thermistor resistance R_t and the fixed resistor R is given by

$$R_p = \frac{RR_t}{R + R_t} \tag{5.36a}$$

The sensitivity of the combination is

$$\frac{dR_p}{dT} = \frac{R^2}{(R + R_t)^2} \frac{dR_t}{dT} \tag{5.36b}$$

Let us analyze the sensitivity characteristic of a thermistor with $R_0 = 10\,\text{k}\Omega$ at 25°C and $\beta = 3965\,\text{K}$. If we plot R_t for temperature from 0°C to 100°C, the nonlinearity is evident from the plot shown in Figure 5.5a. Now, if the thermistor is shunted by a resistor having the same value as that of the thermistor resistance at 20°C, the nonlinearity is improved, however, the sensitivity is reduced (Figure 5.5b).

Although we have assumed a value for the fixed shunting resistor, an optimal value of R can be chosen. To do so, we take three equivalent points in the resistance temperature characteristic so that [4]

$$T_2 - T_1 = T_3 - T_2$$

$$\text{and } R_{p2} - R_{p1} = R_{p3} - R_{p2}$$

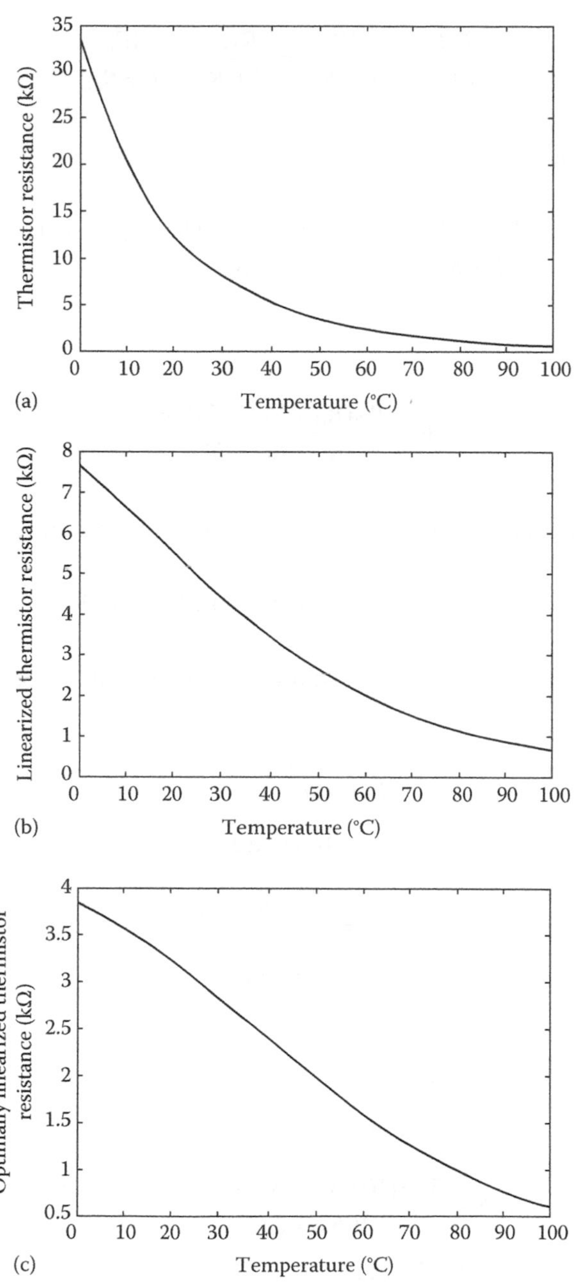

FIGURE 5.5
Thermistor characteristics: (a) nonlinear, (b) linearized, and (c) optimally linearized.

From Equation 5.28

$$\frac{RR_{t2}}{R+R_{t2}} - \frac{RR_{t1}}{R+R_{t1}} = \frac{RR_{t3}}{R+R_{t3}} - \frac{RR_{t2}}{R+R_{t2}}$$

By rearranging

$$\frac{R(R_{t1}+R_{t2})}{(R+R_{t1})} = \frac{RR_{t2}(R+R_{t3})-(R+R_{t2})(RR_{t2})}{(R+R_{t2})(R+R_{t3})}$$

we get

$$R = \frac{R_{t2}(R_{t1}+R_{t3})-2R_{t1}R_{t3}}{R_{t1}+R_{t3}-2R_{t2}} \tag{5.37}$$

For this thermistor, the three specified equidistant temperature points are

$$T_1 = -20°\text{C}, \quad T_2 = 40°\text{C}, \quad \text{and } T_3 = 100°\text{C}.$$

The corresponding thermistor resistances (using Equations 5.35 and 5.37) are

$$R_{t1} = 10e^{3965\left(\frac{1}{253}-\frac{1}{298}\right)}$$
$$= 106.6 \text{ k}\Omega$$

$$R_{t2} = 10e^{3965\left(\frac{1}{313}-\frac{1}{298}\right)}$$
$$= 5.28 \text{ k}\Omega$$

$$R_{t3} = 10e^{3965\left(\frac{1}{373}-\frac{1}{298}\right)}$$
$$= 0.68 \text{ k}\Omega$$

$$\therefore R = \frac{5.28(106.6+0.68)-2\times 106.6\times 0.68}{106.6+0.68-2\times 5.28}$$
$$= 4.35 \text{ k}\Omega$$

When the thermistor is shunted by the resistor with the optimal value of 4.35 kΩ, the linearization is better than that of Figure 5.5b. The new

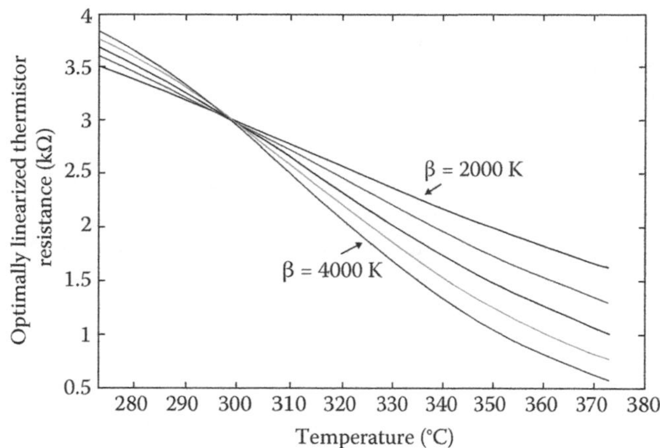

FIGURE 5.6
Thermistor linearization characteristics with different values of β.

characteristic is shown in Figure 5.5c. The constant β has a role to play in the linearization of the thermistor.

Note that in temperature-voltage characteristic (Figure 5.6), lower values of β show better linearity over a wider temperature range, while higher values of β produce increased sensitivity over a narrower temperature range. The MATLAB® program for the thermistor linearization is shown below:

```
%This program simulates a thermistor with a resistance of 10K
ohm at 25 Deg. Centigrade and % Beta of 2000K and changed to
4000K. The thermistor before and after linearization is
plotted. % The linearization for different values of beta are
also plotted. Temperature is changed from 0% to 100Degree
Centigrade
clc;
close all;
clear all;
ro=10; % thermistor resistance at 25 Deg C
b=2000:500:4000; % Change Beta
t1=0:100;% Change temperature
t=t1+273;
r =ro*(exp(b.*((1./t)-(1/298)))) % Thermistor resistance
before linearization
figure(1);
plot(t1,r)
xlabel('Temperature,^o C')
ylabel('Thermistor resistance, K ohm')
rl=r.*ro./(r+ro); % Thermistor resistance with shunted
resistance
figure(2)
plot(t1,rl)
```

```
xlabel('Temperature,^o C')
ylabel('Linearized thermistor resistance, K ohm')
for i=1:length(b)
  j=1:length(t)
  b1=b(i)
  t11=t(j)
% Thermistor resistance by changing Beta
  r(i,j)=ro*(exp(b1.*((1./t11)-(1/298))))
% Thermistor resistance shunted by optimal resistance
  ro1=r.*(4.35)./(r+4.35);
end
figure(3)
plot(t11,ro1)
xlabel('Temperature,^o C')
  ylabel('Optimally linearized thermistor resistance, K ohm')
```

A thermistor can be linearized in voltage mode by connecting a resistance in series with the thermistor, to form a voltage divider circuit as shown in Figure 5.7a. For the thermistor, as discussed for analysis earlier, with $R_{25} = 10\,\text{k}\Omega$ and $\beta = 3965\,\text{K}$, we take the series resistance $R = 10\,\text{k}\Omega$ and a voltage of $V_s = 5.0\,\text{V}$.

Example 5.4

A thermistor with $R_0 = 5\,\text{k}\Omega$ at 25°C and $\beta = 3000\,\text{K}$ is linearized by two different methods as described below over a temperature range of –20°C to 100°C.

Method A: By shunting with a resistor of optimal value R. The parallel combination $(R_t \| R)$ is connected to branch-1 of an unbalanced Wheatstone bridge with a supply voltage of 5V. The resistors of the other three branches are made

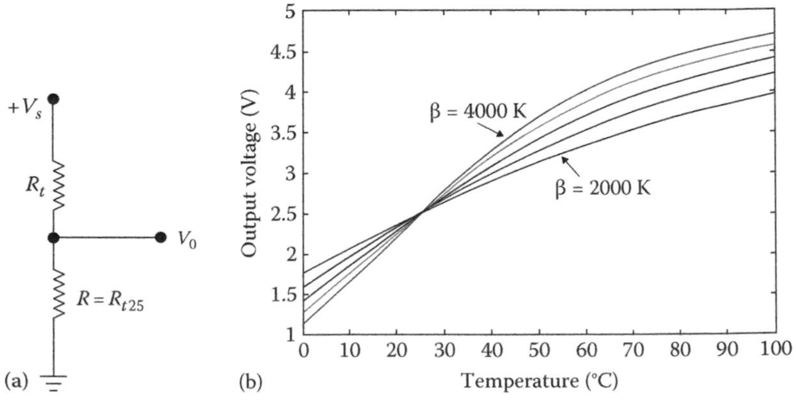

FIGURE 5.7
(a) Voltage divider linearization of thermistor and (b) temperature–voltage characteristics for different values of β.

equal to ($R_0 \| R$). The output voltage of the bridge is the linearized response of the thermistor.

Method B: By forming a voltage divider circuit with a fixed resistance of R_{25} in series with the thermistor (Figure 5.7a). The supply voltage V_s is 5 V. The output voltage V_0 is the linearized response of the thermistor.

Analyze and compare the performance of the two linearization methods using three equidistant temperature points: −20°C, 40°C and 100°C.

Solution

Given data: $R_{25} = 5\,\text{k}\Omega$, $\beta = 3000\,\text{K}$, supply voltage $V_s = 5\,\text{V}$

Temperature range: −20°C to 100°C.

For the analysis of the linearization, resistances at these temperatures are

At −20°C,

$$R_{-20} = 5e^{3000\left(\frac{1}{253} - \frac{1}{298}\right)} = 29.96\,\text{k}\Omega$$

At 40°C,

$$R_{40} = 5e^{3000\left(\frac{1}{313} - \frac{1}{298}\right)} = 3.086\,\text{k}\Omega$$

At 100°C,

$$R_{100} = 5e^{3000\left(\frac{1}{373} - \frac{1}{298}\right)} = 0.66\,\text{k}\Omega$$

From Equation 5.37, the optimal value of shunt resistance

$$R = \frac{3.086(29.96 + 0.660) - 2 \times 29.96 \times 0.660}{29.96 + 0.660 - 2 \times 3.086}$$

$$= 2.24\,\text{k}\Omega$$

The parallel combination of the thermistor and shunt resistance of 2.24 kΩ are

$$\text{At } -20°\text{C}, R_{P-20} = 29.96 \,\|\, 2.24 = 2.084\,\text{k}\Omega$$

$$\text{At } 40°\text{C}, R_{P40} = 3.086 \,\|\, 2.24 = 1.290\,\text{k}\Omega$$

$$\text{At } 100°\text{C}, R_{100} = 0.660 \,\|\, 2.24 = 0.509\,\text{k}\Omega$$

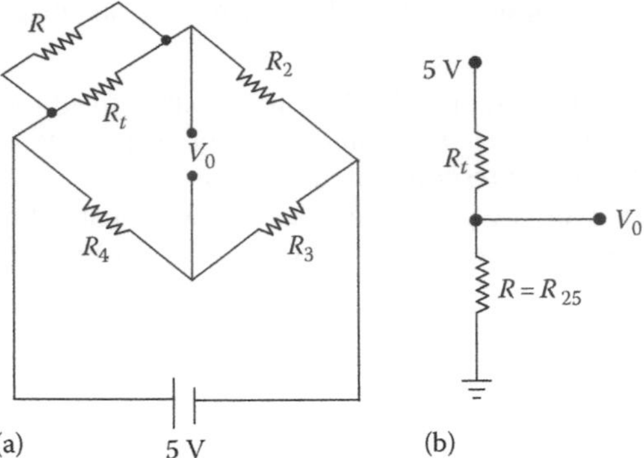

FIGURE E5.4
Circuit diagrams for Example 5.4: (a) Wheatstone bridge and (b) voltage divider.

For finding out the unbalanced output voltage from the Wheatstone bridge circuit shown in Figure E5.4a, the output voltage equation is

$$V_0 = V_s\left[\frac{R_p}{R_p + R_{25}} - \frac{1}{2}\right]$$

In Figure E5.4, $R_2 = R_3 = R_4 = R_0 \parallel R$.
where

$$R_0 = 5e^{3000\left(\frac{1}{275} - \frac{1}{298}\right)} = 12.57\ \text{k}\Omega$$

Therefore, $R_2 = R_3 = R_4 = R_0 \parallel R$.

$$= \frac{12.57 \times 2.24}{12.57 + 2.34}$$

$$= 1.901\,\text{k}\Omega.$$

The output voltages are

$$\text{At } -20°\text{C, } 5 \times \left[\frac{1}{2} - \frac{2.084}{2.084 + 1.901}\right] = -114.80\ \text{mV}$$

$$\text{At } 40°\text{C, } 5 \times \left[\frac{1}{2} - \frac{1.290}{1.290 + 1.901}\right] = 478.69\ \text{mV}$$

$$\text{At } 100°\text{C, } 5 \times \left[\frac{1}{2} - \frac{0.509}{0.509 + 1.901}\right] = 1443.98\ \text{mV}$$

The sensitivities

$$-20°\text{C to } 40°\text{C}: \quad \frac{478.80-(-114.80)\text{ mV}}{60°\text{C}}$$

$$= 9.89 \text{ mV/°C}$$

$$40°\text{C to } 100°\text{C}: \quad \frac{1443.98-(478.8)\text{ mV}}{60°\text{C}}$$

$$= 16.08 \text{ mV/°C}$$

Change in sensitivity = 6.19 mV/°C

Method B:
The voltage divider circuit is shown in Figure E5.4b.
The output voltages (V_0) at various temperatures are as follows:

$$\text{At } -20°\text{C}, V_{0,-20} = 5\times\frac{5}{5+29.96} = 0.715 \text{ V}$$

$$\text{At } 40°\text{C}, V_{0,40} = 5\times\frac{5}{5+3.08} = 3.09 \text{ V}$$

$$\text{At } 100°\text{C}, V_{0,100} = 5\times\frac{5}{5+0.66} = 4.14 \text{ V}$$

The sensitivities

$$-20 \text{ to } 40°\text{C}: \quad \frac{(3.09-0.715)}{60°\text{C}}$$

$$= 39.58 \text{ mV/°C}$$

$$40°\text{C}-100°\text{C}: \quad \frac{(4.41-3.09)}{60°\text{C}}$$

$$= 22.08 \text{ mV/°C}$$

Change in sensitivities = 17.50 mV/°C.

Therefore, linearity achieved in method A is higher; however, the sensitivity in method B is higher.

5.4 Higher-Order Linearization Using MOS

Input–output relationships of most sensors are nonlinear in nature and second-order models are most common. Although sensors are approximated by second order models, for more accurate results, higher-order model approximations are required.

Linearization techniques using OPAMP circuits discussed in Sections 5.2 and 5.3 are limited to second-order models only. Third- or higher-order linearization is possible with MOS-based vector multipliers [5] used to implement a linearized equation.

5.4.1 Quadratic Linearization Using Divider Circuit

Let us consider a second-order nonlinear sensor equation given by

$$E_t = at \pm bt^2 \tag{5.38}$$

where
- E_t is the quadratic signal to be linearized
- t is the measurable quantity
- a and b are the constants

Let the measurable quantity t produce a linear output signal given by

$$E_0 = \gamma t \tag{5.39}$$

where γ is the proportionality constant.

From Equations 5.38 and 5.39, we get

$$E_t = a\frac{E_0}{\gamma} \pm b\left(\frac{E_0}{\gamma}\right)^2$$

$$E_t = AE_0 \pm BE_0^2 \tag{5.40}$$

where
- $A = (a/\gamma)$
- $B = (b/\gamma^2)$

Equation 5.40 can further be written in the form

$$BE_0 = \pm\left(\frac{E_t}{E_0} - A\right)$$

or

$$E_0 = \pm\left(\frac{E_t}{E_0} - A\right)\frac{1}{B} \tag{5.41}$$

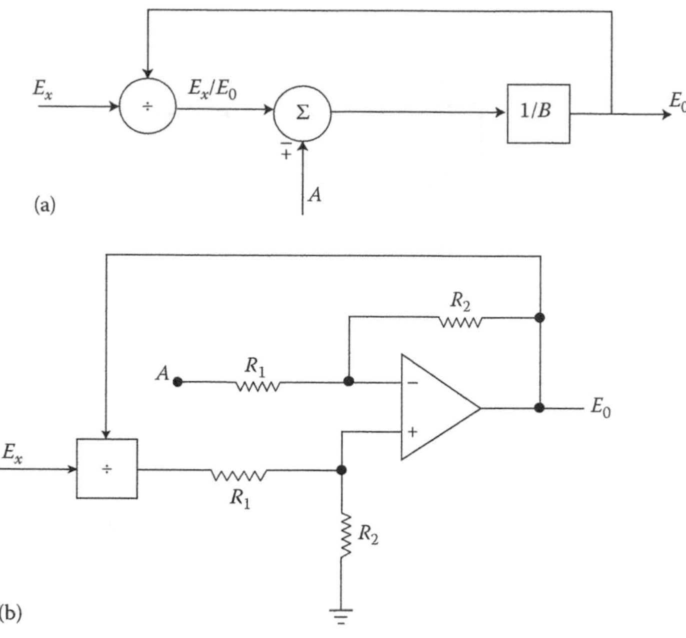

FIGURE 5.8
Second-order linearization: (a) block diagram and (b) circuit diagram. (Reprinted from IEEE, copyright © IEEE. With permission.)

Substituting the expression of E_t and E_0 from Equations 5.38 and 5.41

$$E_0 = \pm\left(\frac{at \pm bt^2}{\gamma t} - A\right)\frac{1}{B}$$

$$E_0 = \pm\left(\frac{a}{\gamma} \pm \frac{b}{\gamma}t - A\right)\frac{1}{B} \tag{5.42}$$

Equation 5.42 is linear with t; however, the constants a and b of the nonlinear equation are replaced by $((b/\gamma) - A)(1/B)$ and zero, respectively, and a constant term $\pm((a/\gamma) - A)(1/B)$ is added. Equation 5.41 can be realized by a feedback system and a corresponding circuit shown in Figure 5.8a and b, respectively.

The output of the differential OPAMP is given by

$$E_0 = -\left(\frac{E_t}{E_0} - A\right)\frac{R_2}{R_1} \tag{5.43}$$

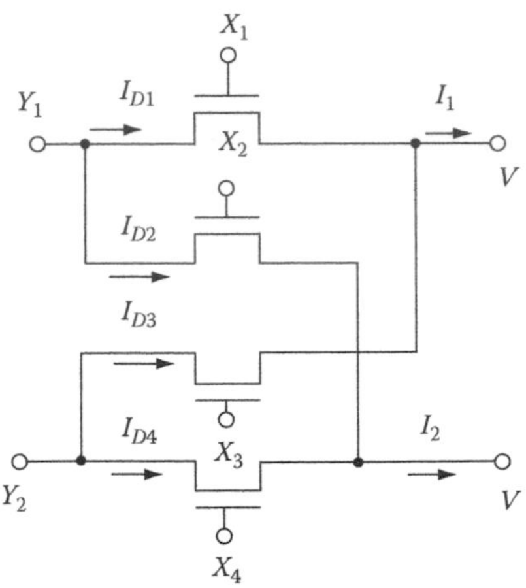

FIGURE 5.9
Four-transistor MOS transconductor. (Reprinted from IEEE, copyright © IEEE. With permission.)

where

$$\frac{R_2}{R_1} = \text{the OPAMP gain (comparing with Equation 5.41)} = \frac{1}{B}$$

The divider unit at the input of the differential amplifier can be realized by the MOS circuits.

The MOS multiplier/divider is based on a four-transistor MOS transconductor, as shown in Figure 5.9. The currents I_{L_1} and I_{L_2} are given by

$$I_{L_1} = I_{LD1} + I_{LD3} = \mu_{cox}\frac{W}{L}[(X_1 - V_{TB})(Y_1 - V) + (X_3 - V_{TB})(Y_2 - V)] \tag{5.44}$$

and

$$I_{L_2} = I_{LD2} + I_{LD4} = \mu_{cox}\frac{W}{L}[(X_1 - X_2)Y_1 + (X_3 - X_4)Y_2 + (X_2 + X_4 - X_1 - X_3)V] \tag{5.45}$$

Now the current differences $I_1 - I_2 = I_{L_1} - I_{L_2}$, therefore

$$I_1 - I_2 = \mu_{cox}\frac{W}{L}[(X_1 - X_2)Y_1 + (X_3 - X_4)Y_2 + (X_2 + X_4 - X_1 - X_3)V] \tag{5.46}$$

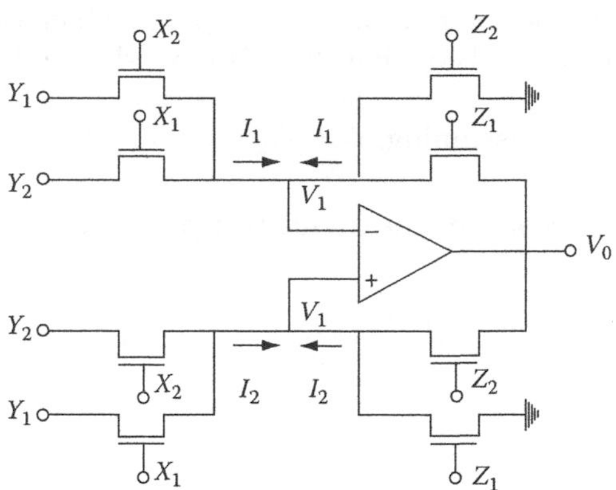

FIGURE 5.10
Multiplier and divider MOS cell. (Reprinted from IEEE, copyright © IEEE. With permission.)

The voltage nodes V are connected to an ideal and balanced OPAMP, therefore to get a linear current difference $I_1 - I_2$, V must be zero. Therefore, forcing V to zero we get

$$I_1 - I_2 = \mu_{cox} \frac{W}{L} [(X_1 - X_2)Y_1 + (X_3 - X_4)Y_2] \tag{5.47}$$

A divider circuit comprising two sets of such 4-MOS transconductors, one set connected to the input and the other to the output of the OPAMP, is shown in Figure 5.10. The current difference of the input and output transconductance is given by

$$I_1 - I_2 = -\mu_{cox} \frac{W}{L} [(Y_1 - Y_2) + (X_1 - X_2)] \tag{5.48}$$

and

$$I_1' - I_2' = \mu_{cox} \frac{W}{L} [V_0(Z_1 - Z_2)] \tag{5.49}$$

The output voltage of the cell is

$$V_0 = \left[\frac{(W/L)i}{(W/L)o} \right] \frac{\Delta Y \, \Delta X}{\Delta Z} \tag{5.50}$$

where

subscripts i and o denote input and output transconductors, respectively
$\Delta X = X_1 - X_2$
$\Delta Y = Y_1 - Y_2$
$\Delta Z = Z_1 - Z_2$

The circuit performs the computation ((ΔYΔX)/ΔZ), which can be utilized for either multiplication or division. For the cell to work as a divider, we make

$$X_1 = V_{C1};\quad X_2 = V_{C2} \text{ and assuming, } Y_1 = -Y_2 = y,\quad Z_1 = Q_0 + z \text{ and } Z_2 = Q_0 - z$$

where Q_0 is a dc shift voltage. The divider output voltage is given by

$$V_{0d} = \left[\frac{(W/L)_i(V_{C1} - V_{C2})y}{(W/L)oZ}\right]$$

$$= d_0 \frac{y}{z} \tag{5.51}$$

where $d_0 = [(W/L)_i(V_{C1} - V_{C2})/(W/L)o]$, which is the cell gain factor.

For the implementation of the 2×4 MOS transistor cell in the form of the circuit of Figure 5.8b, the following equalities are required:

$$Y_1 = -Y_2 = E_t,\quad Z_1 = Q_0 + E_0,\quad \text{and}\quad Z_2 = Q_0 - E_0$$

so that we get

$$V_{0d} = d_0 \frac{E_t}{E_0} \tag{5.52}$$

Another set of 4-MOS transconductors are used for implementing the differential amplifier (Figure 5.11) where

$$Y_1 = V_{0d},\quad Y_2 = A,\quad X_1 = V_{C1},\quad X_2 = V_{C2},\quad Z_1 = V'_{C1},\quad \text{and}\quad Z_2 = V'_{C2}$$

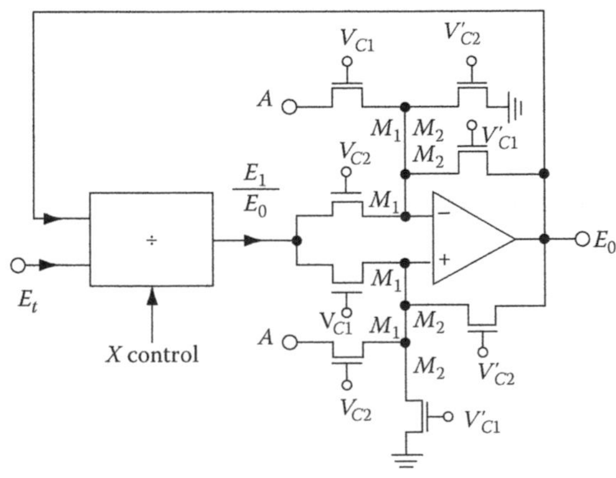

FIGURE 5.11
Quadratic linearizing MOS circuit. (Reprinted from IEEE, copyright © IEEE. With permission.)

The linearized output is given by

$$I_0 = \frac{(W/L)_i (V_{C1} - V_{C2})}{(W/L)_0 (V'_{C1} - V'_{C2})} \left[d_0 \frac{E_t}{E_0} - A \right]$$

$$= K_0 \left[d_0 \frac{E_t}{E_0} - A \right] \tag{5.53}$$

Example 5.5

A MOS linearizing cell shown in Figure 5.10 is used to linearize a pressure sensor signal given by

$$E_p = 0.5p + 0.25p^2$$

The linearized output is given by

$$E_p = 5p$$

where p ranges from 0 to 1 N/m^2. The parameters of the input–output MOS transconductors are

$$\frac{(W/L)_i}{(W/L)_0} = 50 \quad \text{and} \quad V_{C1} - V_{C2} = 2\ \text{V}$$

Determine $(V'_{C1} - V'_{C2})$ to be applied to the cell.

Solution
From the nonlinear equation
$E_p = 0.5p + 0.25p^2$ comparing with Equation 5.38

$$\therefore a = 0.5; \quad b = 0.25$$

From the linearized equation

$$E_0 = 5p$$

$$\therefore \gamma = 5$$

$$A = \frac{a}{\gamma} = \frac{0.5}{5} = 0.1$$

$$B = \frac{b}{\gamma^2} = \frac{0.25}{5^2} = 0.01$$

From Equation 5.53

$$K_0 = \frac{1}{B} = \frac{(W/L)_i (V_{C1} - V_{C2})}{(W/L)_0 (V'_{C1} - V'_{C2})}$$

Substituting the values of the *W*/*L* ratio, $(V_{C1} - V_{C2})$ and *B*

$$\frac{1}{0.01} = 50\frac{2}{V'_{C1} - V'_{C2}}$$

$$\therefore V'_{C1} - V'_{C2} = 2\times 50\times 0.01 = 1\text{ V}$$

5.4.2 Third-Order Linearization Circuit

Similar to the second-order linearization equations, the third-order nonlinear equation can be written as

$$E_t = at \pm bt^2 \pm ct^3 \tag{5.54}$$

The linearized equation

$$E_0 = \gamma t \tag{5.55}$$

$$\therefore E_t = AE_0 \pm BE_0^2 \pm CE_0^3 \tag{5.56}$$

where
$A = (a/\gamma)$
$B = (b/\gamma^2)$
$C = (c/\gamma^3)$

Equation 5.56 can be rearranged to get

$$E_0 = \frac{1}{A}\left(E_t \mp BE_0^2 \mp CE_0^3\right) \tag{5.57}$$

The possible block diagram representation for the third-order linearization and the equivalent MOS circuit are shown in Figure 5.12. There are two multiplies in the circuit of Figure 5.12: first, a vector multiplier with $n = 3$ (Figure 5.12a) and a gain of $1/A$, and the other as a squarer with a gain of *C*. The gain factor $1/A$ of the vector multiplier for all input transistor and for equal (W/L) is

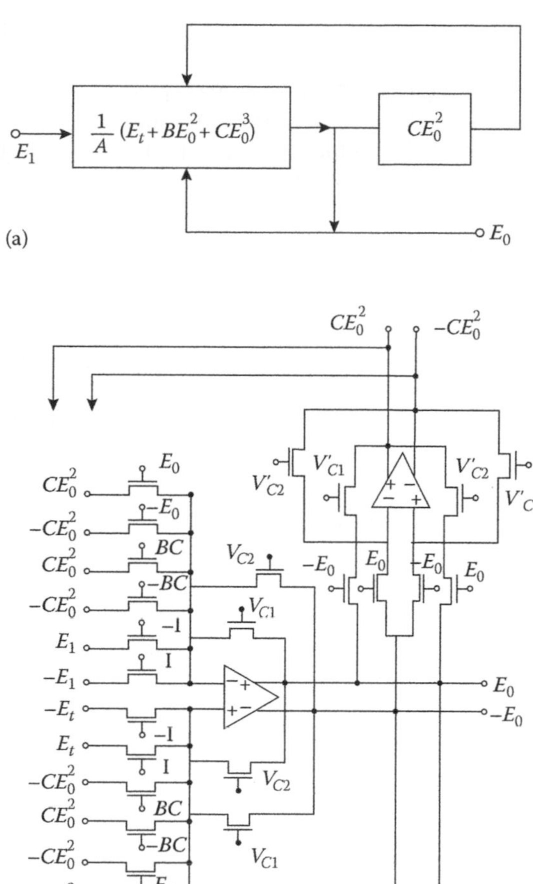

FIGURE 5.12
Third-order linearization using vector multiplier block (a) with gain $1/A$ and (b) with squarer circuit and gain C. (Reprinted from IEEE, copyright © IEEE. With permission.)

$$\frac{1}{A} = \frac{(W/L)_i}{(W/L)_0}(V_{C1} - V_{C2})^{-1} \tag{5.58}$$

The gain factor C for the squarer circuit

$$\frac{1}{A} = \frac{(W/L)_i}{(W/L)_0}(V'_{C1} - V'_{C2})^{-1} \tag{5.59}$$

Example 5.6

The voltage divider circuit (Figure E5.6) for an RTD produces output voltage of 2.5 V with $R_t = 100\ \Omega$ at 0°C and fixed resistance $R = 100\ \Omega$. The output voltage V_t for the temperature range 0°C–100°C follows the equation

$$V_t = V_0[1 + \alpha_1 T + \alpha_2 T^2 + \alpha_3 T^3]$$

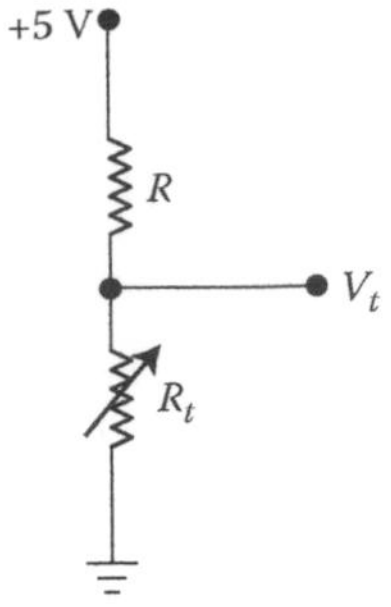

where the output voltage signal V_t is to be linearized using MOS transconductors by the linearized equation $E_0 = 0.05T$ and the temperature coefficients are

FIGURE E5.6
Voltage divider circuit for Example 5.6.

$$\alpha_1 = 3.24 \times 10^{-3}\ \text{V/°C}$$

$$\alpha_2 = 2.33 \times 10^{-5}\ \text{V/°C}$$

$$\alpha_3 = 1.88 \times 10^{-7}\ \text{V/°C}$$

The transconductors have the width-to-length ratio as

$$\frac{(W/L)_i}{(W/L)_0} = 5$$

Determine the parameters $(V_{C1} - V_{C2})$ and $(V'_{C1} - V'_{C2})$.

Solution
The output voltage at $T = 0$°C

$$V_t = 5 \times \frac{R_0}{R_0 + R}$$

$$= 5 \times \frac{100\,\Omega}{100\,\Omega + 100\,\Omega}$$

$$= 2.5\ \text{V}$$

$$\therefore V_0 = 2.5\ \text{V}$$

The nonlinear output equation

$$V_t = 2.5\left[1 + 3.24 \times 10^{-3} T + 2.63 \times 10^{-5} T^2 + 1.88 \times 10^{-7} T^3\right]$$

$$= 2.5 + 8.1 \times 10^{-3} T + 6.57 \times 10^{-5} T^2 + 4.7 \times 10^{-7} T^3$$

Let $V_t' = V_t - 2.5$

$$\therefore V_t' = 8.1\times10^{-3}T + 6.57\times10^{-5}T^2 + 4.7\times10^{-7}T^3$$

Here $a = 8.1\times10^{-3}$, $b = 6.57\times10^{-5}$, and $c = 4.7\times10^{-7}$.

From the given linearized equation

$$E_0 = 0.05T$$

$$\gamma = 0.05$$

$$\therefore A = \frac{a}{\gamma} = \frac{8.1\times10^{-3}}{0.05} = 0.162$$

$$B = \frac{b}{\gamma^2} = \frac{6.57\times10^{-5}}{0.05^2} = 0.026$$

$$C = \frac{c}{\gamma^3} = \frac{4.7\times10^{-7}}{0.05^3} = 3.76\times10^{-3}$$

Using Equation 5.58

$$\frac{1}{A} = \frac{(W/L)_i}{(W/L)_0}(V_{C1} - V_{C2})^{-1}$$

$$\therefore 5(V_{C1} - V_{C2})^{-1} = \frac{1}{0.162}$$

$$\Rightarrow V_{C1} - V_{C2} = 0.81\text{ V}$$

Using Equation 5.59

$$\frac{1}{C} = \frac{(W/L)_i}{(W/L)_0}(V'_{C1} - V'_{C2})^{-1}$$

$$\therefore 5(V'_{C1} - V'_{C2})^{-1} = \frac{1}{3.76\times10^{-3}}$$

$$\Rightarrow V'_{C1} - V'_{C2} = 0.02\text{ V}$$

5.5 Nonlinear ADC- and Amplifier-Based Linearization

Since most of the sensor signal processing involves ADC, analog linearization could be avoided provided a nonlinear ADC is utilized in the digital processing stage.

5.5.1 Nonlinear Counting–Type ADC

Recall a linear counter–type ADC algorithm where an n-bit binary counter produces digital outputs sequentially, driven by a clock. The digital output is converted to analog by a digital-to-analog converter (DAC) and the analog equivalent of the digital number is compared with the analog input signal (V_{in}) by a comparator. If the analog equivalent is smaller than V_{in}, the counter is incremented. When the two analog matches (approximated to a threshold error), the counter output is outputted as the digital output. On completion of the conversion, the ADC control logic outputs an "End of Conversion" signal so that the next analog samples enter the ADC for the next conversion. Since the binary counter is linearly incremented and the counter outputs are linearly converted to analog by the linear DAC, the ADC is said to perform a linear conversion. However, if either the binary counter outputs are converted to a nonlinear binary function or the binary counter outputs are converted to analog by a nonlinear DAC, the ADC is said to perform a nonlinear conversion.

In the first method, the output of the binary counter addresses a LUT stored in an erasable programmable read-only memory (EPROM). The EPROM stores the nonlinear digital functions as per a specific characteristic equation. The total memory space of the EPROM is divided into 2^m segments, each of which can store 2^r sample data. Each segment stores sample data for a specific characteristic equation. Therefore, the address length for the EPROM is $p + m$ where r address lines come from binary counter while m address lines are user specific. The block diagram representation of the nonlinear ADC is shown in Figure 5.13 [6].

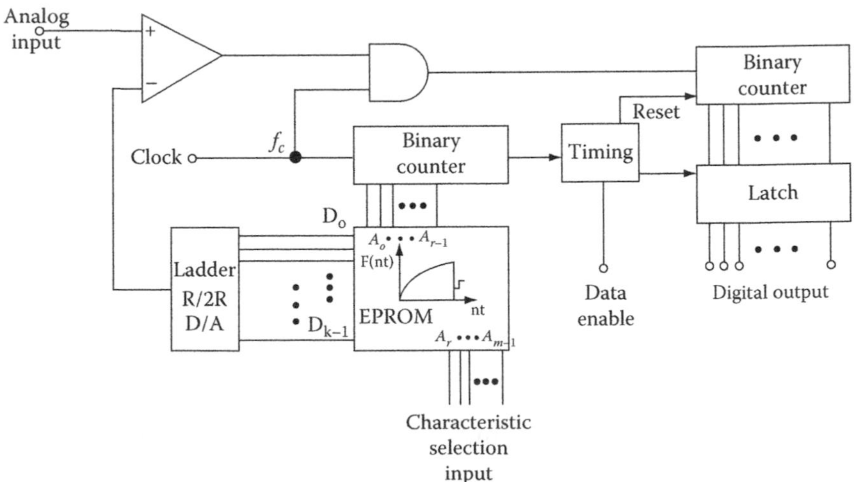

FIGURE 5.13
A nonlinear ADC. (Reprinted from IEEE, copyright © IEEE. With permission.)

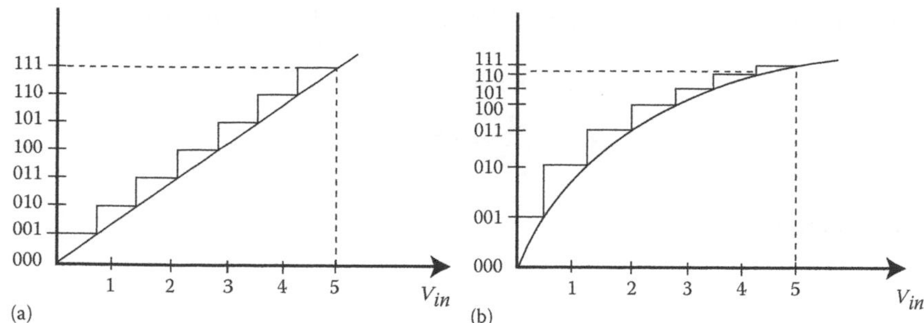

FIGURE 5.14
ADC characteristics: (a) linear ADC and (b) nonlinear ADC.

The difference between linear and nonlinear ADC is shown by their conversion characteristics in Figure 5.14 for a 3-bit ADC, with $V_{max}=5\,V$. For example, an analog input of 3.7 V will be converted to a digital value of 110 by this linear ADC shown in Figure 5.14a. The conversion process for converting an analog input of 3.7 V is shown in Table 5.1.

The registered digital output is shown in bold in the table.

Now let us see how a nonlinear ADC works, which is different from a linear ADC.

Let the ADC follow a nonlinear characteristic given by

$$N = KV_{in}^2 \tag{5.60}$$

where

N is the digital code value
V_{in} is the analog input
K is a constant

TABLE 5.1
Conversion Steps for $V_{in}=3.7\,V$

Clock	V_{in}	Counter Output	Equivalent Analog DAC Output (V)	Decision in Comparator
Start		000	0	Lower
1		001	0.625	Lower
2		010	1.250	Lower
3		011	1.875	Lower
4		100	2.50	Lower
5		101	3.125	Lower
6		**110**	3.750	Higher (stop)

For a 3-bit ADC with V_{in} (max) = 5 V, Equation 5.60 takes the form

$$7 = K \times 5^2$$

$$\therefore K = \frac{7}{25} = 0.28$$

From Equation 5.60

$$V_{in} = \sqrt{\frac{N}{K}}$$

$$= 1.88\sqrt{N}$$

$$\therefore V_{in} = 1.88\,N^{1/2} \tag{5.61}$$

Equation 5.61 is a nonlinear characteristic equation, which will be used to form a LUT for storing in the EPROM (Table 5.2). Table 5.4 shows the analog values for the corresponding 3-bit binary values for Equation 5.61.

The EPROM can be used store more than one type of nonlinear characteristic LUT depending on its size, which can be selected using a characteristic selection address. Therefore, let us consider another nonlinear equation given by

$$V_{in} = KN^2 \tag{5.62}$$

Taking the highest ranges of V_{in} and N, the value of

$$K = \frac{5}{7^2} = 0.102$$

$$\therefore V_{in} = 0.120N^2 \tag{5.63}$$

Table 5.3 shows the values for Equation 5.63.

TABLE 5.2

Characteristic Table for Equation 5.61

Digital Code Address $A_2A_1A_0$	N	Analog Value (V)
000	0	0
001	1	1.88
010	2	2.65
011	3	3.25
100	4	3.76
101	5	4.20
110	6	4.60
111	7	4.97

TABLE 5.3

Characteristic Table for Equation 5.63

Digital Code Address $A_2A_1A_0$	N	Analog Value (V)
000	0	0
001	1	0.102
010	2	0.408
011	3	0.918
100	4	1.632
101	5	2.550
110	6	3.672
111	7	5.000

Since the EPROM stores the LUTs for two characteristic equations and each LUT contains $2^3 = 8$ sample data, as stated before, here $m = 3$ and $r = 1$. This ensures that there is $2^r = 2^1 = 2$ characteristic LUTs and each LUT contains $2^m = 2^3 = 8$ data. The EPROM address lines are formatted as

A_3	A_2	A_1	A_0
Characteristic equation selection	**Data address**		

Table 5.4 shows the combined LUT for the two characteristic equations.

Note that an analog voltage of 3.7 V will be converted to digital as 100 as per Equation 5.61, while it is coded as 110 for each Equation 5.63.

TABLE 5.4

LUT for the Two Characteristic Equations

Address $A_3A_2A_1A_0$	Data V_{in} (V)
0000	0
0001	1.88
0010	2.65
0011	3.25
0100	3.76
0101	4.20
0110	4.60
0111	4.97
1000	0
1001	0.102
1010	0.408
1011	0.918
1101	1.632
1110	2.550
1111	5.000

Example 5.7

A 12-bit counter–type nonlinear ADC is configured with an EPROM where data for five different characteristic equations are stored. Determine the EPROM address format and EPROM size.

Solution

The number of bits in ADC $= m = 12$.

Total number of characteristic equation $= 2^r = 5$

$$\therefore r \geq 3$$

Total address length $= m + r = 12 + 3 = 15$

$$\therefore \text{EPROM size} = 2^{15} = 32\,\text{kB}$$

The address format is shown in Figure E5.7.

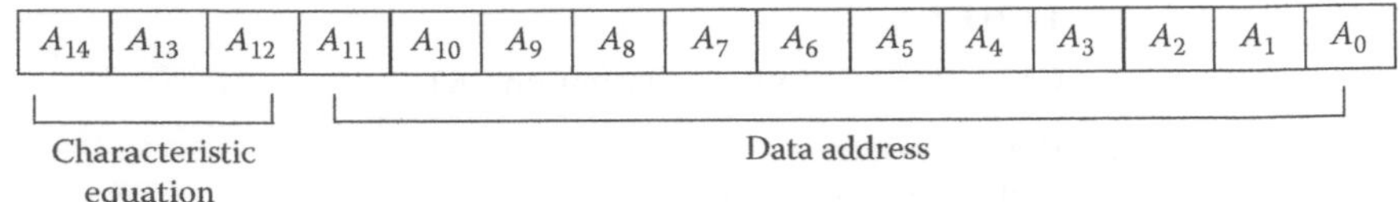

FIGURE E5.7
Address format for Example 5.7.

5.5.2 Nonlinear Successive Approximation ADC

Recall that in a linear successive approximation ADC (SA-ADC), the first clock pulse sets a '1' in the most significant bit (MSB) of the counter and the counter output is converted to an equivalent analog signal by a DAC. The DAC output is then compared with the analog input signal V_{in}. If it is higher than V_{in}, the MSB is made a '0' and the next MSB is made a '1', but if it were lower, the next MSB is also set to a '1'. Then process advances in each clock pulse till the counter output is equal to V_{in}. For a 3-bit SA counter, the conversion path is shown in Figure 5.15a and the bit weightage for a V_{max}=5 V is shown in Figure 5.15b.

The conversion steps involved in converting an analog input of V_{in}=3.7 V in a linear SA-ADC are shown in Table 5.5.

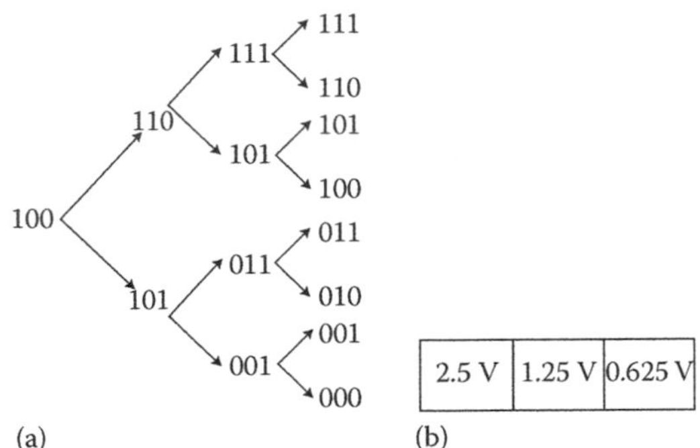

FIGURE 5.15
(a) Conversion tree diagram of a 3-bit SA-ADC and (b) bit weightages for V_{max}=5 V.

TABLE 5.5
Conversion Steps for Converting V_{in}=3.7 V

Clock	V_{in} (V)	Counter Output	Equivalent Analog (V)	Comparator Decision
Start	3.7	100	2.5	Lower
1	3.7	**110**	3.75	Higher (stop)

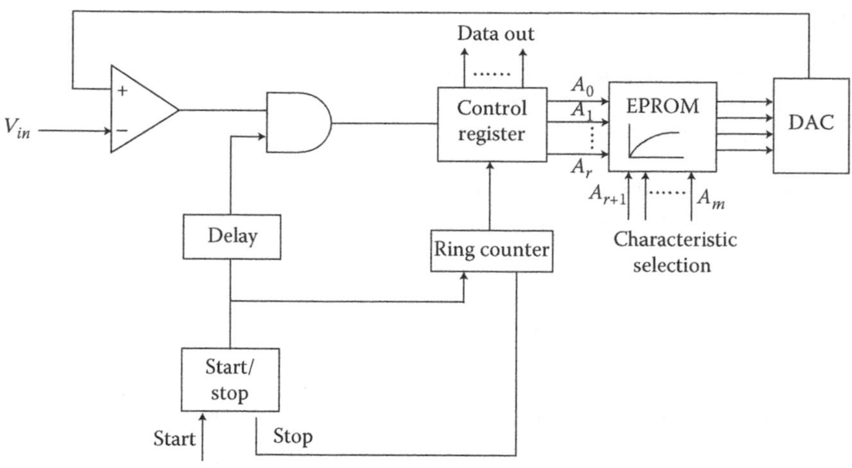

FIGURE 5.16
Block diagram of a nonlinear SA-ADC.

The registered digital output is shown in bold in the table, which is 110. Note that it takes only one clock pulse while six clock pulses are required in a linear counter–type ADC, as shown in Table 5.1.

The linear SA-ADC can be used in the nonlinear mode similar to a nonlinear counter–type ADC. The SA counter addresses an EPROM where nonlinear data as per some characteristic equations are stored. The schematic block diagram of the counter is shown in Figure 5.16.

Example 5.8

A temperature-measuring circuit employing an RTD generates a nonlinear voltage as per the following equation:

$$V_0 = \alpha_1 T + \alpha_2 T^2, \quad \text{where} \quad \alpha_1 = 4.5\times10^{-2}\ \text{V/°C} \quad \text{and} \quad \alpha_2 = -10^{-4}\ \text{V °C}^2$$

The circuit is used for a temperature range of 0°C–200°C. Develop a LUT to be stored in EPROM for designing an 8-bit nonlinear ADC (show the significant codes only).

Solution
The maximum voltage of the ADC at 200°C

$$V_{max} = 4.5\times10^{-2}\times200 - 10^{-4}\times200^2 = 5.0\ \text{V}$$

Temperature value of 1 least significant bit (LSB) = 200/256 = 0.78°C.

For each increment of 1 LSB in digital code, the temperature values and the nonlinear analog voltages are calculated

For example:

Digital code: 0000 0001; decimal equivalent: 1
Temperature: $1 \times 0.78°C = 0.78°C$
Analog voltage = $V_0 = 4.5\times10^{-2}\times(0.78) - 10^{-4}\times(0.78)^2 = 0.0350$ V

Digital code: 0000 1000; decimal equivalent = 8
Temperature: $8 \times 0.78 = 6.24°C$
Analog voltage = $V_0 = 0.2769$ V

Digital code: 1000 0000; decimal equivalent = 128
Temperature: $128 \times 0.78 = 99.84°C$
Analog voltage = $V_0 = 3.49$ V

The analog voltages at a few significant codes are shown below:

Code	N	Temperature (°C)	Voltage (V)
0000 0000	0	0	0
0000 0001	1	0.78	0.035
0000 1000	8	6.24	0.2769
0001 0000	32	24.96	1.0609
1000 0000	128	99.84	3.4960
1000 1000	136	106.08	3.6483
1111 1111	255	198.90	4.9944

5.5.3 Pulse Width Modulation Nonlinear ADC

The basic principle of pulse width modulation (PWM) of a signal is converting the signal to a pulse, width of which is proportional to the signal amplitude; however, the pulse amplitude is considered as constant. Figure 5.17a shows a technique that can produce a PWM signal and Figure 5.17b shows the output waveforms of the circuit.

The sawtooth waveform $C(t)$ applied to the comparator modulates a train of pulse at the output, the width of which are proportional to the amplitude

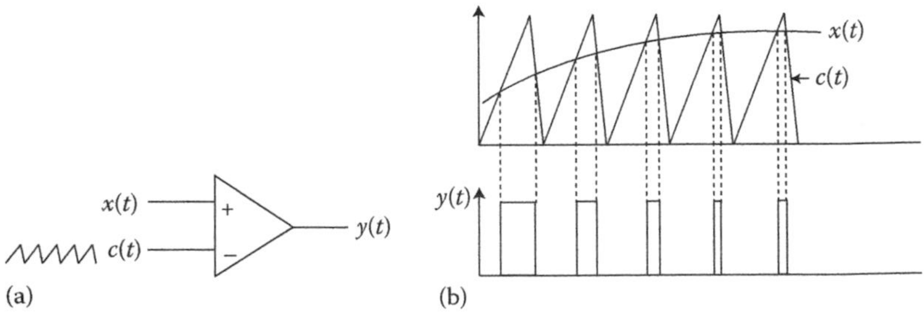

FIGURE 5.17
(a) A basic comparator of PW modulator and (b) input sawtooth and output pulse waveform.

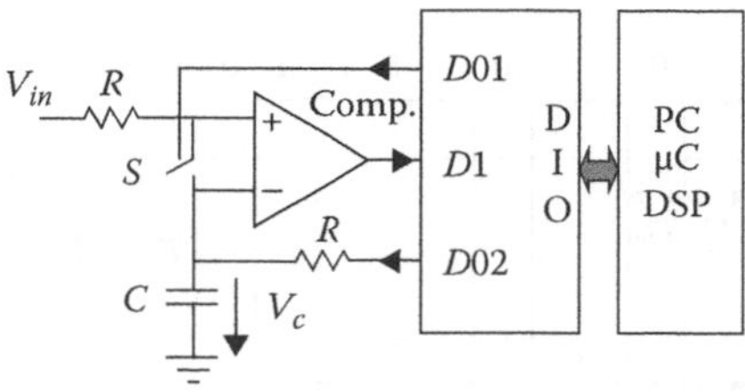

FIGURE 5.18
Block diagram of PWM ADC. (Reprinted from Dias Pereira, J.M. et al., A discrete and cost effective ADC solution based on pulse-width modulation technique, in *Proceedings of the CONFETELE 2001*, Figuera da Fiz, Portugal, April 2001. With permission.)

of the message signal. By applying digital pulses of various levels and duration to the inverting terminal of the comparator, a nonlinear ADC can be accomplished. Figure 5.18 shows the block diagram of a basic PWM ADC [7]. The system consists of a digital I/O interface from a PC or a microprocessor or a DSP unit, a comparator, a low pass RC filter, and an analog switch (S). Before the start of the conversion process, the switch S is closed by $D01$ and the capacitor (C) is charged to the input analog voltage V_{in}. After this initialization operation, the analog switch is opened. A digital pulse from $D02$ determines the comparator output. The capacitor output voltage variation with a constant voltage V_F is given by

$$V_C(t) = V_F + (V_{CO} - V_F)e^{-t/RC} \tag{5.64}$$

where
V_{CO} represents the capacitor initial voltage
RC is the time constant of the circuit

Say the $D02$ outputs a digital pulse train of positive pulses of amplitude V_H and duration T_{P1}, and negative pulse of amplitude V_L and duration T_{P0}, then the capacitor voltage is given by

$$V_C = V_{in} + n(V_H - V_{CO})\left[1 - e^{-T_{P1}/RC}\right] + m(V_L - V_{CO})\left[1 - e^{-T_{P0}/RC}\right] \tag{5.65}$$

where m and n are the number of positive and negative pulse, respectively, delivered by $D02$. Since the total of the positive and negative pulses, i.e., $n+m$ is the bit number of the ADC

$$2^B = n + m$$

Assuming the RC time constant to be much higher than the pulse period (T_P) delivered by $D02$, i.e.,

$$RC \gg T_P$$

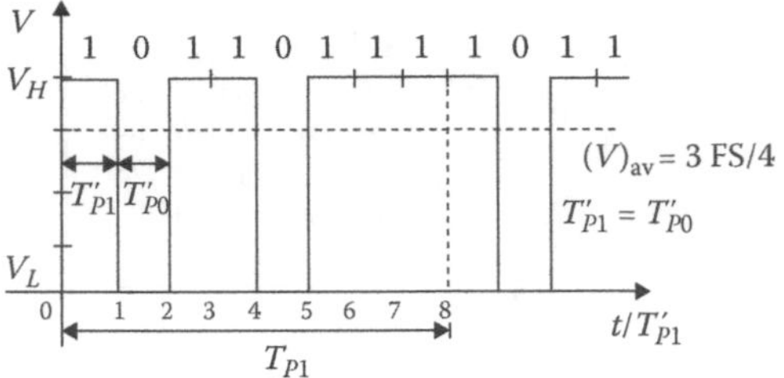

FIGURE 5.19
A PWM signal generated by a PWM ADC. (Reprinted from IEEE, copyright © IEEE. With permission.)

The following relation between V_{in} and ADC output code (n) can be expressed

$$V_{in} = \frac{m}{m + nk^{-1}} V_H + \frac{n}{mk + n} V_L \tag{5.66}$$

where k is a nonlinear factor determined by T_{P1} and T_{P0} given by

$$k = \frac{1 - e^{-T_{P1}/RC}}{1 - e^{-T_{P0}/RC}} \tag{5.67}$$

The nonlinearity of the input voltage V_{in} can be corrected by taking a proper value of k, which depends on the ratio of T_{P1} and T_{P0}. Figure 5.19 shows an example of the PWM signal for an interval of eight units of total pulses where six are positive pulses while two are negative or low pulses. If the reference voltage to the ADC is V_H, the V'_H and V'_L levels set through $D02$ by the microprocessor are given by [8]

$$V'_H = \frac{n_{HH} V_H + n_{HL} V_L}{n_{HH} + n_{HL}} \tag{5.68a}$$

$$V'_L = \frac{n_{LH} V_H + n_{LL} V_L}{n_{LH} + n_{LL}} \tag{5.68b}$$

For $V_H = 5$ V, $V_L = 0$ V, $n_{HH} = 6$, $n_{LL} = 0$, $n_{LH} = 2$, and $n_{HL} = 2$

$$\therefore V'_H = 3.75\text{ V}, \quad V'_L = 1.25\text{ V}$$

For an 8-bit ADC with $V'_H = 3.75$ V, $V'_L = 1.25$ V, and a linear factor $K = 1$, analog input voltage corresponding to a few values of PWM code (n) is calculated using Equation 5.66, as shown in Table 5.6.

TABLE 5.6

Linear Relationship between V_{in} and m

Number of Positive Pulse (m)	Number of Negative Pulse (n)	Analog Voltage (V_{in}) (V)
0	256	1.75
1	255	1.7578
2	254	1.7656
10	246	1.8281
20	236	1.9063
30	226	1.9844
100	156	2.5312
128	128	2.75
255	1	3.7422
256	0	3.75

The steps for the calculation of $m = 100$ and $n = 156$ for $k = 1$ is shown below:

$$V_{in} = \frac{100}{100 + 156 \times 1} \times 3.75 + \frac{156}{100 + 156} \times 1.75$$
$$= 2.5312 \text{ V}$$

Table 5.6 shows that corresponding to the analog values from 1.75 to 3.75 V, the number of positive pulse (m) increases linearly. It is possible to adjust the factor k such that the relationship between V_{in} and PWM code n becomes nonlinear. To do so, let us take the following relationship between T_{P1}, T_{P0}, and RC:

$$T_{P1} : T_{P0} : RC = 1:2:1000$$

Hence

$$k = \frac{1 - e^{-T_{P1}/RC}}{1 - e^{-T_{P0}/RC}}$$
$$= \frac{1 - e^{-1/1000}}{1 - e^{-2/1000}} \cong 0.5 \tag{5.69}$$

The values for V_{in} corresponding to the values of m similar to Table 5.6 the case of $k = 0.5$ are shown in Table 5.7. An example calculation for $m = 100$ and $n = 156$ for $k = 0.5$ is shown below:

$$V_{in} = \frac{100}{100 + 156 \times 2} \times 3.75 + \frac{156}{100 \times 0.5 + 156} \times 1.75$$
$$= 2.2354 \text{ V}$$

TABLE 5.7
Nonlinear Relationship between V_{in} and m

Number of Positive Pulse (m)	Number of Negative Pulse (n)	Analog Voltage, V_{in} (V)
0	256	1.75
1	255	1.7539
2	254	1.7578
10	246	1.7898
20	236	1.8313
30	226	1.8744
100	156	2.2354
128	128	2.4166
255	1	3.7344
256	0	3.75

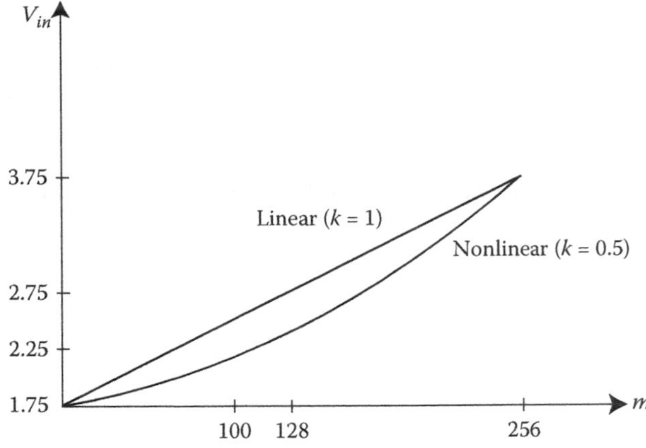

FIGURE 5.20
V_{in}–m characteristics for $k=1$ and $k=0.5$.

It follows from the Table 5.7 that the relationship between V_{in} and m is nonlinear. Figure 5.20 illustrates the linear and nonlinear relationship between V_{in} and m.

Example 5.9

A nonlinear ADC of Figure 5.19 produces a PWM 8-bit code for analog voltage ranging from 1.75 to 3.75 V. For an equal number of positive and negative pulses of the PWM signal, the equivalent analog voltage is 3.0 V. The RC time constant is 1 s and the width T_{P1} is 1 ms.

1. Determine the width T_{P0}.
2. Determine the analog nonlinearity by verifying the voltages for m starting from 0 to 256 at an interval of 32.

Solution
Given values:

$$V_{in} = 3.0\text{ V}$$

$$B = 8, \quad V_H = 3.75\text{ V}; \quad V_L = 1.75\text{ V}$$

1. For $m + n = 2^B = 2^8 = 256$
$m = n = (256/2) = 128$, given value of $V_{in} = 3.0\text{V}$

Using Equation 5.66

$$\frac{128 \times 3.75}{128 + (128/K)} + \frac{128 \times 1.75}{128K + 128} = 3.0$$

$$\Rightarrow \frac{224 + 480K}{128 + 128K} = 3$$

$$\Rightarrow 96K = 160$$

$$\Rightarrow K = 1.66$$

Using Equation 5.67

$$\frac{1 - e^{-T_{P1}/RC}}{1 - e^{-T_{P0}/RC}} = 1.66$$

$$\Rightarrow e^{-T_{P0}/RC} = 0.994$$

$$\Rightarrow T_{P0}/RC = 0.006$$

$$\therefore T_{P0} = 0.006RC = 6\text{ ms}$$

2. For $m = 0$, using Equation 5.66

$$V_{in} = \frac{0}{0 + (256/1.66)} \times 3.75 + \frac{256}{0 + 256} \times 1.75 = 1.75\text{ V}$$

For $m = 32$; i.e., $n = 224$

$$V_{in} = \frac{32 \times 3.75}{32 + 224/1.66} + \frac{224 \times 1.75}{32 \times 1.66 + 224} = 2.1334\text{ V}$$

Similarly, the values for various m are

$$m = 64, \quad V_{in} = 2.4624 \text{ V}$$

$$m = 96, \quad V_{in} = 2.7480 \text{ V}$$

$$m = 128, \quad V_{in} = 2.9981 \text{ V}$$

$$m = 160, \quad V_{in} = 3.2190 \text{ V}$$

$$m = 192, \quad V_{in} = 3.4156 \text{ V}$$

$$m = 224, \quad V_{in} = 3.5915 \text{ V}$$

$$m = 256, \quad V_{in} = 3.75 \text{ V}$$

5.5.4 Nonlinear Resistor–Based Amplifier

In linear applications, a field effect transistor (FET) is operated in the constant current region of the output characteristics. In the region before pinch off, where V_{DS} is small ($V_{DS} < (V_{GS} - V_{GS(off)})$), the channel resistance reduces to [8]

$$R_{DS} = \frac{R_{DS(ON)}}{1 - (V_{GS}/V_p)} \tag{5.70}$$

where the minimum resistance for $V_{GS} = 0$ is given by

$$R_{DS(ON)} = \frac{V_P}{2I_{DSS}} \tag{5.71}$$

Using Taylor's series expansion on Equation 5.70

$$R_{DS} = R_{DS(ON)}\left[1 + \frac{V_{GS}}{V_P} - \frac{1}{2}\left(\frac{V_{GS}}{V_P}\right)^2 + \frac{1}{6}\left(\frac{V_{GS}}{V_P}\right)^3 + \cdots\right] \tag{5.72}$$

Since R_{DS} changes with V_{GS} nonlinearly as per Equation 5.72, linearization is possible if nonlinear voltages are applied as V_{GS} as per Equation 5.72. Figure 5.21 shows a nonlinear OPAMP with an array of voltage-controlled resistors (VCRs) in the feedback path of the OPAMP. The OPAMP feedback branch comprises of a parallel combination of VCR, where the resistance of a single VCR is given by

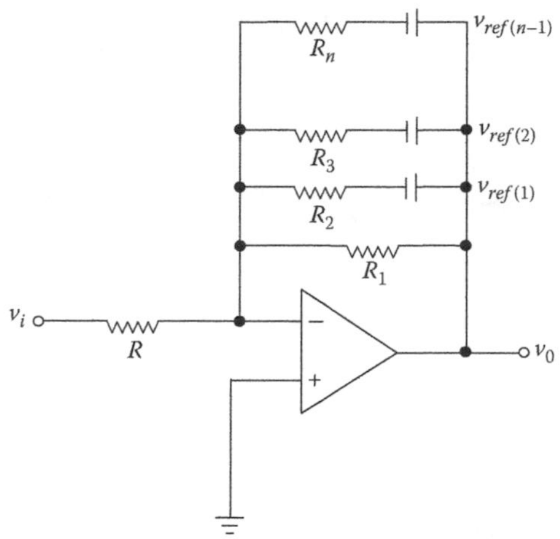

FIGURE 5.21
A nonlinear OPAMP. (Redrawn from IEEE, copyright © IEEE. With permission)

$$R_{VCR} = R_0(1+\alpha V_i) = \frac{R_{DS}(\text{ON})}{1-\dfrac{V_O}{V_P}} \tag{5.73}$$

where α is a coefficient and $R_0 = R_{DS}(\text{ON})$. The amplifier can be used for n different gains for different intervals given by $A_v(i) = -\frac{R_1 \| R_2 \ldots \| R_i}{R}$ for $i = 1, 2, \ldots n$. Therefore, the amplifier provides piece-wise linearization of a nonlinear characteristics.

5.5.5 Logarithmic Amplifier and FET-Based Linearization

Sensors with response characteristics exponentially decaying can be linearized using a log-amplifier. The basic log-amplifier circuit is shown in Figure 5.22.

The base of the transistor is grounded and placed in the feedback path of the OPAMP. Since the base is grounded and the collector is connected to the virtual ground, the transistor behaves like a diode giving the emitter and collector current as

$$I_C = I_E = I_S\left(e^{\frac{qV_E}{KT}} - 1\right) \tag{5.74}$$

where
- I_S is the emitter saturation current
- K is the Boltzmann's constant
- T is the absolute temperature (K)

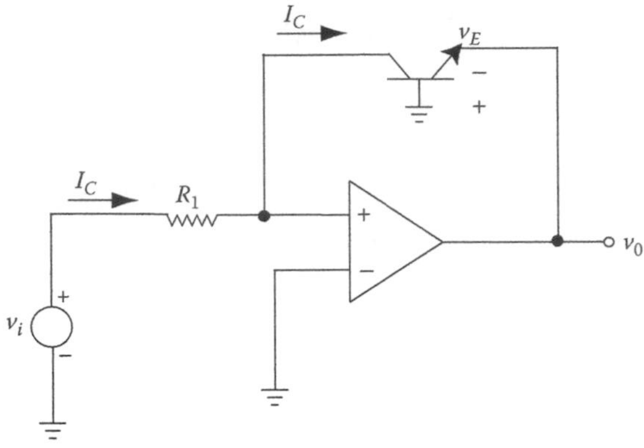

FIGURE 5.22
A basic logarithmic amplifier circuit.

Since $I_C/I_S \gg 1$, Equation 5.74 reduces to

$$e^{\frac{qV_E}{KT}} \approx \frac{I_C}{I_S} \tag{5.75}$$

Taking natural log on both sides

$$V_E = \frac{KT}{q}\ln\left(\frac{I_C}{I_S}\right) \tag{5.76}$$

Since in the OPAMP, $I_C = (V_i/R_1)$ and $V_E = -V_0$, the output voltage is given by

$$V_0 = -\frac{KT}{q}\ln\left(\frac{V_i}{I_S R_1}\right) \tag{5.77}$$

A slight modification of the basic circuit is shown in Figure 5.23, where a negative temperature coefficient (NTC) thermistor is connected through an inverting OPAMP to the log amplifier [9]. The output voltage of the first OPAMP is given by

$$V_1 = \frac{VR}{R_T} \tag{5.78}$$

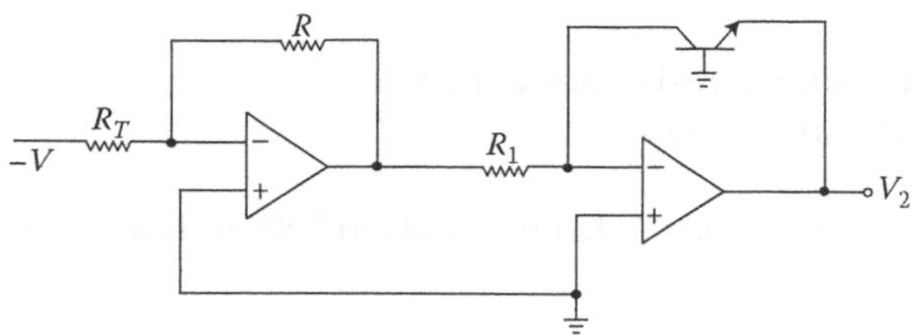

FIGURE 5.23
A thermistor-coupled log-amplifier. (Redrawn from IEEE, copyright © IEEE. With permission.)

which we can replace for V_i in Equation 5.77 to get

$$V_2 = \frac{KT}{q}\ln\left(\frac{VR}{I_S R_T R_1}\right) \tag{5.79}$$

The characteristic equation of a thermistor is given by

$$R_T = R_{T0} e^{\beta\left(\frac{1}{T}-\frac{1}{T_0}\right)} \tag{5.80}$$

and substituting Equation 5.80 in Equation 5.79, we get

$$\begin{aligned} V_2 &= \frac{KT_a}{q}\left[\ln\frac{VR}{I_S R_{T0} R_t} - \ln\left(e^{\beta\left(\frac{1}{T}-\frac{1}{T_0}\right)}\right)\right] \\ &= \frac{KT_a}{q}\ln\frac{VR}{I_S R_{T0} R_1} - \frac{KT_a}{q}\frac{\beta}{T_0} + \frac{KT_a}{q}\frac{\beta}{T} \end{aligned} \tag{5.81}$$

$$\therefore V_2 = \frac{KT_a}{q}\left[\ln\frac{VR}{I_S R_{T0} R_1} - \frac{\beta}{T_0}\right] + \frac{KT_a}{q}\frac{\beta}{T} \tag{5.82}$$

The output voltage V_2 can be applied to the gate-source as V_{gs} of an FET to obtain a linear output V_0 proportional to the temperature T. The FET linearizing circuit is shown in Figure 5.24 [9]. The drain-source resistance of an FET is given by

$$R_{DS} = \frac{R_0}{1-(V_{gs}/V_p)} \tag{5.83}$$

where

R_0 is the drain-source resistance with $V_{GS}=0$
V_p is the pinch-off voltage

Putting the expression for V_{GS} from Equation 5.82, in Equation 5.83

$$R_{DS} = \frac{R_0}{1-\frac{KT_a}{qV_p}\left[\ln\frac{VR}{I_S R_{T0} R_1} - \frac{\beta}{T_0}\right] + \frac{KT_a}{q}\frac{\beta}{T}} \tag{5.84}$$

In Figure 5.24, the output voltage is given by

$$V_0 = V_R \frac{R_{DS}}{R_{DS}+R_r} \tag{5.85a}$$

Substituting R_{DS} from Equation 5.84 in Equation 5.85a, we get

$$V_0 = \frac{R_0 V_R}{R_r\left[1-\frac{KT_a}{qV_p}\left[\ln\frac{VR}{I_S R_{T0} R_1} - \frac{\beta}{T_0}\right] + \frac{KT_a\beta}{qV_pT}\right] + R_0} \tag{5.85b}$$

Now letting $1-\frac{KT_a}{qV_p}\left[\ln\frac{VR}{I_S R_{T0} R_1} - \frac{\beta}{T_0}\right] = 0$, we get

$$R_{DS} = \frac{R_0}{1-\frac{K\beta T_a}{TqV_p}} \tag{5.86}$$

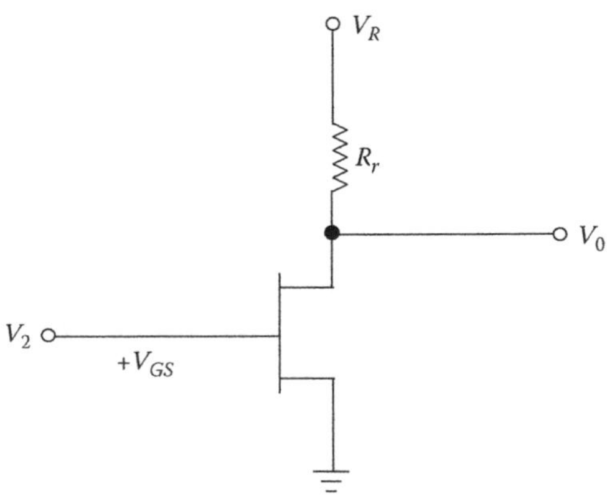

FIGURE 5.24
An FET linearizing circuit.

$$V_2 = \frac{KT_a}{q}\left[1+\frac{\beta}{T}\right]$$

$$V_0 = \frac{R_0 V_R}{\dfrac{R_r K T_a \beta}{T q V_p} + R_0} \tag{5.87}$$

Since $((R_r K T_a \beta)/Tq) \gg R_0$, we get

$$V_0 = \frac{V_R V_p R_0 T q}{K\beta R_r T_a} = CT \tag{5.88}$$

where

$$C = \frac{V_R V_p R_0 q}{K\beta R_r T_a} \tag{5.89}$$

Equation 5.88 shows that the output voltage V_0 is proportional to the temperature T, which the NTC thermistor is subjected to. For proper operation of the FET as a voltage controlled variable resistor, the condition, $V_{gs} \leq (1/2) V_p$ should be fulfilled. Therefore, if required, V_2 should be conditioned using a biasing OPAMP as suggested in [9]. Further, for a practical realization of the assumption of Equation 5.88, an OPAMP conditioning circuit is suggested [9], as shown in Figure 5.25.

The constant C for the circuit of Figure 5.25 is found as

$$C = \frac{R_6 V_R V_p R_0 q}{R_C K\beta R_7 T_a} \tag{5.90}$$

where $R_C = R_2/(1 + (R_5/R_4))$.

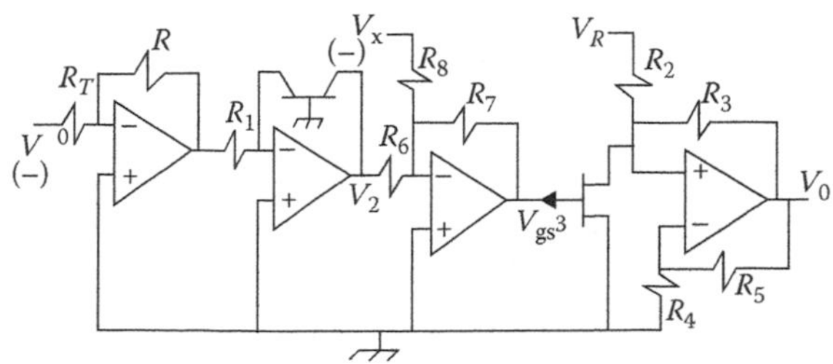

FIGURE 5.25
Practical thermistor linearization circuit. (Reprinted from IEEE, copyright © IEEE. With permission.)

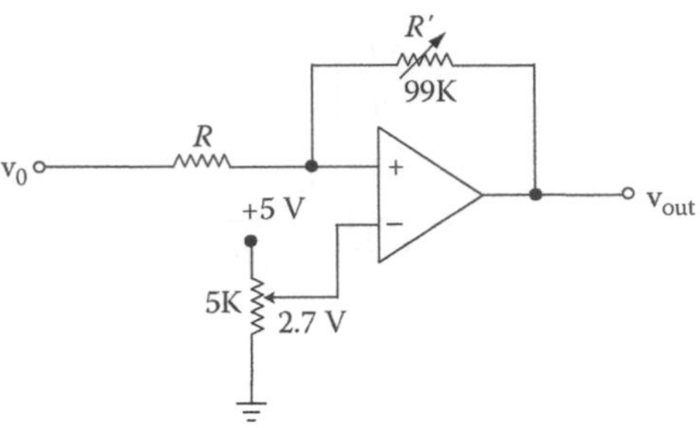

FIGURE E5.10
OPAMP calibration circuit.

Example 5.10

A thermistor having $\beta = 4000\,K$ and $R_{T0} = 5\,k\Omega$ is linearized using the circuit of Figures 5.18 and 5.19 over a temperature range of 0°C–100°C. The FET has a pinch-off voltage of 2.6 V and R_0 of 500 Ω. The FET circuit uses a reference voltage (V_R) of 5 V and the pull up resistor (R_r) of 500 Ω. Take 25°C as reference temperature.

1. Determine the output voltage at 0°C, 25°C, 50°C, and 100°C.
2. Suggest an OPAMP to calibrate the output voltage in the range 0–5 V and find the linearized and calibrated voltages.

$$\left[q = 1.602 \times 10^{-19}\,\text{C}, K = 1.381 \times 10^{-23}\right]$$

Solution
Given values:

$$V_p = 2.6\,\text{V}, \quad R_0 = 500\,\Omega, \quad T_a = 25°\text{C}, \quad R_r = 500\,\Omega, \quad V_R = 5\,\text{V}$$

$$q = 1.602 \times 10^{-19}\,\text{C}$$

$$K = 1.381 \times 10^{-23}$$

The values of V_2, R_{DS}, and V_0 for the given temperature are calculated using Equations 5.87, 5.83, and 5.85a, respectively, as

1. At $T = 0°\text{C} = 273\,\text{K}$, $T_a = 25°\text{C} = 273 + 25 = 298\,\text{K}$

$$V_2 = \frac{KT_a}{q}\left[1 + \frac{\beta}{T}\right]$$

$$= \frac{1.381 \times 10^{-23} \times 298}{1.602 \times 10^{-19}}\left[1 + \frac{4000}{273}\right]$$

$$= 0.391\,\text{V}$$

$$R_{DS} = \frac{R_0}{1-(V_{gs}/V_p)} = 588.5\ \Omega$$

$$V_0 = V_R \frac{R_{DS}}{R_{DS}+R_r} = 2.70\ \text{V}$$

Similarly, the values at other temperatures are calculated and found as

$$\text{At } 25°\text{C:}\quad V_2 = 0.37\ \text{V};\quad R_{DS} = 583.06\ \Omega;\quad V_0 = 2.69\ \text{V}$$

$$\text{At } 50°\text{C:}\quad V_2 = 0.3435\ \text{V};\quad R_{DS} = 576.164\ \Omega;\quad V_0 = 2.675\ \text{V}$$

$$\text{At } 100°\text{C:}\quad V_2 = 0.301\ \text{V};\quad R_{DS} = 565.48\ \Omega;\quad V_0 = 2.65\ \text{V}$$

2. The variation of output voltage over 0°C–100°C is = (2.70 − 2.65)V = 0.05 V

$$\text{Amplified range} = 5\ \text{V}$$

$$\therefore \text{Amplification required} = \frac{5}{0.05} = 100$$

The offset required to be compensated is the voltage at 0°C = 2.70 V.
The amplification factor = 100 = (1 + (R'/R))

$$\therefore \frac{R'}{R} = 99$$

Taking $R' = 99$ K and $R = 1$ K
Now the calibrated voltages (V_{out}) for the given temperature values are

$$\text{At } 0°\text{C:}\quad V_{out} = (V_0 - 2.70)\times A = (2.70 - 2.70)\times 100 = 0\ \text{V}$$

$$\text{At } 25°\text{C:}\quad V_{out} = (2.70 - 2.69)\times 100 = 1\text{V}$$

$$\text{At } 50°\text{C:}\quad V_{out} = (2.70 - 2.675)\times 100 = 2.5\text{V}$$

$$\text{At } 100°\text{C:}\quad V_{out} = (2.70 - 2.65)\times 100 = 5\text{V}$$

Example 5.11

The thermistor of Example 5.10 is to be linearized using the circuit of Figure 5.25 where the following condition has to be fulfilled:

$$1 - \frac{V_x R_7}{V_p R_8} - \frac{K T_a}{q V_p}\frac{R_7}{R_6}\left[\ln \frac{VR}{R_{T0} R_1 I_S \alpha} + \frac{\beta}{T_0}\right] = 0 \quad \text{and} \quad (R_2/R_3) = (R_4/R_5)$$

The transistor has an $I_S = 0.2$ nA, $\alpha = 0.9$, the circuit resistances and voltage are

$$R_6 = R_8 = 0.5\ \text{k}\Omega,$$

$R_7 = 100\ \text{k}\Omega$, $R = 1\ \text{k}\Omega$, $R_1 = 10\ \text{k}\Omega$, $R_2 = 120\ \Omega$, $R_5 = 9\ \text{k}\Omega$, $R_4 = 1\ \text{k}\Omega$, $V = 10$ V.
Determine

1. The conditioning voltage V_x
2. The output voltages at 0°C and 100°C

Solution

1. Substituting the values of (KT_a/q), V_p, β, and T_0 in the equation

$$1 - \frac{V_x R_7}{2.6R_8} - \frac{25.6\times10^{-3}}{2.6}\times\frac{R_7}{R_6}\left[\ln\frac{VR}{R_{T0}R_1I_s\alpha} + \frac{4000}{298}\right] = 0$$

$$\Rightarrow \frac{25.68\times10^{-3}}{2.6}\frac{R_7}{R_6}\left[\ln\frac{VR}{R_{T0}R_1I_s\alpha} + 13.42\right] = 1 - \frac{V_x R_7}{2.6R_8}$$

$$\Rightarrow 0.0098\frac{R_7}{R_6}\ln\left(\frac{VR}{R_{T0}R_1I_s\alpha}\right) = 1 - \frac{V_x R_7}{2.6R_8} - 0.0098\frac{R_7}{R_6}\times13.42$$

$$\Rightarrow \ln\frac{VR}{R_{T0}R_1I_s\alpha} = \frac{2.6R_6^2R_8 - V_xR_6R_7 - 0.3419R_6R_7R_8}{0.02548R_6R_7R_8} \quad \text{(E5.11)}$$

Given values:

$V = 10$ V, $R = 1\text{k}\Omega$, $R_1 = 10\text{k}\Omega$, $R_2 = 120\ \text{k}\Omega$, $I_s = 0.2$nA, $R_6 = R_8 = 0.5\text{k}\Omega$,

$R_7 = 100\ \text{k}\Omega$

For Equation E5.11

$$\therefore \text{LHS} = \ln\frac{10\times10^3}{5\times10^3\times10\times10^3\times0.2\times10^{-9}\times0.9} = 13.92$$

$$\therefore 2.6R_6^2R_8 - V_xR_6R_7 - 0.3419R_6R_7R_8 = 0.3547R_6R_7R_8$$

$$\Rightarrow 2.6R_8R_6 - V_xR_7 + 0.0128R_7R_8 = 0$$

$V_x = 12.9$ V (putting the values of resistance in the above equation)

2. The constant C from Equation 5.90

$$C = \frac{R_6V_RV_pR_0q}{R_CK\beta R_7T_a} = \frac{5\times0.5\times10^3\times500\times2.6}{25.68\times10^{-3}\times4000\times100\times10^3\times R_C}$$

where, $R_C = \dfrac{R_2}{1+(R_5/R_4)} = 12.0$

Putting the value of R_C and other values in the equation of C we get

$$C = 0.0264$$

The output voltage using Equation 5.88

$$\therefore \text{At } 0°\text{C}, V_0 = 0.0264 \times (0 + 273) = 7.2072 \text{ V}$$

and at 100°C, $V_0 = 0.0264 \times (100 + 273) = 9.8472$ V

5.6 Interpolation

Interpolation is a mathematical technique of using known data to estimate unknown values. Although there are various methods of interpolation, linear interpolation is the simplest one. Briefly speaking, it requires the knowledge of two points and a constant rate of change between them to estimate another point.

5.6.1 Linear Interpolation

The concept of linear interpolation assumes that the rate of change between two known values is constant, for estimating a value other than the known two values. If the unknown point lies inside the two known points, it is called interpolation, whereas if the unknown point lies outside, it is called extrapolation. The linear interpolation can be expressed in the form

$$p_u = p_0 + u(p_1 - p_0) \tag{5.91}$$

where p_0 and p_1 are the points that we interpolate between and u is a gradient between points p_0 and p_{-1} or p_0 and p_1. Let us express the points in X–Y coordinates as (Figure 5.26)

$$p_{-1}: \quad (x_{-1}, y_{-1})$$

$$p_0: \quad (x_0, y_0)$$

$$p_1: \quad (x_1, y_1)$$

When we interpolate points between p_0 and p_1, it is given by the equation

$$y_u = y_0 + u_y(x_1 - x_u) \tag{5.92}$$

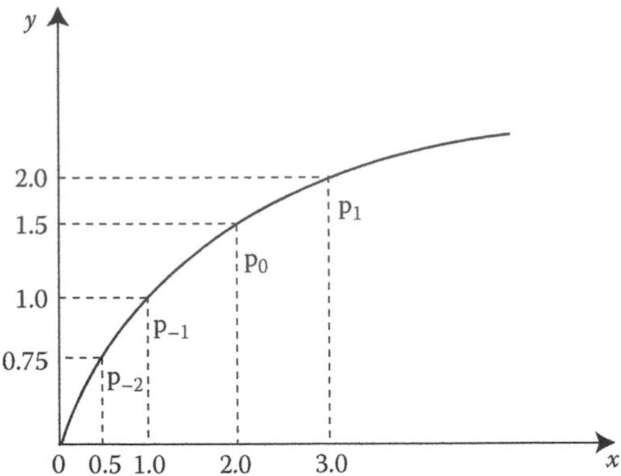

FIGURE 5.26
Linear interpolation characteristics.

where

$$u_y = \frac{dy}{dx} = \frac{(y_0 - y_{-1})}{(x_0 - x_{-1})} \tag{5.93}$$

where x_u is the X-coordinate of the unknown point. The gradient u_y gives the indication about the slope, i.e., the rate of change of y with respect to x. Therefore, the slope between p_{-1} and p_0 is considered for u_y. In Figure 5.26, point p_1 can be obtained by interpolating linearly from p_{-1} and p_0. Let us take the coordinates of p_{-1} and p_0 as (1,1) and (2,1.5), respectively, then the gradient u_y can be obtained as

$$u_y = \frac{1.5 - 1.0}{2.0 - 1.0} = 0.5$$

and using Equation 5.92, the interpolated point is

$$\begin{aligned} y_1 &= y_0 + u_y(x_1 - x_0) \\ &= 1.5 + 0.5(3 - 2) \\ &= 2.0 \end{aligned}$$

Let us take another point with $x = 0.5$ for which the gradient prior to it is not known. Since the gradient beyond the point (0,0) is not available, a point with $x = 0.5$ can be interpolated backward using the equation

$$p_u = p_{-1} + u_y(x_1 - x_u)$$

Here, the gradient can be found as

$$u_y = \frac{(y_0 - y_{-1})}{(x_0 - x_{-1})} = 0.5$$

$$p_u = 1.0 - 0.5(1 - 0.5) = 0.75$$

Example 5.12

The output voltages of a thermocouple were measured using a millivoltmeter within a temperature range of 0°C–400°C. The output voltages were recorded as shown below:

T (°C):	0	100	200	300	400
V_0 (mV):	0	44	71	105	135

1. Determine the voltage at 250°C by linear interpolation.
2. What is the sensitivity of the thermocouple if a linear characteristic is considered with a span of 0°C–300°C?
3. Find the projected values at 250°C and 450°C.

Solution

The unknown point is between 200°C and 300°C, therefore, we find gradient as

$$u = \frac{71 - 44}{200 - 100} = 0.27 \text{ mV/°C}$$

1. The interpolated point for $x = 250$°C

$$y_u = 71 + 0.27(250 - 200) = 84.50 \text{ mV}$$

2. Sensitivity for 0°C–300°C range

$$S = \frac{105 \text{ mV}}{300} = 0.35 \text{ mV/°C}$$

3. With this sensitivity, the voltage at

$$250\text{°C} = 0.35 \times 250 = 87.5 \text{ mV}$$

Voltage at 450°C = 0.35 × 450 = 157.5 mV.

5.6.2 Gradient Acceleration Factor

In some sensors, the nonlinear characteristic progresses with an increase in the gradient. In such cases, an acceleration factor α has to be considered while interpolating points. Let us consider the following data of a J-type thermocouple. The first row is for the temperature values and the second row is for the corresponding output voltage of the thermocouple:

T (°C):	0	10	20	30	40	50	60	70	80	90	100
V_0 (mV)	0	0.507	1.019	1.536	2.058	2.585	3.115	3.649	4.186	4.725	5.268

Now we will examine the sensitivity of the thermocouple at each temperature value using the equation

$$u = \frac{v_1 - v_0}{\Delta T} \tag{5.94}$$

where

v_1 and v_0 are the output voltages at two consecutive temperature values
ΔT is the difference between the two temperature values

The sensitivity of the thermocouple is observed to increase from one temperature point to another by a factor given by the equation

$$\alpha = \frac{u_1 - u_0}{\Delta T} \tag{5.95}$$

and the gradient value can be interpolated by the equation

$$u_1 = u_0 + \alpha(T_1 - T_0) \tag{5.96}$$

where u_1 and u_0 are the sensitivities at two consecutive temperature points T_1 and T_0. An example to show how the values for u and α can be calculated is given below:

Between 0°C and 10°C

$$u_1 = \frac{0.507 - 0}{10 - 0} = 0.0507 \text{ mV/°C}$$

Between 10°C and 20°C

$$u_2 = \frac{1.019 - 0.507}{20 - 10} = 0.0512 \text{mV/°C}$$

Since we have obtained the sensitivities for two consecutive temperature readings, the gradient acceleration factor is calculated from

TABLE 5.8

Gradient (mV/°C) and the Acceleration Factors

Temperature Range (°C)	Gradient (u) mV/°C	Acceleration Factor (α), mV/°C/°C × 10^{-5}
0–10	0.0507	—
10–20	0.0512	5
20–30	0.0517	5
30–40	0.0522	5
40–50	0.0527	5
50–60	0.0533	6
60–70	0.0534	1
70–80	0.0537	2
80–90	0.0539	3
90–100	0.0543	4

At 20°C:

$$\alpha = \frac{0.0512 - 0.0507}{20 - 10} = 5 \times 10^{-5}$$

The values for u and α for the entire data length are shown in Table 5.8, and Figure 5.27 shows the graph for u.

Now we will apply the acceleration factor to calculate the output voltage at 55°C, 75°C, and 85°C. Before applying the factor α for interpolation, let us find the values by simple linear interpolation for comparison.

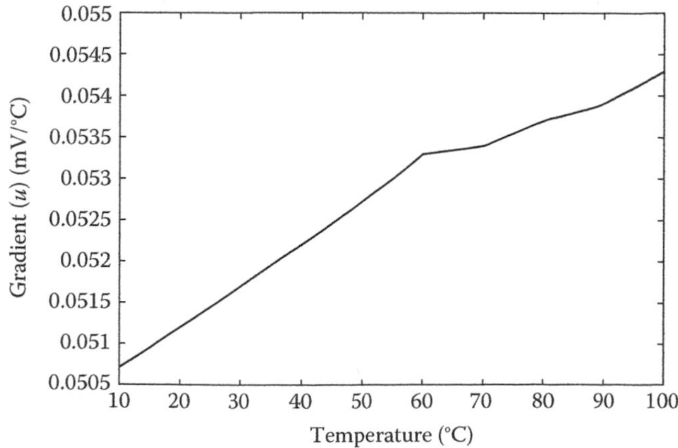

FIGURE 5.27
Interpolation characteristics with acceleration of gradient.

By simple linear interpolation

$$V_{55} = V_{50} + u_{40-50}(55-50) = 2.84\,\text{mV}$$

Similarly

$$V_{75} = 3.649 + 0.0534 \times 5 = 3.916\,\text{mV}$$

$$V_{85} = 4.186 + 0.0537 \times 5 = 4.4545\,\text{mV}$$

By gradient acceleration interpolation

Using Equation 5.96, the gradients are calculated as

$$u_{55} = u_{40-50} + \alpha u_{50-60} \times \Delta T = 0.05295$$

$$\therefore V_{55} = 2.585 + 0.05295 \times 5 = 2.849\,\text{mV}$$

Similarly

$$V_{75} = 3.9167\,\text{mV}$$

$$V_{85} = 4.455\,\text{mV}$$

Note that according to the datasheet of J-type thermocouple, the voltages at 55°C, 75°C, and 85°C are 2.849, 3.917, and 4.455 mV, which are more closely obtained by acceleration interpolation rather than simple interpolation.

One important point to be noted in the above analysis is that for interpolation of the thermocouple data, we have considered temperature as the parameter for the X-ordinate and output voltage for the Y-ordinate. But this is not the case practically when the system linearizes the data; it uses voltage data that is interpolated to the corresponding temperature value. In this situation, one has to consider voltage in X-ordinate and temperature in the Y-ordinate. We will examine this now for an output voltage of 1.797 mV of the J-type thermocouple.

Recalling Equation 5.92 and writing the equation for X-ordinate as

$$x_u = x_0 + u_x(y_1 - y_u) \tag{5.97}$$

Here u_x is given by

$$u_x = \frac{(x_0 - x_{-1})}{(y_0 - y_{-1})} \tag{5.98}$$

When the gradient (u_y) increases as parameter on X-ordinate increases, the gradient u_x, on the other hand, decreases. Similar to Equation 5.96, a new gradient can be obtained by interpolation as

$$u_{x1} = u_{x0} + \beta(v_1 - v_0) \tag{5.99}$$

where

$$\beta = \frac{u_{x1} - u_{x0}}{\Delta v} \tag{5.100}$$

Now let us apply Equations 5.97 and 5.98 to find the temperature corresponding to a voltage of 1.797 mV by simple interpolations.

For the given voltage, the gradient in the preceding data points is

$$u_x = \frac{30 - 20}{1.536 - 1.019} = 19.34\ °\text{C/mV}$$

Note that the unit of sensitivity is reciprocal of that found earlier for the thermocouple. Using Equation 5.91, for interpolating temperature values

$$T_{1.797} = 30 + 19.34(1.797 - 1.536) = 35.048°\text{C}$$

Note that this is closer to the data sheet value of the thermocouple, which is 1.797 mV at 35°C. We expect that the accuracy improves if we use Equations 5.99 and 5.100. Similar to Table 5.8, we prepare a table (Table 5.9) to show all values of u_x and β. Here we take from the table

u_{x0} = 19.15(°C/mV) and β = 0.19(°C/mV/mV) and using Equation 5.99

$$\therefore u_{x1} = 19.15 + 0.19(1.797 - 1.536)$$
$$= 19.199(°\text{C/mV})$$

Using Equation 5.97

$$T_{1.797} = 30 + 19.199(1.797 - 1.536) = 35.01\,\text{mV}$$

This value is much closer to the value in the data sheet (35 mV) than that obtained by simple interpolation (35.048 mV) earlier.

TABLE 5.9

Gradient (°C/mV) and Decelerating Factors (β)

Voltage Range (°C)	Gradient (°C/mV)	Deceleration Factor β (°C/mV/mV)
0–0.507	19.72	—
0.507–1.019	19.53	0.19
1.019–1.536	19.34	0.19
1.536–2.058	19.15	0.19
2.058–2.585	18.97	0.18
2.585–3.115	18.86	0.11
3.115–3.649	18.72	0.14
3.649–4.186	18.62	0.10
4.186–4.725	18.55	0.07
4.725–5.268	18.41	0.14

5.6.3 Gradient Deceleration Factor

In some sensors, the nonlinearity of the characteristic shows a declining gradient at higher values of inputs. This is the most typical trends of sensor characteristics. Consider the following data of a pressure sensor:

Pressure (psi):	100	200	300	400	500
V_0 (mV):	15	22.5	30	32.5	35

If we calculate the gradient u_y between any two consecutive pressure intervals using Equation 5.93, we will find that the gradient decreases, however the gradient u_x of Equation 5.98 increases. The values of u_y and β calculated are shown in Table 5.10.

TABLE 5.10

Gradient and Deceleration Factor for Pressure Sensor

Pressure Range (psi)	Gradient (mV/psi)	Deceleration Factor (β)
0–100	0.15	—
100–200	0.075	-7.5×10^{-4}
200–300	0.075	0
300–400	0.025	-5.0×10^{-4}
400–500	0.025	0

If we want to interpolate data points without considering the gradient deceleration factor to get the value of output voltage for a pressure of 312 psi, we proceed like this

The gradient: $u_{312} = 0.075\,\text{mV/psi}$
Hence, $V_{312} = 30 + 0.075(312 - 300) = 30.9\,\text{mV}$

If we want to interpolate considering the gradient deceleration factor, we proceed like this

$$\text{The deceleration factor } (\beta) = -5\times10^{-4}$$

The gradient

$$u_{312} = 0.075 - 5\times10^{-4}(312 - 300) = 0.069$$

Hence, $V_{312} = 30 + 0.069(312 - 300) = 30.82\,\text{mV}$.

5.6.4 Hardware Implementation of Linear Interpolation

For implementing linear interpolation in hardware, the following units are required:

1. Floating point-to-integer converter
2. EPROM for LUT
3. Adders and multipliers (floating point)

The hardware implementation is illustrated in Figure 5.28. The floating point value of the sampled input $x(k)$ is first converted to an integer index n by the floating point to integer converter. The integer index n is used to address the LUT, which is stored in the EPROM. The LUT stores the characteristic values $f(x_0)$, the gradient $f'(x_0)$, and $n\Delta x$. The term $n\Delta x$ is the

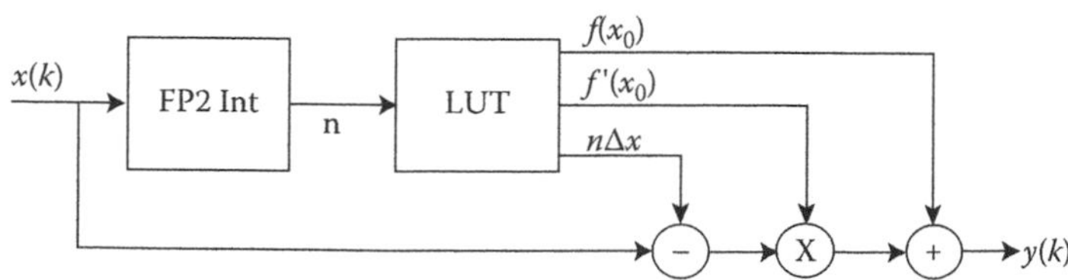

FIGURE 5.28
Hardware schematic of linear interpolation.

multiplication of the index n with the difference between two consecutive input data points, i.e.,

$$\Delta x = x(k+1) - x(k).$$

It requires three EPROM to store the values of $f(x_0)$, $f'(x_0)$, and $n\Delta x$, which are addressed simultaneously by the same index n. The output from LUT is then processed by two adders and one multiplier. The structure of the LUT for the following X–Y data is shown in Figure 5.29.

X:	0	0.1	0.5
Y:	0	3	5

An index (n) with a range of 0–10 is used for the X-data range of 0–0.5. The $f'(x)$ values are calculated using the equation

$$f'(x) = \frac{f(x_1) - f(x_0)}{n_1 - n_0} \tag{5.101}$$

The interpolated value at the system output is given by

$$y(k) = f(x_0) + f'(x_0)[n\Delta x_1 - x_1] \tag{5.102}$$

For example, say $x_1(k) = 0.2$. The floating point to integer converter converts it to an integer as

$$n = \frac{10}{0.5} \times 0.2 = 4$$

For interpolating, $n = 4$, $f(x_0) = 3$, $f'(x_0) = 1.5$, and $\Delta x = 0.1$. The LUT is shown in Figure 5.29.

Index (n)	$f(x)$	$f'(x)$	Δx
0	0	–	–
2	3	$\frac{3}{2} = 1.5$	0.1
10	5	$\frac{2}{8} = 0.25$	0.4

FIGURE 5.29
An example of the LUT for hardware-based interpolation.

Hence,

$$y(k) = 3 + 1.5[4\times 0.1 - 0.2]$$
$$= 3 + 1.5\times 0.2$$
$$= 3.3$$

Example 5.13

Prepare a LUT for the hardware implementation of the linear interpolation of the following load-cell data:

W (kg):	0	0.2	0.4	0.6
V_0 (mV):	0	10.2	18.5	25.7

Determine the interpolated value of load for a voltage of 15 mV.

Solution

The input voltage is converted to an integer index (n) of range 0–257. Therefore, the conversion factor is

$$\frac{257}{25.7} = 10.$$

The gradient values are calculated by

$$f'(x) = \frac{\Delta W}{\Delta V_0}$$

The Δx values are calculated by

$$\Delta x = x_1 - x_2$$

The LUT for the example is shown in Figure 5.30.

Index (n)	$f(x)$	$f'(x)$	Δx
0	0	–	–
102	0.2	1.96×10^{-3}	10.2
185	0.4	2.40×10^{-3}	8.3
257	0.6	2.77×10^{-3}	7.2

FIGURE 5.30
LUT for Example 5.13.

Now for a voltage of 15 mV

$$n = 15 \times 10 = 150$$

$$f(x) = 0.2; \quad f'(x) = 1.96 \times 10^{-3}; \quad \Delta x = 10.2$$

$$\therefore W = 0.2 + 1.96 \times 10^{-3}[150 - 102]$$

$$= 0.29\,\text{kg}$$

5.6.5 Progressive Polynomial Interpolation

Progressive polynomial interpolation method is generally applied for the compensation or correction of sensor output data. Let a sensor result in an output that is either higher or lower than the ideal. If the ideal characterization is known, the corrected output reading corresponding to a measured value can be compensated by using progressive polynomial method. This is a step by step linearization technique such that in each step one correction is done and a correction coefficient is developed.

The correction coefficient (a_1) is developed based on the difference between corrected output ($h(x)$) and previous ideal output (y_2). This coefficient is used to upgrade the corrected output $h_1(x)$ by adding a factor with a ratio of a_1 on the difference between corrected $h_1(x)$ and previous ideal output (y_1). Following are the steps involved in the correction process [10]:

Step 1: Develop correction coefficient $a_1 = y_1 - f(x_1)$

Step 2: Develop corrected output $h_1(x) = f(x) + a_1$

Step 3: Develop $a_2 = \dfrac{y_2 - h_1(x_2)}{h_1(x_2) - y_1}$

Step 4: Develop $h_2(x) = h_1(x) - a_2[h_1(x) - y_2]$

Step 5: Develop $a_3 = \dfrac{y_3 - h_2(x_3)}{[h_1(x_2) - y_1][h_2(x_3) - y_2]}$

Step 6: Develop $h_3(x) = h_2(x) + a_3[h_1(x) - y_1][h_2(x) - y_2]$

Step n: Develop $a_n = \dfrac{y_n - h_{n-1}(x_n)}{\Pi_{i=1}^{n-1}[h_i(x_n) - y_i]}$

Step $n+1$: Develop $h_n(x) = h_{n-1}(x) + a_n \Pi_{i=1}^{n-1} h_i[h_i(x) - y_i]$

In the above steps, the terms used are defined as follows:

a_n = correction coefficient for nth reading

$h_1(x_1)$ = first corrected output based on correction on first reading

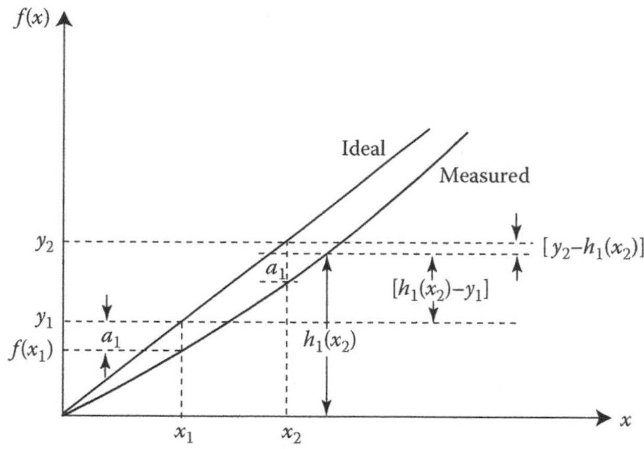

FIGURE 5.31
Correlation characteristics of progressive polynomial.

$y_2 - h_1(x_1)$ = difference between ideal and corrected value for second reading

$h_1(x_1) - y_1$ = difference between corrected values of second reading with ideal of first reading

The characteristic of the progressive correction method is shown in Figure 5.31.

5.6.6 Cressman Interpolation

Cressman interpolation algorithm was developed by George Cressman in 1959. The technique interpolates data of a coordinate point to a user-defined *X–Y* grid. Multiple progresses are made through the grid at consecutively smaller radii of influence, achieving better to best precision. The radius of influence is defined as the maximum radius from grid point to a target point so that observed point value may be weighted to estimate the value at the grid point (Figure 5.32a). At each progress, a correction factor is developed and a new value is calculated at each grid point. A distance weighted value is used with the errors. The weight is given by

$$W_k^m = \frac{R^2 - r^2}{R^2 + r^2} \tag{5.103}$$

where

R is the radius of influence

r is the distance between grid point and observed point

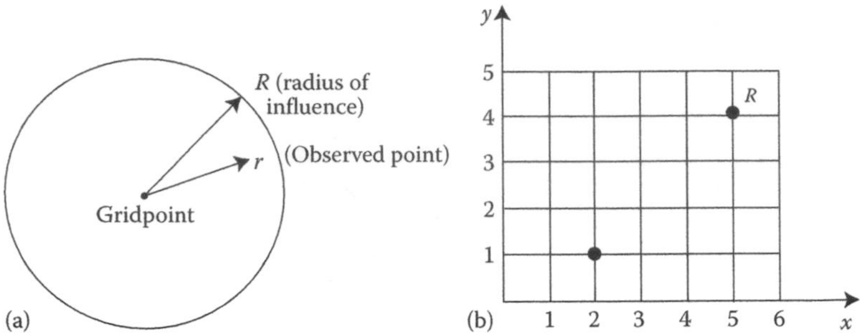

FIGURE 5.32
Cressman polynomial. (a) Radius of influence and (b) grid.

The interpolated updated point is given by

$$p_i^{m+1} = p_i^m + \frac{\sum_{k=1}^{k} w_{ik}^m (o_k - p_k^m)}{\sum_{k=1}^{k} w_i^m k + e^2} \tag{5.104}$$

The subscripts used in the above equation are
m is the number of iterations
k is the observation point
i is the grid point

and

$$e = \text{error} = (R - P_k) \tag{5.105}$$

Now, we consider an *X–Y* grid coordinate (Figure 5.32b) where the radius of influence is at (5,4) and grid point is at (2,2). The possible observed points are

$$[(5,4),(5,3),(5,2)], \quad [(4,4),(4,3),(4,2)] \quad \text{and} \quad [(3,4),(3,3),(3,2)]$$

For interpolation in *X*-direction, let us take grid point at $p=2$. In the first iteration for the weights and errors using Equations 5.103 and 5.105, we get

$$W_1' = \frac{5^2 - 3^2}{5^2 + 3^2} = 0.47$$

$$W_2' = \frac{5^2 - 2^2}{5^2 + 2^2} = 0.72$$

$$W_3' = \frac{5^2 - 1^2}{5^2 + 1^2} = 0.92$$

$$e = 5 - 2 = 3$$

The observed *X*-ordinate values are (5, 4, 3)
The interpolated value in *X*-ordinate is (using Equation 5.104)

$$p_i = 2.0 + \frac{0.47(5-2)+0.72(4-2)+0.92(3-2)}{(0.47+0.72+0.92)+3^2}$$
$$= 2.33$$

Second iteration

$$W_1^2 = \frac{5^2-(5-2.33)^2}{5^2+(5-2.33)^2} = 0.55$$

$$W_2^2 = \frac{5^2-(4-2.33)^2}{5^2+(4-2.33)^2} = 0.79$$

$$W_3^2 = \frac{5^2-(3-2.33)^2}{5^2+(3-2.33)^2} = 0.96$$

$$e = (5-2.33) = 2.67$$

$$p_2 = 2.33 + \frac{0.55(5-2.33)+0.79(4-2.33)+0.96(3-2.33)}{(0.55+0.79+0.96)+2.67^2}$$
$$= 2.69$$

Third iteration

$$W_1^3 = \frac{5^2-(5-2.69)^2}{5^2+(5-2.69)^2} = 0.64$$

$$W_2^3 = \frac{5^2-(4-2.69)^2}{5^2+(4-2.69)^2} = 0.87$$

$$W_3^3 = \frac{5^2-(3-2.69)^2}{5^2+(3-2.69)^2} = 0.99$$

$$e = (5-2.69) = 2.31$$

$$p_2 = 2.69 + \frac{0.64(5-2.69)+0.87(4-2.69)+0.99(3-2.69)}{(0.64+0.87+0.99)+2.31^2}$$
$$= 3.06$$

Fourth iteration

$$W_1^4 = \frac{5^2 - (5-3.06)^2}{5^2 + (5-3.06)^2} = 0.73$$

$$W_2^4 = \frac{5^2 - (4-3.06)^2}{5^2 + (4-3.06)^2} = 0.93$$

$$e = (5-3.06) = 1.94$$

$$p_2 = 3.06 + \frac{0.73(5-3.06) + 0.93(4-3.06)}{(0.73+0.93) + 1.94^2}$$

$$= 3.48$$

Fifth iteration

$$W_1^5 = \frac{5^2 - (5-3.48)^2}{5^2 + (5-3.48)^2} = 0.83$$

$$W_2^5 = \frac{5^2 - (4-3.48)^2}{5^2 + (4-3.48)^2} = 0.97$$

$$e = (5-3.48) = 1.52$$

$$p_2 = 3.48 + \frac{0.83(5-3.48) + 0.97(4-3.48)}{(0.83+0.97) + 1.52^2}$$

$$= 3.90$$

The progress characteristic for the above interpolation is shown in Figure 5.33.

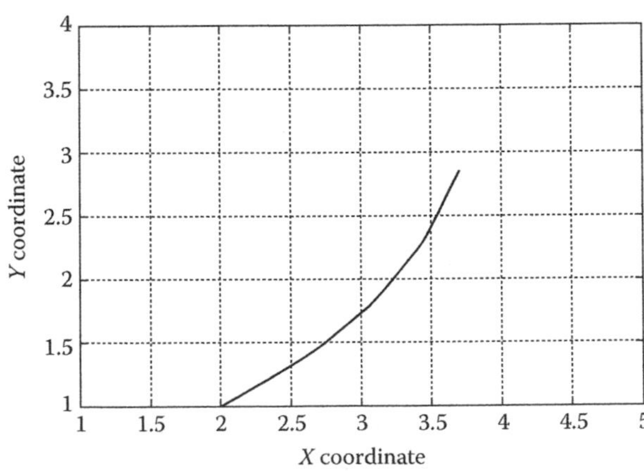

FIGURE 5.33
Progress characteristic of Cressman interpolation.

5.7 Piecewise Linearization

In the piecewise linearization technique, the nonlinear sensor's data over a working operating range is segmented into a certain number of pieces of ranges where each range is fitted to a known characteristic equation. The selection of the number of segments depends on

1. The accuracy of linearization required
2. The length of the operating range

On the other hand, the characteristic equation for each segment depends on the trends on the characteristic of the sensor data for a particular range. The basic idea about piecewise linearization is illustrated in Figure 5.34. In Figure 5.34, the *X*–*Y* characteristic of a nonlinear sensor is segmented in to *n*-segments for the entire input–output ranges as $(0-x_1, 0-y_1)$; $(x_1-x_2, y_1-y_2)\ldots(x_{n-1}-x_n, y_{n-1}-y_n)$. The reason why the range is segmented into *n* is that the linearization technique can assign known characteristic equations based on the sensor's experimental data. Figure 5.34 shows that each segment or piece is approximated by a straight line indicated by extended straight lines, which intercept on the *Y*-axis.

To understand the segmentation process, we take the input–output characteristic of a pressure sensor shown in Figure 5.35. For simplicity, the characteristic showing input pressure in unit of psi, V_s output voltage in volt, is segmented into four pieces of ranges up to 100 psi only and the rest of the range can similarly be segmented. Following are the data points for the four segments:

$$P_1(0-25)\text{ psi},\quad V_{01}(0-4.32)\text{ mV}$$

$$P_2(25-50)\text{ psi},\quad V_{02}(4.32-4.52)\text{ mV}$$

$$P_3(50-75)\text{ psi},\quad V_{03}(4.52-4.67)\text{ mV}$$

$$P_4(75-100)\text{ psi},\quad V_{04}(4.67-4.77)\text{ mV}$$

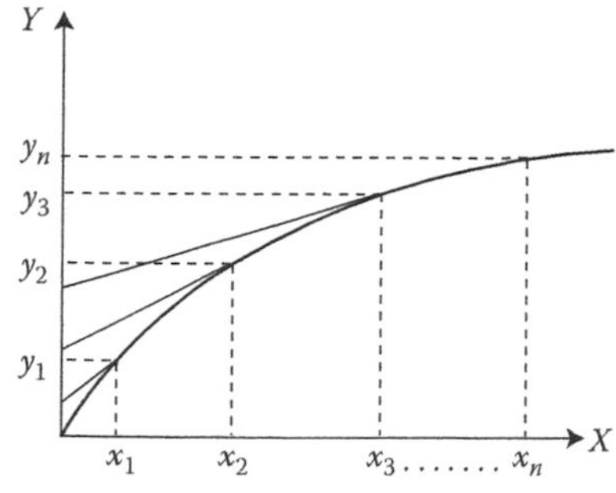

FIGURE 5.34
Characteristics of piecewise linearization.

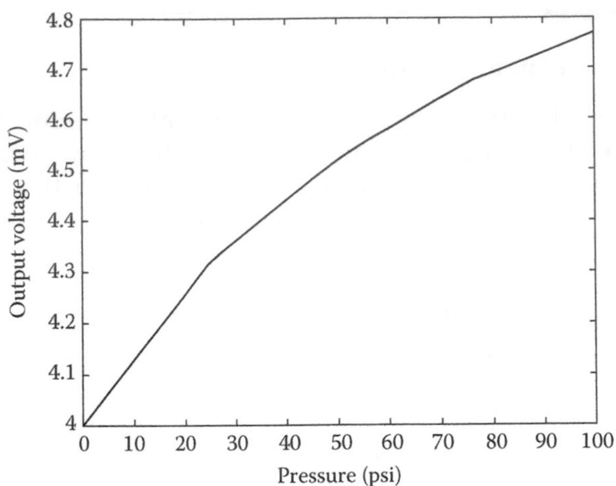

FIGURE 5.35
Nonlinear characteristic of a pressure sensor.

Each segment of the characteristic can be expressed by a straight line equation for which slopes (m) and intercepts (c) are different. The dotted line indicates the straight lines for the segmented characteristics. The equation of the straight lines can be determined either graphically or using the coordinates of the two end points of the segments. In the first method, the equation of a straight line is given by

$$y_{n-1,n} = m_n x_{n-1,n} + c_n \tag{5.106}$$

where

subscript n indicates the segment number

$y_{n-1,n}$ and $x_{n-1,n}$ indicate values within the range $(n - 1)$ to n

For example, Equation 5.106 for the first segment is

$$4.32 = m_1 \times 25 + 4.0$$

$$\therefore m_1 = 0.0128$$

Therefore, the equation of the first segment is

$$y = 0.0128x + 4.0 \tag{a}$$

On the other hand, in the second method using coordinates of the end points of first segment as (0, 0) and (25, 4.32), the straight line equation is given by

$$\frac{y_2 - y_1}{x_2 - x_1} = \frac{y - y_1}{x - y_1} \tag{5.107}$$

and substituting the coordinate values

$$\frac{4.32 - 4}{25 - 0} = \frac{y - 0}{x - 0}$$

we get

$$y = 0.0128x$$

Similarly for the second segment, if graphical values are used, we get (using Equation 5.106)

$$4.52 = m_2 \times 50 + 4.1$$

$$\therefore m_2 = 0.0084$$

The segment equation is

$$y = 0.0084x + 4.1$$

Again if the coordinates of the end points of the segment are used in Equation 5.107, we get

$$\frac{4.32 - 4.52}{25 - 50} = \frac{y - 4.32}{x - 25}$$

$$\Rightarrow y = 0.008x + 4.12$$

There is obviously a difference in accuracy of the two equations $y = 0.0084x + 4.1$ and

$$y = 0.008x + 4.12 \tag{b}$$

Strictly speaking, the second equation is more accurate because it is based on sensor data points, whereas the first equation is based on graphical measurement of the intercept. Therefore, we determine the equations for the rest of the segments by the second method as

$$y = 0.006x + 4.22 \tag{c}$$

$$y = 0.004x + 4.37 \tag{d}$$

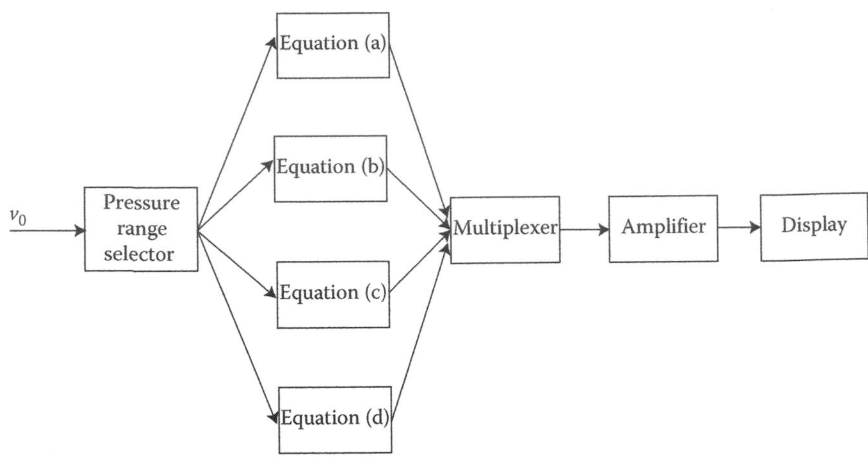

FIGURE 5.36
Block diagram of piecewise linearization of pressure sensor.

However, in a practical situation, the linearization process uses output y to determine input x rather than the reverse, as discussed above. Hence the four Equations (a), (b), (c), and (d) are converted using the engineering notation as

$$p = 5.787v_0 \quad \text{for} \quad 0 \le v_0 \le 4.32\,\text{mV} \tag{e}$$

$$p = 1257v_0 - 515 \quad \text{for} \quad 4.32\,\text{mV} \le v_0 \le 4.52\,\text{mV} \tag{f}$$

$$p = 166.66v_0 - 703.3 \quad \text{for} \quad 4.52\,\text{mV} \le v_0 \le 4.67\,\text{mV} \tag{g}$$

$$p = 250v_0 - 1092.5 \quad \text{for} \quad 4.67\,\text{mV} \le v_0 \le 4.77\,\text{mV} \tag{h}$$

The piecewise linearization technique that uses the above four equations is shown in the block diagram of Figure 5.36.

The linearization operation shown in Figure 5.36 can be performed in various ways depending on the application such as

1. Analog processing
2. Digital processing
3. Microcontroller

5.7.1 Analog Processing

In analog processing, diode clipper and operational amplifiers can be employed for the realization of the range selector as well as the analog computation of the straight line characteristic equation. If we write the piecewise linearization equation in generic form, the following equations result:

$$
\begin{aligned}
v_{x_1} &= k_1 v_i - V_0 \quad \text{for} \quad v_0 \le v_i \le v_1 \\
v_{x_2} &= k_2 v_i - V_1 \quad \text{for} \quad v_1 < v_i \le v_2 \\
v_{x_3} &= k_3 v_i = V_2 \quad \text{for} \quad v_2 < v_i \le v_3 \\
v_{x_k} &= k_k v_i = V_k \quad \text{for} \quad v_{k-1} < v_i \le v_k \\
&\vdots \\
v_{x_n} &= k_n v_i - V_n \quad \text{for} \quad v_{n-1} < v_i \le v_n
\end{aligned}
\tag{5.108}
$$

where $V_0, V_1, \ldots, V_n$ are offset voltages.

In the above equation, v_i is the sensor output voltage, which is input to the linearizer and v_x is an equation for linearized voltage proportional to a physical variable x.

Figure 5.37 shows a diode-clipper-OPAMP-based realization circuit in a general form for n-segmented characteristic equations. The diode-clipper circuit works as a range selector. In the kth diode-clipper circuit output, the input voltage v_i falling within v_{k-1} to v_k appears while the rest of the diode-clipper circuits will output the reference voltage. The outputs of the

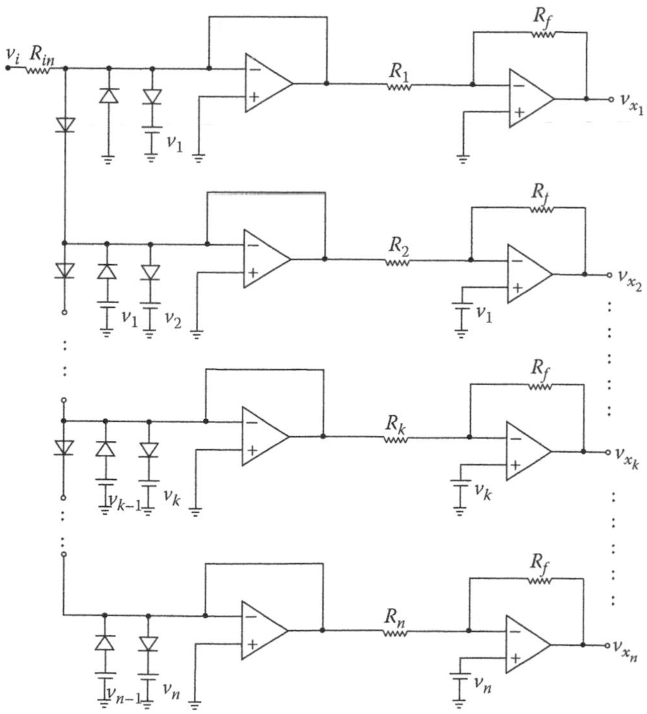

FIGURE 5.37
Diode clipper and OPAMP-based piecewise linearizing circuit.

diode-clipper circuits are applied to n numbers of non-inverting OPAMPs to implement characteristic equation. The circuit works as follows.

The sensor voltage v_i is connected to each of the n numbers of diode-clipper circuits. This sensor voltage will appear at the output of the corresponding circuit in which the range is set by reference voltages v_{k-1} and v_k, i.e., if v_i falls within the range 0–v_1, v_i will appear at the output of the first circuit only, if v_i falls within the range v_1–v_2, v_i appears at the output of second circuit only and so on. When one circuit passes v_i, the output of the rest of the circuit becomes equal to the reference voltage. The output of each diode-clipper circuit is connected to an array of n-OPAMPs through inverting buffers. Each amplifier is configured with gain (K_k) and offset (V_K) corresponding to Equation 5.108. The gain is adjusted as

$$K_1 = \frac{R_f}{R_1}, \quad K_2 = \frac{R_f}{R_2}, \ldots, K_n = \frac{R_f}{R_n} \tag{5.109}$$

and the offset voltages $V_1, V_2, V_3, \ldots, V_n$ are applied to the inverting terminal of the OPAMPs. The gain of an amplifier is set such that the sensor signal voltage is amplified to the level of the corresponding characteristic equation. The outputs $V_{x1} \ldots V_{x_n}$ represents the segmented linearized outputs of v_i. Note that there is only one output in $V_{x1} \ldots V_{x_n}$, which relates to the corresponding segment. Now we have to design a circuit to identify the output that corresponds to the exact range of v_i.

Analog multiplexer (MUX) switches can be used for multiplexing the outputs $V_{x1} \ldots V_{x2}$ by applying the proper selecting address. The schematic diagram of the analog MUX is shown in Figure 5.38a. The channel selector

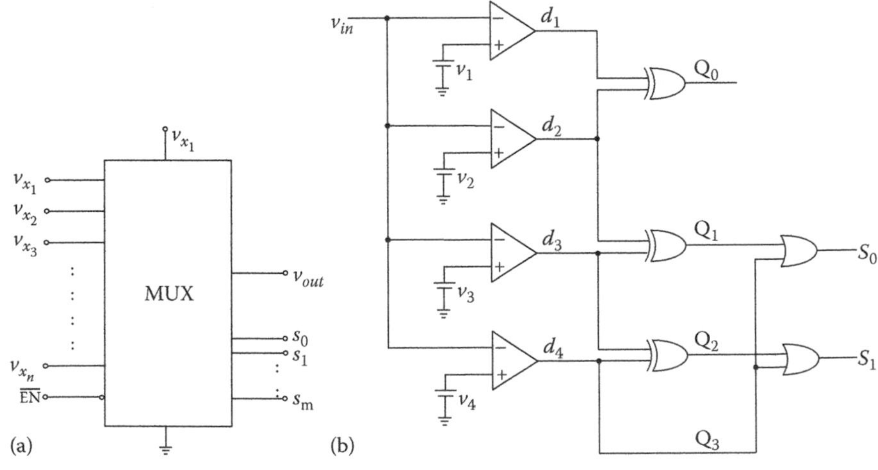

FIGURE 5.38
(a) Block diagram of a MUX and (b) channel selector circuit.

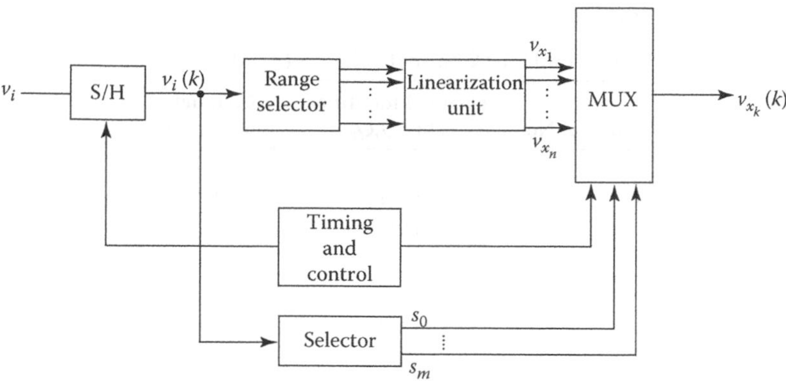

FIGURE 5.39
Block diagram of the piecewise linearization technique.

for the MUX can be realized by an OPAMP comparator-gate combination, as shown in Figure 5.38b. When the linearization circuit accommodates n-segmented pieces, the number of select lines (m) is given by $n=2^m$. Commercially available MUXs (8:1 MUX) can accommodate three select lines. For simplicity, let us examine how a channel selector can be realized for $n=4$. Figure 5.38b shows an array of four comparators that receive the same input voltage v_i, which is triggered gradually from lowest to highest segment of voltage, i.e., first comparator goes HIGH ($d_1=1$) when $0 \le v_i \le v_1$, second comparator triggers when $0 < v_i \le v_2$, and so on. Four outputs Q_0, Q_1, Q_2, and Q_4 are produced, any one of which goes HIGH when v_i falls within the particular range, i.e., only Q_0 goes HIGH when $0 \le v_i \le v_1$, only Q_1 goes HIGH when $v_1 < v_i \le v_2$, and so on. Using a common truth table for the MUX and the channel selector circuit, a circuit can be designed as shown in Figure 5.39.

Since there are only four channels, two select lines are sufficient. Using Table 5.11, the logic equation for the select lines are found using K-map simplification method as

$$S_0 = Q_1 + Q_3$$

$$S_1 = Q_2 + Q_3$$

The circuit shown in Figure 5.38a and b are, truly speaking, not practical circuits since the circuits are unable to synchronize the linearization process with each other. Truly practical circuits will result when timing and control units are introduced. Figure 5.39 shows the schematic diagram of the complete system that includes a sample and hold, and a timing control unit.

TABLE 5.11
Truth Table for Channel Selector ($n=4$)

Channel No.	Channel Identifier $Q_3Q_2Q_1Q_0$	Channel Selector S_1S_0
×	0000	××
1	0001	00
2	0010	01
×	0011	××
3	0100	10
×	0101	××
×	0110	××
×	0111	××
4	1000	11
×	1001	××
×	1010	××
×	1011	××
×	1100	××
×	1101	××
×	1110	××
×	1111	××

Example 5.14

Take the J-type thermocouple data used in Section 5.6 to linearize the thermocouple output between 0°C and 100°C. Perform piecewise linearization in five segments and design the range selector, linearizing amplifier and the MUX selector.

Solution

The given thermocouple data is reduced to the following form for five-segment piecewise linearization:

T (°C):	0	20	40	60	80	100
V_0 (mV):	0	1.019	2.058	3.115	4.186	5.268

Using Equation 5.107, the characteristic equation of the five segments is determined as follows:

Segment 0°C–20°C:

$$\frac{0-1.019}{0-20}=\frac{y-0}{x-0} \tag{E5.14.1}$$

$$\Rightarrow x = 19.62y$$

where

x = temperature, °C(T)
y = voltage, mV(v_i)

Segment 20°C–40°C:

$$\frac{1.019-2.058}{20-40}=\frac{y-1.019}{x-20} \tag{E5.14.2}$$
$$\Rightarrow x = 19.24y + 0.384$$

Similarly for

$$40°\text{C} - 60°\text{C}: \quad x = 18.92y + 1.06 \tag{E5.14.3}$$

$$60°\text{C} - 80°\text{C}: \quad x = 18.67y + 1.83 \tag{E5.14.4}$$

$$80°\text{C} - 100°\text{C}: \quad x = 18.48y + 2.62 \tag{E5.14.5}$$

The amplifier gains are

$$K_1 = 19.62, \quad K_2 = 19.24, \quad K_3 = 18.92, \quad K_4 = 18.67, \quad \text{and} \quad K_5 = 18.48$$

The reference voltages for clipper circuits and amplifier offset are

$$v_0 = 0\text{ V}, \quad v_1 = 0.384\text{ V}, \quad v_2 = 1.06\text{ V}, \quad v_3 = 1.83\text{ V}, \quad \text{and} \quad v_4 = 2.62\text{ V}$$

The gains are set as

Taking $R_f = 200\,\text{k}\Omega$

$$K_1 = 19.62 = \frac{R_f}{R_1} = 200\text{ k}\Omega/R_1$$

$$\therefore R_1 = 10.19 \cong 10.2\,\text{k}\Omega$$

Similarly

$$R_2 = 10.39 \cong 10.4\,\text{k}\Omega$$
$$R_3 = 10.57 \cong 10.6\,\text{k}\Omega$$
$$R_4 = 10.71 \cong 10.7\,\text{k}\Omega$$
$$R_5 = 10.82 \cong 10.8\,\text{k}\Omega$$

The channel selector circuit will have first stage outputs Q_0, Q_1, Q_2, Q_3, and Q_4 as shown in Figure 5.40. The truth table and the logic circuit for the channel selector

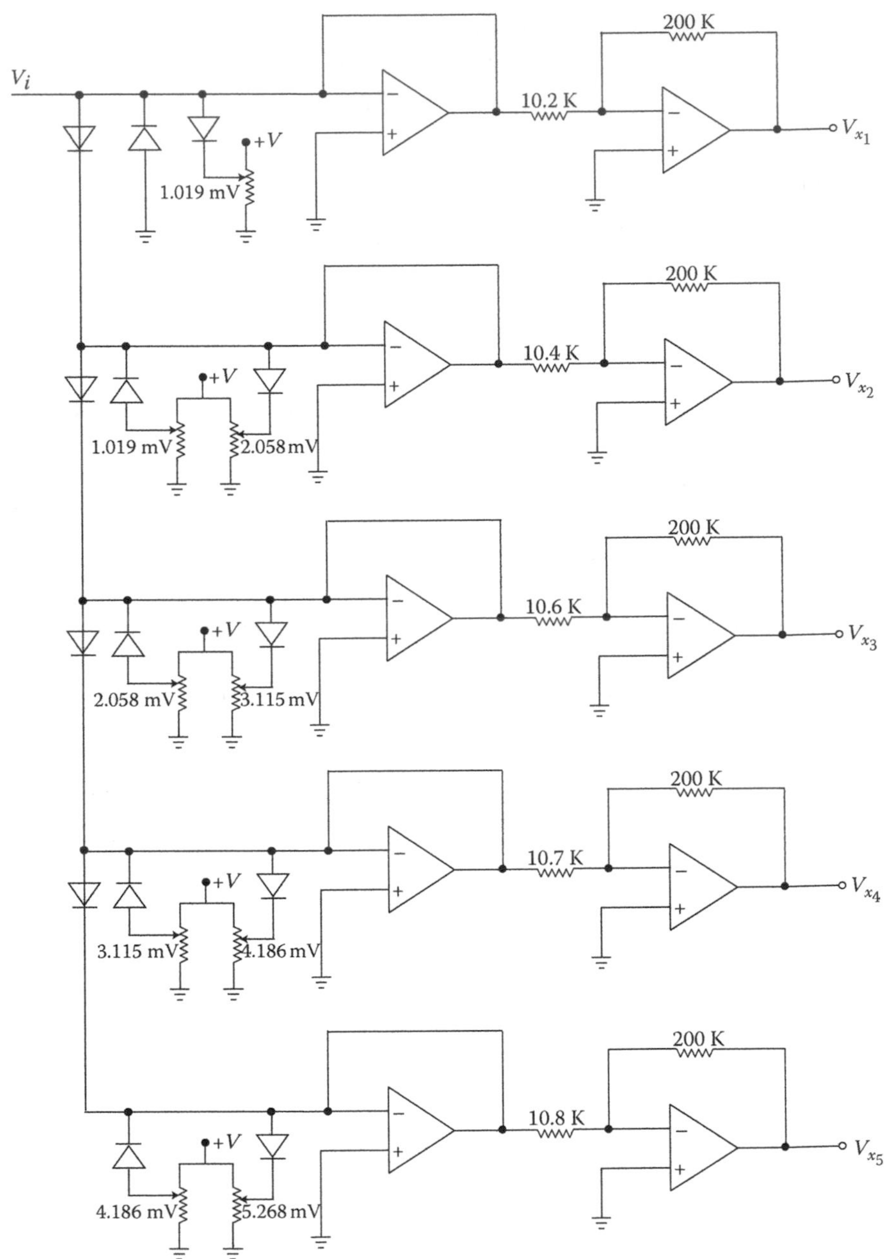

FIGURE 5.40
Piecewise linearization circuit.

TABLE 5.12
Channel Selector Truth Table

Channel	$Q_4Q_3Q_2Q_1Q_0$	$S_2S_1S_0$
×	00000	×××
1	00001	000
2	00010	001
×	00011	×××
3	00100	010
×	00101	×××
×	00110	×××
×	00111	×××
4	01000	011
×	01001	×××
…	…	…
×	01111	×××
5	10000	100
×	10001	×××
…	…	…
×	11111	×××

are shown in Table 5.12 and Figure 5.41, respectively. The truth table is converted to K-map for simplification. The equation for S_0, S_1, and S_2 are found from K-map simplification method (Figure 5.42) as

$$S_0 = Q_1 + Q_3$$

$$S_1 = Q_2 + Q_3$$

$$S_2 = Q_4$$

5.7.2 Digital Processing

The process of piecewise linearization can also be performed by digital processing using an ADC. Similar to nonlinear ADC-based linearization techniques discussed in Section 5.5, the characteristic straight line equation of the sensor for several ranges, or segments can be stored in EPROM, which is addressed by digital codes generated by counters. Based on the method, the algorithm of conversion may be either counting type or successive approximation.

5.7.2.1 Lookup Table–Based ADC

As discussed in Section 5.5, in the SA nonlinear ADC, the EPROM stores the voltages according to nonlinear characteristic equations, where the characteristics can be selected by additional selects bits in the address line of

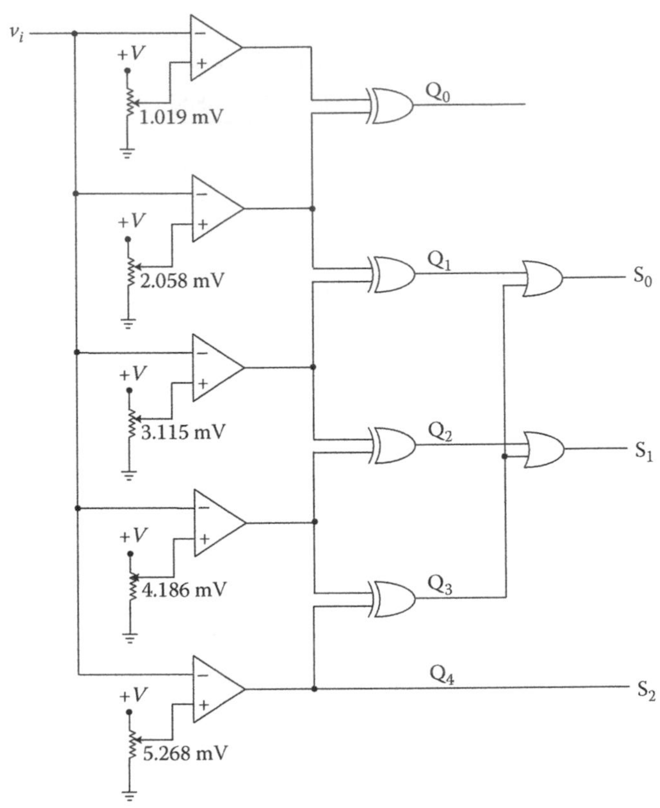

FIGURE 5.41
Channel selector circuit.

the EPROM. Similarly, in the piecewise linearization method using ADC, the characteristic equation for different ranges can be selected by selects bits generated separately by a selector circuit same as that used for MUX in analog processing. The schematic block diagram of such a technique with SA-ADC is shown in Figure 5.43, however in this figure, the counter and control logic circuits are not shown separately. The piecewise linearization process works as follows.

When the ADC starts a conversion cycle, the counter is started by the logic control to output the binary signal (A_r ...A_1A_0). Simultaneously, the selector is triggered to develop the select lines S_m ...S_1S_0. For an 8-bit counter $r=7$ and, therefore, the total coded value is $2^{7+1}=256$. If the number of total select lines $m+1=1$, then the total address length of the EPROM is $(r+1)+(m+1)=8+1=9$. With $m+1=1$, two characteristic segments can be used.

The corresponding digital value from the EPROM is fed to a DAC to convert it to an equivalent analog level. The analog voltage is compared with the amplified sensor output V_{amp} by a comparator, and when the two voltages become equal, the control logic stops the counter and the digital output

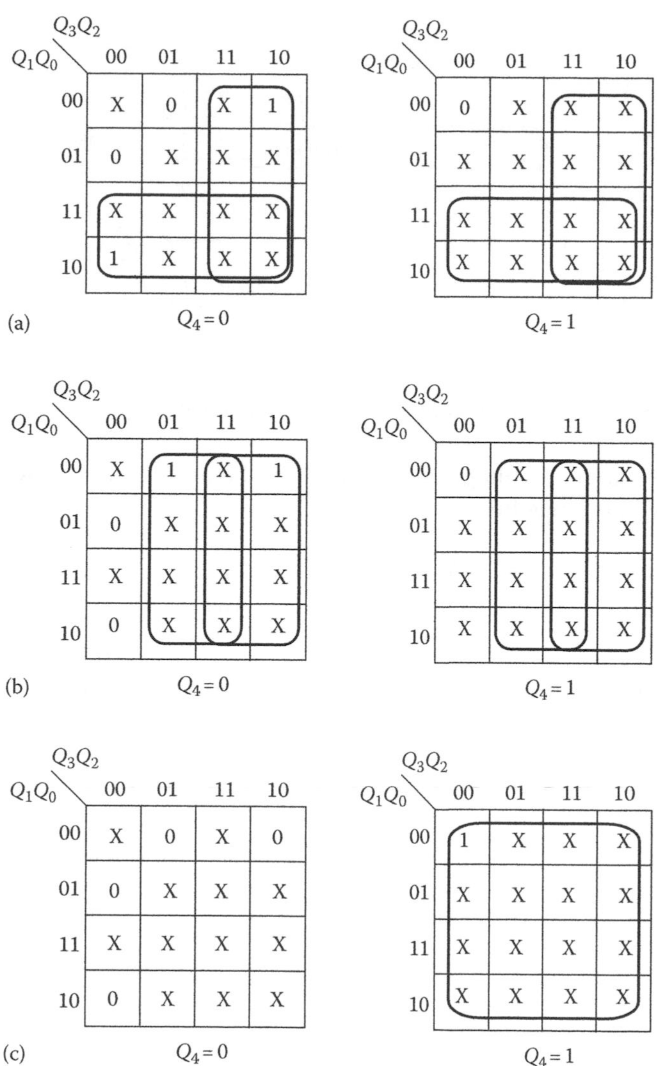

FIGURE 5.42
K-map for select lines for (a) S_0, (b) S_1, and (c) S_2.

(S_m … S_1S_0 A_r … A_1A_0) becomes the digital output of the linearized sensor voltage. Let us consider the characteristic equation of a pressure sensor of Figure 5.35 segmented into two segments only for the simplicity of explaining the piecewise linearization by ADC. Recall the two ranges of the pressure sensor as

P_i(psi):	0	50	100
V_0(mV):	0	4.52	4.77

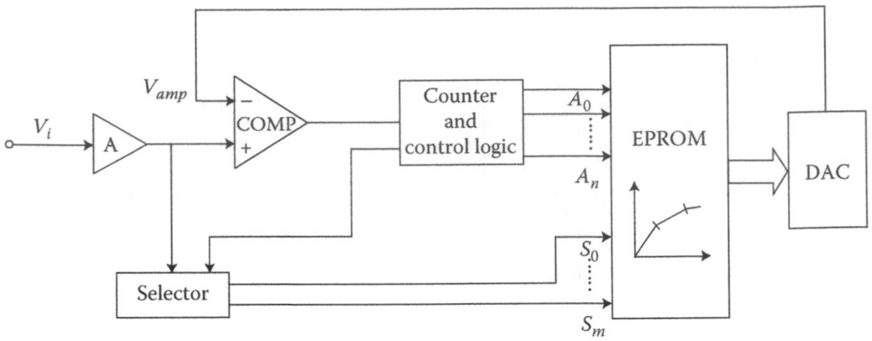

FIGURE 5.43
Digital processing of an SA-ADC-based piecewise linearization.

If the entire range of voltage is amplified by the amplifier at ADC input to a range 0–5 V, the range becomes

P_i(psi):	0	50	100
V_{amp}(V):	0	4.73	5.00

Since the EPROM stores the voltage levels, we develop two characteristic equations in the form $v = mp + C$ using Equation 5.107 as shown below:

For 0–50 psi

$$\frac{0-4.73}{0-50} = \frac{y-0}{x-0}$$

$$\Rightarrow x = 0.094y$$

$$\Rightarrow v = 0.094p \quad \text{for } 0 \le V_{amp} \le 4.73\text{ V}$$

Similarly, for the range 50–100 psi, we get

$$v = 0.0054p + 4.46 \quad \text{for } 4.73 < V_{amp} \le 5\text{ V}$$

Now let us design the selector circuit similar to that used for MUX selector in analog processing. Similar to Figure 5.38b, we use two comparators and one NOR gate as shown in Figure 5.44. Since there are two characteristic equations, the number of selector lines $m = 2 - 1 = 1$. Truth table of Table 5.13 gives the clue to design the circuit for S_0. The equation for S_0 is evident from Table 5.13 that

$$S_0 = Q_0$$

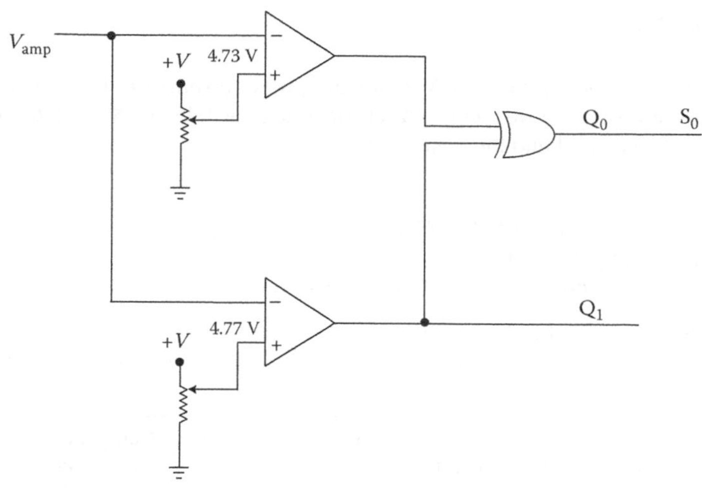

FIGURE 5.44
Selector circuit for ADC linearizer.

Therefore, when the sensor voltage is in the range of first segment $S_0 = 0$ and when the sensor voltage falls in the second segment, $S_0 = 1$ (Table 5.13).

The next requirement of the piecewise linearization circuit is to develop the LUT for the EPROM. Considering the ADC to be an 8-bit one, the counter output is 7-bit and the S_0 bit becomes the MSB bit. The ADC conversion logic is shown in Table 5.14.

TABLE 5.13
Truth Table for the Select Circuit

Characteristic Equation	Q_1Q_0	S_0
First	00	0
Second	01	1
×	10	×
×	11	×

TABLE 5.14
ADC Conversion Logic Table

Characteristic Equation	EPROM Address $S_0A_6A_5A_4A_3A_2A_1A_0$	LUT Voltage (V)	Pressure (psi)	Code (N)
	0 0 0 0 0 0 0 0	0	0	0
$v = 0.094p$	0 0 0 0 0 0 0 1	0.037	0.3937	1
	⋮	⋮	⋮	⋮
	0 1 1 1 1 1 1 0	4.660	49.518	126
	0 1 1 1 1 1 1 1	4.73	50	127
	1 0 0 0 0 0 0 0	4.732	50.3937	128
$v = 0.0054p + 4.46$	1 0 0 0 0 0 0 1	4.734	50.786	129
	⋮	⋮	⋮	⋮
	1 1 1 1 1 1 1 0	4.997	99.60	254
	1 1 1 1 1 1 1 1	5.000	100	255

Example 5.15

The piecewise linearization SA-ADC for the pressure sensor output for the two segments discussed above receives a sensor voltage of 4.45 mV. Show the 8-bit SA conversion steps and find the digital output.

Solution
The range of sensor output: 0–4.77 mV
The range of amplifier output: 0–5 V
Hence the amplified output = 5/4.77 × 4.45 = 4.66 V

Since the voltage of 4.66 V is within the first segment, i.e., 0–4.73 V, the selector output $S_0 = 0$. Hence the digital output in the SA-ADC process is as shown below:

Clock	$S_0A_6A_5A_4A_3A_2A_1A_0$	Code (N)	Voltage (V)	Comparator Decision
1	0 1 0 0 0 0 0 0	64	2.368	Less
2	0 1 1 0 0 0 0 0	96	3.552	Less
3	0 1 1 1 0 0 0 0	112	4.140	Less
4	0 1 1 1 1 0 0 0	120	4.440	Less
5	0 1 1 1 1 1 0 0	124	4.580	Less
6	**0 1 1 1 1 1 1 0**	126	4.662	STOP

The voltage levels are calculated as shown below:

In first clock: Digital output = 0100 0000
Equivalent decimal $(N) = 64$

$$1\,\text{LSB for pressure (psi)} = \frac{50}{2^7 - 1} = \frac{50}{127} = 0.3937\ \text{psi}$$

∴ Equivalent pressure = 0.3937 × 64 = 25.196 psi
Equivalent voltage = $v = 0.094p = 0.094 \times 25.196 = 2.368$ V

In a similar way, the voltages for other digital outputs are calculated.

Hence the digital output = $(01111110)_2$. (Shown in bold in table above)

5.7.2.2 Piecewise Linear ADC

In Section 5.5, we have seen how counting and successive approximation principle can be employed for converting nonlinear analog signals to digital words with the help of LUT stored in EPROMs. Figure 5.14b shows how a 3-bit digital word corresponds to nonlinear analog voltage characteristics. Also recall Tables 5.2 and 5.3 where the analog level that corresponds to two different nonlinear equations are shown against the digital words. In this section, we will see how the nonlinear characteristic equations, say that of Table 5.3, can be considered for piecewise linearization using another type of ADC—a flash ADC.

A conventional linear flash ADC is shown with a basic circuit diagram in Figure 5.45. The circuits comprise of a resistive divider network, sample-and-hold circuit, an array of comparators, and an 8:3 priority encoder. Since equal values of resistances (R) are used in the resistive divider networks, the reference voltages of the comparators are also equally spaced as 0 V (LSB), 5/8 V, …, 35/8 V (MSB). The truth table showing comparator outputs and the 8 to 3 line priority encoder is shown in Table 5.15.

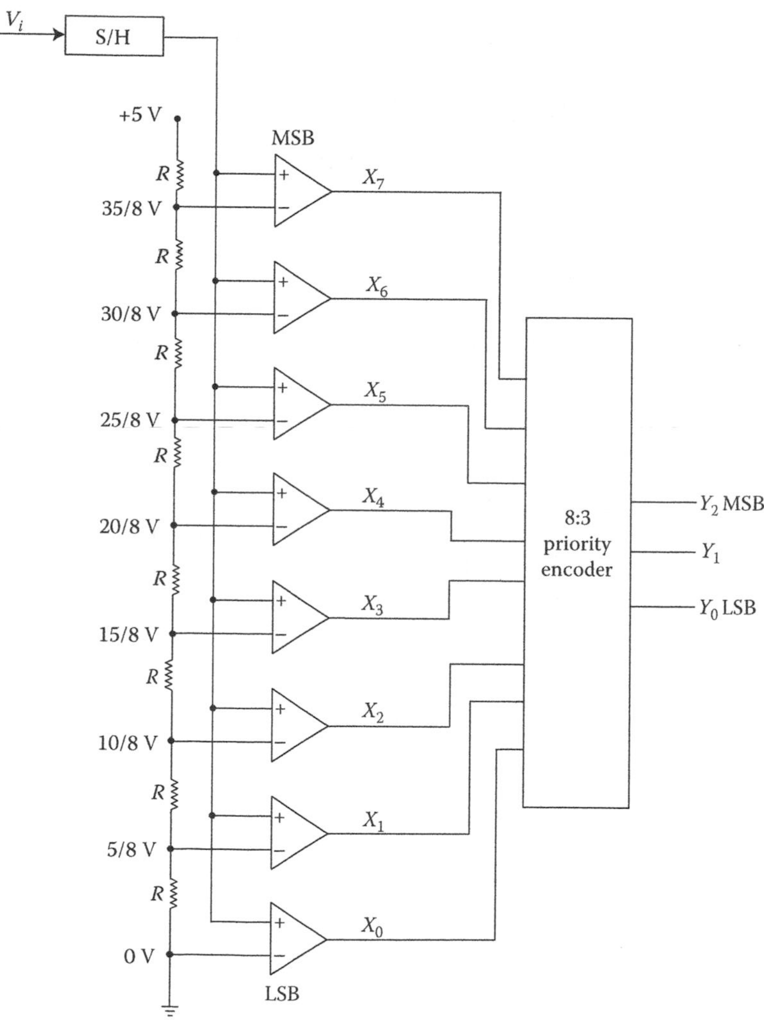

FIGURE 5.45
A basic linear flash ADC.

TABLE 5.15
Truth Table of the Linear Flash ADC

V_i Range (V)	$x_7x_6x_5x_4x_3x_2x_1x_0$	$y_2y_1y_0$
$0 \le v_i \le \frac{5}{8}$	0000 0001	000
$\frac{5}{8} < v_i \le \frac{10}{8}$	0000 0011	001
$\frac{10}{8} < v_i \le \frac{15}{8}$	0000 0111	010
$\frac{15}{8} < v_i \le \frac{20}{8}$	0000 1111	011
$\frac{20}{8} < v_i \le \frac{25}{8}$	0001 1111	100
$\frac{25}{8} < v_i \le \frac{30}{8}$	0011 1111	101
$\frac{30}{8} < v_i \le \frac{35}{8}$	0111 1111	110
$\frac{35}{8} < v_i \le \frac{40}{8}$	1111 1111	111

Now let us consider the nonlinear equation $v_i = KN^2$, which was tabulated for values of $N = 0, 1, 2, \ldots, 7$ in Table 5.3 for the LUT method of linearization. Linearization of such a characteristic equation can be performed by a flash ADC using a nonuniform resistive divider circuit. Therefore, an output over a segment can be called as piecewise linearized digital output. The resistive divider circuit can be designed with the following approach.

In the nonlinear equation $v_i = KN^2$, for a 3-bit ADC, at maximum values of $v_i = 5\,\text{V}$ and $N = 7$, $K = (5/7^2) = (5/49)$.

The reference voltages are

$$V_{ref\,0} = \frac{5}{49} \times 0 = 0\ \text{V}$$

$$V_{ref\,1} = \frac{5}{49}\ \text{V}$$

$$V_{ref\,2} = \frac{5}{49} \times 4\ \text{V}$$

$$V_{ref\,3} = \frac{5}{49} \times 9\ \text{V}$$

$$V_{ref\,4} = \frac{5}{49} \times 16\ \text{V}$$

$$V_{ref5} = \frac{5}{49} \times 25 \text{ V}$$

$$V_{ref6} = \frac{5}{49} \times 36 \text{ V}$$

$$V_{ref7} = \frac{5}{49} \times 49 \text{ V}$$

The resistance values are

$$V_{ref0} = 0\,\text{V}, \quad \text{so} \quad R_0 = 0\,\Omega\,(\text{grounded})$$

$$V_{ref1} = \frac{5}{49}$$

$$= 5 \times \frac{1}{1+48}$$

$$\therefore R_1 = 1\,\text{k}\Omega$$

$$V_{ref2} = \frac{5}{49} \times 4$$

$$= 5 \times \frac{1+3}{(1+3)+45}$$

$$\therefore R_2 = 3\,\text{k}\Omega$$

Similarly, we get

$$R_3 = 5\,\text{k}\Omega, \quad R_4 = 7\,\text{k}\Omega, \quad R_5 = 9\,\text{k}\Omega, \quad R_6 = 11\,\text{k}\Omega, \quad \text{and} \quad R_7 = 13\,\text{k}\Omega$$

The piecewise linearizing flash ADC is shown in Figure 5.46 and the truth table is shown in Table 5.16.

A piecewise linearizing flash ADC with $N=15$ break voltages for linearizing a cosine characteristic (Figure 5.47) of a sensor developed in [11] is shown in Figure 5.48. The ADC follows a two-stage conversion procedure: in the first stage (during control time ϕ_1 and ϕ_2), when a start of conversion (SOC) signal is received by the ADC, the analog input voltage is converted to a 4-bit word ($b_7b_6b_5b_4$) and stored in register A. In this stage, the digital word is first developed by 15 comparators, the reference voltage to which are set by the piecewise linearized resistive dividers. The 15-bit word is then encoded to the 4-bit word by a 4-bit encoder. The bits $b_7b_6b_5b_4$ form the MSB of the digital output. The conversion is further fine-tuned in the

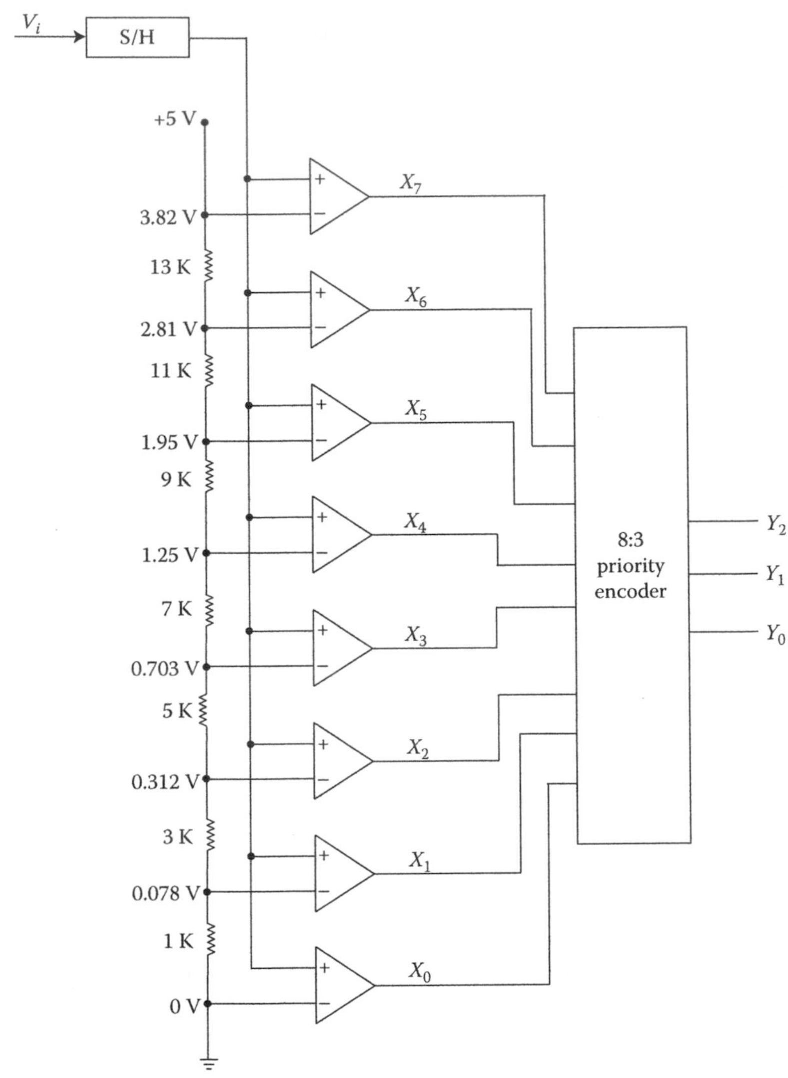

FIGURE 5.46
A piecewise linearizing flash ADC.

second stage when two 16–1 MUX controlled by the MSB bits sets linearly spaced reference voltages to the comparators, which produces the second set of 15-bit word. This 15-bit word is now converted to a 4-bit word ($b_3b_2b_1b_0$) and stored in register B. The second stage is performed during timing ϕ_3 and ϕ_4.

The controller unit is a finite state machine that generates the timing signals ϕ_1, ϕ_2, ϕ_3, and ϕ_4 in the proper sequence. When the data are stored in registers A and B, the controller generates an end of conversion (EOC) signal.

TABLE 5.16
Truth Table of the Piecewise Linearized Flash ADC

V_i Range (V)	$x_7x_6x_5x_4\ x_3x_2x_1x_0$	$y_2y_1y_0$
0–0.078	0000 0001	000
0.078–0.312	0000 0011	001
0.312–0.703	0000 0111	010
0.703–1.25	0000 1111	011
1.25–1.95	0001 1111	100
1.95–2.81	0011 1111	101
2.81–3.82	0111 1111	110
3.82–5.00	1111 1111	111

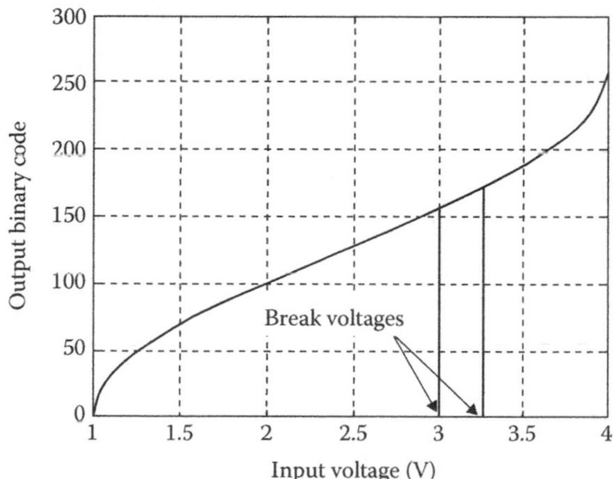

FIGURE 5.47
Cosine characteristic of a sensor. (Reprinted from IEEE, copyright© IEEE. With permission.)

The method of-how the converter resolves the entire range into nonlinear segments will be discussed with the help of the following 6-bit two-stage flash converter.

The 6-bit flash converter is shown in Figure 5.49. The ADC comprises of eight comparators, the output of which are converted to 3-bit words by the 3-bit encoder. The comparator develops the outputs X_7, X_6,…, X_0 by comparing analog input v_i with the piecewise linearized voltage divider circuit. Let us consider that the ADC receives an analog voltage $v_i = 0.50\,\text{V}$. From Table 5.14, it is found that the comparator output at time T_1 will be $X_7X_6X_5X_4X_3X_2X_1X_0 = 0000\ 0111$. This word is converted to a 3-bit word $Y_2Y_1Y_0 = 010$. So at time T_2, register A stores the MSB word $b_5b_4b_3 = 010$. The characteristic of the converter is shown in Figure 5.50a. At time T_3, two operations take place simultaneously:

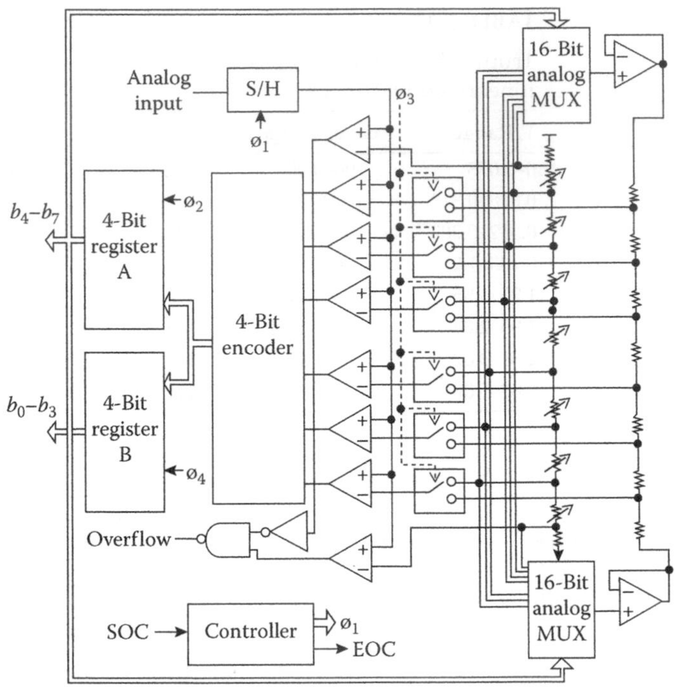

FIGURE 5.48
Piecewise linearizing flash ADC for a sensor with cosine characteristics. (Reprinted from IEEE, copyright © IEEE. With permission.)

1. The 3-bit MSB word (i.e., 011) is used to address an 8:1 MUX-1 through a register C. The MUX multiplexes the lower limit voltage from v_0, $v_1, \ldots, v_7$ so the output becomes 0.312 V.
2. A binary "1" is added to the 3-bit word to get the address of the next higher-limit voltage. So the address becomes $011 + 1 = 100$ and the output becomes 0.703 V.

Between these two limit voltages, there are eight equal segments developed by the *R*–*R* divider circuit to get the break voltages as

$$0.312, 0.36, 0.408, 0.456, 0.504, \ldots, 0.703 \text{ V}.$$

At time T_4, these reference voltages are connected to the comparators using the 8-switch array. Since the input voltage $v_i = 0.5$ V, the comparator output becomes $X_7X_6X_5X_4X_3X_2X_1X_0 = 00011111$. This 8-bit word is converted to a 3-bit word $b_2b_1b_0 = 100$ and at time T_5, it is stored in register B, hence the equivalent 0.5 V is converted to a 6-bit word as

$$b_5b_4b_3 \quad b_2b_1b_0 = 011 \quad 100$$

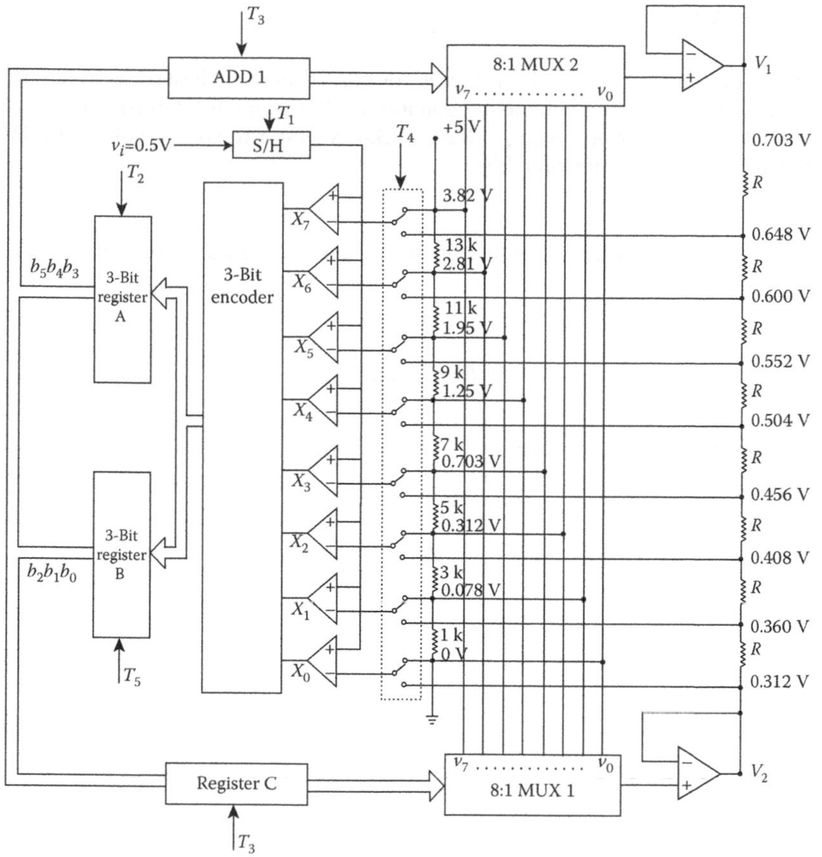

FIGURE 5.49
A 6-bit piecewise linearizing flash ADC.

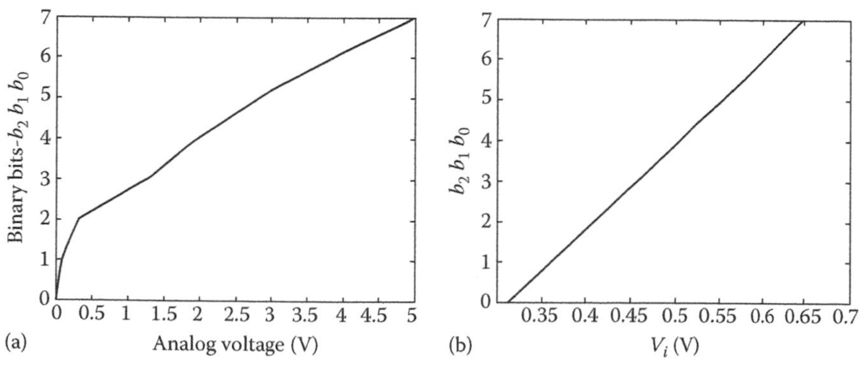

FIGURE 5.50
Conversion characteristics of the 6-bit flash ADC: (a) first stage and (b) second stage.

Example 5.16

For the piecewise linearized flash ADC discussed above, design the reference voltages circuit for the nonlinear equation $v_i = KN^2$ with 8-bit resolution. Find the digital output for an analog voltage of 3.43 V and a reference voltage of 5 V. Prepare the ADC conversion table.

Solution

Since the ADC is 8-bit, each of the registers, A and B, should store 4-bit words as

$$A = b_7b_6b_5b_4$$

$$B = b_3b_2b_1b_0$$

Total number of comparators = 16

For $v_i = 5$ V and $N = 16$

$$K = \frac{5}{16^2} = \frac{5}{256}$$

The reference voltages are

$$V_{ref0} = \frac{5}{256} \times 0 = 0 \text{ V}$$

$$V_{ref1} = \frac{5}{256} \times 1 = 19.53 \text{ mV}$$

$$V_{ref2} = \frac{5}{256} \times 2^2 = 78.12 \text{ mV}$$

$$V_{ref3} = \frac{5}{256} \times 3^2 = 0.175 \text{ V}$$

$$V_{ref4} = \frac{80}{256} = 0.312 \text{ V}$$

$$V_{ref5} = \frac{125}{256} = 0.488 \text{ V}$$

$$V_{ref6} = \frac{180}{256} = 0.703 \text{ V}$$

$$V_{ref7} = \frac{245}{256} = 0.957 \text{ V}$$

$$V_{ref8} = \frac{320}{256} = 1.25 \text{ V}$$

$$V_{ref9} = \frac{405}{256} = 1.582 \text{ V}$$

$$V_{ref9} = \frac{405}{256} = 1.582\text{ V}$$

$$V_{ref10} = \frac{500}{256} = 1.953\text{ V}$$

$$V_{ref11} = \frac{605}{256} = 2.363\text{ V}$$

$$V_{ref12} = \frac{720}{256} = 2.812\text{ V}$$

$$V_{ref13} = \frac{845}{256} = 3.3\text{ V}$$

The resistance values are

$$R_0 = 0\ \Omega$$

$$R_2 = 1\text{k}\Omega$$

$$R_3 = 3\text{ k}\Omega$$

$$R_4 = 5\text{ k}\Omega$$

$$R_5 = 7\text{ k}\Omega$$

$$R_6 = 9\text{ k}\Omega$$

$$R_7 = 11\text{k}\Omega$$

$$R_8 = 13\text{ k}\Omega$$

$$R_9 = 15\text{ k}\Omega$$

$$R_{10} = 17\text{ k}\Omega$$

$$R_{11} = 19\text{ k}\Omega$$

$$R_{12} = 21\text{k}\Omega$$

$$R_{13} = 23\text{ k}\Omega$$

$$R_{14} = 25\text{ k}\Omega$$

$$R_{15} = 27\text{ k}\Omega$$

$$R_{16} = 29\text{ k}\Omega$$

$$R_{17} = 31\text{k}\Omega$$

Since $v_i = 3.43$ V at first stage, the 4-bit digital word $b_7b_6b_5b_4 = 1101$ is stored in register A. in the second stage, the multiplexed outputs are

$$V_2 = 3.3\text{ V}$$

$$V_1 = 3.828\text{ V} \quad \text{Hence } \Delta V = 3.828 - 3.3 = 0.528\text{ V}$$

TABLE E5.16A

ADC Conversion Table

Range of V_i (V)	$x_{15}x_{14}x_{13}x_{12}x_{11}x_{10}x_9x_8$ $x_7x_6x_5x_4x_3x_2x_1x_0$	$y_3y_2y_1y_0$
0–0.0195	0000 0000 0000 0001	0000
0.0195–0.0781	0000 0000 0000 0011	0001
0.0781–0.175	0000 0000 0000 0111	0010
0.175–0.312	0000 0000 0000 1111	0011
0.312–0.488	0000 0000 0001 1111	0100
0.488–0.703	0000 0000 0011 1111	0101
0.703–0.957	0000 0000 0111 1111	0101
0.957–1.25	0000 0000 1111 1111	0110
1.25–1.582	0000 0001 1111 1111	1000
1.582–1.953	0000 0011 1111 1111	1001
1.953–2.363	0000 0111 1111 1111	1010
2.363–2.812	0000 1111 1111 1111	1011
2.812–3.3	0001 1111 1111 1111	1100
3.3–3.828	0011 1111 1111 1111	1101
3.828–4.394	**0111 1111 1111 1111**	**1110**
4.394–5.00	1111 1111 1111 1111	1111

This range is shown in bold in Table E5.16a. The uniform reference voltages of second stage are

$$v_{y0} = 3.3 \text{ V}$$

$$v_{y1} = 3.3 + \frac{\Delta V}{16} = 3.3 + 0.033 = 3.333 \text{ V}$$

Similarly we get

$$v_{y2} = 3.366 \text{ V}$$

$$v_{y3} = 3.399 \text{ V}$$

$$v_{y4} = 3.432 \text{ V}$$

$$v_{y5} = 3.465 \text{ V}$$

$$v_{y6} = 3.498 \text{ V}$$

$$v_{y7} = 3.531 \text{ V}$$

$$v_{y8} = 3.564 \text{ V}$$

$$v_{y9} = 3.597 \text{ V}$$

TABLE E5.16B

Break Voltages and Digital Codes

Range of V_i (V)	$x_{15}x_{14}x_{13}x_{12}x_{11}x_{10}x_9x_8$ $x_7x_6x_5x_4x_3x_2x_1x_0$	$y_3y_2y_1y_0$
3.3–3.333	0000 0000 0000 0001	0000
3.333–3.366	0000 0000 0000 0011	0001
3.366–3.399	0000 0000 0000 0111	0010
3.399–3.432	0000 0000 0000 1111	0011
3.432–3.465	0000 0000 0001 1111	0100
3.465–3.498	0000 0000 0011 1111	0101
3.498–3.531	0000 0000 0111 1111	0101
3.531–3.564	0000 0000 1111 1111	0110
3.564–3.597	0000 0001 1111 1111	1000
3.597–3.63	0000 0011 1111 1111	1001
3.63–3.663	0000 0111 1111 1111	1010
3.663–3.696	0000 1111 1111 1111	1011
3.696–3.729	0001 1111 1111 1111	1100
3.729–3.726	0011 1111 1111 1111	1101
3.726–3.795	0111 1111 1111 1111	1110
3.795–3.828	1111 1111 1111 1111	1111

$$v_{y10} = 3.63\text{ V}$$

$$v_{y11} = 3.663\text{ V}$$

$$v_{y12} = 3.696\text{ V}$$

$$v_{y13} = 3.729\text{ V}$$

$$v_{y14} = 3.762\text{ V}$$

$$v_{y15} = 3.795\text{ V}$$

$$v_{y16} = 3.828\text{ V}$$

The break voltages between 3.3 and 3.828 V and the corresponding digital codes are shown in Table E5.16b.
The 4-bit word stored in register B is $b_3b_2b_1b_0 = 0011$.
Hence the final digital output is 11010011.

Example 5.17

The relation between pressure and output voltage of a pressure sensor when amplified by an amplifier is given by a nonlinear equation $p = K_1v + K_2v^2$, where K_1 and K_2 are two constants. The output voltage of the sensor at maximum pressure of 100 psi is 5 V and it is 3 V at mid range.

1. Derive an equation for output voltage.
2. Segment the pressure range into 16 segments and find the voltages.
3. Using the voltage ranges between segments as reference voltages, design the resistive divider circuit.

Solution

1. Applying the pressure and voltage values: at $p=100$ psi, $v=5$ V and $p=50$ psi, $v=3$ V

$$\therefore 3K_1 + 9K_2 = 50$$

$$5K_1 + 25K_2 = 100$$

By solving we get

$$K_1 = \frac{35}{3} = 11.66$$

$$K_2 = \frac{5}{3} = 1.6$$

The equation can be written as

$$11.66v + 1.66v^2 - p = 0$$

The solution of the equation gives

$$v = \frac{-11.66 \pm \sqrt{(11.66)^2 + 4 \times 1.66 \times P}}{2 \times 1.66}$$

$$= \frac{-11.66 \pm \sqrt{(135.95) + 6.64P}}{3.32}$$

2. If the pressure range is segmented in 16 segments, we get

$$\Delta p = \frac{100}{16} = 6.25$$

For $p_0 = 0$; $v_0 = \dfrac{-11.66 + \sqrt{135.95 + 6.64P}}{3.32} = 0$

and for $p_1 = 6.25$; $v_1 = \dfrac{-11.66 + \sqrt{135.95 + 6.64P}}{3.32} = 0.5$ V

Similarly, we get other values, which are tabulated below (Table E5.17):

TABLE E5.17

Pressure–Voltage Relationship

Pressure (psi)	Voltage (V)	Pressure (psi)	Voltage (V)
6.25	0.5	56.25	3.28
12.5	0.94	62.5	3.55
18.75	1.34	68.75	3.81
25.00	1.72	75.00	4.07
31.25	2.07	81.25	4.31
37.5	2.39	87.50	4.55
43.75	2.70	93.75	4.78
50.00	3.00	100.00	5.00

3. Let total resistance of the resistive divider $= R_T = 100\,K$

For $V_{ref1} = 0.5\text{ V} = 5 \times \dfrac{R_1}{R_T}$

$$\therefore R_1 = \frac{0.5 \times 100}{5} = 10\,K$$

$$V_{ref2} = 0.94\text{ V} = 5 \times \frac{R_1 + R_2}{R_T}$$

$$\therefore R_2 = 8.8\,K$$

Similarly, we get

$R_3 = 8\,K,\quad R_4 = 7.6\,K,\quad R_5 = 7\,K,\quad R_6 = 6.4\,K,\quad R_7 = 6.2\,K,\quad R_8 = 6\,K,$

$R_9 = 5.6\,K,\quad R_{10} = 5.4\,K,\quad R_{11} = 5.2\,K,\quad R_{12} = 5.2\,K,\quad R_{13} = 4.8\,K,$

$R_{14} = 4.8\,K,\quad R_{15} = 4.6\,K,\quad R_{16} = 4.4\,K$

5.8 Microcontroller-Based Linearization

The rapid decrease in the cost of microcontrollers and increase in on-chip capabilities have made it possible to perform larger mathematical operations in sensor signal processing. The basic functions that a microcontroller has to perform in sensor signal linearization are

1. To receive the sensor signal from the ADC
2. To perform the mathematical linearization operation to fit or compare the data with a reference value stored in EPROM
3. To output the data to a display unit or any other outside world devices through DAC

This text will emphasize on the second operation only, and the first and third functions related to input/output operation is available in other texts. With respect to sensor signal linearization by microcontroller, two methods are commonly used:

1. Use of an LUT
2. Polynomial fit to the data by mathematical operation

5.8.1 Lookup Table Method

An LUT holds values digitally corresponding to a linear fit of the sensor measurement data. The data is stored in an EPROM from where it is retrieved by the microcontroller at each linearization cycle. When the acquisition speed is critical, LUT methods are faster than the polynomial fit method. Although the LUT method is faster because no calculation is involved, it requires a large EPROM size. The more the number of measurement points used, the better the accuracy of linearization, however at the cost of EPROM size.

Microcontroller-based linearization of sensor signal using a LUT stored in EPROM basically involves the process of converting the analog voltage by an ADC and using the ADC output to address the EPROM where the LUT is stored. The flowchart of Figure 5.51 illustrates the microcontroller actions for the linearization process. Now let us examine how a basic microcontroller can be configured with an ADC and an external EPROM. Most microcontrollers have an on-chip internal EPROM; however, we will consider here a microcontroller without an internal EPROM such as Intel 8031. Figure 5.52 shows the connections of 8031 with a flash ADC, 8 K RAM and a 16 K EPROM (IC 27128). The microcontrollers have four 8-bit ports—port 0, port 1, port 2, and port 3. The digital output of the flash ADC ranges from 00h for 0 V to FFh for +5 V of analog input. The control signals of the ADC are chip select enable ($\overline{CS}$), write strobe ($\overline{WR}$), and read strobe ($\overline{RD}$). These control signals are connected to P3.2, P3.3, and P3.3 of port 3.

The EPROM is connected to port 0 and port 2 for its addressing and port 0 is time multiplexed for both addressing and data. The microcontroller first sends the lower bytes of the address through port 0, which is then latched to an external register to store the lower byte of address. It is stored by applying a clock pulse to address latch enable (ALE). Once it is latched, port 0 can be used as a data bus to read the EPROM. Port 2 is used to connect to the higher

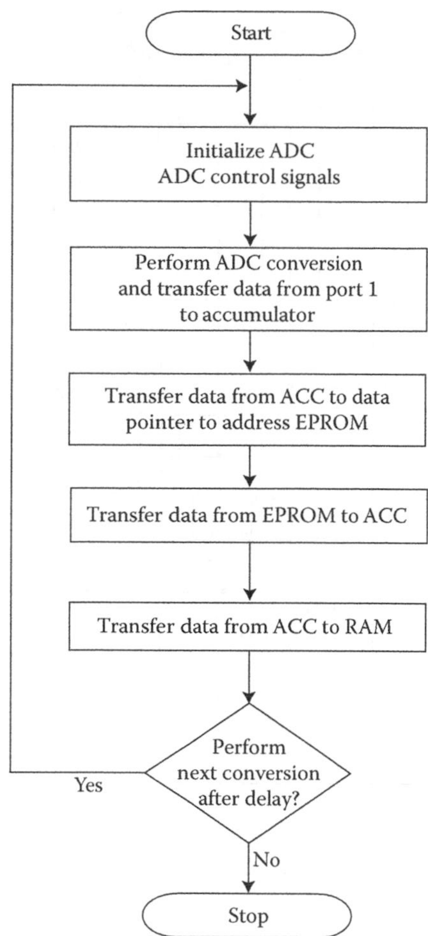

FIGURE 5.51
Flowchart of microcontroller-based linearization using LUT.

byte of the address (AD13-AD8) and ($\overline{PSEN}$) for EPROM addressing. The latched lower byte of the address A7-A0 and higher byte A12-A8 are used to address the RAM. The data byte D7-D0 for both EPROM and RAM is connected to P 0.0–P 0.7.

Now we examine how a microcontroller can be programmed to perform the linearization with the steps shown in the flowcharts of Figure 5.51. We assume that the EPROM stores a LUT for data as per a nonlinear equation $y=x^2$, where y is the sensor input variable and x is the code of the digital output of the ADC. If the sensor circuit generates voltages in the range of 0–5 V for an input variable of range 0–50, for simplicity, we use a 3-bit ADC, as shown in Table 5.17, for linearizing the given characteristic equation in seven segments.

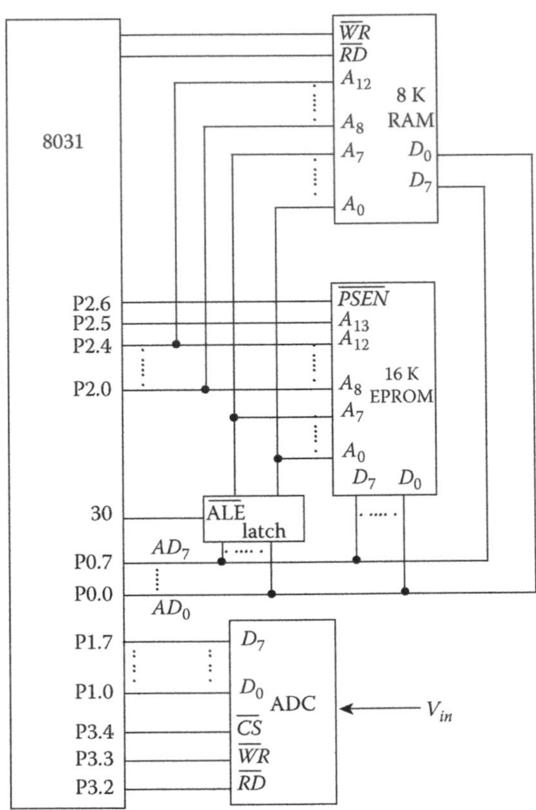

FIGURE 5.52
Connection diagram of ADC, external ROM, and RAM with 8031 microcontroller.

TABLE 5.17
ADC Characteristics Stored in LUT

Analog Voltage (V)	Digital Output $d_2d_1d_0$	Code	x^2 Stored in (HEX) LUT	Decimal Data (LUT)
0	000	0	00h	00^2=00d
0.71	001	1	01h	01^2=01d
1.42	010	2	04h	02^2=04d
2.13	011	3	09h	03^2=09d
2.84	100	4	10h	04^2=16d
3.55	101	5	19h	05^2=25d
4.26	110	6	24h	06^2=36d
4.97	111	7	31h	07^2=49d

The following program stores the data in the EPROM starting from location with address 0000h.

Mnemonics	Comment
.org 0000h	; start address of EPROM is 0000h
.db 00h	; EPROM location 0000h stores 00h
.db 01h	; EPROM location 0001h stores 01h
.db 04h	; EPROM location 0002h stores 04h
.db 09h	; EPROM location 0003h stores 09h
.db 10h	; EPROM location 0004h stores 10h
.db 19h	; EPROM location 0005h stores 19h
.db 24h	; EPROM location 0006h stores 24h
.db 31h	; EPROM location 0007h stores 31h

Let us consider that the ADC completes one conversion cycle in 50 μs, hence the ADC operation must be completed by the microcontroller within this time including the time required for the instruction cycles. Therefore, a delay must be used to synchronize the processor with the ADC. The program that performs the communication of the microcontroller with an ADC, an external RAM, and an EPROM is shown below

```
  .equ begin, 2000h      ; start address of RAM for storing data
  .equ delay,            ;
  .org 0000h
adc:clr p3.2             ; activates CS to ADC (1)
Loop: clr p3.3           ; activates WR pulse
  set b p3.3             ; (1)
  clr p3.4               ; activates RD pulse (1)
  mov a, p1              ; move data from port to Acc (1)
  set b p3.4             ; deactivate RD pulse (1)
  mov 0083, a            ; store data in low byte of DPTR (2)
  mov 0082,# 00h         ; store 00h in high byte of DPTR (2)
  mov a, # 00h           ; (1)
  mov a, @ a+dptr        ; move data form EPROM with address at
                           DPTR (2)
  mov dptr, # begin      ; point to start of RAM (1)
  movx @ dptr, a         ; store data in RAM (2)
  mov r1, # delay        ; generate delay
  cycle: djnz r1, cycle  ; (2)
  sjmp loop              ; (2)
  .end
```

The figures inside parenthesis are the number of instruction cycles.
The total number of cycles in the program = 17
Time for instructions execution = $17 \times 0.75\,\mu s = 12.75\,\mu s$
Since one conversion cycle = $50\,\mu s$
Delay required = $50 - 12.75\,\mu s = 37.25\,\mu s$.

Delay cycles required = 37.25/0.75 = 49.66 and taking it as 49 we get actual conversion time

$$= (49 + 17) \times 0.75 = 49.5 \ \mu s.$$

5.9 Artificial Neural Network–Based Linearization

Artificial neural network (ANN) models have been successfully implemented for estimating sensor performance in soft sensors. The advantage of ANN models in sensor performance estimation has already been discussed in Section 4.4.2.2. The LUT method of linearization using microcontroller or embedded microcontroller is a simple method; however, good resolution cannot be achieved by low memory LUT. ANN-based linearization is a solution where comparatively higher resolution can be achieved with lower capacity processor. In general, large ANN architecture need powerful processor but linearizing ANN can be achieved with simple architecture.

The basic about ANN-based linearization is setting a single-input single-output (SISO) multilayer perceptron (MLP) network where input is the sensor measurement data and output or target is the corresponding linear or theoretical data. Figure 5.53a and b shows a basic block diagram of the ANN-based linearization method. In Figure 5.53b, the sensor output V_{nl} is input to the ANN that produces a corresponding linear output V_l. Figure 5.53a illustrates that the ANN is trained with the nonlinear sensor characteristic data as inputs and the linear characteristic data as output or targets. The structure of an ANN for sensor signal linearization is shown in Figure 5.54.

The output from the hidden neurons are given by

$$0_{1,1} = \frac{1}{1+e^{-net1,1}} \tag{5.110a}$$

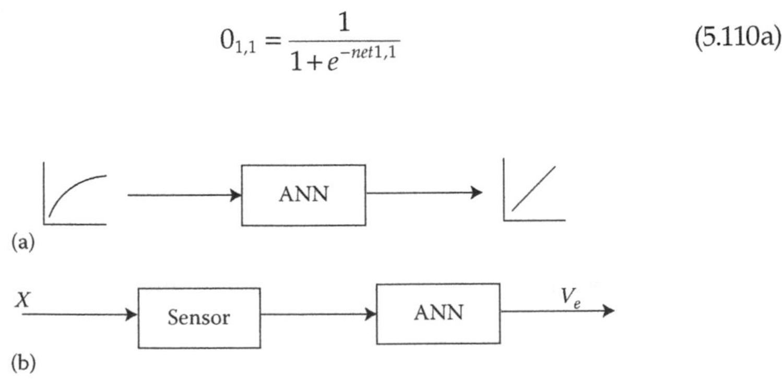

FIGURE 5.53
ANN-based sensor linearization. (a) Training and (b) testing.

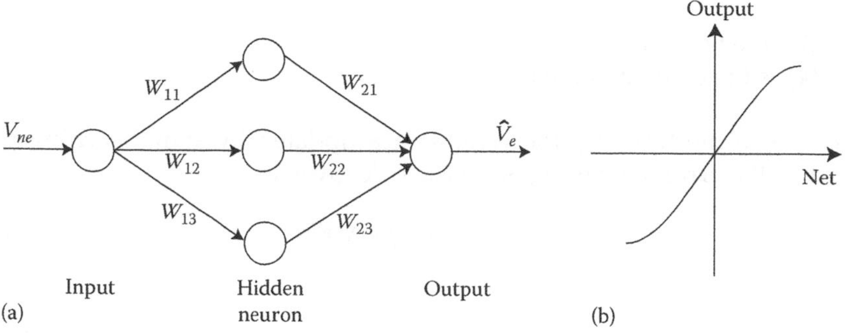

FIGURE 5.54
(a) ANN structure and (b) nonlinear activation function.

$$0_{1,2} = \frac{1}{1+e^{-net1,2}} \tag{5.110b}$$

$$0_{1.3} = \frac{1}{1+e^{-net1,3}} \tag{5.110c}$$

The above equations follow a nonlinear activation function, as shown in Figure 5.54b. The equations for the outputs "net" are given by

$$net_{1,1} = V_{nl}W_{11} \tag{5.111a}$$

$$net_{1,2} = V_{nl}W_{12} \tag{5.111b}$$

$$net_{1,3} = V_{nl}W_{13} \tag{5.111c}$$

The estimated ANN output V_{nl} is given by

$$\hat{V}_e = \frac{1}{1+e^{-net1,2}} \tag{5.112}$$

where

$$net2 = 0_{1,1}W_{21} + 0_{1,2}W_{22} + 0_{1,3}W_{23} \tag{5.113}$$

While training the network, the outputs 0_{11}, 0_{12}, and 0_{13}, and then V_{nl} is estimated using some randomly selected weights W_{11}, W_{12}, W_{13}, W_{21}, W_{22}, and W_{23}. Since the target is set for each input data, an error is calculated given by

$$e(k) = V_l(k) - \hat{V}_l^0(k) \tag{5.114}$$

where
$V_l(k)$ is the target data
$\hat{V}_l^0(k)$ is the estimated data

To converse the training, the weights are updated by some learning rule based on the error. Following is a basic rule given as

$$w(k+1) = w(k) + \Delta w(k) \quad (5.115)$$

where

$$\Delta w(k) = \beta e(k) V_{nl}(k) \quad (5.116)$$

where β is a learning constant. At each epoch, the weights are updated and the error approaches zero.

Example 5.18

A sensor produces output voltage nonlinearly as per the equation $v = ax + bx^2$, where a and b are two constants. The sensor circuit produces a maximum voltage of 5 V at the maximum of 50 units of x, while 25 units of x produce 3 V. Develop the input and target data vector for the entire operating range for training an ANN for linearization.

Solution
Given equation: $v = ax + bx^2$
Applying the given values of x and v in the equation
at $x = 25$, $v = 3.0$ V
therefore, $25a + (25)^2 b = 3$

$$\Rightarrow 25a + 625b = 3 \quad (a)$$

at $x = 50$, $v = 5$ V
therefore, $50a + (50)^2 b = 5$

$$\Rightarrow 50a + 2500b = 5 \quad (b)$$

Solving Equations a and b, we get

$$a = 0.14, b = -8 \times 10^{-4}$$

Hence the nonlinear equation takes the form

$$v = 0.14x - 8 \times 10^{-4} \quad (c)$$

Now we find the linearized equation for the sensor with the values: at $x = 0$, $v = 0$ V and at $x = 50$, $v = 5$ V; therefore, the linear equation takes the form

$$v = \frac{5}{50} x$$

$$\Rightarrow v = 0.1x \quad (d)$$

TABLE E5.18

Linear–Nonlinear Relationship

x	V_{nl}	V_l	x	V_{nl}	V_l
0	0	0	10	1.320	1.0
1	0.139	0.1	.	.	.
2	0.276	0.2	.	.	.
3	0.412	0.3	.	.	.
4	0.547	0.4	20	2.480	2.0
5	0.681	0.5	.	.	.
6	0.811	0.6	.	.	.
7	0.940	0.7	30	3.480	3.0
8	1.068	0.8	.	.	.
9	1.195	0.9	50	5.00	5.0

The nonlinear and linear voltage data for $x=0, 1, 2, \ldots, 50$ are calculated using Equations c and d, respectively, and tabulated below (Table E5.18).

The input vector to the ANN is given by

$$V_{NLj}^{T} = \begin{bmatrix} 0 & 0.139 & 0.276 & \cdots & 5.0 \end{bmatrix}$$

The target vector is given by

$$V_{L,j}^{T} = \begin{bmatrix} 0 & 0.1 & 0.2 & \cdots & 5.0 \end{bmatrix}$$

An ANN-based linearization technique for NTC resistive sensor by the methodology discussed above has been developed by Medrano-Marques et al. [12]. Figure 5.55a shows the schematic of the linearizer and Figure 5.55b is the proposed ANN structure. In Figure 5.55a, the sensor resistive divider output V_T is the nonlinear voltage, which is used to train the ANN. V_{ANN} is the target error calculated by subtracting measurement data from ideal linear data. The linear output V_{LIN} is obtained by adding V_T with V_{ANN}. An NTC with $R_0=10\,\text{k}\Omega$ and $\beta=3750\,\text{K}$ was used to generate V_T for temperature ranging from 268 to 328 K. Each of the input and target vectors were having 240 data. The nonlinear output errors for the NTC and ANN linearizer is shown in Figure 5.56.

5.10 Nonlinear Adaptive Filter–Based Linearization

Filters are almost inevitable parts of signal processing. The key function of a filter is to achieve the desired spectral characteristic of a signal, to reject

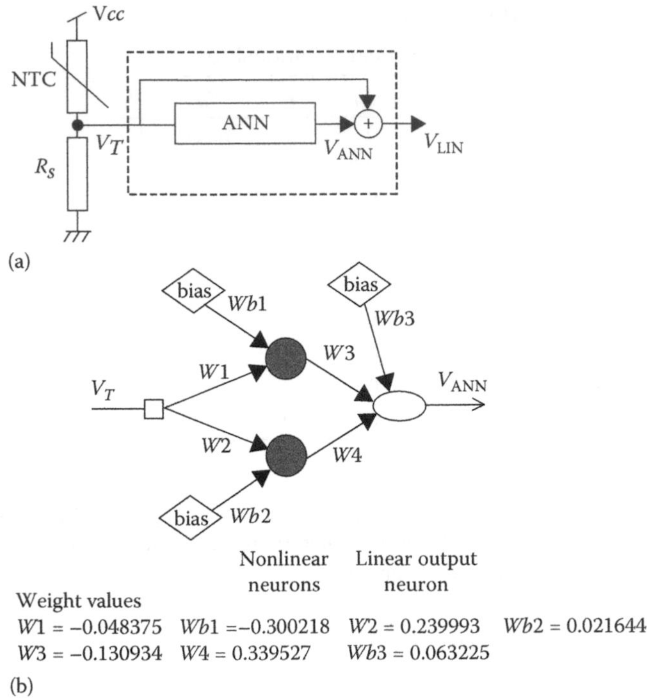

FIGURE 5.55
Linearizer for NTC sensor. (a) Resistive divider circuit and (b) MLP architecture of ANN with weights after training. (Reprinted from IEEE, copyright © IEEE. With permission.)

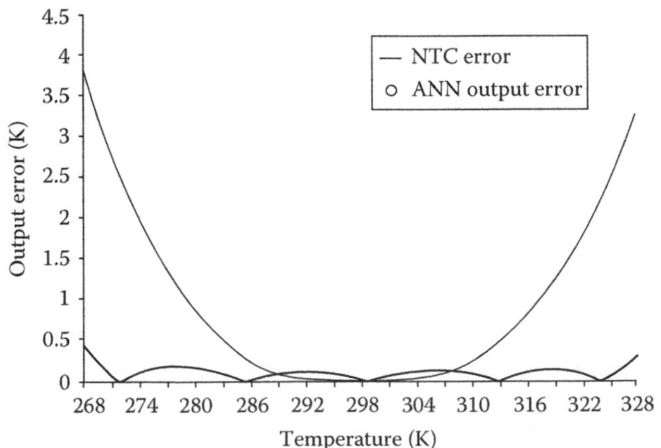

FIGURE 5.56
ANN output and nonlinear error. (Reprinted from IEEE, copyright © IEEE. With permission.)

noise and interference, to reduce the bit rate in signal transmission, etc. Typically a filter is designed with fixed parameters based on a prior knowledge about the spectral characteristic of the signal it handles; however, when situation arises such as when it needs to tackle the problems that cannot be predicted in advance, a filter must be able to change its parameters. For example, when the characteristic of a signal changes, the interference pattern also changes with environment or the circuit behaves differently with time. Adaptive filters that are designed with some adaptive capabilities can solve such problems.

The LMS (least mean square) adaption algorithm is a most common technique, which is an approximation of the steepest descent algorithm. This algorithm uses an instantaneous estimate of the gradient vector of a cost function, which is determined from the system input vector and an error signal. At each iteration of the algorithm, the filter coefficients are updated or improved moving toward the approximation gradient. Figure 5.57 shows the block diagram representation of the adaptive filter. The LMS algorithm works with an error signal given by

$$e[n] = d[n] - C^H[n]\, x[n] \tag{5.117}$$

where $d[n]$ is a reference signal and $C^H[n] = (C^*)^T[n]$ is the Hermitan of vector $C[n]$, which are the filter coefficient giving the filter output as

$$y[n] = C^H[n]\, x[n] \tag{5.118}$$

The LMS algorithm works to find a set of filter coefficient C minimizing the LMS of error signal. In LMS algorithm, the filter coefficients are updated by using an updating equation given by

$$C[n+1] = C[n] + \mu x[n] e^* \tag{5.119}$$

where μ is a step-size parameter, which controls the distance of error the algorithm uses between two steps. The adaptive filter can be used to

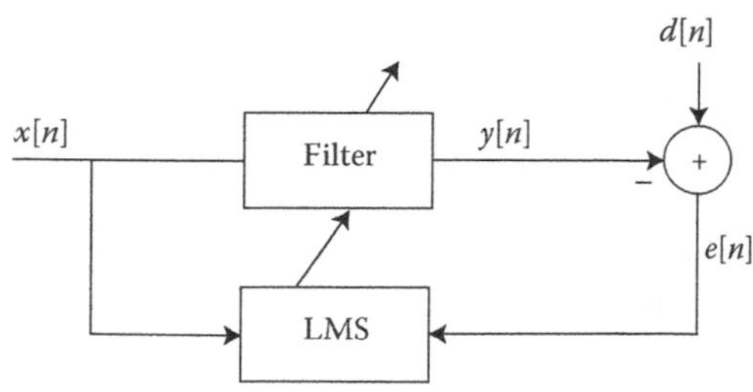

FIGURE 5.57
An LMS adaptive filter.

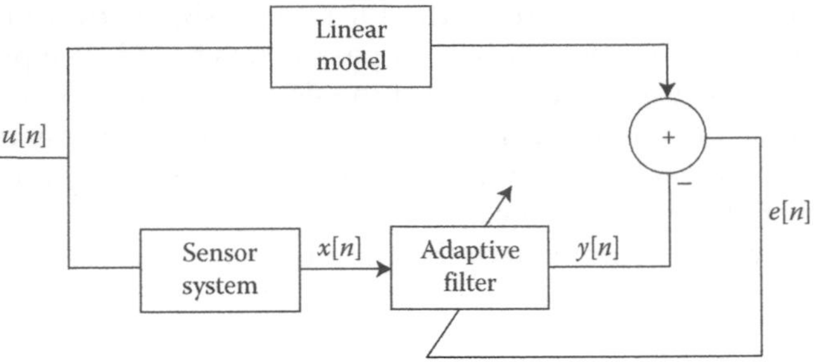

FIGURE 5.58
Adaptive filter–based linearization.

compensate nonlinearity of a sensor or sensor signal processing with the schematic shown in Figure 5.58. The input $u[n]$ when applied to the sensor system produces the nonlinear output $x[n]$. A reference $d[n]$ is generated by applying the input $u[n]$ to a linear model and an error $e[n]$ is applied to the LMS algorithm of the adaptive filter. The filter output $y[n]$ is the linearized output.

5.11 Sensor Calibration

The term calibration has originated from the words *"calibre"* and *"calliper"* used in the eighteenth century to describe an instrument for measuring or comparing the internal diameter or bore of guns. Callipre is the instrument used to measure the internal diameter of a bore. The term calibration later on extended to correction of irregularities in measuring scales of thermometers, barometers, pressure gauge, etc. Nowadays, calibration has a broader scope in the domain of electronic measuring or sensing techniques. As the sensing technology has developed from the simplest to the most complex form, calibration methods have also developed manyfold. With the same input, no sensor can deliver perfect reading, i.e., when a sensor is repeated with the same input, it gives readings with different values. This uncertainty in measurement is inevitable. The key to the process of calibration is the estimation of the uncertainty in measurement. The following sources mainly contribute to the uncertainty:

1. Environmental changes such as variation in temperature, pressure, flow rate, humidity, etc.
2. Low resolution or discrimination of the sensor
3. Low repeatability of the sensor
4. Malfunctioning of the sensor due to the manufacturing defects

5. Loss of cleanliness, accumulation of dirt, etc., on the sensing surface
6. Operator error
7. Processing stage error

A sensor should be made capable of counteracting the uncertainties developed due to the factors listed above and transfer the physical variable correctly to the electrical output signal. Calibration is the method of providing this capability to a sensor system. The calibration process needs costly reference standard equipment and considerable time and attention. Hence, by integrating a programmable calibration circuit on the sensor interface chip, intelligent sensor calibration can be automated.

Although a conventional calibration method adopts manual technique, under the purview of intelligent sensor, this section will emphasis on different techniques developed for automatic calibration. However, we will first examine the convectional calibration techniques before discussing sensor auto-calibration.

5.11.1 Conventional Calibration Circuits

Conventionally, sensor calibration circuits have features added to the sensor voltage–generating circuits for eliminating errors, drifts, interference, offset, etc. Such conventional calibration techniques can be classified into two broad categories—resistor adjustment–based calibration and digitally controlled amplifier–based calibration.

5.11.1.1 Resistor Adjustment–Based Analog Calibration

In most resistive sensor measurement systems, a Wheatstone bridge is commonly used to obtain an unbalanced bridge output corresponding to the variable to be measured. The bare minimum requirement for such a bridge circuit is shown in Figure 5.59a.

The bridge is formed by four resistors R_{s1}, R_{s2}, R_{s3}, and R_{s4}, which are the sensor resistances; however, a bridge can also be formed with a single sensor (R_{s1}) and the remaining three resistors are fixed resistors; or two sensors with opposite resistance variation polarity (R_{s1} and R_{s2}) and other two are fixed resistors. Figure 5.59a shows only one active branch where R_1 is the sensor resistance and the resistances of the bridge can be considered as $R_1=R_2=R_3=R_4=R_0$ to get a zero output at resting conditions of the sensor. If the sensor is affected by temperature variation, an offset voltage will be available at output even if there is no input to the sensor. The circuit of Figure 5.59b can be used to compensate the offset temperature component. Compensating resistor R_{tc} can be adjusted to get zero output when the output is drifted by any temperature variation. R_{tc} can compensate a sensor either at R_1 or at R_4 by changing the switch position.

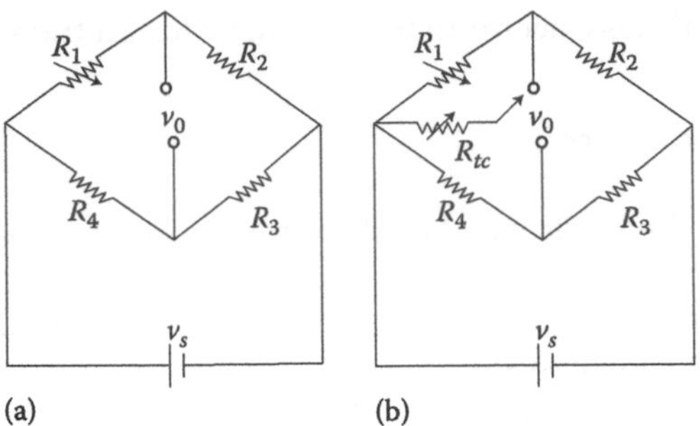

FIGURE 5.59
(a) A basic Wheatstone bridge and (b) temperature offset compensation.

Example 5.19

A sensor with resistance 100 Ω at normal condition is connected to branch R_1 of Figure 5.59b. The resistances of the other three resistors are 50 Ω each. A variable resistor R_{tc} of 100 Ω is connected parallel to R_1 as shown in the figure to compensate the offset due to temperature. If the resistance of the sensor changes by 0.3 Ω due to a temperature variation, find the value of resistance to which R_{tc} should be adjusted for temperature offset compensation.

Solution

Sensor resistance at branch 1, $R_1 = 100\ \Omega$, $R_2 = R_3 = R_4 = 50\ \Omega$
Change in resistance in sensor due to temperature = 0.3 Ω
Hence, sensor resistance at changed temperature = 100 + 0.3 = 100.3 Ω
For balance of the bridge, the parallel combination of R_1 and $R_{tc} = 50\ \Omega$.

$$\therefore \frac{R_1 R_{tc}}{R_1 + R_{tc}} = 50\ \Omega$$

$$\Rightarrow \frac{100.3 \times R_{tc}}{100.3 + R_{tc}} = 50$$

$$\Rightarrow 100.3R_{tc} - 50R_{tc} = 50 \times 100.3$$

$$\Rightarrow R_{tc} = 99.7\ \Omega$$

Resistive sensors such as strain gauges configured in Poisson's configuration are very common where four active strain gauges are used, as shown in Figure 5.60 [13]. For the balanced condition of the bridge, the sensor resistances are normally selected as $R_1 = R_2 = R_3 = R_4 = R_0$. Due to manufacturing defect or wrong installation, the bridge balance may not be achieved normally. To achieve bridge balance by compensating any offset present due to problems in the sensor or installations,

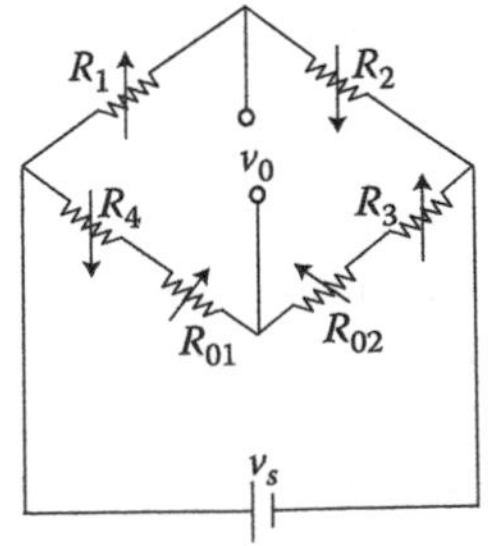

FIGURE 5.60
Resistive bridge–based calibration circuit.

two resistors R_{01} and R_{02} are connected as shown in the Figure 5.60. With the sensor unexcited by external signal; the two resistors R_{01} and R_{02} are adjusted to achieve balance, i.e., zero output voltage. The offset developed in the sensors due to reasons described above can be compensated by adjusting R_{01} and R_{02} until the output becomes zero. To what values the resistors R_{01} and R_{02} have to be adjusted can be determined from the following analysis:

1. Offset due to imbalance in R_4:
   Let the resistance R_4 change by an amount ΔR_4 to give the following balanced condition:

$$R_1\left(R_3+R_{02}\right)=R_2\left(R_4+\Delta R_4+R_{01}\right)$$

   Since $R_1=R_2=R_3=R_4=R$,

$$RR+RR_{02}=R\left(R+\Delta R_4\right)+RR_{01}$$

   Equating for compensating

$$RR_{02}=R\left(R+\Delta R_4\right)$$

$$\Rightarrow R_{02}=R+\Delta R_4$$

   and $RR_{01}=RR$

$$\Rightarrow R_{01}=R$$

2. Offset due to imbalance in R_3:
   The balanced condition is

$$R\left(R+\Delta R_{03}\right)+RR_{02}=RR+RR_{01}$$

   Similarly, we get

$$R_{01}=R+\Delta R_3;\quad R_{02}=R$$

3. Offset due to imbalance in both R_3 and R_4:
   Writing the balance equation

$$R\left(R+\Delta R_3\right)+RR_{02}=R\left(R+\Delta R_4\right)+RR_{01}$$

$$\Rightarrow RR+R\Delta R_3+RR_{02}=RR+R\Delta R_4+RR_{01}$$

   Similarly, we get

$$R_{01}=\Delta R_3;\quad R_{02}=\Delta R_4$$

4. Offset due to imbalance in R_1:
   Similarly, we can write

$$(R+\Delta R_1)R+(R+\Delta R_1)R_{02}=RR+RR_{01}$$

and we get

$$R_{02}=\frac{RR}{(R+\Delta R_1)};\quad R_{01}=(R+\Delta R_1)$$

5. Offset due to imbalance in R_2:
   We can write

$$RR+RR_{02}=(R+\Delta R_2)R+(R+\Delta R_2)R_{01}$$

and we get

$$R_{01}=\frac{RR}{(R+\Delta R_2)};\quad R_{02}=(R+\Delta R_2)$$

6. Offset due to imbalance in all four resistances:

$$\begin{aligned}&(R+\Delta R_1)(R+\Delta R_3)+(R+\Delta R_1)R_{02}\\&=(R+\Delta R_2)(R+\Delta R_4)+(R+\Delta R_2)R_{01}\end{aligned}$$

we get

$$R_{01}=\frac{(R+\Delta R_1)(R+\Delta R_3)}{(R+\Delta R_2)}$$

$$R_{02}=\frac{(R+\Delta R_2)(R+\Delta R_4)}{(R+\Delta R_1)}$$

Example 5.20

In the offset compensation bridge circuit of Figure 5.60, the sensor resistances at normal condition are 100 Ω each; however, due to improper installation, the shift in resistance values for R_1, R_2, R_3, and R_4 are 0.1%, 0.5%, 0.2%, and −0.3%, respectively. To what values R_{01} and R_{02} should be adjusted to restore balance for calibration? Verify the compensation after the adjustment of R_{01} and R_{02}.

Solution

$$R_1 = R_2 = R_3 = R_4 = 100\,\Omega$$

$$\Delta R_1 = \frac{0.1}{100} \times 100 = 0.1\,\Omega$$

$$\Delta R_2 = \frac{0.5}{100} \times 100 = 0.5\,\Omega$$

$$\Delta R_3 = \frac{0.2}{100} \times 100 = 0.2\,\Omega$$

$$\Delta R_4 = \frac{-0.3}{100} \times 100 = -0.3\,\Omega$$

From the condition (f)

$$R_{01} = \frac{(100+0.1)(100+0.2)}{(100+0.5)} = 99.80\,\Omega$$

$$R_{02} = \frac{(100+0.5)(100-0.3)}{(100+0.1)} = 100.09\,\Omega$$

After the adjustment of R_{01} and R_{02} to the above values, the balance equation

$$R_1(R_3 + R_{02}) = R_2(R_4 + R_{01})$$

LHS

$$\Rightarrow 100.1 \times (100.2 + 100.09) = 20{,}049$$

RHS

$$\Rightarrow 100.5 \times (99.7 + 99.8) = 20{,}049$$

So LHS = RHS and the offset gets canceled.

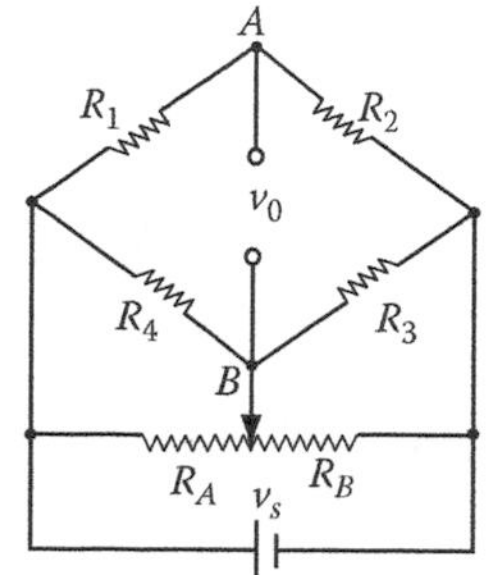

FIGURE 5.61
Combined sensor mismatch correction circuit.

Another circuit for compensating any offset present due to the mismatch of sensor resistive characteristic is shown in Figure 5.61. Offset voltage can be zeroed precisely using a variable resistor of resistance $(R_A + R_B)$ connected in shunt to the power supply V_S and terminal *B*. Let us assume that the sensor resistances are shifted due to the offset present in them.

The equation for the output voltage V_0 with sensor offset ΔR_1, ΔR_2, ΔR_3, and ΔR_4

$$V_0 = V_s\left[\frac{R_4 \parallel R_A}{(R_4 \parallel \Delta R_4)+(R_3+\Delta R_3)\parallel R_B} - \frac{R_1+\Delta R_1}{R_1+\Delta R_1+R_2+\Delta R_2}\right]$$

Since $R_1 = R_2 = R_3 = R_4 = R$ we can write

$$V_0 = V_s\left[\frac{\dfrac{(R+\Delta R_4)}{(R+R_A+\Delta R_4)}}{\dfrac{(R+\Delta R_4)}{(R+R_A+\Delta R_4)}+\dfrac{(R+\Delta R_3)R_B}{(R+R_B+\Delta R_3)}} - \frac{R+\Delta R_1}{R+\Delta R_1+R+\Delta R_2}\right]$$

$$= V_s\left[\frac{1}{1+\dfrac{(R+R_A+\Delta R_4)(R+\Delta R_3)R_B}{(R+\Delta R_4)R_A(R+R_B+\Delta R_3)}} - \frac{R+\Delta R_1}{2R+\Delta R_1+\Delta R_2}\right] \quad (5.120)$$

If the sensors are free from offset, then $\Delta R_1 = \Delta R_2 = \Delta R_3 = \Delta R_4 = 0$ and the above equation of V_0 should fulfill the condition that when $V_0 = 0$, $R_A = R_B$. This is proved in the following steps for $V_0 = 0$:

$$\frac{1}{1+\dfrac{(R+R_A+\Delta R_4)(R+\Delta R_3)R_B}{(R+\Delta R_4)R_A(R+R_B+\Delta R_3)}} = \frac{R+\Delta R_1}{2R+\Delta R_1+\Delta R_2}$$

Dividing RHS by $R + \Delta R_1$

$$\frac{1}{1+\dfrac{(R+R_A+\Delta R_4)(R+\Delta R_3)R_B}{(R+\Delta R_4)R_A(R+R_B+\Delta R_3)}} = \frac{1}{1+\dfrac{R+\Delta R_2}{R+\Delta R_1}}$$

$$\Rightarrow \frac{(R+R_A+\Delta R_4)}{(R+\Delta R_4)R_A}\frac{(R+\Delta R_3)R_B}{(R+R_B+\Delta R_3)} = \frac{R+\Delta R_2}{R+\Delta R_1}$$

$$\Rightarrow \frac{(R+R_A+\Delta R_4)}{R_A}\frac{R_B}{(R+R_B+\Delta R_3)} = \frac{(R+R_2)(R+\Delta R_4)}{(R+\Delta R_1)(R+\Delta R_3)} \quad (1)$$

When $\Delta R_1 = \Delta R_2 = \Delta R_3 = \Delta R_4 = 0$

$$\frac{(R+R_A)R_B}{(R+R_B)R_A} = 1$$

$$\Rightarrow R_A = R_B$$

which is obtained as expected. For a particular set of offset in the sensor resistances, the ratio of potentiometer resistance R_A and R_B can be found from the following equations:

From Equation (1) we can write

$$\frac{RR_B + R_AR_B + \Delta R_4R_B}{RR_A + R_AR_B + \Delta R_3R_B} = K \tag{5.121a}$$

$$\text{where } K = \frac{(R+R_2)(R+\Delta R_4)}{(R+\Delta R_1)(R+\Delta R_3)} \tag{5.121b}$$

Hence, $RR_B + R_AR_B + \Delta R_4R_B = K(RR_A + R_AR_B + \Delta R_3R_B)$
From which the resistance R_B can be written as

$$R_B = \frac{KR_A(R+\Delta R_3)}{R+R_A(1-K)+\Delta R_4} \tag{5.122}$$

From Equation 5.122, again we can prove that
when $\Delta R_1 = \Delta R_2 = \Delta R_3 = \Delta R_4 = 0$,
$K=1$, and putting $K=1$ in Equation 5.122, we get

$$R_B = R_A$$

Example 5.21

Determine the value of R_B and R_A when using a 10 K variable resistance for compensating the following sensor mismatches: $\Delta R_1 = 0\ \Omega$, $\Delta R_2 = 10\ \Omega$, $\Delta R_3 = 0\ \Omega$, and $\Delta R_4 = 7\ \Omega$. Take sensor resistances as 350 Ω each.

Solution
Given: $\Delta R_1 = 0\Omega$, $\Delta R_2 = 10\Omega$, $\Delta R_3 = 0\Omega$ and $\Delta R_4 = 7\Omega$

$$R = 350\ \Omega, \quad R_B + R_A = 10\ \text{k}\Omega$$

Using Equation 5.121b and putting the given values

$$K = \frac{(R+R_2)(R+\Delta R_4)}{(R+\Delta R_1)(R+\Delta R_3)} = 1.05$$

From Equation 5.122 and putting the given values

$$R_B = \frac{KR_A(R+\Delta R_3)}{R+R_A(1-K)+\Delta R_4} = \frac{367.5R_A}{357-0.05R_A}$$

Since the pot resistance $R_B + R_A = 10\,\text{k}\Omega$

$$\therefore R_A = 10 - R_B$$

Putting $R_A = 10 - R_B$ in the equation of R_B

$$R_B = \frac{367.5(10-R_B)}{357-0.05(10-R_B)}$$

Simplifying, we get

$$0.05R_B^2 + 724R_B - 3675 = 0$$

Solving the above equation, we get

$$R_B = \frac{-724 \pm \sqrt{(724)^2 - 4\times 0.05\times(-3675)}}{2\times 0.05}$$

$$= 5.074\,\text{k}\Omega \text{ (Taking the positive value)}$$

$$\therefore R_A = 10 - 5.074$$

$$= 4.926\,\text{k}\Omega$$

5.11.1.2 Digitally Programmable Resistor

Remember that while performing manual calibration, the resistor R_{tc} in Figure 5.59b, R_{01} and R_{02} in Figure 5.60, and R_{AB} in Figure 5.61 are manually trimmed using mechanically adjustable potentiometers. When a sensor calibration circuit in IC technology has to be set permanently, laser-trimmed resistors are used where a laser beam is used to adjust the properties of the resistor deposited on the IC surface. In addition to mechanically adjustable or laser-trimmed resistor, there is another type of variable resistor—programmable resistor array. Programmable resistor array is the key to auto-calibration technique since the calibrating resistors can be varied by a digitally coded signal. The basic form of a programmable resistor array is shown in Figure 5.62. The total resistance across terminals A and B of the resistor array is given by

$$\frac{1}{R_T} = \frac{d_0}{R} + \frac{d_1}{2R} + \frac{d_2}{4R} + \cdots + \frac{d_n}{2^n R} \tag{5.123}$$

$$= \left[\frac{1}{2^0}d_0 + \frac{1}{2^1}d_1 + \frac{1}{2^2}d_2 + \frac{1}{2^3}d_3 + \cdots + \frac{1}{2^{n-1}}d_{n-1} + \frac{1}{2^n}d_n\right] \tag{5.124}$$

where

$(n+1)$ is the array length

d is the digital multiplier equal to analog zero for binary '0' and analog one for binary '1'

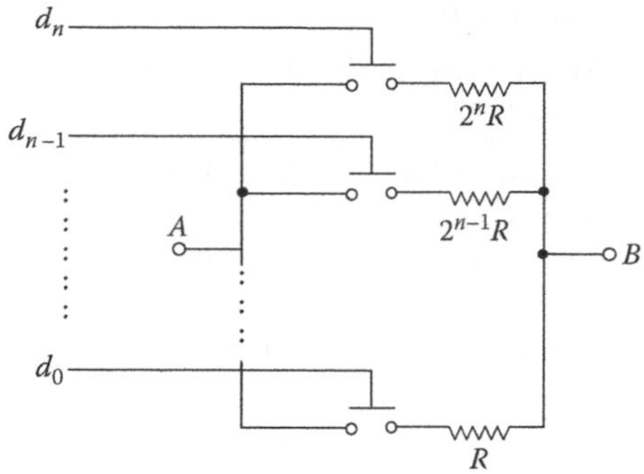

FIGURE 5.62
Digitally programmable resistor array.

Note that the numerical value of the quantity in parenthesis in Equation 5.124 can be written as

$$\left[2-\frac{1}{2^n}\right] \tag{5.125}$$

Hence, the equation of minimum resistance R_T takes the form

$$R_{T,\min} = \frac{R}{[2-(1/2^n)]} \tag{5.126}$$

From Equation 5.126, the LSB value of resistance in 4-bit array is

$$R_{T,\min} = \frac{R}{[2-(1/2^3)]} = \frac{8R}{15} \tag{5.127}$$

Hence, we can say that the least significant bit value of resistance is

$$1\text{ LSB resistance} = \frac{R}{[2-(1/2^n)]} \tag{5.128}$$

The denomination of resistance for a 4-bit array is shown in Table 5.18. Similarly, the array resistances for the significant programmed positions for an 8-bit system are shown in Table 5.19.

Therefore, the general equation for deciding the maximum and minimum array resistances can be written as

TABLE 5.18
Resistances for a 4-Bit Array System

$d_3\,d_2\,d_1\,d_0$	Array Resistance
0000	$8/0=\infty$
0001	$8R/1=8R$
0010	$8R/2=4R$
0011	$8R/3=2.66R$
0100	$8R/4=2R$
0101	$8R/5=1.60R$
0110	$8R/6=1.33R$
0111	$8R/7=1.14R$
1000	$8R/8=R$
1001	$8R/9=0.88R$
1010	$8R/10=0.80R$
1011	$8R/11=0.72R$
1100	$8R/12=0.66R$
1101	$8R/13=0.615R$
1110	$8R/14=0.57R$
1111	$8R/15=0.53R$

TABLE 5.19

Resistances for Significant Bits of 8-Bit Array System

$d_7d_6d_5d_4d_3d_2d_1d_0$	Array Resistance
0 0 0 0 0 0 0 0	∞
0 0 0 0 0 0 0 1	$128R$
⋮⋮	⋮
0 0 0 0 1 0 0 0	$128R/8=16R$
⋮⋮	⋮
0 0 0 0 1 1 1 1	$128R/15=8.53R$
0 0 0 1 0 0 0 0	$128R/16=8R$
⋮⋮	⋮
1 0 0 0 0 0 0 0	$128R/128=R$
⋮⋮	⋮
1 1 1 1 0 0 0 0	$128R/240=0.53R$
⋮⋮	⋮
1 1 1 1 1 1 0 1	$128R/253=0.506R$
1 1 1 1 1 1 1 1	$128R/255=0.501\text{R}$

$$R_{T,max} = 2^{n-1}R$$

$$R_{T,min} = \frac{2^{n-1}}{2^n - 1}R$$

where n is the number of bits in the array. The following example shows how the resistance values can be programmed in an array resistor system.

Example 5.22

In a programmable resistor array system, the resistance has to be varied from a minimum of 50 Ω to a maximum of 13 kΩ approximately. The minimum resolution of variation needed is 0.20 Ω approximately. Choose the value of R and number of bits. Determine actual values of maximum and minimum resistances obtained.

Solution

From the equation of maximum and minimum resistance

$$\text{Max}: 2^{n-1}R = 13\ \text{k}\Omega = 13{,}000\ \Omega$$

$$\text{Min}: \frac{2^{n-1}}{2^n - 1}R = 50\ \Omega$$

and the smallest resolution

$$\frac{2^{n-1}R}{2^n-1}-\frac{2^{n-1}R}{2^n-2}=0.2\,\Omega$$

Putting the values of maximum and minimum resistance values in the above equation

$$50-\frac{13{,}000}{2^n-2}=0.20$$

$$\therefore \frac{13{,}000}{2^n-2}=49.8$$

$$\Rightarrow 2^n=263$$

Since $2^7=128$, $2^8=256$, and $2^9=512$
we choose $n=8$
Putting $n=8$ in equation for maximum resistance

$$2^{8-1}R=13{,}000$$

$$\because R=\frac{13{,}000}{128}=101.56\ \Omega$$

Again putting $n=8$ in equation for minimum resistance

$$\frac{2^7}{2^8-1}R=50$$

$$\Rightarrow R=\frac{255\times 50}{128}=99.60\,\Omega$$

So we choose $R=100\ \Omega$

$$\text{Maximum resistance} = 2^{n-1}R = 128\times 100 = 12.8\text{ k}\Omega$$

$$\text{Minimum resistance} = \frac{2^{n-1}}{2^n-1}R=\frac{128\times 100}{255}=50.19\ \Omega$$

$$\text{Minimum resolution} = \frac{2^{n-1}}{2^n-1}R-\frac{2^{n-1}}{2^n-2}R$$

$$=\frac{12{,}800}{255}-\frac{12{,}800}{254}=0.20\,\Omega$$

The programmable resistor array discussed above can be used for calibration of sensor signal using circuits of Figure 5.59b, 5.60, and 5.61. The following example shows such an application.

Example 5.23

In the resistive bridge calibration circuit of Figure 5.60, the resistive sensors have the following specification: (Same as Example 5.20)

$$\text{Normal resistance} = R_1 = R_2 = R_3 = R_4 = 100\ \Omega.$$

The variation in sensor resistance due to manufacturing error:

$$\Delta R_1 = 0.1\,\Omega, \quad \Delta R_2 = 0.5\ \Omega, \quad \Delta R_3 = 0.2\ \Omega, \quad \text{and} \quad \Delta R_4 = -0.3\ \Omega,$$

The mismatch compensating resistors R_{01} and R_{02} are realized by combining a fixed resistance and an array resistor in parallel and series, respectively. Assume suitable value of the fixed resistor and design the array.

Solution
The mismatch compensating resistor R_{01} and R_{02} are given by (See solution of Example 5.20)

$$R_{01} = \frac{(R_1 + \Delta R_1)(R_3 + \Delta R_3)}{(R_2 + \Delta R_2)}$$

$$R_{02} = \frac{(R_2 + \Delta R_2)(R_4 + \Delta R_4)}{(R_1 + \Delta R_1)}$$

The ranges of R_{01} and R_{02} can be found as

$$R_{01} = \frac{(100 + 0.5)(100 - 0.2)}{(100 + 0.5)}$$

$$= 99.8\,\Omega$$

$$R_{02} = \frac{(100 + 0.5)(100 - 0.3)}{(100 + 0.1)}$$

$$= 100.09\,\Omega$$

Configuration of R_{01}: (Figure E5.23a through c)

Choosing $R = 100\ \Omega$

$$\therefore R_T = \infty \text{ to } R_{T(\min)}$$

$$R_{T(\min)} = \frac{100 R_T}{100 + R_T}$$

$$= 99.8\,\Omega$$

By solving, $R_T = 4.99\,\text{k}\Omega$

$$\therefore R_{T(\max)} = \infty \quad \text{and} \quad R_{T(\min)} = 4.99\ \text{k}\Omega$$

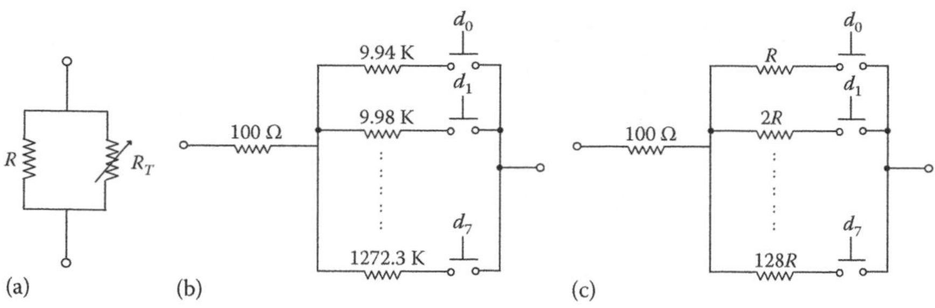

FIGURE E5.23
Compensating array (a) parallel (b) series-parallel (c) programmable array.

We know $R_{T(\min)} = (2^{n-1} R)/(2^n - 1)$ for array resistor

Taking $n = 8$ bits

$$\therefore \frac{2^7 R}{2^8 - 1} = 4.99\text{ k}\Omega$$

$$\Rightarrow R = 9.94\text{ k}\Omega$$

For R_{02}:
Choosing $R = 100\,\Omega$

$$\therefore R_T = 0 - 0.09\,\Omega$$

$$R_{T(\min)} = 0$$

$$R_{T(\max)} = 0.09\,\Omega$$

We know

$$R_{T(\max)} = 2^{n-1} R$$

$$\therefore 2^7 R = 0.09\,\Omega$$

$$\therefore R = 7.03 \times 10^{-4}\,\Omega$$

5.11.2 Multiplying DAC Calibration

In multiplying DAC calibration, the sensor voltage is digitally controlled to the levels of interest using an array of digital signal. Typically, the voltage is converted to a current signal, which is allowed to divide through an *R*–2*R* ladder network as shown in Figure 5.63 [10]. The current through each 2*R* resistor is either sinked to ground or summed to the output using a digital switch. The output current I_{out} is the scaled version of the input voltage v_{in} and I_{out} is again converted back to a voltage using a current-to-voltage converter. The output current I_{out} can be expressed by the equation

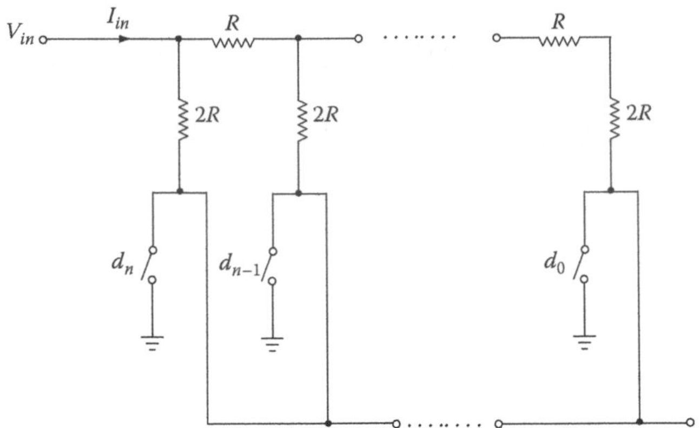

FIGURE 5.63
Multiplying DAC calibration circuit.

$$I_{out} = I_{in}\left[d_0\frac{1}{2^n} + d_1\frac{1}{2^{n-1}} + \cdots + d_{n-1}\frac{1}{2^2} + d_n\frac{1}{2^1}\right] \tag{5.129}$$

where $I_n = (V_{in}/R)$.

Equation 5.129 is equivalent to multiplying the sensor voltage V_{in} by a weighted digital byte, hence the technique is called a multiplying DAC calibration. The digital switches can be implemented by complementary metal oxide semiconductor (CMOS) voltage switches or bipolar current switches [10]. For an 8-bit multiplying DAC with $V_{in} = 5\,\text{V}$ and $R = 1\,\text{k}\Omega$, Table 5.20 shows the outputs for a few significant digital control signals.

TABLE 5.20
Output Currents for Few Digital Control Signals

$d_7d_6d_5d_4\ d_3d_2d_1d_0$	I_{out} (nA)	Equivalent Voltage (mV)
0000 0000	0	0
0000 0001	19.53	19.53
⋮	⋮	⋮
0000 1111	292.95	292.95
0001 0000	312.48	312.48
⋮	⋮	⋮
1111 0000	4687.20	4687.20
⋮	⋮	⋮
1111 1111	4980.15	4980.15

Example 5.24

The output signal of a linear pressure sensor after amplification is to be calibrated to a full-scale range of 20 mA current output for each of the following pressure ranges: 0–20, 0–40, 0–60, 0–80, and 0–100 psi. The amplifier outputs a voltage of 5 V for an input pressure of 100 psi.

1. Design a 4-bit multiplying DAC calibration circuit and determine the digital control signal for each of the calibration ranges.
2. Determine the calibration accuracy for 0–20 and 0–100 psi ranges.
3. How much accuracy can you get for 100 psi range if you use an 8-bit DAC multiplier?

Solution
Given values:
Calibration ranges:

$$P_1 = 20\text{ psi},\quad P_2 = 40\text{ psi},\quad P_3 = 60\text{ psi},\quad P_5 = 80\text{ psi},\quad P_5 = 100\text{ psi}$$

$$I_{out} = 20\text{ mA}, (n-1) = 4$$

1. The corresponding sensor amplified voltages for the calibration ranges are

$$V_1 = \frac{5}{100} \times 20 = 1\text{ V}$$

$$V_2 = \frac{5}{100} \times 40 = 2\text{ V}$$

Similarly, $V_3 = 3$ V, $V_4 = 4$ V, and $V_3 = 5$ V

Taking $V_1 = 1$ V

$$I_{out} = \frac{V_1}{R}$$

$$\therefore R = \frac{V_1}{I_{out}} = 50\,\Omega$$

The current equation of the 4-bit multiplier DAC is

$$I_{out} = I_{in}\left[d_0\frac{1}{16} + d_1\frac{1}{8} + d_3\frac{1}{4} + d_3\frac{1}{2}\right] = I_{in}[D]$$

For $V_1 = 1$ V

$$20\text{ mA} = 20\text{ mA}\,[D_1]$$

$$\therefore D_1 = 1$$

TABLE E5.24
Digital Control Signals

$d_3d_2d_1d_0$	D
0000	0
0001	0.0625
0010	0.125
0011	0.1875
0100	0.25
0101	0.3125
0110	0.375
0111	0.4375
1000	0.50
1001	0.5625
1010	0.625
1011	0.6875
1100	0.75
1101	0.8125
1110	0.875
1111	0.937

The values of D for the 4-bit digital signals are shown below:

Hence the digital word $d_1 = [1111]$

For $V_2 = 2\,V$

$$20\,mA = \frac{V_2}{R}[D_2] = \frac{2}{50[D_2]}$$

$$\therefore D_2 = 0.5$$

Hence the digital word

$$d_2 = [1000]$$

Similarly we get

$$d_3 = [0101]$$

$$d_4 = [0100]$$

$$d_5 = [0011]$$

2. Calibration accuracy:
   For 0–20 psi, the actual current output

$$I_{out} = \frac{V_1}{R}[D_1] = 20\,mA \times 0.937$$

$$= 18.74\,mA$$

$$\text{Percentage accuracy} = 1 - \left|\frac{20 - 18.74}{20}\right| \times 100\% = 93.7\%$$

For 0–100 psi

$$I_{out} = \frac{V_5}{R}[D_5] = 18.75\,mA$$

$$\text{Percentage accuracy} = 1 - \left|\frac{20 - 18.75}{20}\right| \times 100\% = 93.75\%$$

3. For an 8-bit DAC multiplier

$$D = \left[\frac{d_0}{256} + \frac{d_1}{128} + \frac{d_2}{64} + \frac{d_3}{32} + \frac{d_4}{16} + \frac{d_5}{8} + \frac{d_6}{4} + \frac{d_7}{2}\right]$$

When $d_5 = 1111\ 1111$

$$D = 0.996$$

So we get 99.6% accuracy.

5.11.3 Offset Calibration

The multiplying DAC calibration circuit discussed above refers to full-scale calibration of the sensor. In Example 5.24, it is shown how a sensor signal can be calibrated to a full-scale output current (20 mA in the example) for different input ranges. This is similar to how we can use the same scale of a voltmeter for different input voltage ranges by changing the attenuator resistance. But what we generally do when the voltmeter reading has an offset? In a voltmeter, the offset can be nullified by adjusting a screw placed near the pointer, which is connected to a spring controlling the pointer movement. In a similar way, in the full-scale calibrating circuit, an appropriate current of proper polarity is added to the output current of the multiplying DAC [10]. Figure 5.64 shows how the two multiplying DACs are combined to nullify the offset present in the sensor signal. While V_{in} is converted to a current I_{in} and calibrated to deliver a current I_{01} in the upper multiplying DAC of Figure 5.64, a reference voltage V_{ref} is used to calibrate with an offset correction with I_{02} using the lower multiplying DAC. The polarity of I_{ref} is chosen based on the polarity of the offset. At the output, the two currents I_{01} and I_{02} are summed and converted to a voltage

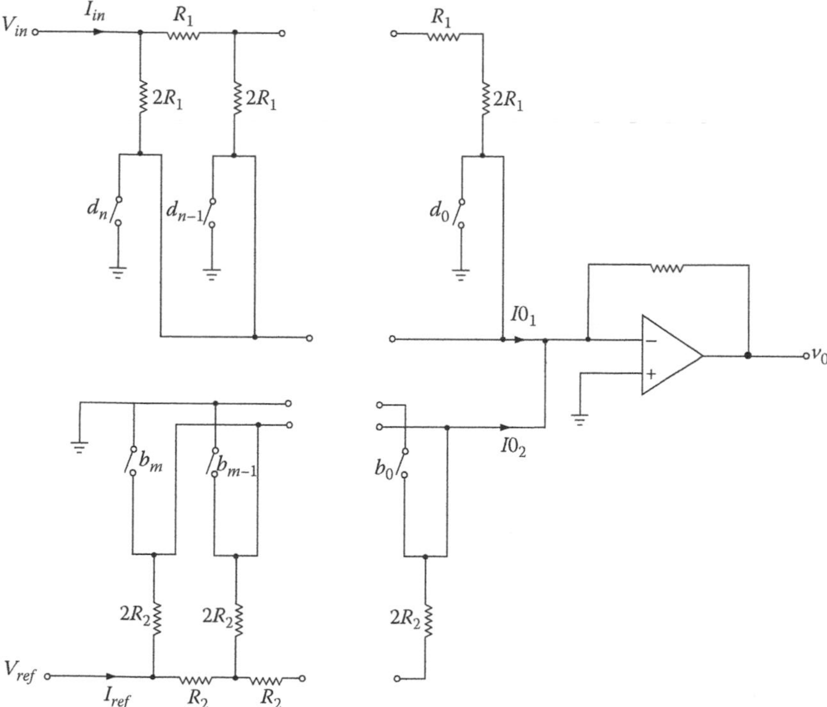

FIGURE 5.64
Offset calibrating multiplying DAC circuit.

V_0 using the current-to-voltage OPAMP converter. While the sensor voltage can be scaled or calibrated by applying the weighted digital control signal $[d_n \ldots d_1 d_0]$, the offset component can be scaled by the digital signal $[b_n \ldots b_1 b_0]$.

The equation of the calibrating and offset currents can be written as

$$I_{01} = I_{in}\left[d_0 \frac{1}{2^n} + d_1 \frac{1}{2^{n-1}} + \cdots + d_{n-1} \frac{1}{2^2} + d_n \frac{1}{2^1}\right] \tag{5.130}$$

and

$$I_{02} = I_{ref}\left[b_0 \frac{1}{2^n} + b_1 \frac{1}{2^{n-1}} + \cdots + b_{n-1} \frac{1}{2^2} + b_n \frac{1}{2^1}\right] \tag{5.131}$$

where

$$I_{in} = \frac{V_{in}}{R_1} \tag{5.132}$$

and

$$I_{ref} = \frac{V_{ref}}{R_1} \tag{5.133}$$

The I–V converter converts the current outputs using the relation

$$V_0 = K\left(I_{0_1} + I_{0_2}\right) \tag{5.134}$$

By keeping a constant value of V_{ref}, the offset current can be controlled by the digital signal $[b_n \ldots b_1 b_0]$; however, the direction of I_{02} can be changed (i.e., a negative current $-I_{02}$) by changing the polarity of $+V_{ref}$ to $-V_{ref}$. The pressure sensor calibration problem of Example 5.24 will be discussed again with the additional feature of offset calibration.

Example 5.25

In the pressure sensor calibration problem of Example 5.24, the sensor signal has to be calibrated for an offset that may range from −0.05 to +0.05 V.

1. Design the circuit for offset calibration using a multiplying DAC.
2. Determine the output current when pressure is 15 psi and offset is +0.05 V.

Solution

Since full-scale pressure calibration is done for 0–20 mA current range, the offset calibration should also have full-scale range of 0–20 mA. Hence

1.

$$I_{O_2} = \frac{V_{ref}}{R_2}$$

$$\therefore R_2 = 250\,\Omega$$

To get smaller offset calibration currents, we use an 8-bit multiplying DAC. The output current is given by

$$I_{O_2} = 20 \times \left[\frac{b_0}{256} + \frac{b_1}{128} + \frac{b_2}{64} + \frac{b_3}{32} + \frac{b_4}{16} + \frac{b_5}{8} + \frac{b_6}{4} + \frac{b_7}{2}\right] \text{mA}$$

2. The offset current for an offset voltage of +0.05 V

$$I_{O_2} = -\frac{I_{O_2}(\text{full scale})}{\text{offset range}} \times \text{offset voltage}$$

$$= -\frac{20}{5} \times 0.05 = -0.2\,\text{mA}$$

We know

$$I_{O_1} = \frac{I_{O_1}(\text{full scale})}{\text{pressure}(\text{full scale})} \times \text{Input pressure}$$

$$= \frac{20}{100} \times 15 = 3\,\text{mA}$$

$$\therefore I_O = 3 - 0.2 = 2.8\,\text{mA}$$

The digital control for generating 0.2 mA offset

$$0.2 = 20\,[B]$$

$$\therefore [B] = 0.01$$

So $b = [0000\ 0011]$

5.11.4 Pulse-Modulated Calibration

The performance of the multiplying DAC–based calibration circuit discussed above strictly depends on two factors—the matching between the resistor array and the digital switches and the number of bits. Any mismatch of the resistors and the digital switching transistors causes a nonlinearity in the output with increasing digital code. On the other hand, good resolution of the output can be achieved only with higher bits, say, not below 8 bits. In Example 5.25, it was observed that 0.2 mA output current cannot be achieved in a 4-bit DAC; hence, an 8-bit DAC was used.

Pulse-modulated DAC circuits are a solution to such problems. In pulse-modulated circuits, the current is attenuated by binary-scaled time slots instead of binary-scaled division [10]. PWM is such a technique where the analog input signal is turned on and off by a higher-frequency pulse signal, the width of which can control the magnitude of the analog signal. The output signal is the time average of the chopped analog signal. A PWM circuit repeatedly produces a fixed period of time and outputs an analog signal that is active for a certain percentage of the time, which is called the duty cycle of the signal. Figure 5.65 shows a PWM signal, the duty cycle of which has been varied from 0% to 100% linearly. Figure 5.65b shows five pulses "high" for half of the PWM cycle and five "low" pulses for next half of the cycle. The frequency of the "on/off" PWM signal is

$$F = \frac{1}{10\,\mu S} = 100\text{ kHz} \tag{5.135}$$

Distributed PWM (Figure 5.65c) uses five "high" and five "low" pulses, but in a distributed manner. Hence, the frequency of the PWM signal in this case is

$$F = \frac{1}{2\,\mu S} = 500\,\text{kHz}$$

The above frequencies are obtained for a 50% duty cycle, but the signal frequency gradually decreases as the duty cycle increases.

The PWM frequency can also be expressed by

$$F = \frac{f}{2^N} \tag{5.136}$$

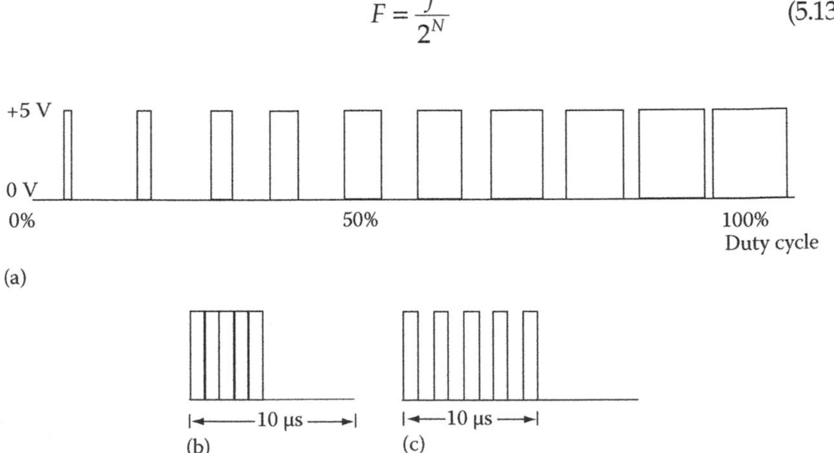

FIGURE 5.65
PWM signals: (a) with varying duty cycle, (b) ON/OFF PWM signal, and (c) distributed PWM signal.

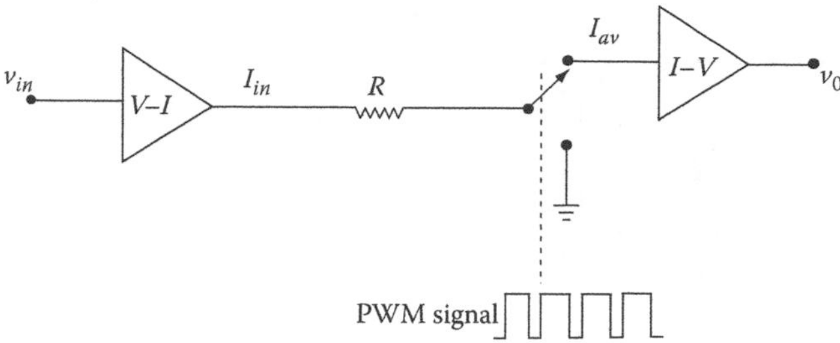

FIGURE 5.66
A PWM-based calibration circuit.

where
- f is the clock frequency
- N is the number of stages in the PWM counter chain, which determines the resolution of the DAC

When this clock frequency is 50 MHz, an 8-bit PWM DAC generates outputs of

$$F = \frac{50 \times 10^4}{2^{12}} = 12.20\,\text{kHz}$$

Equations 5.135 and 5.136 show that neither high speed nor high resolution can be achieved simultaneously in a PWM-DAC. PWM calibration of a sensor signal can be achieved using a circuit as shown in Figure 5.66. The sensor output voltage V_i is first converted to current I_{in} by a V–I converter and an average current I_{av} is obtained by modulating the input current I_{in} by the PWM train of pulses with the help of a switch. The average current obtained from the multiplying DAC is then converted back to a voltage V_0 by a I–V converter.

5.11.5 ADC Calibration

An ADC is inevitable in any sensor signal processing circuit, which makes the sensor signal ready for interfacing with the digital world. The calibration of a signal to the full-scale range of an ADC output can be performed by applying an approximate reference voltage to the ADC. Figure 5.67 shows a classical ADC, the reference voltage of which can be attenuated to any of the values 1, 2, 3, 4, and 5 V. The 8-bit digital output will obviously show different resolutions for different calibration ranges. The minimum and maximum resolutions that can be achieved from this calibrating ADC are, respectively

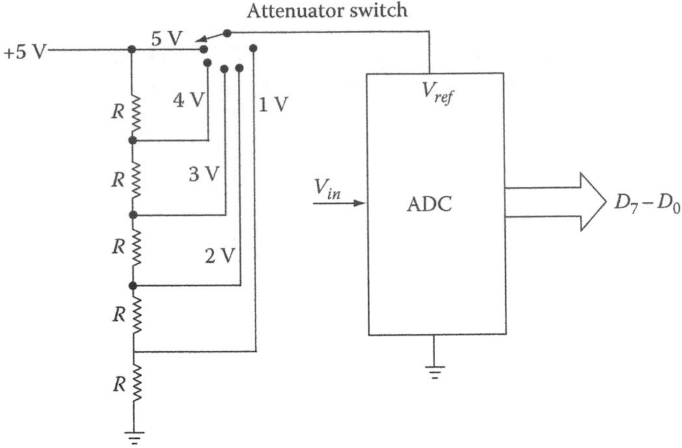

FIGURE 5.67
ADC reference voltage attenuator.

$$r_{min} = \frac{1\,\text{V}}{2^8} = 3.90\,\text{mV}$$

$$r_{max} = \frac{5\,\text{V}}{2^8} = 19.53\,\text{mV}$$

In the above circuit, the calibrating attenuator is placed in the reference voltage, but the same function can be obtained when the attenuator is used for the analog input signal keeping the reference voltage fixed (Figure 5.68). In this case, voltage positions are reversed and the reference voltage is made equal to 1 V. According to the input analog voltage range needed for calibration, the position of the calibrator is changed. By using voltage division rule, the resistances of the attenuator are calculated as

Taking $R_5 = 10\,\text{k}\Omega$ and for $V_s = 5\,\text{V}$ range,

$$V_{in} = V_s \times \frac{R_5}{R_T}$$

where R_T is the total resistance.

$$\therefore 5 \times \frac{R_5}{R_T} = 1$$

$$\Rightarrow R_T = 50\,\text{k}\Omega$$

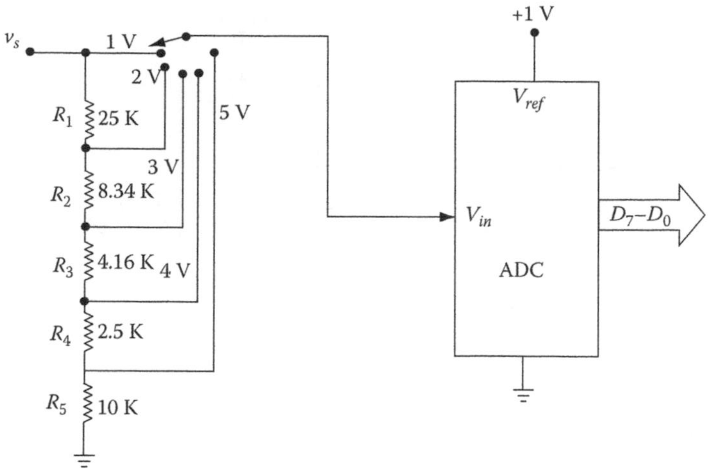

FIGURE 5.68
ADC input voltage attenuator.

For $V_s = 4\,\text{V}$

$$4 \times \frac{R_5 + R_4}{R_T} = 1$$

$$\Rightarrow 4 \times \frac{10 + R_4}{50} = 1$$

$$\therefore R_4 = 2.5\,\text{k}\Omega$$

Similarly $R_3 = 4.16\,\text{k}\Omega$, $R_2 = 8.34\,\text{k}\Omega$, and $R_1 = 25\,\text{k}\Omega$.

Example 5.26

The output voltage of a thermocouple is linearized and then amplified to a level of 0–5 V covering a temperature range of 0°C–500°C. Design an input voltage attenuator for an 8-bit ADC for calibrating the ADC output in two input ranges: 0°C–500°C and 0°C–200°C. Also determine their ratio of resolutions.

Solution

Figure E5.26 shows the circuit diagram.
First temperature range: 0°C–500°C
Voltage range: 0–5 V
Second temperature range: 0°C–200°C
Voltage range: 0–2.0 V
Since, 2.5 V is the smallest input voltage range, we choose $V_{ref} = 2.0\,\text{V}$.
When $V_s = 5\,\text{V}$, $V_{in} = 2.0\,\text{V}$

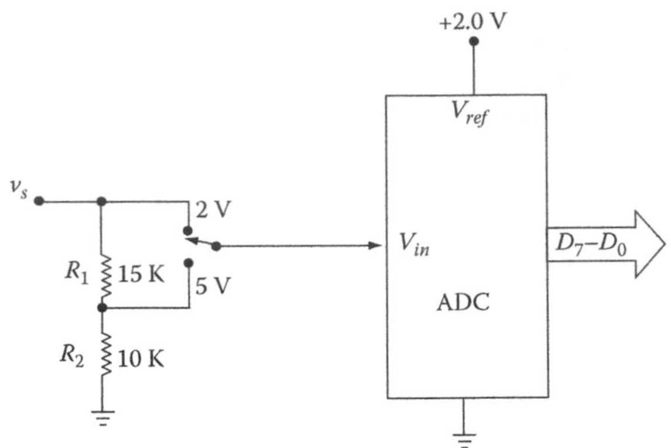

FIGURE E5.26
ADC input voltage attenuator for Example 5.26.

$$\therefore 5 \times \frac{R_2}{R_1 + R_2} = 2$$

$$\Rightarrow 5R_2 = 2R_1 + 2R_2$$

$$\Rightarrow R_1 = \frac{3}{2R_2}$$

Taking $R_2 = 10\,\text{k}\Omega$, $R_1 = 15\,\text{k}\Omega$
Resolution at 500°C range and 200°C range are

$$r_{500} = \frac{500}{2^8}$$

$$= 1.95°\text{C}$$

$$r_{200} = \frac{200}{2^8}$$

$$= 0.758°\text{C}$$

$$\therefore \quad \frac{r_{500}}{r_{200}} = 2.5$$

5.11.5.1 Sigma–Delta ADC Calibration

Classical ADCs such as the successive approximation type used for sensor signal calibration have their limitations in resolution, integration, and cost. Recently, sigma–delta ADC has drawn attention of intelligent system designers for their capability to provide high accuracy and resolution at relatively low sampling rates. A first-order (1-bit) sigma-delta ADC is shown in the block diagram of Figure 5.69.

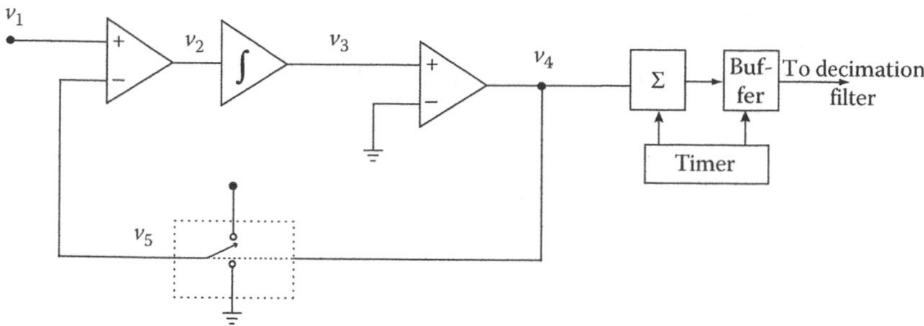

FIGURE 5.69
A first-order sigma–delta ADC

It comprises of a difference amplifier, integrator, comparator, and a 1-bit DAC in the feedback loop. The 1-bit DAC selects either a 'high' or a 'low' using the switch and the voltage V_5 is applied to the negative terminal of the difference amplifier. The difference voltage V_2 is integrated by the integrator to give V_3. This voltage is threshold by a comparator that generates a pulse V_4 when V_3 crosses the threshold. This pulse (V_4) triggers the DAC and increments the counter. As the timer signals when the prefixed time expires, the counter is strobed into the buffer and the counter resets. For an increasing input, the ADC generates a greater number of "1"s. Here, the integration works as low pass filter to the signal and high pass filter to the quantization noise. Hence the quantization noise is pushed toward the high-frequency side, thereby making the noise easier to be filtered out by a low pass filter. The output of the sigma–delta modulator is a 1-bit stream of digital data. This data is fed to a decimation filter, which converts the data stream into a binary code. For a binary stream of 1,1,0,0,1,0,1,1, there are five "1"s out of 8 bit, so the density of "1" is 62%. The following example shows how the DAC produces the bit-stream.

Example 5.27

For a 1-bit sigma–delta ADC, analyze the sequence of operation and determine the first two beat streams when the DAC polarities are ±2.5 V and input analog voltage is 1.0 V. Consider that the integrator and the DAC were reset to zero initially.

Solution

Using the notations of Figure 5.69, at the first sequence

$$V_1 = 1.0\ \text{V},\quad V_3 = 0,\quad V_5 = 0$$

hence $V_2 = V_1 - V_5 = 1.0 - 0 = 1.0\ \text{V}$

$$V_3 = \sum V_2 = 0 + 1.0 = 1.0\ \text{V}$$

The comparator output V_4 becomes a "1"
hence, $V_4 = 1$ (first bit stream)
A "high" input (i.e., "1") to DAC (V_4) produces a +2.5 V
hence, $V_5 = 2.5$ V
Next sequence:

$$V_1 = 1.0\,\text{V}, \quad V_5 = 2.5\,\text{V}$$

$$\therefore V_2 = 1.0 - 2.5 = -1.5\,\text{V}$$

Integrator output

$$\sum V_2 = 0 + 1.0 - 1.5 = -0.5\,\text{V}$$

The comparator output becomes a "0"

$$\therefore V_4 = 0 \text{ (second bit stream).}$$

A zero input to DAC produces −2.5 V. Hence, $V_S = -2.5$ V and the bit stream $b_s = 10\ldots$, this sequence continues to produce a bit stream.

The bit stream developed by the ADC is fed to a decimation filter. The function of the decimation filter is to produce a digital code equivalent to the bit stream density based on the DAC full-scale range. The decimation filter is a combination of digital and analog circuit that performs the following steps:

1. It generates a number (d_S) that represents the bit stream density of "1"s for a prefixed time period.
2. It scales d_S to an analog level.

The bit stream sequence is converted to d_S by using the following equation:

$$d_S = \frac{n}{N} \tag{5.137}$$

where
n is the number of "1"s
N is total bits in the bit stream

The scaling circuit produces a mean analog voltage, which is given by

$$V_m = d_S \times V_{FS}\,(\text{unipolar}) \tag{5.138}$$

$$V_m = d_S \times V_{FS} - \frac{V_{FS}}{2}\,(\text{bipolar}) \tag{5.139}$$

Now let us examine how a bit stream sequence can be decimated to an analog level and a digital code. Let us take a bit-stream output 11001111 obtained from a sigma–delta ADC with a reference voltage of ±2.5 V. The bit stream density is

$$d_S = \frac{6}{8} = 0.75$$

Since the ADC is a bipolar one, using Equation 5.139 the mean value of voltage is

$$V_m = 0.75 \times 5 - 2.5 = \pm 1.25\,\text{V}$$

If an 8-bit digital code is used with full-scale range of ±2.5 V, this value of +1.25 V can be coded as

$$1\,\text{LSB} = \frac{2.5}{2^7} = 19.53\,\text{mV}$$

Hence, decimal equivalent of +1.25 V = 1.25 V/19.53 mV = 64

The 8-bit digital code = $(01000000)_2$

Let us take another bit stream that is produced by negative analog voltage say −10000011. In this case

$$d_S = \frac{3}{8} = 0.375$$

$$V_m = 0.375 \times 5 - 2.5 = -0.625\,\text{V}$$

$$\text{Decimal equivalent} = \frac{-0.625\,\text{V}}{19.53\,\text{mV}} = -32$$

The digital code = $(10100000)_2$

Similar to using classical ADCs for sensor signal calibration, sigma–delta ADCs can also be programmed for tuning to different scaled levels of the full-scale range. MAXIM has developed a highly integrated sigma–delta ADC-based calibration IC-MAX1402, a block diagram of which is shown in Figure 5.70 [14].

The IC draws a low quiescent current of 250 μA in operating mode and 2 μA in power down mode, and the IC provides a 16-bit resolution at a sampling rate of 480 Hz and 12-bit resolution at 4.8 kHz. The IC comprises of an input MUX that can receive three differential analog inputs or five pseudo differential inputs. The multiplexed input is applied to two chopper amplifiers and then to a programmable gain amplifier (with a gain from 1–128). A DAC receives already stored offset values from the interface and adds the analog values to the PGA output for offset correction. The analog output from the PGA is applied to the sigma-delta ADC, which is finally decimated

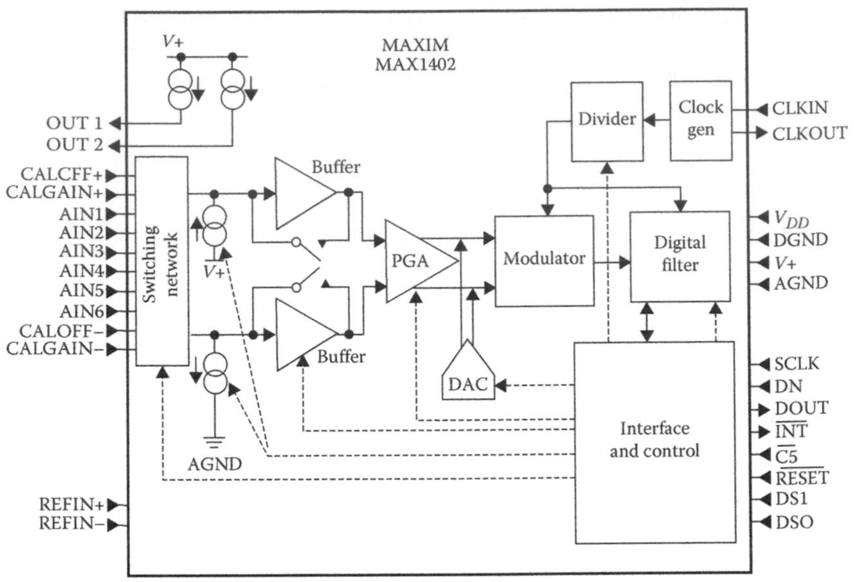

FIGURE 5.70
Sigma–delta ADC-based calibration IC-MAX 1402. (Reprinted from MAXIM, copyright © www.maxim-ic.com. With permission.)

by a digital filter and made available via an SPITM-/QSPITM-compatible 3-wire serial interface. The chip provides both offset calibration (CALOFF) and gain calibration (CALGAIN), which is user controlled. In addition to these calibrations, the PGA can also be used for the calibration of the input analog voltage. Hence the current input range is established by V_{ref}, PGA gain, CALGAIN, CALOFF, and DAC code. Further, the range can be scaled using a $U/\bar{B}$ bit (Figure 5.71). For example, MAXIM have provided an interpolation equation for the input voltage range as

$$\text{voltage} = \frac{V_{ref} \times [\text{DAC Code} - \text{CALOFF Code}]}{[\text{CALGAIN Code} - \text{CALOFF Code}] \times 2^{\text{PGACODE}}} \tag{5.140}$$

For example, with $V_{ref} = 2.5\,\text{V}$, DAC Code $= (1110)_2 = 13$, CALGAIN $= 7$, CALOFF $= 1$, PGA Code $= (000)_2 = 0$ and $U/\bar{B} = 0$, the full-scale voltage that the IC can cover is from equation

$$\text{voltage} = \frac{2.5 \times (13 - 1)}{(7 - 1) \times 2^{\circ}} = 5\,\text{V},$$

which can be obtained from Figure 5.71 also.

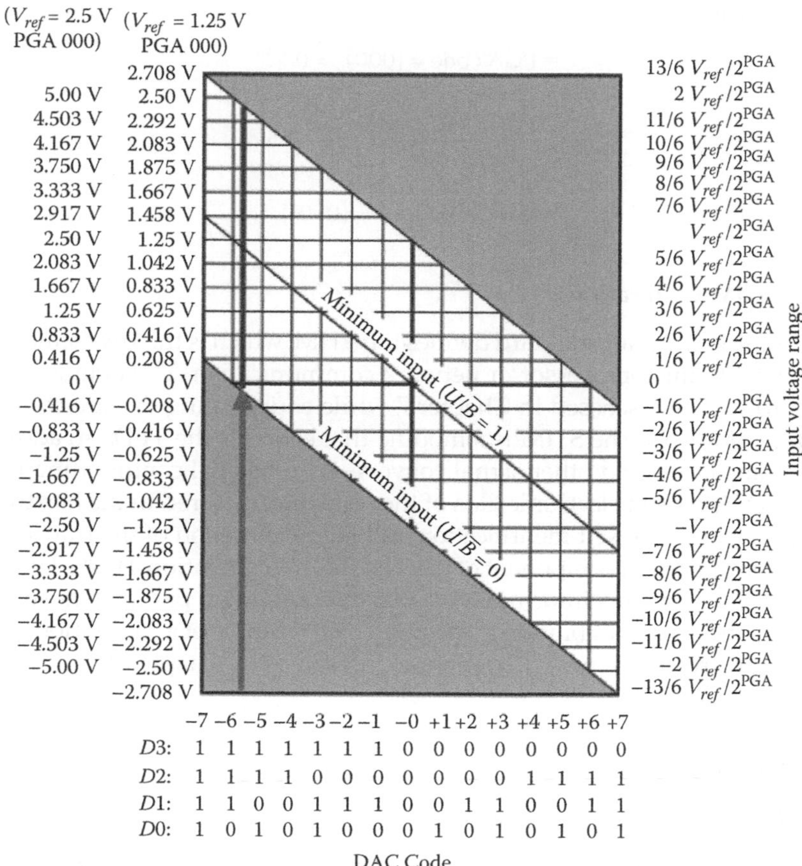

FIGURE 5.71
Input voltage range settings for MAX 1402. (Reprinted from MAXIM, copyright © www.maxim-ic.com. With permission.)

Example 5.28

For what PGA code the range in the MAX-1402 IC can be selected as 2.5 V with DAC code = 1100, CALGAIN = 7, CALOFF = 1, $U/\bar{B} = 0$, and V_{ref} = 2.5 V? If PGA code is 000, what DAC code is to be used?

Solution

First part:

From Equation 5.140

$$\frac{2.5 \times (13 - 1)}{(7 - 1) \times 2^{PGA}} = 2.5$$

$$\Rightarrow 2^{PGA} = 2$$

$$\Rightarrow PGA = 1 = (001)_2$$

Second part:

$$\text{PGA code} = (000)_2 = 0$$

$$\therefore \frac{2.5 \times (\text{DAC Code} - 1)}{(6) \times 2^0} = 2.5$$

$$\Rightarrow \text{DAC Code} = 5 = (0101)_2$$

5.11.6 STIM Calibration

The IEEE-P1451.2 draft standard defines the STIM, which is the key element for transducer to microprocessor or network communication. The details of this standard will be discussed in Chapter 7, while enumerating the smart sensor interface protocols. The STIM is a module that contains the TEDS in addition to the transducers and other signal conversion and signal conditioning modules. TEDS is a set of electronic data of the transducers stored in the hardware. This data is used for self-identification and auto-calibration of the transducer. One of the important component of the TEDS is the calibration TEDS. The calibration TEDS contains information about the calibration parameters such as last calibration data, calibration interval, etc. An example of a calibration TEDS for a single pressure sensor [15] is shown in Table 7.1 of Chapter 7.

5.12 Offset Compensation

One of the major problems in precision application of sensors is the presence of an offset, i.e., presence of an additive error at the output voltage even in the absence of any input. The reason for this offset in a sensor is either due to mismatch in design or hysteresis. For example, in a Hall sensor with a supply voltage of 12 V, an offset voltage as high as 100 mV may appear across the electrodes even when no magnetic field acts. The causes of such offsets in hall sensor may be due to eccentric positioning of electrodes or nonuniformity in composition of the material. In another example, in a linear variable differential transformer (LVDT), the output voltage at null position is ideally zero; however, harmonics in the excitation voltage or stray capacitance between primary and secondary windings produces a small offset voltage. For ordinary applications, small offset voltage may be accepted but for precision measurements, the offset needs elimination. The elimination of offset can be addressed by the following methods [16].

5.12.1 Improved Manufacturing Process

Defect in the manufacturing process is one of the most common causes of offset. In the fabrication of integrated sensors, the defects may be inhomogeneous diffusion, epitaxy, oxidation inaccurate mask alignment, etc. In bridge-type sensor configurations, a small error in the fabrication process can imbalance the bridge giving rise to an offset. There are, however, technological and cost limitations to adopting efficient fabrication process. On the other hand, improved sensor structure design is also important for offset elimination. Geometrical ratios of sensors are mostly responsible for offset development.

5.12.2 Counter Bias Inputs

Wherever possible, a counter bias input in physical form can be applied to the sensor to eliminate the offset; however, all kinds of offsets cannot be eliminated by this technique. For example in a strain gauge based load cell, due to any imbalance in the surface of loading structure, the sensor produces an offset even when no load is applied. The offset can be eliminated by applying a pre-strained load opposite to the direction of the offset. This will counter balance the offset.

5.12.3 Chopper Amplifier

Chopper is used to eliminate the offset and $1/f$ noise introduced by OPAMP and dc amplifiers. The same technique can be used for sensor offset elimination. Figure 5.72 shows a basic chopper amplifier. The chopper amplifier modulates the non-alternative input V_{in} by changing the polarity alternately using the switch at the input. This operation is equivalent to modulating a square wave of input T by the input signal V_{in}. Therefore, V_{in} becomes a band-limited signal with frequency components up to a maximum of $1/T$. After the signal is amplified, it is demodulated by the same square wave, resulting in the output V_0. Since the offset is usually or nearly dc, the original signal gets filtered out from the input.

The chopping action can also be performed in the physical stage, like in optical-based sensors; the light intensity is periodically interrupted by

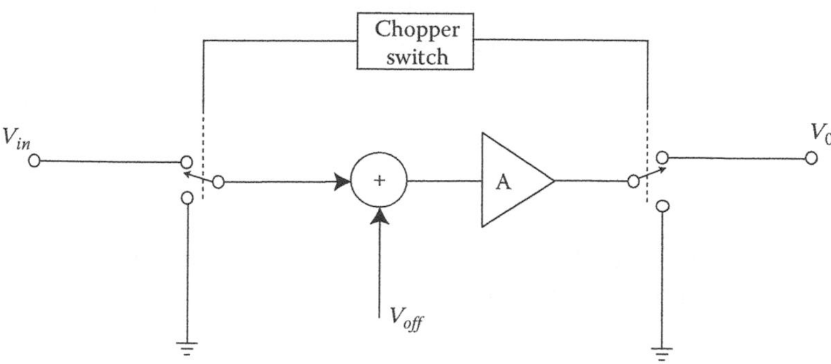

FIGURE 5.72
A basic chopper amplifier.

a rotating electromechanical device. The light intensity signal having an offset is filtered out from the offset in a similar way as that of a chopper amplifier.

5.12.4 Matched Sensor Methods

In micro sensors, two matched sensors exhibiting the same diffusion and oxidation irregularities, mask misalignment, etc., are deposited on the same silicon wafer [16]. It is expected that the two sensors exhibit nearly the same sensitivity and offset. The connections are so made that the desired outputs get added while the offsets get canceled out. This method is successfully used in Hall plates for solid-state keyboards [17]. There may be two situations:

1. Only one sensor is exposed to the measurand and the second sensor is not. This gives the following output voltage equations

$$V_1 = mx + V_0 \tag{5.141}$$

$$V_2 = V_0 \tag{5.142}$$

   where
   m is the sensitivity of the sensor
   V_0 is the offset voltage

   From Equations 5.141 and 5.142, we get

$$x = \frac{(V_1 - V_2)}{m} \tag{5.143}$$

   which is free from offset.

2. Both sensors are exposed to the measurand. In this case, the following voltage equation can be written as

$$V_1 = mx + V_0 \tag{5.144}$$

$$V_2 = mx - V_0 \tag{5.145}$$

   Note that the second sensor produces an offset with negative polarity as that of first sensor. From Equations 5.144 and 5.145, we get

$$x = \frac{(V_1 + V_2)}{2m} \tag{5.146}$$

Example 5.29

Two matched piezoelectric pressure sensors with identical sensitivity of 20 μV/psi and matched offset are deposited on the same wafer. When one of the sensors is applied with a pressure while the other is not exposed to the pressure, the subtracted resultant output voltage is 2 mV. When both sensors are applied with the same pressure and the outputs are added, determine the applied pressure and the output voltage.

Solution

$$m = 20\ \mu\text{V/psi}$$

Case I:

$$V_1 = 20x + V_0$$

$$= 2\,\text{mV} = 2000\,\mu\text{V}$$

$$V_2 = V_0$$

$$V_1 - V_2 = 20x$$

$$\Rightarrow x = \frac{2000}{20} = 100\,\text{psi}$$

Case II:

$$V_1 = 20x + V_0$$

$$V_2 = 20x - V_0$$

$$V_1 + V_2 = 40x = 40 \times 100 = 4000\,\mu\text{V} = 4\,\text{mV}$$

5.13 Error and Drift Compensation

Any difference between the true or ideal value of a sensor for the measured quantity and the sensor reading is called an error. Various definitions and modeling of error have been discussed in Sections 2.2.2 and 2.2.7. This section will deal with the techniques of compensation of error and drifts.

When a sensor gives varying outputs for a constant input, it is said that the sensor has drifts. If the sensor output varies when the input is zero, it is called zero drift, while the output variation during the operating range is called scale factor drift.

5.13.1 Drift Simulation

It is difficult to model sensor drift because it is difficult to obtain actual drift information of a sensor under a specific operating condition [18]; however, sensor drift can be considered under general mathematical model. The sensor drift can be considered as a second-order, nonstationary random process. Such processes generate signals that follow characteristic curves with saturation given by

$$y = (a + bn_j)x^n + cn_j x \tag{5.147a}$$

where $n_j = n_{j-1} + d \cdot j + e \cdot j$ randn.

The expression random denotes a random signal of Gaussian distribution. Two other expressions for drift of sensor are given in [18].

$$y = f \sin(gx + n_j) \tag{5.147b}$$

$$y = h \sin(x + n_j) \tag{5.147c}$$

where f, h, and g are drift coefficients. The above equations can be combined to represent sensor drift, e.g., [1]

$$y = 0.2 + (0.1 + 0.2n_j)\sqrt{\frac{x}{n_j}} + 0.25\sin(0.3x + n_j) + 0.2\,\text{rand} + 0.15\sin(x - n_j) + 0.1\,\text{randn} \tag{5.147d}$$

This corresponds to drift of K-type thermocouple at operating temperature of 1000°C–1100°C inside an air electric furnace.

5.13.2 Compensation by Complementary Sensors and Filters

Wherever possible, a complementary sensor with low performance characteristics such as slow performance but with low or no drift can be used in conjunction with the main sensor that suffers from drift problems. Figure 5.73 shows such a complementary arrangement for drift correction [19]. Sensor-1 is the measuring sensor that has a drift problem while Sensor-2 is drift free but corrupted with noise. The drift of Sensor-1 is filtered out by the high pass filter and the high-frequency noise of Sensor-2 by low pass filter. The filter outputs are added and applied to a Kalman filter for better estimation.

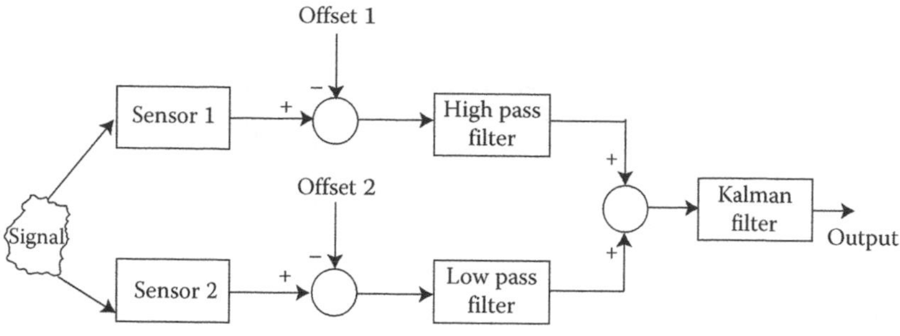

FIGURE 5.73
Complementary filtering scheme.

5.14 Lead Wire Compensation

When a sensor is connected to the processing circuit, the connecting lead wires may have different lengths, sizes, and materials depending upon the distance to be covered, current to be handled, and the durability required, respectively. The lead wires account for the additional resistance to the sensor circuits, which cannot be ignored for precise measurements. If the sensor is a resistive sensor, this accountability becomes higher since the resistance of the lead wires will alter the signal amplitude.

In case of RTDs, which is a resistive temperature sensor, there are three ways to use the sensors—two lead wires, three lead wires, and four lead wires. Three-wire connection is the most advantageous to avoid the lead wire problem whereas four-wire connection is not economic. Therefore, two-wire connection is the most economic if lead wire compensation is done.

5.14.1 Self-Heating Method

An indirect method of temperature measurement using self-heating of two-wire RTD has been proposed in [20], where the lead wire error of RTD is automatically compensated. The schematic circuit diagram of the scheme is shown in Figure 5.74. The two-wire RTD with lead wire resistances r_e and RTD resistance R_T is supplied with two constant current sources I and I_n alternately, where I is the normal measuring current and I_n is a higher current applied for a short duration τ. The technique uses a differential temperature method where, with normal current I, the voltage across RTD at a normal temperature T_1 is given by

$$\begin{aligned} V_1 &= I\left[2r_l + R_{T1}\right] \\ &= I\left[2r_e + R_0(1+\alpha T_1)\right] \end{aligned} \tag{5.148}$$

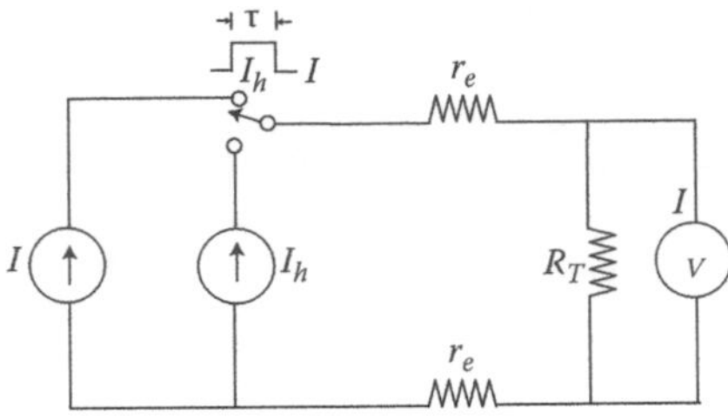

FIGURE 5.74
A self-heating RTD. (Reprinted from IEEE, copyright © IEEE. With permission.)

Again with normal current I, the voltage across RTD at temperature T_2 is given by

$$V_2 = I\left[2r_l + R_0(1+\alpha T_2)\right] \tag{5.149}$$

where
R_0 is the resistance of the RTD at 0°C
α is the temperature coefficient of RTD

The differential voltage

$$V_1 - V_2 = \Delta V = I[R_0\alpha\Delta T] \tag{5.150}$$

where $\Delta T = T_2 - T_1$.

Now the sensor is triggered by a higher current I_h for a short duration τ, thereby providing a self-heating to the RTD. The heat developed will not be dissipated to the surrounding and will increase the sensor's resistance. The heat energy developed is given by

$$Q = I_h^2 R_{aV}\tau \tag{5.151}$$

where R_{aV} is the average resistance of the sensor in the temperature interval ΔT, which is given by

$$R_{aV} = \frac{1}{\Delta T}\int_{T_1}^{T_2} R_T \, dT \tag{5.152}$$

By simplifying to the first-order approximation

$$R_{aV} = R_{T_1} + \frac{1}{2}R_0\alpha\Delta T \tag{5.153}$$

The calorimetric equation of the sensor can be written as

$$Q = m\Delta T \tag{5.154}$$

where m is heat capacity (J/°C).

From Equations 5.151, 5.153, and 5.154, we get

$$m\Delta T = I_h^2\tau\left[R_{T1} + \frac{1}{2}R_0\alpha\Delta T\right]$$

From Equation 5.150

$$R_{T1} = \frac{(\Delta V/IR_0\alpha)\left(m - (I_h^2\tau R_0\alpha/2)\right)}{I_h^2\tau}$$

$$= \frac{\Delta V\left(2m - I_h^2\tau R_0\alpha\right)}{2I\, I_h^2 R_0\alpha\tau} \tag{5.155}$$

Once the value of R_{T1} is calculated, the temperature T_1 can be determined from

$$T_1 = \frac{R_{T1} - R_0}{R_0\alpha} \tag{5.156}$$

Example 5.30

An RTD has the following specifications:
$R_0 = 1000\ \Omega$, $\alpha = 3.9083 \times 10^{-3}\ \Omega/°C$, heat capacity $(m) = 5.25 \times 10^{-3}$ J/°C. The RTD is used in the self-heating mode for its lead wire compensation. The sensor is first excited by a constant current of 5 mA, which is heated to a temperature of T_1. The RTD is then self-heated by a higher current of 20 mA applied for a 5 s duration. The differential voltage measured in the two consecutive measurements is 0.0996 V. Determine the temperature T_1.

Solution
Given values:

$R_0 = 1000\ \Omega$,
$\alpha = 3.9083 \times 10^{-3}\ \Omega/°C$,
$m = 5.25 \times 10^{-3}$ J/°C,
$\tau = 5$ s,
$\Delta V = 0.096$ V,
$I = 5$ mA,
$I_h = 20$ mA

Substituting the values in Equation 5.155, we get

$$R_{T1} = \frac{0.096[2\times 5.25\times 10^{-3} - (20\times 10^{-3})^2 \times 5\times 100\times 3.9083\times 10^{-3}]}{2\times 5\times 10^{-3}\times (20\times 10^{-3})^2 \times 100\times 3.9083\times 10^{-3}\times 5}$$

$$= 119.39\,\Omega$$

From Equation 5.156

$$T_1 = \frac{R_{T1} - R_0}{R_0\alpha}$$

$$= \frac{119.39 - 100}{100\times 3.9083\times 10^{-3}}$$

$$= 49.61°\text{C}$$

5.14.2 Bidirectional Current Mode RTD

Lead wire compensation in RTD has been proposed in [21] where the RTD is used in bidirectional current mode. Figure 5.75 shows the schematic block diagram of the proposed compensation technique.

The RTD is supplied with constant current generated in the CCS. The four switches, SW_1–SW_4, driven by the clock (both high and low) controls the direction of current through the lead wires, diodes (D_1 and D_2) and then the RTD (R_T). An array of sample and hold, SH_1–SH_4, driven by the same clock is used to sample the voltages (V_1, V_1', V_2, V_2') across the RTD circuits in different directions. During high clock pulse, SW1 directs the current I through the lead wire r_{L1}, diode D_1, R_T, lead resistance r_{L2}, and then to ground through SW_3. During this mode SH_1 samples and hold voltage at node 1, i.e., V_1 given by

$$V_1 = r_{L1}I + V_{D1} + R_T I + r_{L2}I + V_{SW_3} \tag{5.157}$$

where V_{D1} and V_{SW_3} are the voltage drops across diode D_1 and SW_3 during high clock pulse. Similarly, SH_3 samples and holds voltage at node 2 (V_1') where

$$V_1' = V_{SW_3} \tag{5.158}$$

Again during low clock pulse, SW_4 and SH_2 participate to give

$$V_2 = r_{L2}I + V_{D_2} + r_{L1}I + V_{SW_2} \tag{5.159}$$

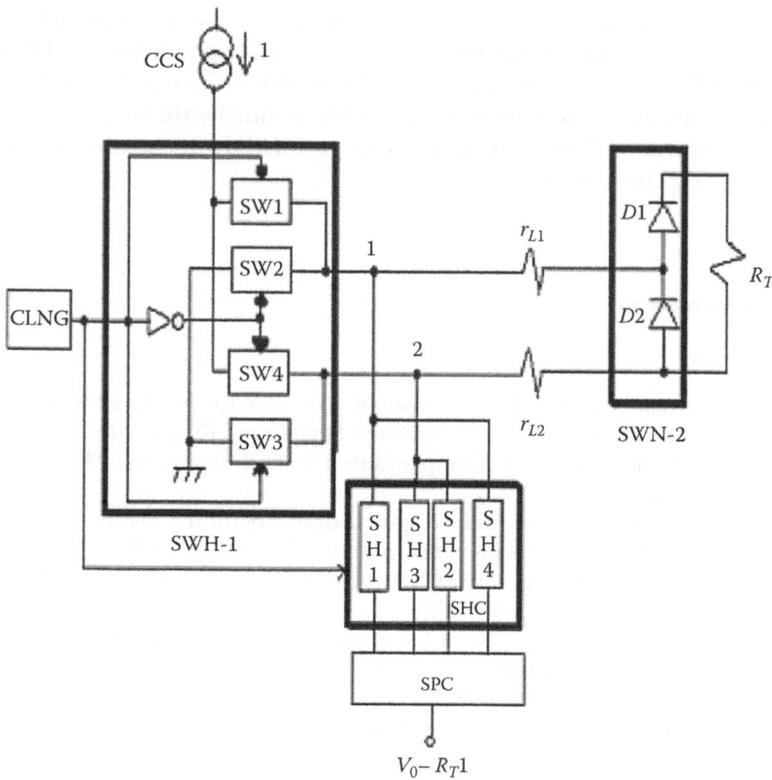

FIGURE 5.75
Block diagram of bidirectional current mode RTD. (Reprinted from IEEE, copyright© IEEE. With permission.)

and

$$V_2' = V_{SW_2} \tag{5.160}$$

These two sets of voltages V_1, V_1', V_2, V_2' are applied to a summing amplifier to get V_3 as

$$\begin{aligned} V_3 &= (V_1 + V_2') - (V_2 + V_1') \\ &= R_T I + V_{D_1} - V_{D_2} \end{aligned} \tag{5.161}$$

For matched diodes, $V_{D_1} = V_{D_2}$ and, therefore

$$V_3 = R_T I \tag{5.162}$$

The output voltage V_3 is free from the lead wire resistances r_{L1} and r_{L2} and it only depends on R_T.

The scheme discussed in Section 5.14.1 needs necessary data-acquisition systems to provide the switching between current triggers for I and I_h, computation of R_{T1} and T_1 using Equations 5.155 and 5.156. On the other hand, the scheme discussed in Section 5.14.2 is entirely hardware based and proper linearization and calibration amplifier can perform the measurement of temperature using the output voltage V_3.

References

1. Mami Konyan, K., Modeling of negative resistance for resistive transducer linearization, *IEEE, Physics Conference*, Saint Petersburg, Russia, 2003.
2. Positive Analog Feedback Compensates PT100 Transducers, MAXIM, 2009, www.maxim-ic.com
3. Nobbs, J.H., Linearization of the response from a platinum resistance thermometer, *Journal of Physics E: Science, Instruments*, 15, 716–718, 1982.
4. Pallas-Areny, R. and Webster, J.G., *Sensors and Signal Conditioning* (2nd edn.), Wiley Interscience, New York, 2001.
5. Khachab, N.I. and Ismail, M., Linearization techniques for nth order sensor models in MOS-VLSI technology, *IEEE Transactions on Circuits and Systems*, 38(12), 1439–1450, December 1991.
6. Lygouras, J.N., Non linear ADC with digitally selectable quantizing characteristic, *IEEE Transactions on Nuclear Science*, 35(5), 1088–1091, 1988.
7. Dias Pereira, J.M., Postolache, O., Girao, P.S., and Serra, A.C., A discrete and cost effective ADC solution based on pulse-width modulation technique, *Proceedings of the CONFETELE 2001*, Figuera da Fiz, Portugal, April 2001.
8. Jafairipanah, M., Al-Hashimi, B.M., and White, N.W., Design consideration and implementation of analog adaptive filters for sensor response correction, *Proceedings of the ICEE 2004*, Mashhad, Iran, May 11–13, 2004.
9. Patranabis, D., Ghosh, S., and Bakshi, C., Linearizing transducer characteristic, *IEEE Transactions on Instrumentation and Measurement*, 37(1), 66–69, March 1988.
10. Horn, G. and Huijsing, J.H., *Integrated Smart Sensors: Design and Calibration*, pp. 93–112, Kluwer Academic Publishers, Boston, MA, 1998.
11. Lopez-Martin, A.J., Zuga, M., and Carlosena, A., A CMOS piecewise linear A/D converter for linearizing sensor characteristic, 0-7803-7059 IEEE, 2001.
12. Medrano-Marques, N.J. and Martin-del-Brio, B., Sensor linearization with neural network, *IEEE Transaction on Industrial Electronics*, 48(6), 1288–1290, December 2001.
13. Johnson, C.D. and Chen, C., Bridge to computer data acquisition with feedback nulling, *IEEE Transactions on Instrumentation and Measurement*, 39, 531–534, 1990.
14. Demystifying Sigma Delta ADC, MAXIM, Application note 1870, 2009, www.maxim-ic.com
15. Woods, S.P., IEEE P1451.2 Smart transducer interface module, Sensors Plug and play, National Instruments, 2009. (http://www.ni.com/teds)

16. Xing, Y.-Z., Kordic, S., and Middlehoek, S., A new approach to offset reduction in sensors: The sensitivity variation method, *Journal of Physics E: Science Instrumentation*, 17, 657–663, 1984.
17. Maupin, J.T. and Geske, M.L., The hall effect in silicon circuits, *The Hall Effect and Its Applications*, Chien, C.L. and Westgate, C.R. (Eds.), Plenum, New York, pp. 421–445, 1980.
18. Sachenko, A., Kochan, V., Kochan, R., Turchenko, V., Tsahouridis, K., and Laopoulos, Th., Error compensation in intelligent sensing instrumentation system, *IEEE Instrumentation and Measurement Technology Conference*, Budapest, Hungary, pp. 869–874, May 21–23, 2001.
19. Lee, H.-J.K. and Jung, S., Gyro sensor drift compensation by filter to control a mobile inverter pendulum robot system, IEEE, *International Conference on Industrial Technology*, ICIT 2009.
20. Sorin-Dan, G., Constantin, I., and Brandusa, P., A sensor self heating method for lead's resistance compensation in two wire RTD's measurements, *Proceedings of the Conference on Precision Electromagnetic Measurements*, vol. 27, IEEE, Boulder, CO, pp. 174–175, June–July 1, 1994.
21. Maiti, T.K., A novel lead-wire resistance compensation technique using two-wire resistance temperature detector, *IEEE Sensors Journal*, 6(6), 1454–1458, 2006.

6

Sensors with Artificial Intelligence

6.1 Introduction: Artificial Intelligence

Artificial intelligence (AI) is the study and design of intelligent systems components, which perceives its environmental conditions and accordingly can take action to maximize its chances of success. The basic characteristics of AI are

1. Reasoning
2. Knowledge
3. Planning
4. Learning

In brief, AI can be defined as the systems that can think and act rationally like humans.

6.1.1 Reasoning

The task of AI is to imitate the step-by-step reasoning methods used by humans while solving puzzles, playing board games, or making logical decisions. However, some researchers say that humans solve most of their problems using fast intuitive judgments rather than the conscious step-by-step deductions. Step-by-step reasoning is possible to be modeled in AI but intuitive judgments are difficult to be modeled and implemented. In most AI-based systems, mimicking of human reasoning is step by step in local but intuitive in global. A problem of reasoning always needs an initial point, a goal, a set of feasible actions, and a set of constraints. The task of AI system is to find the best sequence of permissible actions that can transform the initial point to a goal.

6.1.2 Knowledge

Knowledge representation and knowledge engineering are two important aspects of AI. Most AI-based systems need to explore and use extensive knowledge about the world. The knowledge representation mainly deals with objects, properties, categories, relation between objects, situations, events, states, time, causes, effects, etc.

6.1.3 Planning

Intelligent systems always set goals and try to achieve them by planning. In classical planning problems, the system can assume that it is the only thing acting on the world and the consequence of its actions are known. However, this is not always true and the system must check if the world matches its predictions.

Planning is a role of AI, which increases its autonomy and flexibility through the construction of sequences of actions to achieve its goals.

6.1.4 Learning

An intelligent system should be able to learn from its environment or the user. Machine learning is the key learning in the AI systems. Machine learning allows the AI system to change behavior based on the data from sensors or databases. The major focus on machine learning is providing the ability to the AI system to learn to recognize the complex patterns and draw inference from it intelligently.

6.2 Sensors with Artificial Intelligence

It is difficult to sharply partition the two classes of sensors—"intelligent sensors" and "sensors with AI," since there is a debate about what is "smart," "intelligent," or even "sentient" (which technically means having the five senses) in the context of AI [1].

In generic form, intelligent sensors perform typical metrological operations (such as calibration, error compensation, linearization, offset elimination, etc.), but sensors with AI are able to perform complex operations comparable to human or animal intelligence. Apparently, a self-adaptive sensor looks like possessing the intelligence of human or any other animal since it can adapt to the environment, but in the context of AI, such sensors do not have the capacity of reasoning, knowledge, planning, and learning. AI systems can mimic the human or animal intelligence; however, sensors can be enabled to perform the tasks of reasoning and learning by human.

6.2.1 Multidimensional Intelligent Sensors

Human perception of the sense of audio, visual, olfaction, and touch is multidimensional in nature. The sensory systems of our eyes, ears, nose, and skin accommodate a large number of sensory receptors that sense the signal distributed in a wider space. Single-point sensor technology is very common; however, it has limitations or lacks capabilities of artificial intelligence. Therefore, there have been increasing interests in using multidimensional array-based technology that can investigate spatial distribution of a quantity. The most recently explored area of multidimensional sensor systems is gas sensing.

Multidimensional gas sensors

The major objective of multidimensional gas sensors are

1. Gas classification and recognition
2. Localization and spatial distribution

6.2.1.1 Gas Classification and Recognition

Odors are sensed by the human olfactory system when volatile organic compound (VOC) stimulates the olfactory receptors located in the human olfactory epithelium at the upper surface of the nasal cavity (Figure 6.1). The area of the olfactory receptors in each nasal passage is about 2.5 cm^2 with 50 million receptor cells. The odor receptors can detect and differentiate up to 10,000 different VOCs. Each olfactory receptor neuron holds 8–20 cilia that receive the VOCs. The VOC produces a specific neuron activity pattern and, since patterns are distinctive for each VOC, the pattern allows us to discriminate between a large number of different odors.

VOCs are tiny molecules with molecular weights of 20–300 Da. The human olfactory receptors can detect VOCs with concentrations below 1 ppb. The detection levels of different materials are different such as green leaves, 0.32 ppm; rose, 0.29 ppm; thyme, 86 ppb; lemon, 10 ppb; off flavor in fish, 0.01 ppb; green pepper, 0.001 ppb; and grapefruit, 0.00002 ppb.

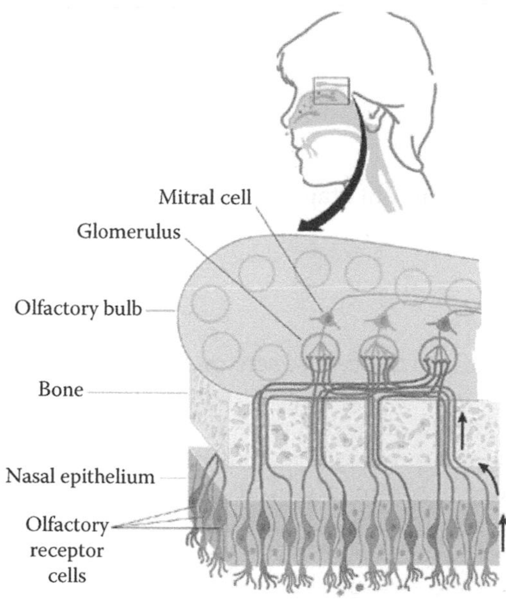

FIGURE 6.1
Human olfactory system.

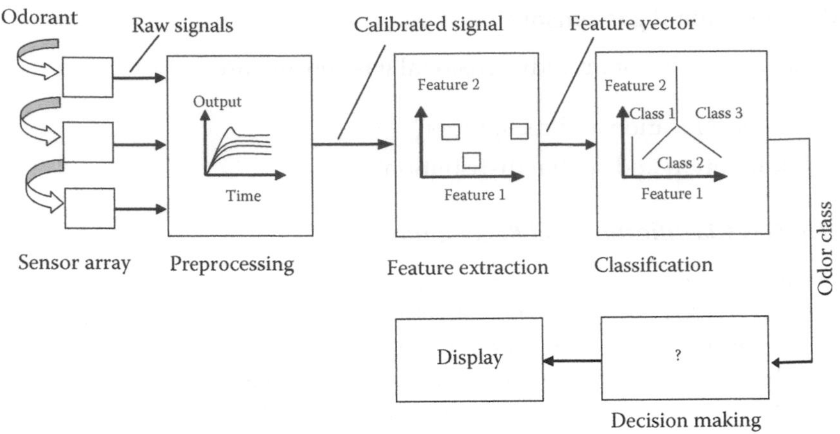

FIGURE 6.2
Block diagram of various stages of E-nose signal processing. (Redrawn from Bhuyan, M., *Measurement and Control in Food Processing*, Taylor & Francis, CRC Press, 208, 2006. With permission.)

The E-nose technology is a mimic of the human olfactory system and is similar to the array of olfactory receptors, signal excitation by neurons, and finally pattern classification in the brain.

The E-nose technology adopts the following basic signal processing and classification stages:

1. Preprocessing
2. Feature extraction
3. Classification

Figure 6.2 shows the different stages of the signal processing of the E-nose system.

6.2.1.1.1 Preprocessing

On application of VOCs to the E-nose sensor array, the E-nose sensors experience a change in conductivity in case of chemoresistive sensors, or a variation in mass and resonant frequency in case of piezoelectric or surface acoustic sensors. This variation is processed to obtain a variation in voltage or current. A Wheatstone bridge is commonly used in case of a chemoresistive sensor, while an oscillator circuit can be used in case of quartz crystal microbalance (QCM) and surface acoustic wave (SAW) sensors. Noise reduction, normalization, drift correction, etc., are other important signal processing of this stage. The preprocessed signal is then applied to a computer using a data-acquisition system and the data is stored in the memory for pattern recognition.

6.2.1.1.2 Feature Extraction

This process uses statistical or other mathematical techniques to find some relevant, but hidden, information from the data. One most common technique of feature extraction is principle component analysis (PCA). PCA is a linear combination of orthogonal vectors accounting for a certain amount of variance in the sensor data. The variance in the data in each principle component is the result of each eigenvalue with varying importance. This technique removes any redundancy and dimensionality of the data.

6.2.1.1.3 Classification

In the classification stage, the class of an unknown sample is determined from a class assignment. On projecting the E-nose data on an appropriate low-dimensional space, the classification stage is used to identify the patterns that are representative of each odor. The classification stage assigns a class to the data by comparing its pattern with the trained patterns. The common mathematical tools used for classification are K-nearest neighbors (KNN), Bayesian classifiers, and artificial neural network (ANN).

6.2.1.2 Localization and Spatial Distribution

When it is necessary to localize the correct source or position of a gas source, the measurement of the spatial gas distribution and its temporal change is important [2]. Since the gas distribution changes its distribution profile due to variation in airflow and diffusion convection, the measurement system should use intelligent and adaptive techniques to fit with actual environments. A two-dimensional array of gas sensors [2] is shown in Figure 6.3.

The gas sensors captures the spatial gas distribution pattern where sensor $S(x, y)$ senses the gas concentration $g(x, y)$. The gas concentration $g(x, y)$ is converted to an image pixel value $p(x, y)$ so that $g(x, y) \leftrightarrow p(x, y)$. For an $N \times N$ pixel image, an $(M \times M)$ image of spatial gas distribution is generated by using linear and bilinear interpolation. Figure 6.4 shows the expansion of a 2×2 pixel kernel to a 4×4 kernel by linear interpolation.

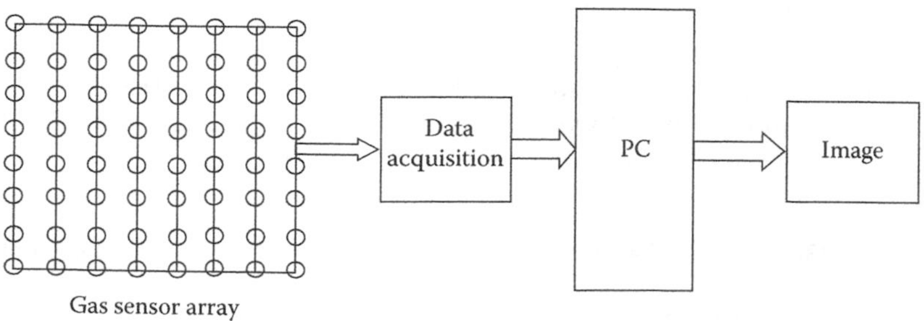

FIGURE 6.3
A two-dimensional gas sensor array. (Redrawn from IEEE, copyright © IEEE. With permission.)

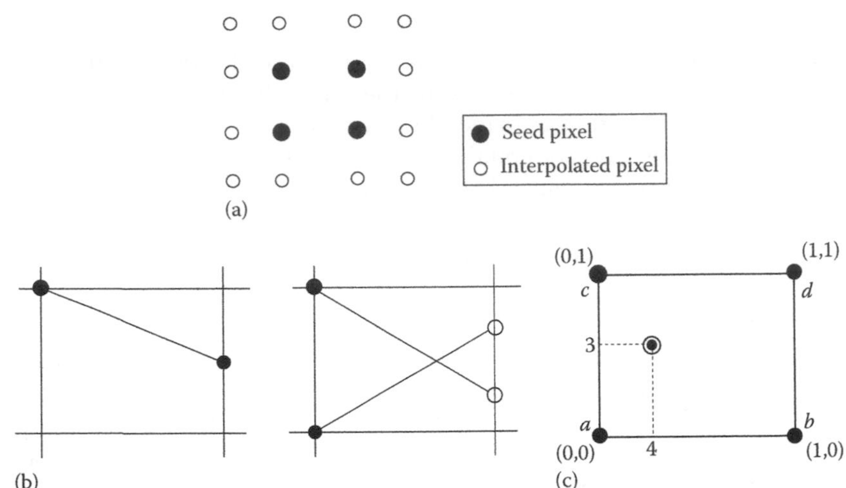

FIGURE 6.4
Linear interpolation in 2D sensor array (a) expansion of resolution, (b) interpolation between 2 points and 4 points, and (c) bilinear interpolation.

In linear interpolation, a known point can be extended to an unknown grid point by a straight line and with the same value. If two known points are to be extended to two unknown points, the farthest points are connected. This linear interpolation scheme is shown in Figure 6.4b.

Bilinear interpolation is used to approximate a value within a rectangular grid, the vertices of which carries the four known points. The bilinear approximation equation is given by [2]

$$p(\zeta,\eta) = a + (b-a)\zeta + (c-a)\eta + (a+d-b-c)\zeta\eta \quad \text{for } 0 \le \zeta \le 1 \text{ and } 0 \le \eta \le 1 \quad (6.1)$$

where
- ζ and η are normalized x, y coordinates in each element rectangle
- a, b, c, and d are the values at the vertices (Figure 6.4c)

Let the pixel values (normalized over 0–255) for the vertices be $a=150$, $b=230$, $c=270$, and $d=190$. If we need to interpolate a pixel at a coordinate of $\zeta=0.5$ and $\eta=0.5$, i.e., at the center of the rectangle, we get

$$p = 150 + (230-150)\times 0.5 + (270-150)\times 0.5 + (150+190-230-270)\times 0.5\times 0.5$$

$$= 150 + 80\times 0.5 + 120\times 0.5 - 40$$

$$= 210$$

Example 6.1

A metal-oxide semiconductor (MOS) gas sensor array is organized in square grids at distances of 20 cm apart to detect gas distribution over the array. Four gas sensors placed on the vertices of a rectangle transmit gas concentration as follows (refer to Figure E6.1): $a = 199$ ppm, $b = 150$ ppm, $c = 142$ ppm, and $d = 120$ ppm. Determine the gas concentration at the mid point of each grid line and at the center using bilinear interpolation.

Solution

Given values:

$a = 199$ ppm
$b = 150$ ppm
$c = 142$ ppm
$d = 120$ ppm

For mid of ab grid line: $\zeta = 0.5$, $\eta = 0$

$$p_{ab} = 199 + (150 - 199) \times 0.5 + (142 - 199) \times 0 + (199 + 120 - 150 - 142) \times 0.5 \times 0$$
$$= 199 + (-24.5)$$
$$= 174.5 \text{ ppm}$$

For mid of ac: $\zeta = 0$, $\eta = 0.5$

$$p_{ac} = 199 + (142 - 199) \times 0.5$$
$$= 170.5 \text{ ppm}$$

For mid of bd: $\zeta = 1.0$, $\eta = 0.5$

$$p_{bd} = 199 + (150 - 199) \times 1.0 + (142 - 199) \times 0.5 + (199 + 120 - 150 - 142) \times 1.0 \times 0.5$$
$$= 135 \text{ ppm}$$

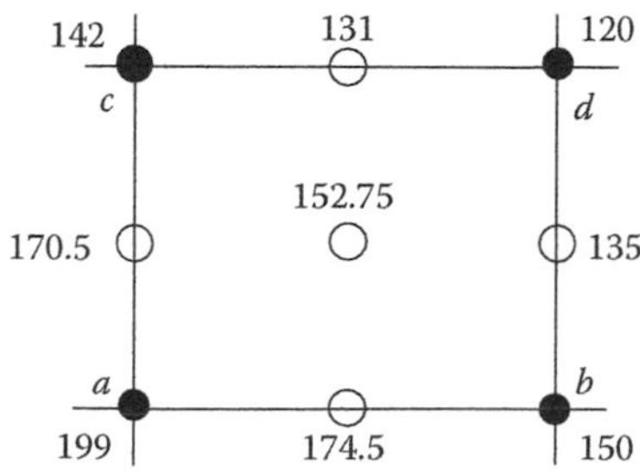

FIGURE E6.1
Bilinear interpolation of gas concentration.

For mid of cd: $\zeta = 0.5$, $\eta = 1.0$

$$p_{cd} = 199 + (150 - 199) \times 0.5 + (142 - 199) \times 1.0 + (199 + 120 - 150 - 142) \times 0.5 \times 1.0$$
$$= 131\,\text{ppm}$$

For center of rectangle: $\zeta = 0.5$, $\eta = 0.5$

$$p_c = 199 + (150 - 199) \times 0.5 + (142 - 199) \times 0.5 + (199 + 120 - 150 - 142) \times 0.5 \times 0.5$$
$$= 152.75\,\text{ppm}$$

An 8×8 array of MOS gas sensor was used in [2] for spatial gas distribution where the gas sensor signals were transformed into visual images.

6.2.2 AI for Prognostic Instrumentation

Traditional condition monitoring for a plant deals with detection, reporting, and response to fault in the system; however, it cannot predict future conditions of the system. Prognostic instrumentation is intended to support condition-based maintenance (CBM), which has the ability to predict a future condition. Prognostic instrumentation is different from diagnostic instrumentation because prognostic can predict a fault allowing reduction in impact to the system, while diagnostic facilitates repairing only. The prognostic instrumentation intelligently utilizes the diagnostic results, knowledge-based data and statistical forecasting so that the remaining life or fault probability of the system can be predicted.

Prognostic instrumentation typically consists of a sensing, interpreting, and reporting module supported by physical experience and knowledge model of the system. The prognostic instrumentation is illustrated in Figure 6.5.

Prognostic instrumentation is useful in high-end and complex systems where system maintenance logistics are high. Some of the applications of prognostic instrumentation are

1. Aircraft engine health
2. Vibrations and mechanical degradation of motor-driven heavy machines

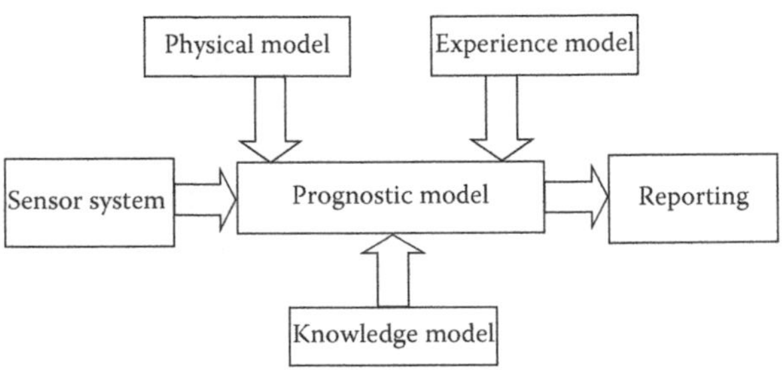

FIGURE 6.5
A typical prognostic instrumentation system.

3. Gas turbines
4. Power plants
5. Human health

The basic components required for prognostic methodology are [3]

1. Identification of failure predictors
2. Sensor for measurement of conditions
3. Data storage, processing, and analysis
4. Fault and degradation models
5. Estimation of fault and degradation levels

6.2.2.1 Identification of Failure Predictors

Before designing the prognostic instrumentation, the failure mechanism must be identified and prioritized. Failure mechanism identification can be done based on the comparison of measurement conditions under various states and ideal ones.

6.2.2.2 Sensors for Measurement of Conditions

The conditions that indicate failure or degradation of the system should be reliably measured using sensors. Identification of measurement variables again depend on how efficiently the failure predictors are identified. Following are the examples of sensors for prognostic instrumentation of few applications:

1. Temperature, humidity, and airflow sensors for environment monitoring
2. Accelerometer and torque sensors for structural and vibration monitoring of rotating machines
3. Flow temperature and pressure sensors for compressor degradation monitoring of gas turbine

Recently, AI models have also been implemented in medical science for prognostic monitoring of nerve injury recovery [4].

6.2.2.3 Data Storage, Processing, and Analysis

The sensor data acquired either online or off-line are stored, processed, and analyzed based on the identified prognostic model algorithm. The speed of data collection from the sensors must compromise with the necessary resolution and capacity of algorithm.

6.2.2.4 Fault and Degradation Models

Prognostics is based on a model that can monitor impending fault in future. The model can be developed from the following sources:

- Experience, knowledge, and historical data
- First principle or empirically derived
- Dependency model linking conditions with failures
- Probabilistic predictions

6.2.2.5 Estimation of Failure and Degradation

For estimation of failure and degradation, the conditional sensor data of the system are fed to the model and then analysis techniques are used to determine the progress of degradation. The prognostic analysis techniques use the following approaches for solving different problems [5]:

1. Experience-based prognostics
2. Evolutionary/statistical trending prognostics
3. State estimator prognostic
4. Model-based or physics failure–based prognostics
5. AI-based prognostics

6.2.2.6 Artificial Intelligence–Based Prognostics

ANN is a dominating tool for classification and recognition of data patterns in array-based sensor systems such as in electronic nose. The basic strength of ANNs is their ability to model nonlinear systems where input–output systems are sole system identifiers. This section describes case studies where ANN has been successively used for the prognostic and predictive estimation of system health.

Case Study 6.1

Battery banks are integral parts of power back-ups of many power plants, solar energy systems, and computer networks. Battery failure in such installations can lead to loss of operations, reduced capability, and down time. A prognostic method to predict the battery state of charge (SOC), capacity (amp-h), and remaining charge cycles (RUL) has been developed in [6]. The prognostic system involves model identification, feature extraction, and data fusion of the sensor data.

6.2.2.6.1 *The Battery Model*

A battery is a electrochemical cell arrangement, each cell of which comprises of two electrodes to facilitate charge transfer under oxidation and reduction reactions. Changes in the electrode surface, diffusion layer, and solution are difficult to detect without dismantling the cell; however, the measurable variables like voltage, current, and temperature can be used to indirectly measure the cell health. The cell SOC is defined as the ratio of the remaining capacity to total capacity where capacity of the cell means the time integral of the current delivered to the load till the voltage of the cell drops to

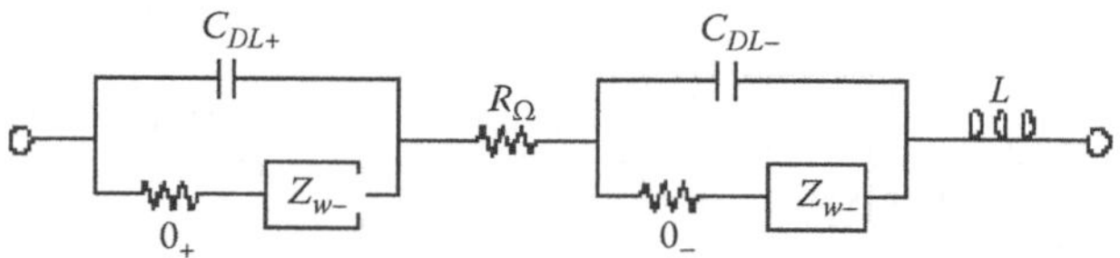

FIGURE 6.6
Electrical model of a battery. (Reprinted from IEEE, copyright © IEEE. With permission.)

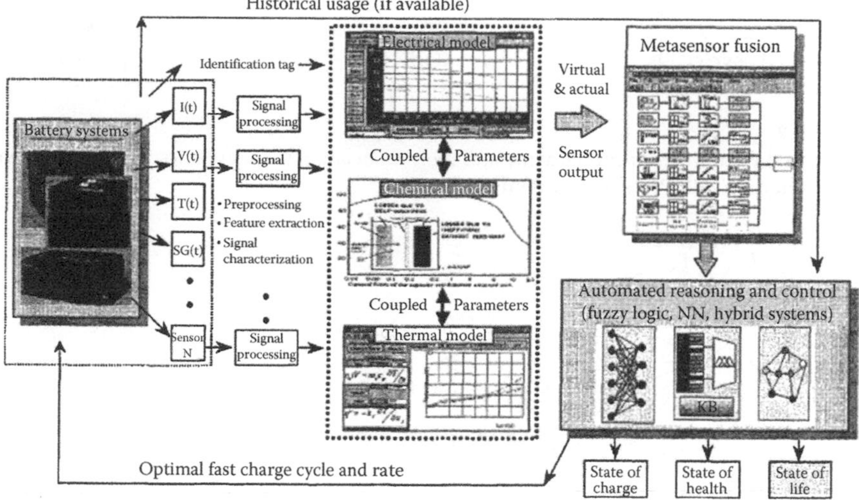

FIGURE 6.7
Instrumentation scheme of battery prognostic monitoring. (Reprinted from IEEE, copyright © IEEE. With permission.)

a lowest threshold level. Another term "state-of-health" (SOH) can also be used to describe the physical condition like the loss of rate capacity or corrosion. The remaining life of the battery is termed as the "state-of-life" (SOL).

The electrical model of a battery can be described by a mathematical model that originates from the Randle circuit shown in Figure 6.6.

6.2.2.6.2 *The Battery Prognostic Instrumentation*

The prognostic instrumentation for battery SOC, SOH, and SOL monitoring is shown in Figure 6.7. The approach involves six stages: sensors (physical and virtual sensors), array system, feature extraction, mapping, fault estimation, and prognostic decisions. Figure 6.8 shows the physical and virtual sensor inputs array and their association to diagnostics [6].

6.2.2.6.3 *ANN-Based Prognostics for Battery SOC*

Lithium batteries of sizes C and 2/3 A under different loading conditions were used for training the ANN. A multilayer radial basis function (RBF)

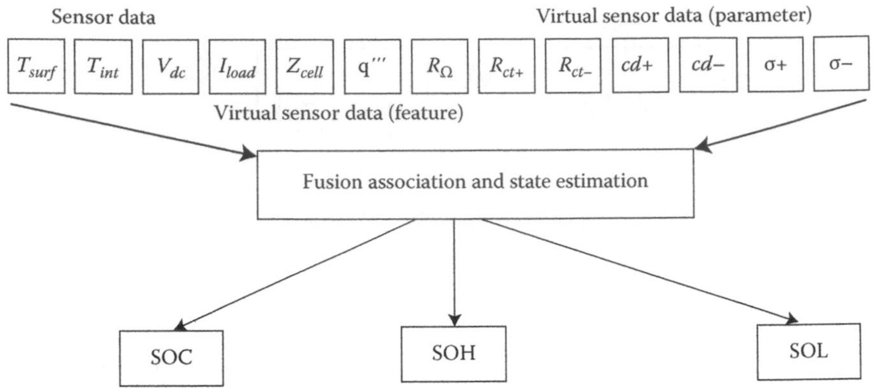

FIGURE 6.8
Physical and virtual sensor inputs array and their association to diagnostics. (Reprinted from Taylor & Francis, CRC Press. With permission.)

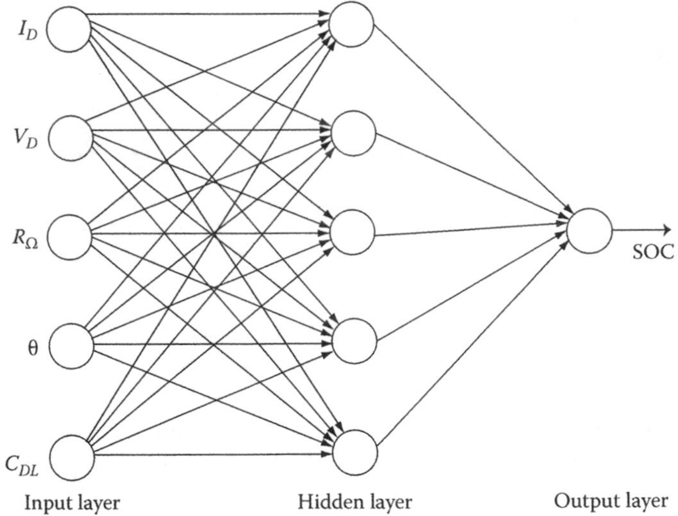

FIGURE 6.9
ANN structure of the SOC prediction of battery.

for ANN structure was used with gradient descent learning algorithm, as shown in Figure 6.9. The inputs to the ANN were subsets of I_D, V_D, R_Ω, θ, and C_{DL}, and the output was SOC [7].

where
- I_D is the discharge current
- V_D is the discharge voltage
- R_Ω is the electrolyte resistance
- θ is the charge transfer resistance
- C_{DL} is the double-layer capacitance

The ANN was structured with 6 and 12 hidden neurons and training errors of 0.6% and 0.8%. The SOC of the battery was evaluated based on the following definition [7]:

1. SOC—The amount of useful capacity remaining in a battery during a discharge.
2. 100% SOC—Full useful capacity. This is the capacity of a battery at the instant a load is applied.
3. 0% SOC—No useful capacity remaining. This is the capacity of a battery when it reaches its end-of-discharge voltage (EDV), which is usually defined by the manufacturer.

The SOC is mathematically represented by

$$\mathrm{SOC}(t) = \frac{C_{total} - \int_0^t I\,dt}{C_{total}} \tag{6.2}$$

where $C_{total} = \int_0^{t_{EDV}} I\,dt$.

The ANN-based SOC estimators were trained with the battery data and impedance parameters were extracted to predict SOC directly. Different training methods can be used such as back propagation of errors with or without momentum, back propagation of errors with or without learning rate adaptation, and back propagation of errors with the Levenberg–Marquardt (LM) method.

6.2.3 ANN-Based Intelligent Sensors

Uncertainty and nonlinearity in sensor dynamic behavior is a major problem in designing signal processing for those sensors. In many sensors, an exact mathematical model showing the relationship between the measurand and the response and its dependency on the environmental parameters is not available. On the other hand, increasing computational complexity in signal processing also makes it difficult to design the signal processing module. Basically, the above two reasons lead to the applications of ANN based sensors in intelligent signal processing. The ANN-based sensors can be classified into the following categories:

1. Linearization and calibration
2. Compensation of errors
3. Soft sensing
4. Sensor failure detection
5. Pattern classification

6.2.3.1 Linearization and Calibration by ANN

Although linear input–output relationship of a sensor is desirable, in practice, most sensors do not show a linear relationship. The linearization of such sensors are carried out by additional hardware circuits or mathematical linearization algorithms implemented in microcontrollers and computers.

Classical software linearization techniques are based on the use of accurate sensor models. The models are used in circuit or software to improve the linearity. Such model-based techniques have many disadvantages—linearization is not general, modeling may not be accurate, and hardware implementation is difficult.

ANNs are alternative approaches for modeling the sensor characteristics when model-based approach is not possible. The basic principle in ANN-based linearization is that an ANN is trained with a set of sensor generated nonlinear output data as input to the network and the corresponding estimated linear output as target output of the network. In Section 5.9, an example of network training data for a nonlinear equation of the type $v = ax + bx^2$ was illustrated for a particular input–output range of data.

Let us consider another nonlinear equation

$$v = 0.2855x + 0.0033x^2$$

while the linear equation is $v = 0.1x$. Taking a data set of x ranging from 0 to 25, the nonlinear and linear outputs v_{nl} and v_l, respectively, are generated.

The two sets of data are used to train a two-layered feed-forward back propagation ANN with an error goal of 0.01. After training the ANN, the whole set of nonlinear data (v_{nl}) can be used to estimate linear outputs $\hat{v}_l$. The variables-nonlinear input (v_{nl}) and linear target (v_l) are shown in Figure 6.10.

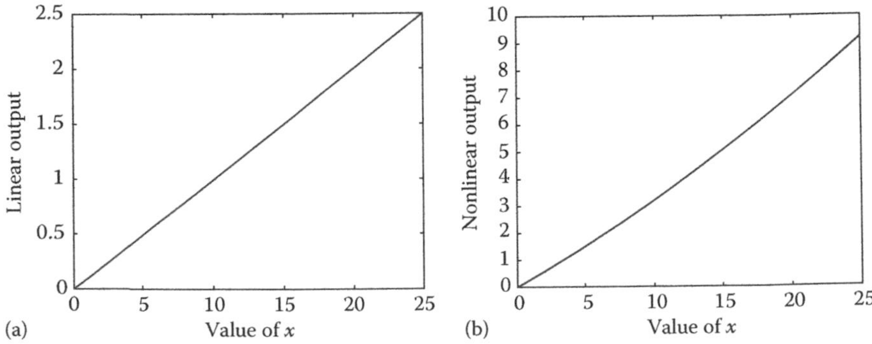

FIGURE 6.10
Plots of ANN inputs and outputs.

Example 6.2

A thermistor with $R_0 = 10\,\text{k}\Omega$ at 25°C and $\beta = 3965\,\text{K}$ gives a nonlinear output given by the equation

$$R_T = R_0 e^{\beta\left(\frac{1}{T} - \frac{1}{T_0}\right)}$$

The thermistor should be linearized in the temperature range of 0°C–100°C by a hardware circuit that produces R_T by the linear equation

$$R_{T,l} = R_0(1 - \alpha T)$$

where α is a linear constant.

Write a MATLAB® program to linearize the thermistor resistance.

Solution

Given: $\beta = 3965\,\text{K}$,
$R_0 = 10\,\text{k}\Omega$ at 25°C
The thermistor resistance at 0°C is given by

$$R_{T,0} = 10e^{3965\left(\frac{1}{273} - \frac{1}{298}\right)}$$
$$= 33.8\,\text{k}\Omega$$

The thermistor resistance at 100°C is given by

$$R_{T,100} = 10e^{3965\left(\frac{1}{373} - \frac{1}{298}\right)}$$
$$= 0.68\,\text{k}\Omega$$

Putting $R_{T,100}$ in the linear equation

$$R_{T,l} = R_0(1 - \alpha \times 100)$$

$$33.8(1 - \alpha \times 100) = 0.68$$

$$\therefore \alpha = 0.0097$$

Now we have two equations:

Linear:

$$R_{T,l} = 33.8(1 - 0.0097T), \quad T \text{ in °C}$$

Nonlinear:

$$R_{T,nl} = 10e^{3965\left(\frac{1}{T} - \frac{1}{298}\right)}, \quad T \text{ in K}$$

Now we generate two data sets for the temperature range 0°C–100°C:

$$R_{T,l} = \begin{bmatrix} 33.8 & \cdots & 0.68 \end{bmatrix}$$

$$R_{T,nl} = \begin{bmatrix} 33.8 & \cdots & 0.68 \end{bmatrix}$$

These two data sets each of 101 data are shown below:

$R_{T,l} =$

[33.8000 33.4688 33.1375 32.8063 32.4750 32.1438 31.8126 31.4813
31.1501 30.8188 30.4876 30.1564 29.8251 29.4939 29.1626 28.8314
28.5002 28.1689 27.8377 27.5064 27.1752 26.8440 26.5127 26.1815
25.8502 25.5190 25.1878 24.8565 24.5253 24.1940 23.8628 23.5316
23.2003 22.8691 22.5378 22.2066 21.8754 21.5441 21.2129 20.8816
20.5504 20.2192 19.8879 19.5567 19.2254 18.8942 18.5630 18.2317
17.9005 17.5692 17.2380 16.9068 16.5755 16.2443 15.9130 15.5818
15.2506 14.9193 14.5881 14.2568 13.9256 13.5944 13.2631 12.9319
12.6006 12.2694 11.9382 11.6069 11.2757 10.9444 10.6132 10.2820
9.9507 9.6195 9.2882 8.9570 8.6258 8.2945 7.9633 7.6320
7.3008 6.9696 6.6383 6.3071 5.9758 5.6446 5.3134 4.9821
4.6509 4.3196 3.9884 3.6572 3.3259 2.9947 2.6634 2.3322
2.0010 1.6697 1.3385 1.0072 0.6760]

$R_{T,nl} =$

[33.8191 32.0731 30.4291 28.8803 27.4206 26.0445 24.7465 23.5219
22.3659 21.2743 20.2432 19.2687 18.3476 17.4764 16.6523 15.8723
15.1339 14.4346 13.7721 13.1442 12.5490 11.9845 11.4489 10.9407
10.4582 10.0000 9.5648 9.1512 8.7581 8.3843 8.0287 7.6904
7.3685 7.0620 6.7702 6.4921 6.2272 5.9747 5.7340 5.5044
5.2854 5.0764 4.8769 4.6865 4.5046 4.3309 4.1649 4.0062
3.8545 3.7094 3.5707 3.4379 3.3109 3.1893 3.0728 2.9613
2.8545 2.7521 2.6540 2.5599 2.4698 2.3833 2.3003 2.2207
2.1443 2.0709 2.0005 1.9328 1.8678 1.8054 1.7454 1.6877
1.6323 1.5789 1.5276 1.4783 1.4308 1.3851 1.3411 1.2988
1.2580 1.2187 1.1808 1.1444 1.1092 1.0753 1.0427 1.0112
0.9808 0.9515 0.9232 0.8959 0.8696 0.8441 0.8196 0.7959
0.7730 0.7508 0.7295 0.7088 0.6888]

With these two data sets, we can simulate the ANN using MATLAB.

Figure E6.2a shows the plots of the thermistor resistances: $R_{T,nl}$, $R_{T,l}$, and the estimated linear resistance $\hat{R}_{T,l}$ in Figure E6.2b.

6.2.3.2 Compensation of Error by ANN

ANNs have potential application for mitigating adverse effects of environmental parameters on the response of sensors. Such applications of ANNs are advantageous when the effects of environmental parameters are difficult to estimate using signal processing. ANN-based compensation schemes are based on implementing an inverse model of the sensor in ANN, as shown in Figure 6.11.

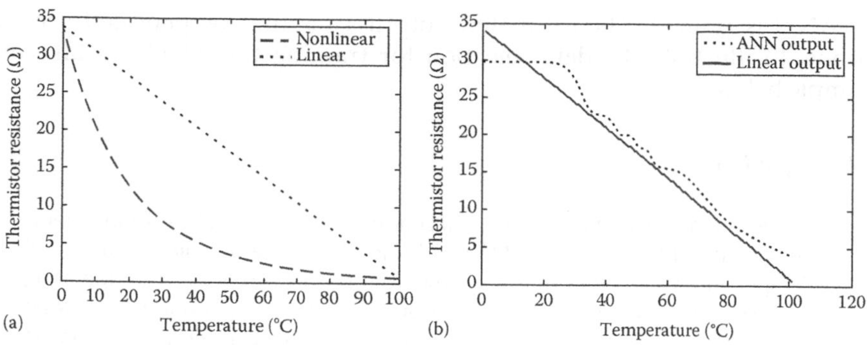

FIGURE E6.2
(a) Plots of linear and nonlinear thermistor resistances and (b) plots of estimated thermistor resistance from ANN output.

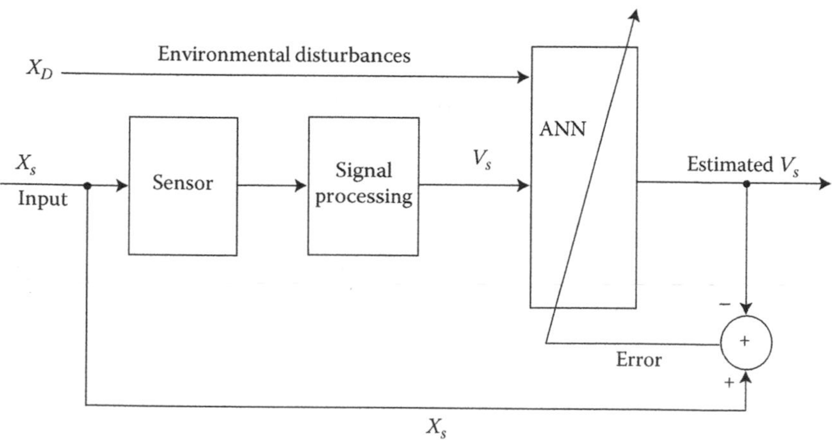

FIGURE 6.11
ANN in training phase of compensation.

As shown in the figure, the ANN is trained with both parts of the input data given by

$$I = [V_S : X_D]$$

where

V_S is the sensor output data
X_D is the environmental disturbance data

Based on the error (e) generated by the equation

$$e = \hat{V}_S - V_S$$

the ANN is trained. The estimated output $\hat{V}_S$ is the compensated sensor output. To illustrate the data selection for training the ANN, we take an example below:

Example 6.3

A metallic strain gauge of resistance 120Ω and gauge factor 2.0, resistance temperature coefficient (α) of 5×10^{-3} Ω/Ω/°C bonded to a beam, which is subjected to a temperature variation from room temperature (25°C) to 50°C. The operating range of gauge is 800 μ strain/s (max). The supply voltage applied to Wheatstone bridge ($R_2 = R_3 = R_4 = 120\,\Omega$) is 12 V. The temperature effect of the gauge is to be compensated by ANN technique. Design the ANN scheme.

Solution

First we find the temperature in output voltage per 1°C per 1 μ strain.

Given gauge parameters:

$$\lambda = 2.0,$$

$$R = 120\ \Omega,$$

$$\alpha = 5 \times 10^{-3}\ \Omega/\Omega/°\text{C},$$

$$V_i = 12\ \text{V}$$

Change in resistance due to α

$$\Delta R_\alpha = R\alpha\Delta T$$

Output voltage due to α

$$V_{0\alpha} = \frac{\alpha \Delta T V_i}{4}$$

Output voltage due to strain (ε)

$$V_{0\varepsilon} = \frac{\lambda \varepsilon V_i}{4}$$

Combined output

$$V_{0\alpha\varepsilon} = \frac{\lambda \varepsilon V_i}{4(1 + \alpha \Delta T)}$$

First, we generate a dataset of V_s at the reference temperature (25°C) for strain values from 0 to 800 μ at an interval or 200 μ (in practical situation, the data acquisition will generate a large number of data during this change of strain). Therefore, at 25°C (means $\Delta T = 0$), the output voltages are

For $\varepsilon = 0$, $V_s = 0$

For $\varepsilon = 100\,\mu$

$$V_{0\varepsilon} = \frac{\lambda \varepsilon V_i}{4} = \frac{1}{4} \times 2.0 \times 100 \times 10^{-6} \times 12 = 0.6\ \text{mV}$$

TABLE E6.3A

Output Voltage (mV) at Different Temperatures

	Output Voltages (mV) at Different Temperatures				
Strain × 10^{-6}	**30°C**	**35°C**	**40°C**	**45°C**	**50°C**
200	1.170	1.142	1.116	1.090	1.066
400	2.340	2.285	2.232	2.180	2.132
600	3.510	3.426	3.348	3.270	3.198
800	4.68	4.568	4.464	4.360	4.264

Similarly, for

$$\varepsilon = 200\,\mu, V_s = 1.2\,\text{mV}$$

$$\varepsilon = 400\,\mu, V_s = 2.4\,\text{mV}$$

$$\varepsilon = 600\,\mu, V_s = 3.6\,\text{mV}$$

$$\varepsilon = 800\,\mu, V_s = 4.8\,\text{mV}$$

So, these four pairs of data (ε_N and V_{ON}) contribute one data set at reference temperature.

Now, we generate the response characteristic of the strain gauge for different ambient temperatures (25°C–50°C at an interval of 5°C), i.e., 30°C, 35°C, 40°C, 45°C, and 50°C using equation $V_0 \alpha \varepsilon$. (In actual practice, the data acquisition will generate a large amount of data during this temperature interval.) This will generate five pairs of data for each strain value, totaling, $5 \times 4 = 20$ data sets. These 20 data sets are shown in Table E6.3a.

The thermistor response characteristics is shown in Figure E6.3a.

It is observed in the plot that there is a wide variation in the sensor response as the ambient temperature changes from 25°C to 50°C. This variation can be applied to the ideal output voltage estimation using ANN. We use here a 2-4-1 MLP architecture, as shown in Figure E6.3b. The input and output matrices for the training of the ANN are shown in Table E6.3b and ANN training error is shown in Figure E6.3c.

The sensor data can be normalized by the equation

$$x_{norm} = \frac{x(t)}{x_{max}}$$

where

$x(t)$ is the sensor data

x_{max} is the maximum data in the dataset

The temperature and voltage data are normalized in the scale 0–1 and the normalized data are shown in Table E6.3b. Similarly, the input and target matrix format is shown below:

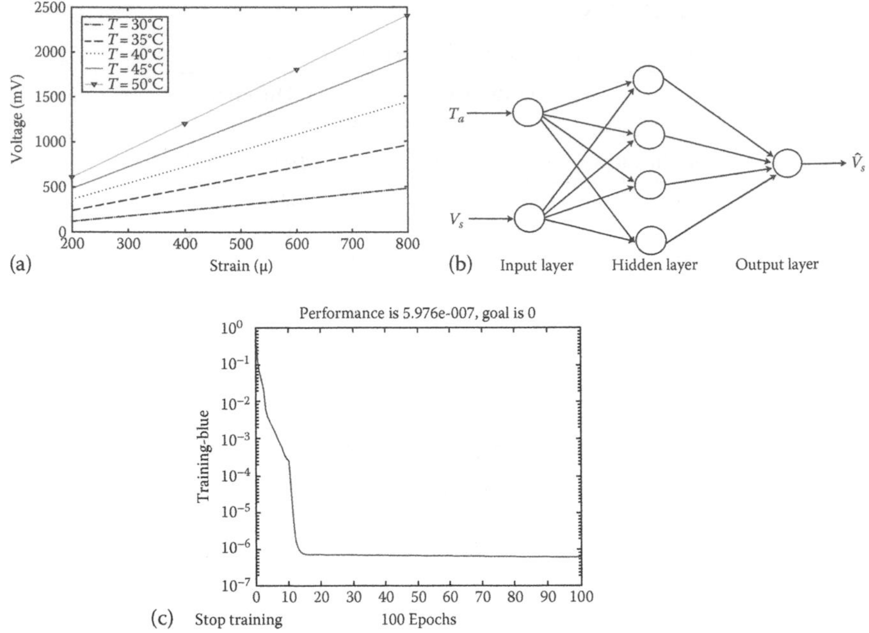

FIGURE E6.3
(a) Response characteristics of the strain gauge at different temperatures, (b) 2-4-1 MLP architecture, and (c) ANN training error.

$$I = \begin{bmatrix} T_a & V_s \end{bmatrix}$$

$$I = \begin{bmatrix} T_{a1}V_{s1} : T_{a2}V_{s2} & \cdots & T_{aN}V_{sN} \end{bmatrix}$$

$$\hat{V}_s = \begin{bmatrix} \hat{V}_{s1}\,\hat{V}_{s2} & \cdots & \hat{V}_{sN} \end{bmatrix}$$

where
T_a is the ambient temperature
V_s is the sensor output voltage

The ANN was tested with a set of test data and the estimated output (temperature compensated) was found as shown in the Table E6.3c.

The first three sets of input data are new data, the outputs of which approximately match with the data of Table E6.3b. The fourth data is taken from the input data set and the output exactly matches the trained data.

6.2.3.3 Soft Sensing by ANN

In Section 4.4.2.2, ANN-based soft sensor design approach for implementing a "dryer status" measurement though Case Study 4.1 was discussed. In this approach, the dryer status (revealing any one of the three drying statuses—underdrying, normal drying, and overdrying) was measured by modeling an ANN experimental measurement data of the dryer (inlet air temperature,

TABLE E6.3B

Normalized ANN Training Data

Temperature	Input Voltage (V_s)				Output Voltage ($\hat{V}_s$)			
0.6	0.0243	0.0486	0.729	0.972	0.25	0.5	0.75	1.0
0.7	0.238	0.476	0.714	0.952	0.25	0.5	0.75	1.0
0.8	0.232	0.464	0.696	0.928	0.25	0.5	0.75	1.0
0.9	0.227	0.454	0.681	0.908	0.25	0.5	0.75	1.0
1.0	0.222	0.444	0.666	0.888	0.25	0.5	0.75	1.0

TABLE E6.3C

Test and Compensated Data (Normalized)

	Temperature		Sensor Measurement Voltage		Compensated Voltage	
Sl No.	Normalized	°C	Normalized	mV	Normalized	mV
1	0.65	32.5	0.72	3.456	0.74801	3.59
2	0.75	37.5	0.25	1.2	0.27989	1.34
3	0.95	47.5	0.92	4.41	0.99988	4.79
4	0.9	45	0.227	1.08	0.24978	1.19

tea feed rate, and exhaust air temperature) and knowledge-based data (drying status). These two data sets for various patterns were used to train the ANN. The trained ANN could estimate the dryer status using the three measurement data.

In this section, the following example is used to explain the basic knowledge about implementing a soft sensor using ANN-based modeling.

Example 6.4

An RLC circuit shown in Figure E6.4a should be simulated as a second-order system to generate response for a step input. The response is to be used to estimate the capacitance (C) of the circuit by training by an ANN. Use rise time of the response as the feature to–train the ANN, the output of which is the estimated capacitance. Use MATLAB® to simulate the system.

Solution

Here, we have to match the feature of the step response of the circuit to the capacitance. Let us consider that the value of R and L are fixed at

$$R = 1000\,\Omega$$

$$L = 0.1\text{H}$$

We vary the capacitance from 0.01 to 0.1 μF and generate the step response. From the response data, we determine the rise time for all values of C. Figure E6.4b shows the plot of the circuit response, Figure E6.4c shows the

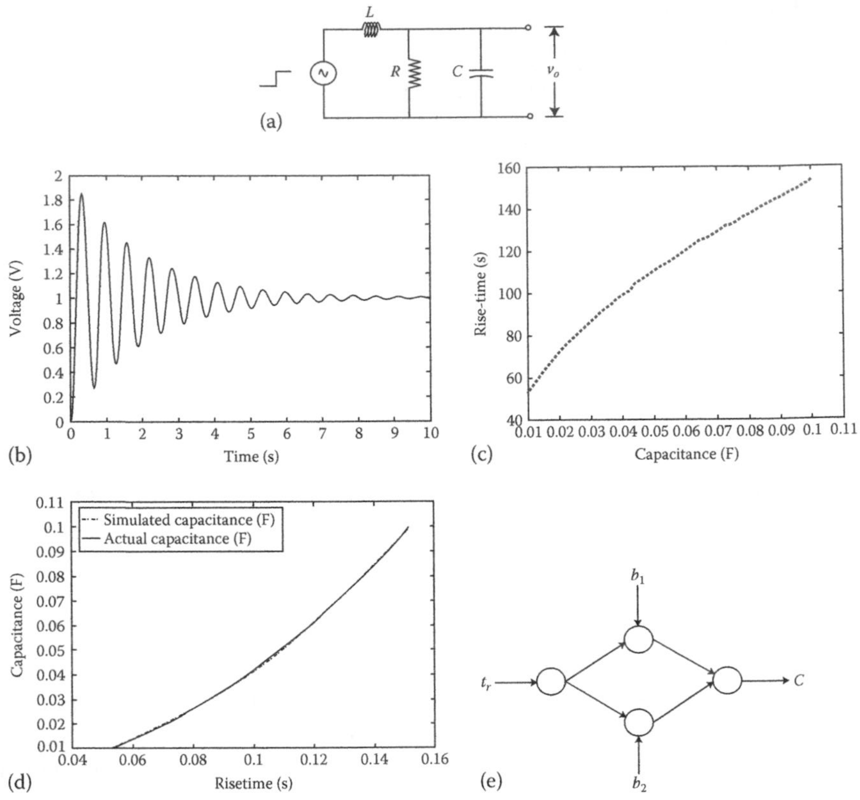

FIGURE E6.4
(a) An RLC circuit, (b) plot of step response of the circuit, (c) plot of the rise time, (d) plot of actual and simulated capacitances, and (e) 1-2-1 layer MLP structure of the ANN.

plot of the rise time for different values of C, and Figure E6.4d shows the actual and simulated capacitances. We use a 1-2-1 MLP architecture, as shown in Figure E6.4e.

The transfer function of the circuit is given by

$$G(s) = \frac{(1/LC)}{s^2 + s(1/RC) + (1/LC)}$$

The transfer function is simulated in MATLAB and the step response is generated. The MATLAB code for the simulation is shown below:

```
% This MATLAB program simulates a RC circuit
% as a first order system and its step response is
% determined for time duration of 0-10sec.
% The capacitance values are varied from 0.01F
% to 0.1F and
% the rise time(tr) of the responses for all the
% capacitance values are determined and plotted.
```

```
clc;
close all;
clear all;
R=10;
L=0.1;
C=0.01:.001:0.1;
csize=size(C);
for i=1:1:csize(2);
   C1=C(i);
   Num=1./(L.*C1);
   Den=[1 1./(R.*C1) 1./(L.*C1)];
   sys=tf(Num,Den);
   t=0:.001:10;
   tsize=size(t);
   y=step(sys,t);
   for j =1:1:tsize(2)
      if (y(j) >= 0.9)
         rise(i)= j;
         break;
      end
   end
end
   plot(C,rise,'b.');
   xlabel('Capacitance, F');
   ylabel('Rise-time, Sec');
   figure(2);
   plot(y);
   xlabel('Time, Sec');
   ylabel('Voltage,Volt');
```

The rise time (t_r) and capacitance (C) data generated from the simulation were used to train an ANN in MATLAB. After training, the train dataset of t_r was used to test the ANN.

6.2.3.4 Fault Detection by ANN

Detection and isolation of faults and failures in plants and measurement systems is of paramount importance due to technology- and safety-related factors. Fault detection has been addressed by various researchers by different techniques, which can be basically classified as

1. Based on parity relations and observers
2. Gross error detection based on first principle methods
3. Statistical correlation–based methods such as PCA, PLS, etc.
4. Intelligent systems such as ANN, Fuzzy logic, and GA

This section describes ANN-based techniques for system and sensor fault detection. Although ANN-based techniques widely vary on a case-to-case basis, they basically fall under two categories:

1. Comparison of features with fault model
2. Detection of measurement consistency

6.2.3.4.1 Comparison of Features with Fault Model

When we use a simple feed forward neural network for comparing the input features of plant variables, the output of the network becomes some features of the fault characteristics. In the simplest form, the input features may be mean, rms value, variance, central moments, etc., while the output features of the network can be thought of a single node with binary levels of "normal" and "faulty." For a specific fault-detection problem, the training data consists of N patterns of input data x^p, $p=1, 2, \ldots, N$ and N patterns of output data y^p, $p=1, 2, \ldots, N$

where

$$y^p = f(W, x^p), \quad p = 1, 2, \ldots, N$$

where W is the weight vector, which is updated by minimizing a cost function

$$E = \frac{1}{2}\sum\left(\hat{y}^p - y^p\right)^2$$

where

$\hat{y}^p$ is the estimated fault characteristic

y^p is the actual fault characteristic

Such a network is shown in Figure 6.12.

The input and the output matrix of the network take the following form:

$$X^p = \begin{bmatrix} f_1^1 & f_1^2 & \cdots & f_1^N \\ f_2^1 & f_2^2 & \cdots & f_2^N \\ f_3^1 & f_3^2 & \cdots & f_3^N \\ \vdots & \vdots & \cdots & \vdots \\ f_k^1 & f_k^2 & \cdots & f_k^N \end{bmatrix} \tag{6.3}$$

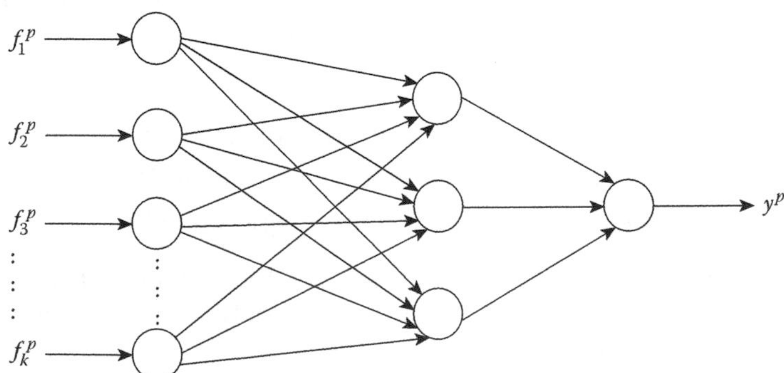

FIGURE 6.12
ANN structure for input feature comparison.

$$Y^p = \begin{bmatrix} b & b^2 & \cdots & b^N \end{bmatrix} \tag{6.4}$$

where b is a binary value ("0" or "1").

Example 6.5

A measurement system comprises of an RLC circuit, as shown in Figure E6.4a of Example 6.4. The circuit is excited by transient signals to get the output voltage V_0. The capacitive component of the circuit is prone to fault due to absorption of moisture from the surrounding medium. In normal sensing condition, the capacitance of the system ranges from 0.01 to 0.05 F, and when the component absorbs moisture, the capacitance suddenly increases beyond 0.05 F. The two respective equations of moisture content versus capacitance are

$$C = 0.001m \quad \text{and} \quad C = m^4 \times 10^{-8}$$

In order to detect fault in the system due to abnormal increase in capacitance, the transient response of the system (step and impulse) are continuously monitored and an ANN is used to detect the fault.

1. Write a MATLAB program to simulate the system and generate two features
2. Design the input–output structure of the ANN
3. Train the ANN and test the validity

Solution

The transfer function of the circuit is given by

$$G(s) = \frac{(1/LC)}{s^2 + s(1/RC) + (1/LC)}$$

Where we take say $R = 10\,\Omega$ and $L = 0.1$ H and vary capacitance from 0.01 to 0.05 F for moisture content of 10%–50% and 51%–100%. For comparing the characteristics of the system with "faulty" and "normal," we need some features of the response.

Two features we can consider are rise time (t_r) in step response case, while maximum overshoot (y_0) in impulse response case. When these two features are generated and plotted against capacitance, it is found that rise time increases with capacitance while maximum overshoot decreases. Since the boundary between "normal" and "fault" is determined by the capacitance value of 0.05 F, we divide the feature vectors into two parts:

$$[f_1, f_2]_1 \quad \text{for } 0\% \le m \le 50\%$$

$$[f_1, f_2]_2 \quad \text{for } 50\% \le m \le 100\%$$

The MATLAB program for this simulation is shown below:

```
% This program simulates a second order RLC circuit to
% generate
% Step and Impulse response for a duration of 10sec when
% capacitance
% values are varied from 0.01 farad to 1 farad due to
% variation of moisture
% from 10% to 100%. The variation of risetime (in step
% response)
% and maximum-overshoot (in impulse response)
% for different values of capacitances are generated and
plotted.
close all;
clear all;
R=10;
L=0.1;
% Vary capacitance linearly with moisture from 10% to 50%
m=10:1:50;
c1=0.001*m;
% Vary capacitance drastically with moisture above 50%
m=51:1:100
c2=(m.^4)*10^(-8);
C=[c1,c2];
csize=size(C);
for i=1:1:csize(2);
  C1 =C(i);
  Num=1./(L.*C1);
  Den=[1 1./(R.*C1) 1./(L.*C1)];
  sys=tf(Num,Den);
  t=0:.001:5;
  tsize=size(t);
  y=impulse(sys,t);
  for j =1:1:tsize(2)
    maxover(i)=max(y);
      rise(i)= t(j);
    break;
  end
end
  for i=1:1:csize(2);
  C1 =C(i);
  Num=1./(L.*C1);
  Den=[1 1./(R.*C1) 1./(L.*C1)];
  sys=tf(Num,Den);
  t=0:0.001:5;
  tsize=size(t);
  y=step(sys,t);
  for j =1:1:tsize(2)
    if (y(j) >= 0.9)
        rise(i)= t(j);
```

```
      break;
    end
  end
end
  plot(C,rise,'b.');
  xlabel('Capacitance, F');
  ylabel('Rise-time, Sec');
  figure(2);
  plot(t,y);
  xlabel('Time, Sec');
  ylabel('Voltage(Step Response),Volt');
  csvwrite('cap.dat',C)
  figure(3)
  plot(C,maxover,'b.');
  xlabel('Capacitance, F');
  ylabel('Maximum overshoot, Volt');
  figure(4);
  plot(t,y);
  xlabel('Time, Sec');
  ylabel('Voltage(impulse response),Volt');
  csvwrite('cap.dat',C)
  input=[rise; maxover];
```

The input matrix can be represented by

$$X = \begin{bmatrix} t_r^1 & t_r^2 & \cdots & t_r^N \\ y_0^1 & y_0^2 & \cdots & y_0^N \end{bmatrix}$$

where superscript N denotes number of data points.

The matrix is divided into two parts:

$$X = [X_n : X_f]$$

where

subscript n denotes "normal"

f denotes "faulty"

Figure E6.5a shows variation of capacitance with moisture content, Figure E6.5b shows the rise time variation with C, and Figure E6.5c shows maximum overshoot variation with C.

The output of the ANN can be denoted by

$$Y = \begin{bmatrix} b^1 b^2 & \cdots & b^N \end{bmatrix}$$

where b is a binary bit, and the bit value is assigned as: normal–1 and faulty–0. The matrix is divided into two vectors—normal and faulty—as

$$Y = [Y_n : Y_f]$$

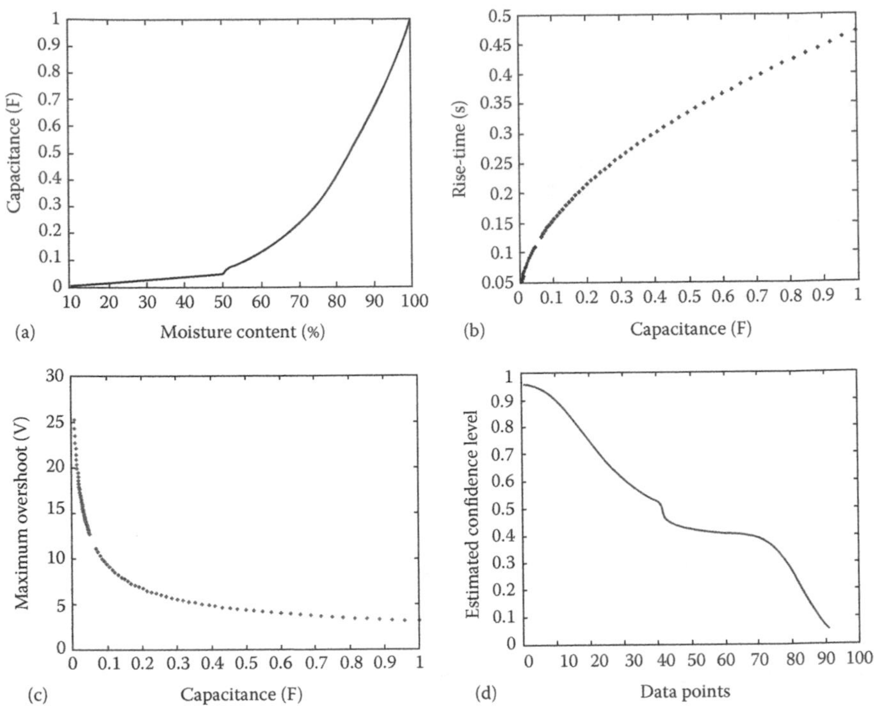

FIGURE E6.5
(a) Plot for variation of capacitance with moisture content, (b) plot for variation of rise-time with capacitance, (c) plot for variation of maximum overshoot with capacitance, and (d) estimated confidence level of fault.

The ANN is trained with data X as input and Y as target. The trained ANN was tested with input X and the estimated output was determined as confidence level of fault (Figure E6.5d).

6.2.3.4.2 Detection of Measurement Consistency

In a multivariable plant, there are some variables that are correlated to one another while some are not correlated. This correlation arises from the transfer function relationship of the plant. When a measurement system includes such correlated variables, the fault detection of sensors can be performed by measuring such variables based on checking the consistency of measurement. Figure 6.13a shows an example of sensor correlation or influence. The sensor influence map (Figure 6.13b) shows that I_2 has influence on both O_1 and O_2, while I_1 and I_3 have influence on O_1 and O_2, only respectively. This sensor influence is first considered to select the set of sensor measurements for which consistency can be detected. In the above example, the sets of sensors that can be considered for fault detection are (I_1, O_1), (I_2, O_1, O_2), and (I_3, O_2).

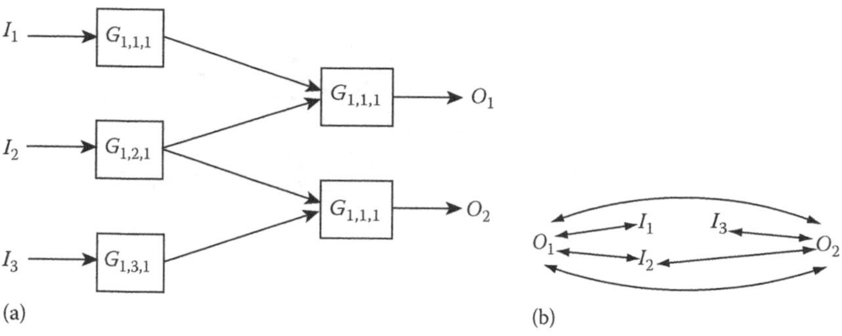

FIGURE 6.13
Variable correlation (a) gain correlation and (b) sensor influence map.

Let us consider any one of the sensor set, say, (I_1, O_1), is used for fault detection of I_1. An ANN is used to map the I_1 and O_1 sensor data sets to estimate a confidence level set C_{I_1} and C_{O_1}. The input to the ANN is given by

$$X = \begin{bmatrix} I_1^1 & I_1^2 & \cdots & I_1^N \\ O_1^1 & O_1^2 & \cdots & O_1^N \end{bmatrix}$$

The corresponding ANN output can be written as

$$O = \begin{bmatrix} C_{I_1}^1 & C_{I_1}^2 & \cdots & C_{I_1}^N \\ C_{O_1}^1 & C_{O_1}^2 & \cdots & C_{O_1}^N \end{bmatrix}$$

where superscript N is the number of data points.

The confidence level is a normalized value ranging from 0 to 1, indicating lowest to highest sensor measurement data confidence. The ANN structures are shown in Figure 6.14. In Figure 6.15, a model of an RC circuit is shown where four variables—frequency (f) of the input voltage, input voltage (V_i), voltage across resistor $R(V_R)$, and voltage across capacitance C (V_C)— are measured by four different sensors. The sensor interface sets are (f, V_i, V_C), (V_i, V_R, V_C), (V_R, V_C), (V_i, V_R), and (V_i, V_C). For simplicity, let us take a set (V_R, V_C) for the corresponding sensor fault detection. To generate the (V_R and V_C) data of the circuit for fault detection, the capacitor fault due to moisture (Example 6.5) is considered. When the moisture absorbed by the capacitor dielectric material changes from 10% to 50%, the capacitance changes linearly with the equation

$$C = 0.001m$$

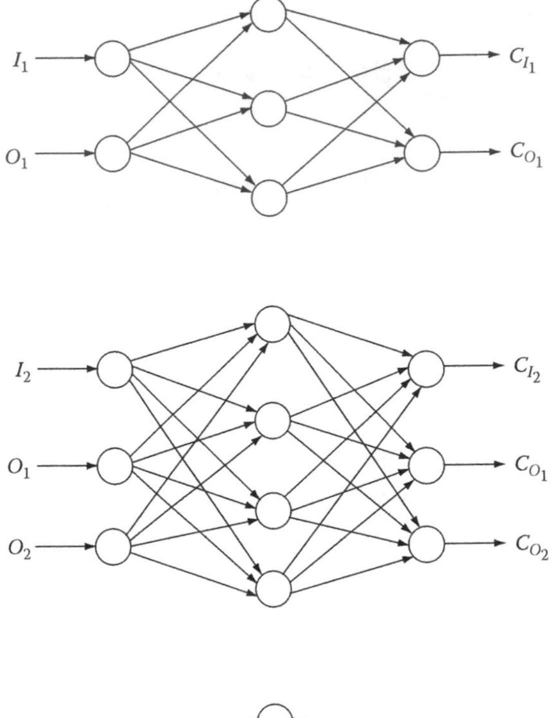

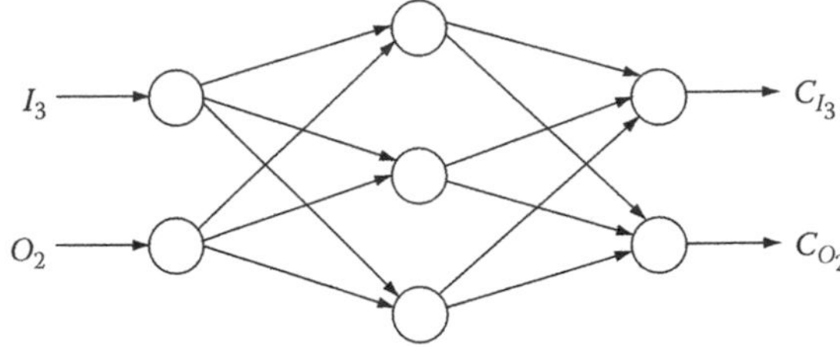

FIGURE 6.14
ANN structure for the three sensor sets.

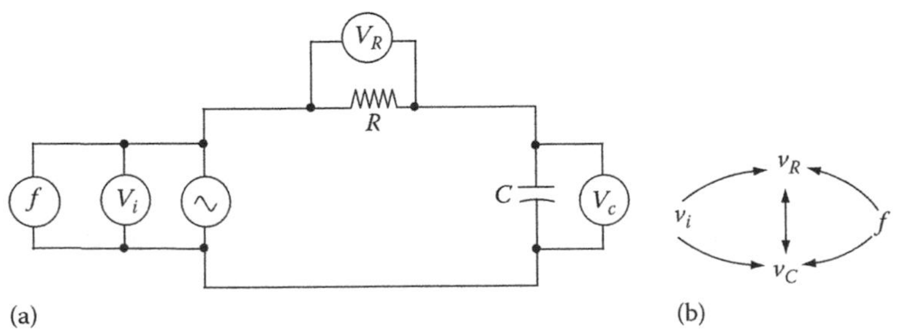

FIGURE 6.15
Sensor correlation in a measurement circuit (a) model of an RC circuit and (b) sensor influence model.

whereas the capacitance abruptly changes with the equation

$$C = m^4 \times 10^{-8}$$

for moisture content of 51%–100%. To train an ANN, the input data is formed with V_R and V_C as

$$X = [V_R : V_C]$$

$$= \begin{bmatrix} V_R^1 & V_R^2 & \cdots & V_R^N \\ V_C^1 & V_C^2 & \cdots & V_C^N \end{bmatrix}$$

The capacitor fault is considered for the moisture absorption by dielectric material and each data of the whole data range is assigned with a fault confidence index in the scale 0–1. A "0" confidence level indicates a faulty system whereas a "1" indicates a "normal" system. Therefore, each data pair of V_R : V_C are assigned a confidence level $CV_R : CV_C$. This confidence matrix is used as the target of the ANN for the fault model. The output can be written as

$$O = [CV_R : CV_C]$$

$$= \begin{bmatrix} CV_R^1 & CV_R^2 & \cdots & CV_R^N \\ CV_C^1 & CV_C^2 & \cdots & CV_C^N \end{bmatrix}$$

If the number of data points is N, the $CV_R : CV_C$ data ranges from 1 to $1/N$. The variation of the capacitance with moisture content and the corresponding confidence levels are illustrated in the following example.

Example 6.6

The capacitance variation due to moisture content (m) variation is described by the following equation:

$$C = 10^{-3}m \quad \text{for } 10\% \le m \le 50\%$$

$$C = 10^{-8}m^4 \quad \text{for } 50\% < m \le 100\%$$

Simulate the circuit by programming in MATLAB to generate $[V_R : V_C]$ and $[CV_R : CV_C]$ data with $R = 1\,\text{k}\Omega$, $f = 10\,\text{kHz}$. The confidence level of $CV_R : CV_C$ changes such that

$$C = 1 \text{ to } \frac{n}{N} \quad \text{for } 10\% \le m \le 50\%$$

$$C = \frac{n+1}{N} \text{ to } \frac{1}{N} \quad \text{for } 50\% < m \le 100\%$$

Take $V_i = 5\,\text{V}$.

Simulate an ANN with $V_R : V_C$ as input and $CV_R : CV_C$ as target.

Solution

From circuit analysis for circuit of Figure 6.15a, we can write the equation for the voltage across the resistor as

$$V_R = V_i \frac{R}{\sqrt{R^2 + X_C^2}}$$
$$= V_i \frac{R}{\sqrt{R^2 + \left(\frac{1}{2\pi fC}\right)^2}}$$

and the voltage across the capacitor as

$$V_C = V_i \frac{X_C}{\sqrt{R^2 + X_C^2}}$$
$$= V_i \frac{X_C}{\sqrt{R^2 + \left(\frac{1}{2\pi fC}\right)^2}}$$

Generation of V_R and V_C: Two segments of V_R and V_C are generated by changing m from 10% to 50% and 51% to 100%. The corresponding changes in capacitance (using the two capacitance equations) are 0.01–0.05 F for first segment and 0.067–1 F for second segment. With the variation of capacitances, we calculate the values of V_R and V_C.

Number of Data Points: If we sample V_R and V_C such that m changes in steps of 1%, the number of data points are

m from 10% to 50% in steps of 1%: 41 data points
m from 51% to 100% in steps of 1%: 50 data points

So, total number of data points are: $41 + 50 = 91$.
Therefore, $n = 41$ and $N = 91$.

Hence, 41 sets for first segments and 50 sets for second segments are generated for ($V_R : V_C$). The $V_R : V_C$ matrix can be shown as

$$X = \begin{bmatrix} V_R \\ V_C \end{bmatrix} = \begin{bmatrix} V_R^1 & V_R^2 & \cdots & V_R^{41} \vdots V_R^{42} & V_R^{43} & \cdots & V_R^{91} \\ V_C^1 & V_C^2 & \cdots & V_C^{41} \vdots V_C^{42} & V_C^{43} & \cdots & V_C^{91} \end{bmatrix}$$

An example V_R and V_C data sets generated by MATLAB programming are shown below:

V_R = [0, 0.0952, 0.1813, 0.2592, 0.3297, 0.3935, 0.4512, 0.5034, 0.5507, 0.5934, 0.6321, 0.6671, 0.6988, 0.7275, 0.7534, 0.7769, 0.7981, 0.8173, 0.8347, 0.8504, 0.8647, 0.8775, 0.8892, 0.8997, 0.9093, 0.9179, 0.9257,

0.9328, 0.9392, 0.9257, 0.9328, 0.9392, 0.9257, 0.9328, 0.9392, 0.9450, 0.9502, 0.9550, 0.9592, 0.9631, 0.9666, 0.9698, 0.9727, 0.9753, 0.9776, 0.9798, 0.9817, 0.9834, 0.9850, 0.9864, 0.9877, 0.9889, 0.9899, 0.9909, 0.9918, 0.9926, 0.9933, 0.9939, 0.9945, 0.9950, 0.9955, 0.9959, 0.9963, 0.9967, 0.9970, 0.9973, 0.9975, 0.9978, 0.9980, 0.9982, 0.9983, 0.9985, 0.9986, 0.9988, 0.9989, 0.9990, 0.9991, 0.9992, 0.9993, 0.9993, 0.9994, 0.9994, 0.9995, 0.9995, 0.9996, 0.9996, 0.9997, 0.9997, 0.9997, 0.9998, 0.9998, 0.9998, 0.9998, 0.9998, 0.9998, 0.9999, 0.999, 0.9997, 0.9997, 0.9998, 0.9998, 0.9998, 0.9998, 0.9998, 0.9998, 0.9999, 0.9999, 0.9999, 0.9999, 0.9999, 0.9999, 0.9999, 0.9999, 0.9999, 0.9999, 0.9999, 1.0000]

V_C = [0, 0.0095, 0.0181, 0.0259, 0.0330, 0.0393, 0.0451, 0.0503, 0.0551, 0.0593, 0.0632, 0.0667, 0.0699, 0.0727, 0.0753, 0.0777, 0.0798, 0.0817, 0.0835, 0.0850, 0.0865, 0.0878, 0.0889, 0.0900, 0.0909, 0.0918, 0.0926, 0.0933, 0.0939, 0.0945, 0.0950, 0.0955, 0.0959, 0.0963, 0.0967, 0.0970, 0.0973, 0.0975, 0.0978, 0.0980, 0.0982, 0.0983, 0.0985, 0.0986, 0.0988, 0.0989, 0.0990, 0.0991, 0.0992, 0.0993, 0.0993, 0.0994, 0.0994, 0.0995, 0.0995, 0.0996, 0.0996, 0.0997, 0.0997, 0.0997, 0.0998, 0.0998, 0.0998, 0.0998, 0.0998, 0.0998, 0.0999 0.0999, 0.0999, 0.0999, 0.0999, 0.0999, 0.0999, 0.0999, 0.0999, 0.0999, 0.0999, 0.1000]

Generation of Confidence Level: The confidence level variation is given as

$$C = 1 \text{ to } \frac{n}{N} \quad \text{for first segment}$$

$$= 1 \text{ to } \frac{41}{91} \quad \text{i.e., 1 to 0.450} \quad \text{for } m = 10\% \;\; 50\%$$

$$C = \frac{n+1}{N} \text{ to } \frac{1}{N} \quad \text{for second segment}$$

$$= \frac{42}{91} \text{ to } \frac{1}{91} \quad \text{i.e., 0.461–0.01} \quad \text{for } m = 51\%\text{–}100\%$$

The equation for the first segment (linear)

$$\frac{y - 0.45}{x - 50} = \frac{0.45 - 10}{50 - 10}$$

$$\Rightarrow \quad y = -0.01375x + 1.1375$$

$$\Rightarrow \quad C_1 = -0.01375m + 1.1375$$

The equation for the second quadrant (power of 4): considering the equation of confidence level for the second segment as

$$y = mx^4 + c$$

Since the capacitance varies as $10^{-8} \times$ (moisture content)^4

From the condition

at $x = 51\%$ moisture content, $y = 0.461$

and $x = 100\%$ moisture content, $y = 0.01$

We get

$$0.461 = (51)^4 m + c$$

and

$$0.01 = (100)^4 m + c$$

Solving we get

$$m = -4.837 \times 10^{-9}$$

$$c = 0.49$$

Therefore, the equation for the second segment

$$C_2 = -4.837 \times 10^{-9} m + 0.49$$

These two confidence level data sets are combined to get the target matrix for the ANN as shown below:

$$O = \begin{bmatrix} C_1^1 & C_1^2 & \cdots & C_1^{41} & C_1^{42} & C_1^{43} & \cdots & C_1^{91} \\ C_2^1 & C_2^2 & \cdots & C_2^{41} & C_2^{42} & C_2^{43} & \cdots & C_2^{91} \end{bmatrix}$$

An example of C_1 and C_2 data sets generated by MATLAB programming is shown below:

C_1 = [1.0000 0.9870 0.9740 0.9610 0.9480 0.9350 0.9220 0.9090
0.8960 0.8830 0.8700 0.8570 0.8440 0.8310 0.8180 0.8050
0.7920 0.7790 0.7660 0.7530 0.7400 0.7270 0.7140 0.7010
0.6880 0.6750 0.6620 0.6490 0.6360 0.6230 0.6100 0.5970
0.5840 0.5710 0.5580 0.5450 0.5320 0.5190 0.5060 0.4930
0.4800]

C_2 = [0.4393 0.4305 0.4218 0.4130 0.4043 0.3956 0.3868
0.3781 0.3693 0.3606 0.3519 0.3431 0.3344 0.3256
0.3169 0.3082 0.2994 0.2907 0.2819 0.2732 0.2645
0.2557 0.2470 0.2382 0.2295 0.2208 0.2120 0.2033

0.1945 0.1858 0.1771 0.1683 0.1596 0.1508 0.1421
0.1334 0.1246 0.1159 0.1071 0.0984 0.0897 0.0809
0.0722 0.0634 0.0547 0.0460 0.0372 0.0285 0.0197
0.0110]

The generation of the matrix of *X* and *O* was performed by MATLAB simulation program as shown below:

```
% This programme simulates an RC circuit to determine the
% voltage
% the resistance(Vr) and the voltage across the
% capacitance(Vc)
% with change in capacitance from 0.01 farad to 0.05 farad
% when the
% moisture content of the capacitor varies from 10% to 50%.
% Again
% when the moisture content changes from 51% to 100% the
% capacitance
% changes from 0.06 farad to 1 farad. The voltage Vr and Vc
% are simulated
% and plotted with capacitance. During the entire data
% range, the
% confidence level of the capacitance fault is changed from
% 1 to 0
% gradually.
clc;
close all;
clear all;
f=10000;
R=1000;
% Vary capacitance linearly with moisture from 10% to 50%
m1=10:1:50;
k=1:1:41;
% generate confidence level for first segment
out1(k)=1.013-k.*0.013;
c1=0.001*m1;
% Vary capacitance sharply with moisture above 50%
m2=51:1:100
k=1:1:50;
% generate confidence level for second segment
out2(k)=0.448-k.*8.74*10^(-3);
c2=(m2.^4)*10^(-8);
C=[c1,c2];
xc=2*(pi)*f.*C;
vr=5*(R./sqrt(R^(2)+(xc).^(2)));
```

```
vc=5*(xc./sqrt(R^(2)+(xc).^(2)));
figure(1)
plot(C,vr)
xlabel('Capacitance, Farad');
ylabel('Resistor voltage, Volt');
figure(2)
plot(C,vc)
xlabel('Capacitance, Farad');
ylabel('Capacitor voltage, Volt');
v=[vr;vc];
nout1=[out1,out2];
nout2=[out1,out2];
nout=[nout1;nout2];
```

Figure E6.6a and b shows the voltages V_R and V_C, respectively, and Figure E6.6c shows the plot of confidence level for the entire data range. An ANN was trained with X as input and O as output. After training, the same input X was used to test the estimator. The estimated confidence levels generated by ANN are shown below:

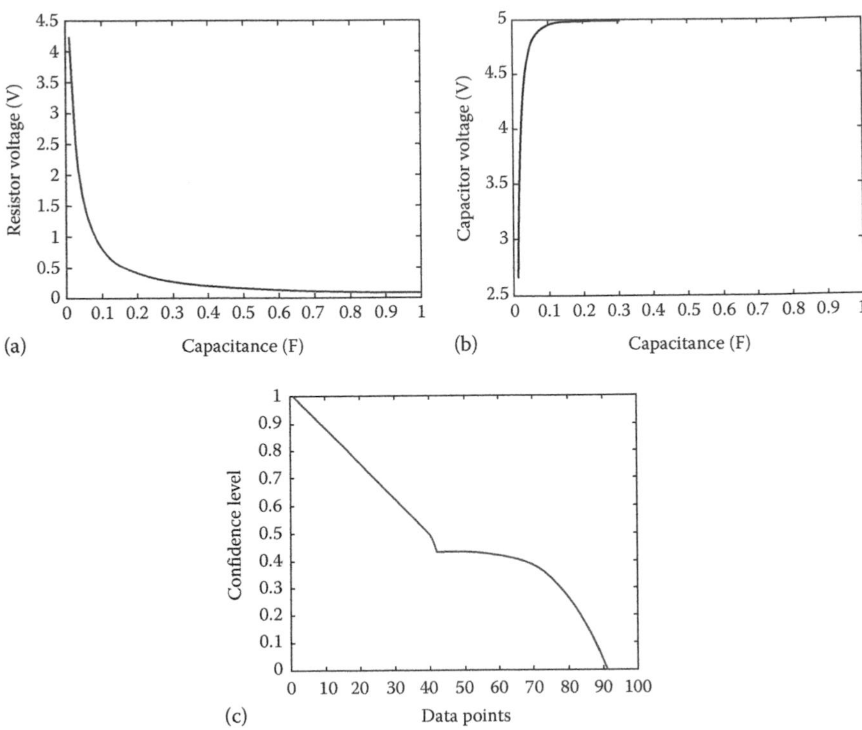

FIGURE E6.6
(a) Voltage across resistor, (b) voltage across capacitor, and (c) confidence level.

$$\hat{C}_1 = [1.0000\ 0.9870\ 0.9740\ 0.9610\ 0.9480\ 0.9350\ 0.9220\ 0.9090\ 0.8960\ 0.8830\ 0.8700\ 0.8570\ 0.8440\ 0.8310\ 0.8180\ 0.8050\ 0.7920\ 0.7790\ 0.7660\ 0.7530\ 0.7400\ 0.7270\ 0.7140\ 0.7010\ 0.6880\ 0.6750\ 0.6620\ 0.6490\ 0.6360\ 0.6230\ 0.6100\ 0.5970\ 0.5840\ 0.5710\ 0.5580\ 0.5450\ 0.5320\ 0.5190\ 0.5060\ 0.4930\ 0.4800\ 0.4393\ 0.4305\ 0.4218\ 0.4130\ 0.4043\ 0.3956\ 0.3868\ 0.3781\ 0.3693\ 0.3606\ 0.3519\ 0.3431\ 0.3344\ 0.3256\ 0.3169\ 0.3082\ 0.2994\ 0.2907\ 0.2819\ 0.2732\ 0.2645\ 0.2557\ 0.2470\ 0.2382\ 0.2295\ 0.2208\ 0.2120\ 0.2033\ 0.1945\ 0.1858\ 0.1771\ 0.1683\ 0.1596\ 0.1508\ 0.1421\ 0.1334\ 0.1246\ 0.1159\ 0.1071\ 0.0984\ 0.0897\ 0.0809\ 0.0722\ 0.0634\ 0.0547\ 0.0460\ 0.0372\ 0.0285\ 0.0197\ 0.0110]$$

$$\hat{C}_2 = [1.0000\ 0.9870\ 0.9740\ 0.9610\ 0.9480\ 0.9350\ 0.9220\ 0.9090\ 0.8960\ 0.8830\ 0.8700\ 0.8570\ 0.8440\ 0.8310\ 0.8180\ 0.8050\ 0.7920\ 0.7790\ 0.7660\ 0.7530\ 0.7400\ 0.7270\ 0.7140\ 0.7010\ 0.6880\ 0.6750\ 0.6620\ 0.6490\ 0.6360\ 0.6230\ 0.6100\ 0.5970\ 0.5840\ 0.5710\ 0.5580\ 0.5450\ 0.5320\ 0.5190\ 0.5060\ 0.4930\ 0.4800\ 0.4393\ 0.4305\ 0.4218\ 0.4130\ 0.4043\ 0.3956\ 0.3868\ 0.3781\ 0.3693\ 0.3606\ 0.3519\ 0.3431\ 0.3344\ 0.3256\ 0.3169\ 0.3082\ 0.2994\ 0.2907\ 0.2819\ 0.2732\ 0.2645\ 0.2557\ 0.2470\ 0.2382\ 0.2295\ 0.2208\ 0.2120\ 0.2033\ 0.1945\ 0.1858\ 0.1771\ 0.1683\ 0.1596\ 0.1508\ 0.1421\ 0.1334\ 0.1246\ 0.1159\ 0.1071\ 0.0984\ 0.0897\ 0.0809\ 0.0722\ 0.0634\ 0.0547\ 0.0460\ 0.0372\ 0.0285\ 0.0197\ 0.0110]$$

6.2.4 Fuzzy Logic–Based Intelligent Sensors

Fuzzy logic–based intelligent sensors are typically modeling of a measurement system using fuzzy logic rules that behaves as human-like reasoning. Fuzzy logic–based measurement models have the following advantages:

1. Complex mathematical model is avoided and uses simple mathematical calculations.
2. Memory requirement is less compared to that of lookup table used in nonlinear measurements.
3. Allows fast computations making the system cost effective in real-time computations.

4. Fuzzy logic–based models are universal approximations and they can model a nonlinear continuous time function.
5. Have less learning time compared to neural networks.

Fuzzy logic–based modeling of measurement systems has already been discussed in Section 4.9.3 under fuzzy logic–based indirect sensing; hence, readers may refer to that section for such nonlinear modeling.

References

1. Swanson, D.C., Chapter 14: Intelligent Sensor Systems, *Signal Processing for Intelligent Sensor Systems*, Marcel Dekker, Inc., New York, pp. 451–519, 2000.
2. Yamasaki, H. and Hiranaka, Y., Multidimensional intelligent sensing systems using sensor array, *Solid State Sensors & Actuators, Digest of Technical Papers, TRANSDUCES' 1991, IEEE*, San Francisco, CA, pp. 316–321, June 24–27, 1991.
3. Chase, L., *Prognostic Health Measurement (PHM): The Basics*, IEEE Reliability Society, Langkawi, Malaysia, www.ieee.org, 2009.
4. Bhuyan, M. and Barthakur, M., Possibility in study of nerve injury by signal processing and artificial intelligent methods on NCV and EMG signals: A future perspective, *Proceedings of the 12th Asian Oceanian Congress of Neurology*, New Delhi, India, October 2008.
5. Byington, C.S., Waston, M., Roemer, M.J., Galie, T.R., and Groarty, J.J., *Prognostic Enhancement to Gas Turbine Diagnostic Systems*, Defence Technical Information Center, Fort Belvoir, VA, www.dtic.mil, 2010.
6. Byington, C.S. and Garga, A.K., Data fusion for developing prognostic diagnostics for electromechanical systems, *Handbook of Multi Sensor Data Fusion Theory and Practice*, Liggins, M.E., Hall, D.L., and Llinar, J. (Eds.), CRC Press, Taylor & Francis, Boca Raton, FL, pp. 702–735, 2009.
7. Kozlowski, J.D. et al., Model based predictive diagnostics for electrochemical energy sources, *Proceedings IEEE Aerospace Conference*, Vol. 6, Big Sky, MT, pp. 63149–63164, March 2001.

7

Intelligent Sensor Standards and Protocols

7.1 Introduction

In the domain of intelligent instrumentation, in particular the industrial instrumentation, interfacing of the intelligent sensors to the processor and the users is a major issue. With the advent of complex data acquisition and control frameworks, the intelligent sensor interfacing protocols have also developed manyfold. In order to achieve higher efficiency, uniformity, and flexibility of the intelligent sensors, various interfacing protocols have been developed either in wireless or in Internet platforms. IEEE 1451 (draft proposal) is such a standard that specifies interfacing of intelligent sensors with fieldbuses or direct coupling to Ethernet-based intranet. In spite of the enormous growth of the protocols, smart sensor applications are being slowed down due to the lack of universal standards in wireless areas [1]. Since the smart sensors and networks are becoming very cost-effective solutions in the application areas, smart sensor interfacing to network has become an economical solution in distributed instrumentation environment. On the other hand, choosing the right network for a particular instrumentation system is also a trade-off in the design criterion of the transducer manufacturer to cut the cost.

This chapter discusses many of the issues related to the standards and protocols related to intelligent instrumentation and their applications.

7.2 IEEE 1451 Standard

In 1993, the smart transducer communication interface standard was first initiated jointly by the National Institute of Standards and Technology (NIST) and IEEE. To initiate widespread applications of smart sensors in a unified approach, the IEEE 1451—the *draft* standard sets a base protocol allowing compatibility between the sensors, networks, and buses. The most important feature of this standard is the open nature of the standard that facilitates different manufacturers to develop transducers that can be self-configured by the manufacturer and can be used by any other manufacturer in a plug-and-play manner. The IEEE 1451 draft standard consists of two basic parts:

- IEEE 1451.1: A network-independent common model for transducers
- IEEE 1451.2: Connection of transducers to network microprocessors

The key features that support this standard are

1. Choosing the best sensor for a selected network
2. Use of a sensor for multiple network
3. Choosing the best suited network
4. Automatic/self-configuration of a network

The basic modules of the IEEE 1451 draft standard are

1. Smart transducer interface module (STIM)
2. Transducer electronic data sheet (TEDS)
3. Network capable application processor (NCAP)
4. Transducer independent interface (TII)

In a nutshell, the STIM communicates the transducer/actuator-generated digital data messages with the NCAP, the NCAP handles the network interface, the TII interfaces to ports of STIM, and TEDS guides the usage of the transducers. The IEEE 1451 draft standard protocol is illustrated by the block diagram of Figure 7.1 [1]. The module functionaries are explained below.

7.2.1 STIM

The STIM is the frontend module of the IEEE 1451 platform that contains TEDS, the logics to interface the transducers to the NCAP, the transducers, actuators, and any other signal-conditioning circuit. The STIM can accommodate up to 255 sensors and actuators and signal conditioners/converters

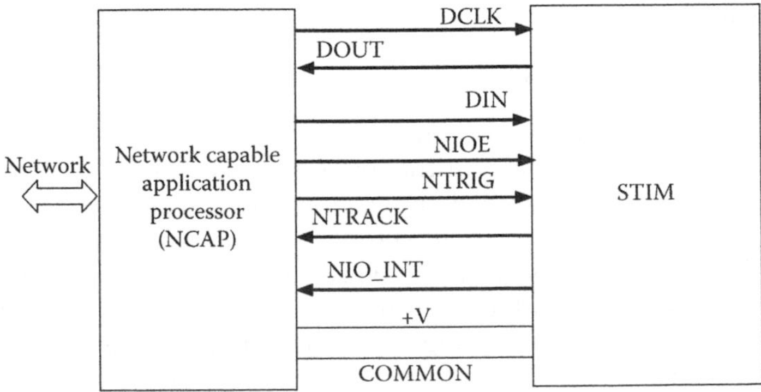

FIGURE 7.1
Block diagram and interface of STIM.

such as ADC/DAC, VOC, V-I/I-V converter, etc. The addressing of the various transducers and signal conditioners are also performed by the STIM. Apart from the basic signal-conditioning circuits, the STIM can accommodate microprocessor/microcontrollers for complex signal-conditioning operations. The STIM makes the transducer information stored in the TEDS available to the NCAP to be communicated to the network.

The STIM can be used for local or distributed measurement and control through network. For local measurement and control, the STIM can use a microprocessor or a microcontroller, and the NCAP directs the STIM locally. For distributed measurement and control the NCAP communicates to the other systems through network. The STIM–NCAP communication signals are listed below:

Driven by STIM:
DOUT: Data from STIM to NCAP
NTACK: Both triggering and data transport acknowledge
NIO_INT: Requests service from NCAP

Driven by NCAP:
DIN: Address and data communication between STIM and NCAP
DCLK: Positive going edge latches data on both DIN and DOUT
NIOE: Enables data transport and delimits data transport framing
NTRIG: Triggering
POWER: 5 V power supply
COMMON: Signal common to ground

7.2.2 TEDS

TEDS is an electronic data sheet of a transducer stored in a chip attached to the transducer or a file. TEDS helps the transducer and the STIM to self-diagnosis of the transducer for operation, maintenance, and calibration. Once a transducer is moved from one location to an another location, the TEDS also gets moved since it is a part of the transducer. Examples of the fields that TEDS stores are manufacturer name, identification number, response time, warm-up time, and sampling period. The TEDS have the following five different parts:

1. Meta-TEDS
2. Channel TEDS
3. Calibration TEDS
4. Application TEDS
5. Industrial and extension TEDS

Table 7.1 shows the example format of the five different TEDS for a single-channel ceramic pressure sensor [1].

TABLE 7.1

Example of TEDS

Meta-TEDS
Data structure–related information
Such as Meta-TEDS Length, IEEE 1451 Standards Family Working Group Number, TEDS Major Version Number, etc.
Timing-related information Worst Case Channel
Worst case channel data repetitions, worst case channel update time worst case, channel write setup time, etc.
Channel grouping–related information channel
Groupings data sub-block length, number of channel groupings, group name length, etc.
Data integrity information
Checksum for Meta-TEDS
Data structure–related information
Meta-identification TEDS length
Identification-related information
Manufacturer's identification length, manufacturer's identification, model, number length, etc.
Data integrity information data sub-block
Checksum for meta-identification TEDS
Channel TEDS
Data structure–related information Channel TEDS length
Calibration key, industry extension key
Transducer-related information
Lower range limit, upper range limit, unit warm up time, etc.
Data converter–related information
Channel data repetitions, series increment, channel update time
Channel write setup time, channel read setup time, etc.
Data integrity information
Checksum for channel TEDS
Data structure–related information
Channel identification TEDS length
Identification-related information
Manufacturer's identification length
Manufacturer's identification, model number length, serial number length, etc.
Data integrity information
Checksum for channel identification TEDS
Calibration TEDS
Data structure–related information
Calibration TEDS length
Calibration-related information
Last calibration date-time, calibration interval, channel degree list, number of segments list, etc.
Data integrity information
Checksum for calibration TEDS

Meta-TEDS: It contains the overall description of the TEDS data structure of all transducers connected to the STIM, worst-case timing parameters and channel grouping information. Since Meta-TEDS stores general information, one Meta-TEDS is provided per STIM.

Channel TEDS: The channel TEDS digitally lists the information on physical units, uncertainty, maximum/minimum rages, warm-up time, presence of self-test, data model, calibration model and triggering parameters, etc. Channel TEDS are specific to the transducer and needs one for each transducer.

Calibration TEDS: The calibration TEDS contains the calibration parameters such as last calibration date/time, calibration interval, correction parameters, number of calibration segments and the offsets, calibration polynomial coefficients, etc. Commonly, calibration TEDSs are specific to the transducer and need one for each transducer; however, the same calibration TEDS can also be used for transducers if need arises.

Application-specific TEDS: Application-specific TEDSs are defined by the end users that are not common to all. Examples of this category of TEDS are temperature compensation parameters for a resistive pressure sensor, ANN parameters (weights, error goal, neuron activation function) for artificial intelligence, etc.

Industrial and Extension TEDS: Future implementation and industry extension informations are stored in this TEDS. This information is received by the IEEE for evaluation and adoption.

The TEDS are normally prepared by the transducer manufacturer and stored in EEPROM. TEDS reading in machine-readable binary format is difficult and hence a compiler has to be used. A TEDS compiler (old version) and 1451.0 format compatibility tester developed by SUNY/Baffalo are shown in Figure 7.2 [2]. The compiler accepts the zip code of the manufacturer so that the universal unique identification number (UUID) and rest of the transducer TEDS data are automatically downloaded to the STIM EEPROM.

7.2.2.1 TEDS Calibration

Inexpensive and speedy calibration of intelligent sensors can be achieved by the unified approach of TEDS calibration in IEEE 1451 standard. This calibration method does not require a large calibration data, which needs to be stored in a lookup table generally.

TEDS is a compact data set stored in either EEPROM embedded within the smart sensor or in file known as virtual TEDS (VTEDS). VTEDS are suitable for storing complex TEDS parameters such as ANN parameters and structures, complex polynomials, large system model matrices, etc. The NI-DAQmx driver software from National Instruments Inc. [3] can accommodate all the features of TEDS. The NI-DAQmx sensor calibration technique using TEDS is discussed below.

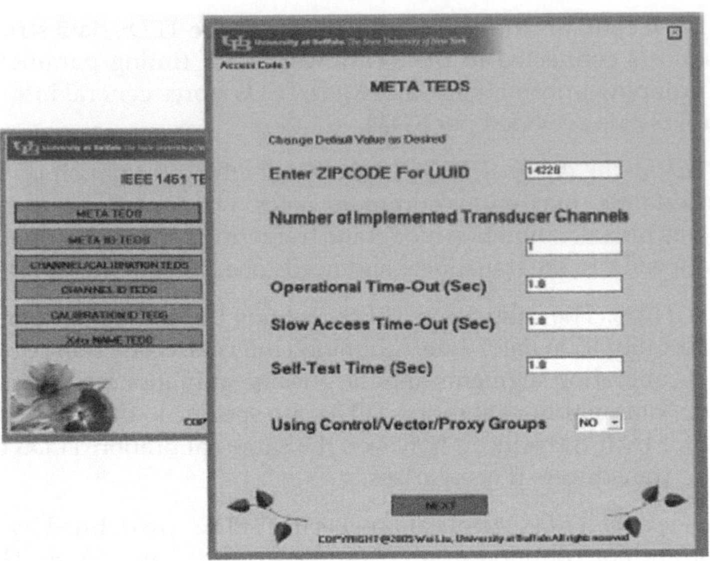

FIGURE 7.2
TEDS compiler window screen for UUID for data downloading. (Reprinted from Eesensors, www.eesensors.com, 2010. With permission.)

The DAQmx driver applies the standard calibration curves and equations specific to the transducers. These calibration parameters are stored in the EEPROM or the VTEDS and the driver chooses a template from the 16 templates (Table 7.2) and applies to the binary data. There are three more templates in addition to the 16 templates, these are calibration table, calibration curve (polynomial), and frequency response table.

7.2.2.1.1 Calibration Table and Curve

The calibration table template refers to a lookup table of the transducer for mapping the electrical outputs to the physical inputs. The table has two basic parameters—*Calpoint_DomainValue* and *CalPoint_RangeValue*, an example of which is shown in Table 7.3.

CalTable_Domain: It is used to select the domain parameter for the calibration table as electrical or physical. For an LVDT, the electrical domain parameter is voltage and the physical domain parameter is ±cm.

CalPoint_DomainValue: This parameter expresses the calibrated value expressed as a percentage of the full-scale range of the transducer. For the LVDT, if the full stroke range is from −5 to +5 cm, then *CalPoint_DomainValue* is 50% if the calibration reference point is 0 cm (null position).

CalPoint_RangeValue: It is the deviation of the range measurement from the expected value expressed as a percentage of the full scale of the transducer.

TABLE 7.2

Standard Templates for Calibration

Type	Template ID	Name of Template
Transducer type template	25	Accelerometer force transducer with constant current amplifier
	26	Charge amplifier (incl. attached accelerometer)
	27	Microphone with built-in preamplifier
	28	Microphone preamplifier with attached user or system
	29	Microphone(capacitive)
	30	High-level voltage output sensors
	31	Current loop output sensors
	32	Resistive sensors
	33	Bridge sensors
	34	AC linear rotary variable differential transformer (LVDT/RVDT) sensors
	35	Strain gage
	36	Thermocouple
	37	Resistance temperature detectors (RTD)
	38	Thermistors
	39	Potentiometric voltage dividers
Calibration template	40	Calibration table
	41	Calibration curve polynomial
	42	Frequency response table
Transducer type template	43	Charge amplifier (incl. attached force transducer)

Source: *Sensor Calibration with TEDS Technology*, National Instruments Inc., Austin, TX, www.ni.com, 2010. With permission.

Let the LVDT gives a reading with a deviation of 0.1 cm from the ideal value, then *CalPoint_DomainValue* is given by

$$CalPoint\_DomainValue = \frac{0.1}{10} \times 100\% = 1\%$$

The above calibration parameters are meant for calibration of the entire operating range with the same coefficients; however, in highly nonlinear characteristics, several segments of characteristics are considered with different polynomial functions. Accordingly TEDS calibration curve template specifies a multi-segment polynomial function. Table 7.4 shows such a template.

Let us take an example to show how the curve polynomial is defined in the TEDS. Let a transducer output is given by the following nonlinear function for an input range of $x = 0-3$,

$$y = x^3 + 3x^2 - 4x + 1$$

TABLE 7.3
Calibration Table Template for ID = 40

Function	Property Command	Description	Access	Bits	Data Type (Range)
ID	TEMPLATE	Template ID	—	8	Integer (0)
Table data	CalTable_Domain	Domain parameter	CAL	1	Enumeration: electrical/physical
	STRUCTARRY CalTable	Number of data sets	CAL	7	Dimension size of 1–127
	CalPoint_ DomainValue	Domain calibration point (% of full scale)	CAL	16	ConRes (0–100 step 0.0015)
	CalPoint_ RangeValue	Range calibration deviation % of full span	CAL	21	ConRes (100–100 step 0.0001)
		Total bits required for TEDS (range) 16–4715 bit			

Source: *Sensor Calibration with TEDS Technology*, National Instruments Inc., Austin, TX, www.ni.com, 2010. With permission.

The TEDS would store the [%*CalCurve_Power*, %*CalCurve_Coeff*] fields as

$$[(3,1),(2,3),(1,-4),(0,1)] \cong \sum_{i=0}^{n} C_i x^{P_i}$$

where
n is the number of coefficient stored in TEDS as *CalCurve_Poly*
C_i is the polynomial coefficient stored in TEDS as *CalCurve_Coef*
P_i is the power of x for coefficient C_i stored in TEDS as *CalCurve_Power*

For each piece of the segments, the TEDS stores two values: $CalCurve\_PieceStart_K$ and $CalCurve\_PieceStart_{K+1}$. The TEDS stores these fields for the example polynomial function as

$$\left[CalCurve\_PieceStart_K, CalCurve\_PieceStart_{K+1}\right] = [0,3]$$

7.2.2.1.2 Frequency Response Table

The calibration table and curve explained above refers to non-varying inputs only and thus the TEDS does not give any information about the frequency response behaviors of the transducer. The frequency response table template provides the frequency response by referring to a lookup table that stores the frequency (*TF_Table_Freq*) and the corresponding amplitude

TABLE 7.4
Calibration Table Template for ID = 41

Function	Property Command	Description	Access	Bits	Data Type (Range)
ID	TEMPLATE	Template ID	—	8	Integer (41)
Curve data	%CalCurve_ Domain	Domain parameter	CAL	1	Enumeration: electrical/ physical
	STRUCTARRY CalTable	Number of calibration curve segments	CAL	8	Dimension size of 1–255
	CalCurve_ PieceStart	Start of segment (array of size CalCurve)	CAL	12	ConRes (10–100 step 0.0123)
	STRUCTARRY CalCurve_Poly	Number of polynomial	CAL	7	Dimension size of 1–127
	%CalCurve_ Power	Power of domain value (2D array with dimension size of CalCurve and CalCurve_Poly)	CAL	7	Con Res (−32–32 step 0.5)
	%CalCurve_ Coeff	Polynomial coefficient (2D array with dimension size of CalCurve and CalCurve_Poly)	CAL	32	Single
		Total bits required for TEDS (range) 17–1,268,132 bit			

Source: *Sensor Calibration with TEDS Technology*, National Instruments Inc., Austin, TX, www.ni.com, 2010. With permission.

(*TF_Table_Ampl*). Table 7.5 shows the frequency response table template for ID = 42.

7.2.3 NCAP

A network capable application processor is a processor module between the STIM and the network that performs communication and data conversion functions such as

- Network communication
- Communication with STIM
- Data conversion
- Application functions

The NCAP structure consists of the following three main blocks (Figure 7.3):

1. STIM driver
2. Application firmware
3. Network communication stacks

TABLE 7.5
Frequency Response Table Template for ID = 42

Function	Property Command	Description	Access	Bits	Data Type (Range)
ID	TEMPLATE	Template ID	—	8	Integer (41)
Table data	STRUCTARRYTF_ Table	Transfer function table	CAL	7	Dimension size of 1–132
	%TF_Table_Freq	Frequency value of size (TF-Table)	CAL	15	ConRes(res 1–1.3E+6, ±0.02%)
	%TF_Table_Ampl	Amplitude (array size TF_table)	CAL	32	ConRes (−100 to 100 step 0.0001)
		Total bits required for TEDS (range): 15–4587 bit			

Source: *Sensor Calibration with TEDS Technology*, National Instruments Inc., Austin, TX, www.ni.com, 2010. With permission.

The NCAP is basically a controller and an interface to the network. For DeviceNet control network, the NCAP can be an 8-bit microprocessor, while for an Ethernet-based control network, it should be a 32-bit microprocessor. The STIM driver performs the following major functions:

1. Software interface driver, which retrieves data across the interface
2. TEDS parser compiles the TEDS data into user-friendly format

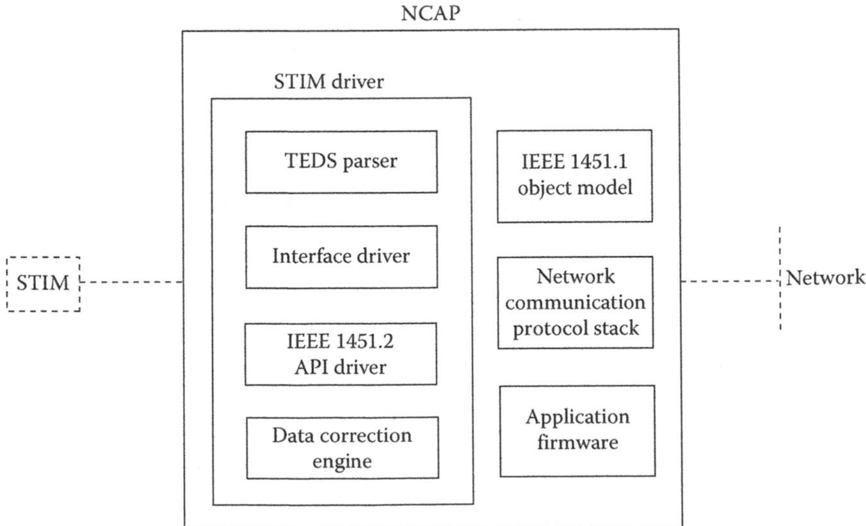

FIGURE 7.3
Block diagram of NCAP.

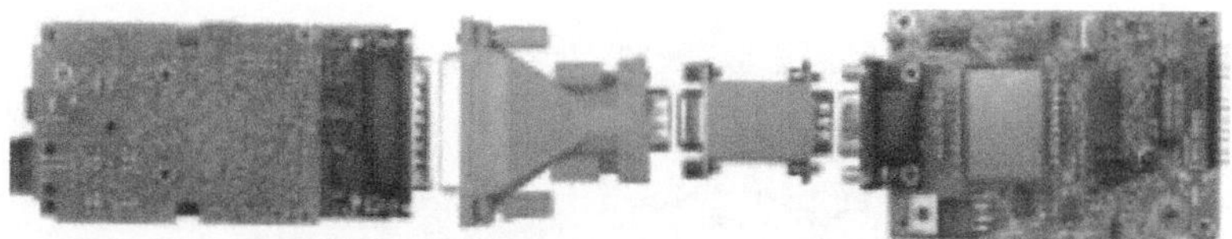

FIGURE 7.4
NCAP (internal version) and TIM with RS232 interface. (Reprinted from Eesensors, www.eesensors.com, 2010. With permission.)

3. Correction engine applies the correction parameters stored in the TEDS and converts raw sensor data received from STIM to calibrated data
4. 1451.2 Application programming interface (API) performs addressing and READ operation of TEDS blocks, transducer readings, actuator control, triggers, and interrupts

When the NCAP is first triggered, it searches for the STIMs and when an STIM is located, it acknowledges by sending a signal to the STIM. On receipt of the signal from the NCAP, the STIM sends the particular TEDS to the NCAP cache area. Using the TEDS, the NCAP then performs the metrological operations on the raw sensor data and converts it to calibrated SI data. The calibrated data is then transmitted to the external network using HTTP protocol and XML.

An NCAP/TIM interface using point-to-point RS-232 connection is shown in Figure 7.4 [2]. The TIM consists of a temperature sensor, a photodiode, and a relay. The TIM data in IEEE 1451.0 format is requested by an Internet browser and encoded in HTTP (TCP/IP). The data is converted to serial format (RS232) and it is received by the TIM, and the resulting data in IEEE 1451.0 format is returned. The interface can also be implemented by other serial buses like RS485 and USB.

For low-power wireless with limited bandwidth, another option is a simplified TEDS-only version (IEEE 1451.4) [2], as shown in Figure 7.5. It uses only analog sensors like accelerometer. This TEDS uses only 8 byte along with the built-in UUID of the 1-wire EEPROM in which the TEDS is stored. The TEDS is used for local wireless transmitted data and the NCAP expands this to the Internet.

7.3 Network Technologies

The networked sensors supported by IEEE 1451.2 protocol have the ability to communicate to a wider range to the network processors. The networked sensors typically have a local processor, which can perform signal transmission, calibration, corrections, and intelligent functions. The only disadvantage

FIGURE 7.5
Wireless sensor with Dot 4 TEDS. (Reprinted from Eesensors, www.eesensors.com, 2010. With permission.)

that networked sensors have complex circuitry, limited bandwidth, queuing delays, and loss of data. Examples of some network technology protocols, their most vital applications, and the respective sponsors are listed below [4]:

Automotive instrumentation:
J-1850: SAE
CAN: Robert Bosch GmbH
MI-Bus: Motorola
D^2B: Philips

Industrial instrumentation:
DeviceNet: Allen Bradley
Smart Distributed Systems: Honeywell
LonTalk/LonWorks: Eschelon Corp.
Hart: Rosemount

Home instrumentation:
Smart House: Smart house LP
CEBus: EIA

Building instrumentation:
Batibus: Merlin Gerin (France)
BACNet: Building Automation Industry

University protocols:
Michigan Parallel Standard: University of Michigan
Integrated Smart-Sensor Bus: Delf University of Technology
Time-Triggered Protocol: University of Wien

Some of the protocols are discussed in the following sections.

7.4 LonTalk

LonTalk is a networking protocol developed by Echelon Corporation aimed at networking devices over various types of channels. The typical application areas of LonTalk are industrial instrumentation and control, home automation, automobile and transportation, building lighting, and HVAC systems. LonTalk is a modified form of 1-Persistant Carrier Sense Multiple Access (1-Persitant CSMA) for peer-to-peer network applications. The flexibility of the protocol is that the services of the protocol can be chosen while programming and compiling.

7.4.1 LonTalk Media Types and Addressing

The protocol supports a variety of media channels through various transreceivers such as twisted pair, power line, RF, infrared (IR), coaxial cable, and fiber optics. The protocol specifies the channel specifications for each media type in distance, bit rate, and topology. The media can support up to a maximum of 32,385 nodes for multiple channels. The LonTalk network specifies two important identifiers—a unique 48-bit domain ID and an address. The ID specifies each individual neuron, which is permanently stored in the chip by the manufacturer. The address refers to source nodes and destination nodes of a LonTalk packet. The addressing modes of the LonTalk protocol are

1. Unicast addressing (subnet/node)—messaging to a single node
2. Multicast addressing (group)—messaging to a group of nodes
3. Broadcast addressing (subnet/domain)—messaging to a subnet or to the complete domain

The address format has three fields:

1. *Domain*—48-bit address of the subsystem of an open media
2. *Subnet*—8-bit address of the subnet of a domain
3. *Node*—7-bit address of the node within a subnet

Following are the major regions where LonTalk has been accepted as a standard protocol:

ANSI 709.1:	Control Networking (United States)
EN 14908:	Building control (European Union)
GB/Z 20177.1:	Control networking and building control (China)
IEEE 1473-L:	Train control (United States)
SEMI E54:	Semiconductor manufacturing sensors and actuators (United States)
IFSF:	International forecourt standards for EU petrol stations

LonTalk has also been recognized by international standardization bodies in the following areas:

ISO/IEC 14908-1	Communication
ISO/IEC 14908-2	Power line signaling
ISO/IEC 14908-3	Twisted pair wire signaling
ISO/IEC 14908-4	IP compatibility (tunneling)

7.5 CEBUS Communication Protocol for Smart Home

The CEBus (customer electronic bus) communication protocol is a standard for home automation and smart home [5] developed by Electronic Industries Association (EIA). The CEBus standard is comprehensive and has the potential to the development of new home appliances and consumer electronic products. The CEBus protocol provides the following features:

- Automation of home appliances
- Compatible to a wide range of appliances
- Low-cost interface embedded in appliances
- Variety of channel types
- Supports wide-band audio and video distribution in different analog and digital formats
- Distributed communication without the need for central controller
- Flexibility and simplicity of addition or removal of appliances providing plug-and-play feature

7.5.1 CEBus Channels

The CEBus supports data communication through the following medias:

- Electric power line
- Twisted pair wires
- Coaxial cable
- IR signal
- Radio frequency signal
- Fiber optics
- Audio–video bus

In home automation, typically, data communication is more convenient via power line layout; however, IR or RF signals are generally used for remote controls. The CEBus home appliance control technology is embedded entirely within the appliances that plug into the conventional outlets or within the remote control units. The CEBus channels carry both the control and the data signals at a rate of about 8000 bit/s.

In case of coaxial cable systems in CEBus, it adopts dual coaxial system where an upstream cable collects in-house generated videos from VCRs and cameras. A head-end unit combines the in-house videos with the external videos and sends to all receivers through a downstream cable.

7.5.2 CEBus Architecture

CEBus supports a flexible architecture and does not specify a topology. A device may be installed at any convenient location connecting to any media. Communication between devices, sensors, and controllers are routed through a unit called *router*. The router need not necessarily be a separate unit, it may be embedded into the appliance. Figure 7.6 shows a typical CEBus network with three media interconnected by routers [5]. The cluster controller controls the appliances such as lighting or energy management. Figure 7.7 illustrates an example of home lighting control in CEBus [5]. The hand-held remote control unit sends IR signal to the IR detector of the TV and then to the processor of the TV for interpreting the signal. If the signal is interpreted as TV control, the TV processor uses it for normal TV operations. When the LIGHTS button on the remote is pressed, the signal is interpreted not as TV control, and then it is passed to the router built in the TV. The power line router transmits this signal to the power line through the power outlet of the TV. This signal is received by the lighting controller and controls the lighting as per the command.

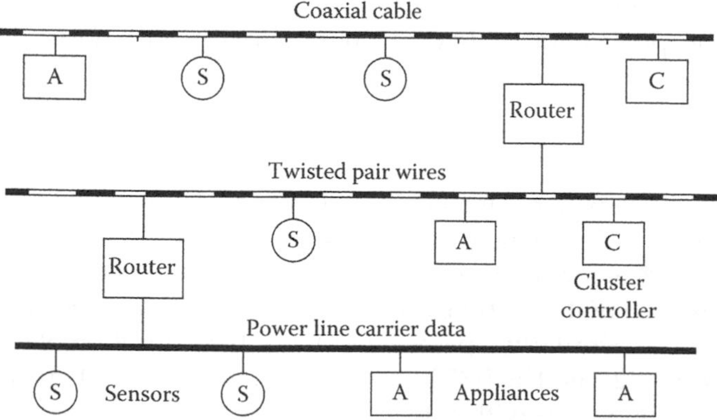

FIGURE 7.6
Example of a CEBus topology. (Wacks, K.P., *Home Automation and Utility Customer Services.* Copyright 1997, all rights reserved. With permission.)

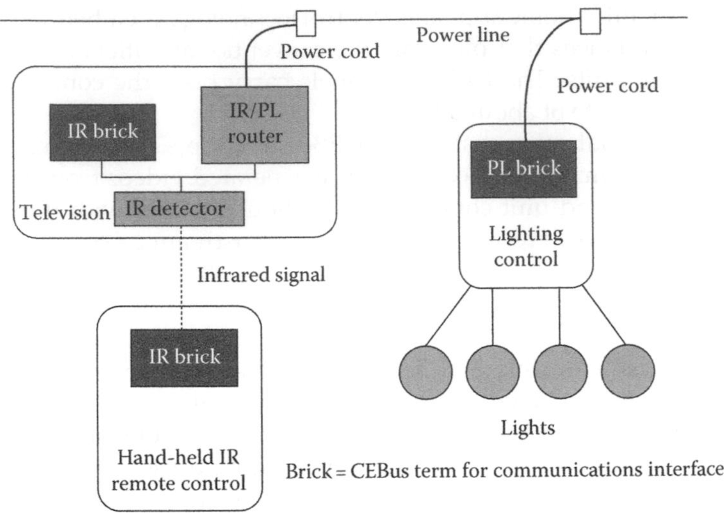

FIGURE 7.7
Example of an application of CEBus in home lighting control. (Wacks, K.P., *Home Automation and Utility Customer Services*. Copyright 1997, all rights reserved. With permission.)

7.6 J1850 Bus

The J1850 Bus protocol developed by SAE is aimed at diagnostic and data sharing in automotive applications. The standard uses the following two modes of signal transmission:

1. Pulse width modulation (PWM) in two-wire differential transmission at 41.6 kbps
2. Variable pulse width (VPW) in single-wire transmission at 10.4 kbps (maximum length of 35 m with 32 nodes)

7.6.1 Signal Logic and Format

A HIGH logic is considered for signals between 4.25 and 20 V, while a LOW is anything below 3.5 V. In single-wire transmission, the HIGH and LOW times are—64 and 128 μs for active and passive conditions, respectively and alternately. A passive HIGH has time duration of 128 μs while an active HIGH has duration of 64 μs. Similarly, a passive LOW has time duration of 64 μs and an active LOW has 128 μs. The J1850 signal format is shown in Figure 7.8.

The signal frame starts with a start of frame (SOF) with a duration of 200 μs, which is followed by a header byte and the data byte. The cyclic redundancy check (CRC) byte follows the data byte then the data is terminated by the end of data (EOD) bit of 200 μs LOW pulse.

SOF	Header	Data	CRC	EOD	IFR	CRC	EOF

FIGURE 7.8
J1850 signal format.

7.7 MI Bus

Motorola Interconnect (MI) is a communication protocol in serial data transfer mode that connects one master with a maximum of eight slaves. The communication network is used to control smart switches, motors, and actuators. The MI bus can also be used as an automotive bus to drive mirrors, seats, window lifts, head light levelers, etc.

The data transfer in the bus follows a push–pull sequence to transfer data between the master and the slaves. The message between the master and the slave contains two fields—a *push* field and a *pull* field (Figure 7.9). The fields contain the data as well as the address of one of the slave. When the master sends a field of "push" message to the slave, the addressed slave responds by transmitting a "pull" field to the master.

The bit lengths of the push–pull message fields are

SOF bit: 3 time slots held LOW
Sync bit (master): 1 bi-phase encoded 0
Data (master): 5 bit of bi-phase encoded data
Address: 3 bit of bi-phase encoded data
Sync bit (slave): 1 bi-phase encoded 1 initiated by the master
Data (slave): 3 bit nonreturn to zero encoded data
EOF: 3 cycles of a 20 kHz square wave

An MI bus interface circuit using microcontroller is shown in Figure 7.10 [6]. The interface unit drives the bus and protects the microcontroller from voltage transients appearing on the line. Similarly, the RX pin of the microcontroller is protected by a protecting circuit. The MI bus has two states:

Dominant (state 0): represented by a maximum of 0.3 V
Recessive (state 1): represented by +5 V via a 10 K pull-up resistor

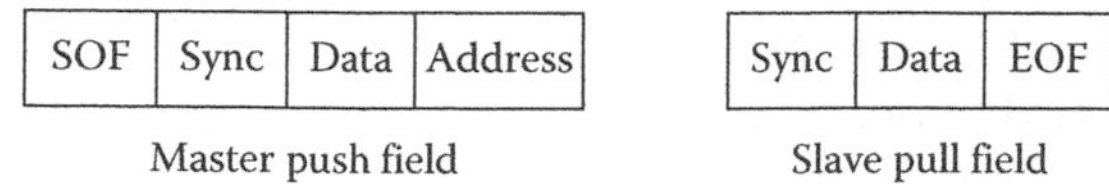

FIGURE 7.9
MI Bus Master–Slave message format.

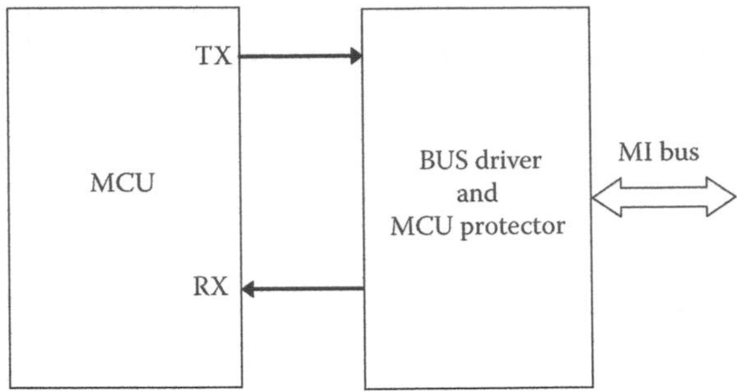

FIGURE 7.10
MI hardware interface bus circuit.

7.8 Plug-n-Play Smart Sensor Protocol [7]

Smart web sensor has become a promising and powerful solution to sensor signal processing in web server browsing or other information-retrieving environment. Typically, data communication using static web pages through HTML or data socket connection that adopts Java applet is common. However, the major drawback with such approaches is that the smart sensor is a proprietary system, and in HTML, the user is simply a reader of the pages formatted as per the manufacturer's criterion. On the other hand, with Java approach, the manufacturer does not supply the protocol information to the users.

To deal with the problems of the above-mentioned approaches, a new architecture based on web service approach for smart sensors has been developed in [7]. In this technique, each sensor defines its IP address and becomes an active node of the network. The advantages of this approach are

1. Faster and easier access to measured data
2. Integration of a large complex web sensors network
3. Realization of flexible custom application
4. Service preprocessing

The block diagram of Figures 7.11 and 7.12 shows the implementation of the smart web sensor hardware. Voltage transducers LV25-P were used to implement the web-based smart sensor technology. The acquired signal from the sensor was preprocessed by the operational amplifier (TL082) and a microcontroller (dsPIC 30F3013) and then sent to the networking embedded system—FOX board to perform final data processing and storage. The 16-bit microcontroller was used to implement the data acquisition and preprocessing. The speed of the microcontroller is 30 million instructions per second (MIPS). The microcontroller has the following features:

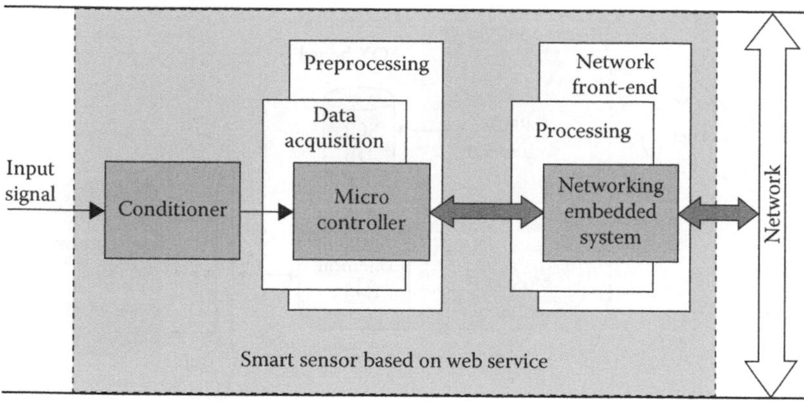

FIGURE 7.11
Block diagram of a smart web sensor hardware. (Reprinted from IEEE, copyright © IEEE. With permission.)

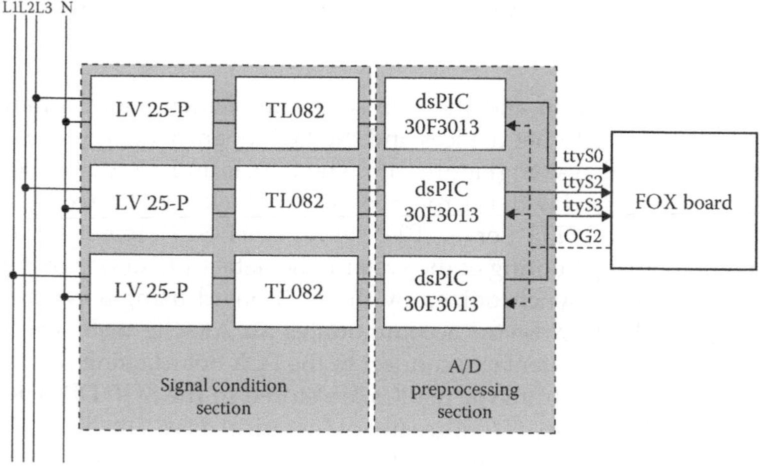

FIGURE 7.12
Linking schematic of the whole system. (Reprinted from IEEE, copyright © IEEE. With permission.)

- 20 I/O ports
- 2048-B static RAM
- 1024-B EEPROM
- 24 kB program flash
- 12 bit ADC
- Sampling rate—100 kS/s

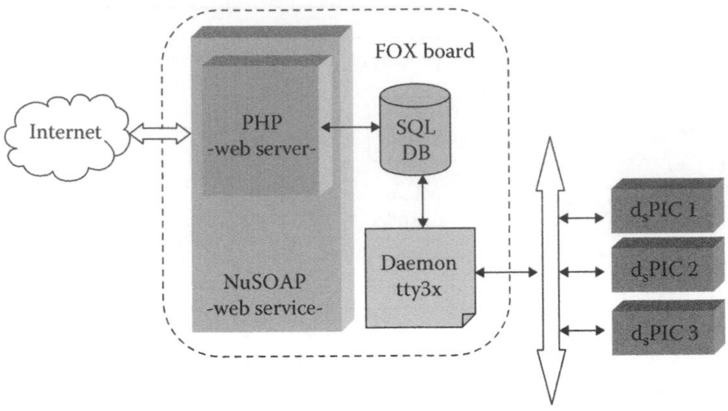

FIGURE 7.13
The web service scheme. (Reprinted from IEEE, copyright © IEEE. With permission.)

- Integrated PLL (4×, 8× and 16×)
- 16-bit timer/input capture/output compare/standard pulsewidth modulator

The network connection and the web server implementation in hardware for developing the Internet devices and the FOX board were performed in a low-cost embedded system (Figure 7.13). The FOX is an ETRAX LX100 MCM processor-based board with microkernel linux. The kernel Linux 2.6 version is compiled for the HTTP server, FTP server, SSH, SCP, Telnet server, and PPP support. During booting of the system, the data acquisitions of the three microcontrollers are synchronized by the FOX board using a synchronization protocol. The FOX board accommodates an Apache web server with NuSOAP library. Any client can connect to the FOX board using the web service standard. Signals from the dsPICs are stored in the SQL DB. The DB is stored in an USB memory for future use of historical data. The NUSOAP web service exports the services from which the information can be retrieved. The details of the techniques are available in the article in [7].

References

1. Woods, S.P. et al., *IEEE-P1451.2 Smart Transducer Interface Module*, Sensors Plug and Play, National Instruments, 2009 (http://www.ni.com/teds)
2. Wobschall, D., *IEEE 1451-A Universal Transducer Protocol Standard*, Eesensors Inc., Amherst, NY, 2010, www.eesensors.com
3. Mary Mentzer, Lebow, *Sensor Calibration with TEDS Technology*, National Instruments Inc., Austin, TX, 2006, www.ni.com

4. Osuna, R.G., *Intelligent Sensor Systems*, Wright State University, Dayton, OH, 2009.
5. Wacks, K.P., Introduction to CEBUS® communication protocol, *Home Automation and Utility Customer Services*, Cutter Information Corporation, http://www.cutter.com, copyright 1997.
6. MI Bus—Motorola Interconnect, 2010, www.motorola.com
7. Ciancetta, F. and Gallo, D., Plug-n-play smart sensor network with dynamic web service, *IEEE Transactions, Instrumentation and Measurement*, 57(10), 2136–2145, 2008.

Questions

Chapter 1

1.1 What is a *process* and a *process parameter*? Give two examples of process parameters in each of the following processes:

1. Our environment
2. An automobile
3. A washing machine
4. An electric motor

1.2 Classify the following process parameters under the two categories—*commonly used* and *rarely used*: flow, pressure, texture, level, RPM, pH, and oxygen content.

1.3 What is the basic difference between a *sensor* and a *transducer*? Give examples.

1.4 Define the four basic classes of transducers with examples.

1.5 Classify the following transducers and sensors under the four basic classes: potentiometer, thermocouple, shaft encoder, strain gauge, photo cell, and pH electrode.

1.6 In a resistive sensor, in what are the ways the change in resistance takes place?

1.7 Explain how a variable area capacitive sensor can be used for the measurement of angular movement? Draw the schematic diagram.

1.8 Can a dc motor be said to be a magnetic transducer? Justify your answer.

1.9 What is the basic difference between an *analog* and a *digital* transducer? How can an analog transducer be converted to a digital one?

1.10 What is the working principle of a radioactive transducer?

1.11 Can *density* and *moisture content* of a material be measured by a radioactive transducer? Explain how?

1.12 What are the significances of *half-life* and *half-distance* in radioactive instrumentation?

1.13 Explain the working principles of a semiconductor thermal sensor and a Hall effect sensor.

1.14 What are the basic stages of pattern-recognition techniques of array-based sensors?

1.15 Give examples of a few actuating devices.

Chapter 2

2.1 What do you mean by *accuracy* and *precision* of a transducer? How can accuracy and precision be calculated?

2.2 A thermocouple with a sensitivity of 40 mV/°C produces the following output voltages corresponding to the temperatures

Temperature (°C)	0	10	20	30	40	50	60	70	80	90	100
Voltage (mV)	0.1	0.4	0.79	1.18	1.55	1.88	2.45	2.67	3.15	3.45	3.85

Determine the *accuracies* of the readings and the *mean error*.

2.3 In Q2.2, the thermocouple produces the following output voltage readings when the applied temperature is 50°C:

Voltage (mV)	1.85	1.88	1.89	1.86	1.86	1.88	1.85	1.88	1.89	1.87	1.87

Determine the *precision* values of the readings.

2.4 What do you mean by *repeatability* of a transducer? How is repeatability of a transducer quantitatively defined?

2.5 In Q2.3, determine the repeatability of the thermocouple?

2.6 In Q2.2, what are the maximum *absolute* and *relative* errors?

2.7 In problem Q2.2, determine

1. Mean sensitivity
2. Offset
3. Maximum nonlinearity

2.8 Why is the *resolution* of a transducer an important characteristic?

2.9 The output voltage of the thermocouple of Q2.2 is read by an analog millivoltmeter with a full-scale range of 10 mV. The millivoltmeter scale has 100 small divisions. Determine the resolution of the thermocouple setup.

2.10 In the thermocouple of Q2.3, determine

1. *Percentage deviation* of each reading
2. *Average deviation*
3. *Standard deviation* and *variance*

2.11 What do you mean by *dynamic sensitivity* and *dynamic error* of a first-order sensor to a step input. Explain with mathematical equations.

2.12 A bare thermocouple behaves as a first-order system with a time constant of 2 min, while a time constant of another 1 min is added when the thermocouple is covered with a sheath, making the system a second-order one. Derive the equation of the output voltage response for a step input of 100°C. Take the sensitivity of the thermocouple as 45 μV/°C.

Chapter 3

3.1 Give two examples each for *periodic, apeariodic,* and *random* signals.

3.2 Indicate the following signals as periodic or random:

1. Wind velocity of a place
2. The sunlight intensity on the surface of the earth on a clear and sunny day
3. Signals generated in the brain
4. The damped movement of a pendulum

3.3 The output of a biosensor is discritized for 5 s with an ADC of 1 kHz sampling frequency. The signal is found to have the following features over a window of 20 ms:

Zero crossings:	4
Peaks:	5
Troughs:	4

Is the sample random over the window?

3.4 What are the distinct advantages of representing signals in Fourier, Laplace, and Z-transform?

3.5 The frequency contents of an ECG cycle is analyzed by discrete Fourier transform. The duration of the ECG cycle is 200 ms and the sampling frequency is 100 Hz. Write the equation of the DFT up to three samples. Consider the discrete ECG signal as $x[0], x[1], \ldots, x[N-1]$.

Chapter 4

4.1 For Case Study 4.7, perform test on sensor fault for the following two cases of sensor readings:
Case A: $T_i = 87°C$; $M_i = 12$ trays/15 min; $T_0 = 75°C$; $X_0 = 1\%$.
Case B: $T_i = 110°C$; $M_i = 9$ trays/15 min; $T_0 = 70°C$; $X_0 = 0.5\%$.

Consider the weight of tea in a tray as 4 kg and tolerance of 0.1% of tea dryness fraction.

4.2 For the measuring system of Example 4.10, develop residuals for the following faults:

1. A 50% sluggishness in ammeter spring
2. A short in voltmeter

(*Hint*: A 50% sluggishness in ammeter spring causes $y_2(t) = 1/2$, a short in voltmeter causes $R = 1\,\text{k}\Omega$ parallel with voltmeter resistance say $100\,\text{k}\Omega$)

4.3 In a digital OR gate system, three meters measure input A, B, and C, and output (Y) is estimated using a digital rule.

1. Write the equation for estimate of Y to be 'low'.
2. In how many cases meter faults for each of A, B, and C can be detected if rest two are known to be healthy?
3. When meter A and B are healthy, what is the probability of fault in the meters C and Y?

Chapter 5

5.1 Determine the following outputs when a sensor output of 3.3 mV is applied to the circuit of Figure 5.40:

1. Linearization amplifier output, i.e., V_x
2. Comparator array output, i.e., d_4, d_3, d_2, d_1, d_0
3. Range selector output, i.e., Q_0, Q_1, Q_2, Q_3, and Q_4
4. MUX selector output, i.e., S_2, S_1, S_0

5.2 Derive the piecewise linearization equation for the thermocouple data of Example 5.14.

5.3 The thermocouple data of Example 5.14 is linearized from 20°C to 60°C and 60°C to 100°C in two linear pieces. Determine the linearity percentage error at 55°C and 85°C if the corresponding output voltages are 2.849 and 4.455 mV, respectively, as per the data sheet of the thermocouple.

5.4 Design a piecewise linearization circuit for the pressure sensor data of Section 5.6.3 using the segments: 100–300 psi, 300–400 psi. Determine the linearity error if the ideal value of pressure at 30.82 mV is 312 psi.

5.5 In Example 5.15, find the digital output for a sensor voltage 4.516 mV. What is the conversion time if the ADC uses a clock of 10 MHz. Show the conversion table.

5.6 Design the selector circuit and the EPROM LUT for the pressure sensor characteristic of Figure 5.35 with the following segments for piecewise linearization:

P_i (psi):	0	25	50	75
V_0 (mV):	0	4.32	4.52	4.67

5.7 Determine the maximum, minimum, and least resolution for a resistor array system to be programmed by a 10-bit ADC. The array has an *R* value of 200 Ω. What is the percentage change in least resolution, if

1. The value of *R* is 100 Ω
2. The ADC has 12-bit resolution

5.8 Prepare a table showing an 8-bit programmable array resistor with value of *R*—500 Ω, when the array is programmed using an 8-bit ring counter.

5.9 A twisted ring counter with 4-bits is used to program a 4-bit array resistor giving a minimum and maximum resistance of 0.53 and 8 kΩ, respectively. Determine the value of *R* and show the resistances in tabular form.

5.10 A 4-bit multiplying DAC calibration circuit for a sensor signal of 0–5 V outputs in the live-zero current standard, i.e., 4–20 mA. Determine

1. The value of *R* and output current when input is 2.5 V and digital control signal is $(1111)_2$
2. The required digital signal for calibrating input voltage of 0–2.5 V at 4–10 mA range

[*Hint*: Use an independent current branch delivering a bias current of 4 mA to the output.]

5.11 A sensor voltage ranging from 0 to 3 V is calibrated in two ranges: 0–0.5 V and 0–3 V.

1. Determine the resistance for a reference voltage attenuator for an 8-bit ADC. Take reference voltage (max) = 3 V.
2. When an input voltage of 0.3 V is applied, find the digital code for the two calibrating ranges.

5.12 Repeat the sequence of the sigma–delta ADC of Example 5.27 and determine the 8-bit stream. What is the density of the "1"s in the bit stream?

5.13 A sigma–delta ADC produces a bit stream 1011101 for an input analog. Determine the 8-bit digital codes and percentage of quantization error for

1. Unipolar mode of 0–5 V, when $V_i = 3.55$ V
2. Bipolar mode of ±5 V, when $V_i = 1.00$ V

5.14 In a sigma–delta ADC, the output bit stream is produced by a clock of 48 MHz for a total duration of 2.082×10^{-7} s. What is the total number of bits? For what bit-stream density the digital code produced by the decimation filter for both bipolar (±5 V) and unipolar (+2.5 V) modes are equal?

5.15 The MAXIM IC1402 can be used for the calibration of an RTD, which produces 20 mV at 0°C and 40 mV at 266°C when an excitation current of 200 μA is applied. What V_{ref} and PGA gain should be used for calibrating the RTD in the range of 0–5 V? Use DAC code = 1100, CALGAIN = 7, CALOFF = 1 and $U/\bar{B} = 0$.

5.16 In Example 5.30, determine the self-heating temperature.

Chapter 6

6.1 Write a MATLAB® program to linearize the following thermocouple data using a FFBP ANN.

T (°C)	0	100	200	300	400
V_0 (mV)	0	44	71	105	135

Determine

1. Voltage at 25°C
2. Temperature corresponding to output voltage of 60 mV

6.2 Write a MATLAB program to linearize the data obtained from the nonlinear equation

$$y = 0.1x + 0.05x^2 + 0.0075x^3$$

The data should be linearized by the equation

$$y_l = 0.15x$$

Take a FFBP ANN and data range as $x = 0{:}0.01{:}10$.

6.3 In the circuit of Figure E6.4a, the values of resistance and capacitance are 10 Ω and 0.01 F, respectively. The inductance is prone to fault by varying its value from 0.01 to 0.1 H. Simulate the circuit in MATLAB to generate the step response and determine time constant (τ) for the whole range. Train an ANN with τ as input and *L* as target to estimate the value of inductance.

6.4 In the circuit of Figure 6.15a, the capacitance is 1.0 F and the resistance is changed with the temperature by the equation

$$R_t = R_0\left(1+\alpha t\right)$$

where

$R_0 = 1000\ \Omega$ at 25°C
$\alpha = 3\times10^{-3}\ \Omega/\Omega°\text{C}$

Take $V_i = 5\,\text{V}, f = 10\,\text{kHz}$ and generate data for $[V_i{:}V_C]$ by simulating the circuit in MATLAB. Consider that the temperature changes from 25°C to 50°C in steps of 1°C. Train an ANN to estimate the confidence level of temperature fault.

Chapter 7

7.1 What are the key features of IEEE 1451 standard?
7.2 What are the various modules of IEEE 1451 standard? Explain briefly.
7.3 Define the various signals communicated between the NCAP and STIM.
7.4 What are the five different parts of TEDS? Explain briefly.
7.5 What is the difference between *CalpointRangeValue* and *CalPOintDomainValue* in TEDS Calibration methods using DAQmx driver of National instruments. Explain with reference to a sensor.
7.6 For the calibration equation: $y = 2x^3 - 2x^2 - 5x$, how the TEDS stores the [*%CalCurve_Power, %CalCurve_Coeff*] fields?
7.7 What are the functions of the NCAP?
7.8 List some network technologies supported by IEEE 1451 standard.
7.9 Give a brief overview of LonTalk.
7.10 What are the CEBus-supported media channels?
7.11 Explain a CEBus architecture briefly.
7.12 Explain the J1850 and MI bus message formats.

Index